Handbook of
PEDIATRIC
TRANSFUSION
MEDICINE

Edited by

Christopher D. Hillyer, MD

Department of Pathology and Laboratory Medicine
Emory University School of Medicine
Atlanta, Georgia

Ronald G. Strauss, MD

Departments of Pathology and Pediatrics
University of Iowa
Iowa City, Iowa

Naomi L.C. Luban, MD

Departments of Pediatrics and Pathology
The George Washington University Medical Center
Washington, D.C.

ELSEVIER
ACADEMIC
PRESS

AMSTERDAM • BOSTON • HEIDELBERG • LONDON
NEW YORK • OXFORD • PARIS
SAN DIEGO • SAN FRANCISCO • SINGAPORE • SYDNEY • TOKYO

Academic Press in an imprint of Elsevier

Elsevier Academic Press
525 B Street, Suite 1900, San Diego, California 92101-4495, USA
84 Theobald's Road, London WC1X 8RR, UK

This book is printed on acid-free paper.

Library of Congress Cataloging-in-Publication Data
Application submitted

British Library Cataloguing in Publication Data
A catalogue record for this book is available from the British Library

ISBN: 0-12-348776-5

For all information on all Academic Press publications
visit our Web site at www.academicpress.com

PRINTED IN THE UNITED STATES OF AMERICA
03 04 05 06 07 08 9 8 7 6 5 4 3 2 1

*To Vivien Strauss, for the incredible love and compassion
she demonstrated throughout her life.*

Table of Contents

Chapter 5 Pretransfusion Compatibility Testing   63

SUSAN T. JOHNSON, MSTM, MT(ASCP)SBB, AND
TINA M. PUGH, MT(ASCP)

Chapter 6 Serologic Investigation of Unexpected Antibodies   73

SUSAN T. JOHNSON, MSTM, MT(ASCP)SBB, AND
TINA M. PUGH, MT(ASCP)

Chapter 7 Leukoreduced Products: Prevention of Leukocyte-Related Transfusion-Associated Adverse Effects   85

LENNART E. LÖGDBERG, MD, PHD

Chapter 8 Preparation of Blood Components to Reduce Cytomegalovirus and Other Infectious Risks   93

JOHN D. ROBACK, MD, PHD

Chapter 9 Irradiated Products   101

EDWARD C.C. WONG, MD

Chapter 10 Washed and/or Volume-Reduced Blood Components   113

S. GERALD SANDLER, MD, AND
JAYASHREE RAMASETHU, MD, FAAP

Chapter 19 Hemoglobinopathies 209
KRISTA L. HILLYER, MD

Chapter 20 The Bleeding Child: Congenital and Acquired Disorders 221
BRIAN M. WICKLUND, MD, CM, MPH

Chapter 21 Transfusion of the Patient with Autoimmune Hemolysis 245
KAREN E. KING, MD

Chapter 22 Platelet Transfusions in the Infant and Child 253
MATTHEW SAXONHOUSE, MD, WILLIAM SLAYTON, MD, AND
MARTHA C. SOLA, MD

Chapter 23 Bone Marrow–Derived Stem Cells 271
GRACE S. KAO, MD, AND STEVEN R. SLOAN, MD, PHD

Chapter 24 Peripheral Blood Stem Cells 283
GRACE S. KAO, MD, AND STEVEN R. SLOAN, MD, PHD

Chapter 25 Umbilical Cord Blood Stem Cells 295
LAURA C. BOWMAN, MD, MICHAEL A. BRIONES, DO,
AND ANN E. HAIGHT, MD

Chapter 26 Transfusion Reactions 301
ANNE F. EDER, MD, PHD

Preface

The *Handbook of Pediatric Transfusion Medicine* is designed as a companion textbook to the *Handbook of Transfusion Medicine,* published in 2001. Both of these handbooks are intended to be eminently readable as a quick reference, with concise paragraphs and multiple headings, as well as comprehensive texts covering blood banking and transfusion medicine in both adults and children. Pediatric transfusion medicine represents a true subspecialty in the field of transfusion medicine, often practiced by "super-subspecialized" pediatric hematologists or adult transfusion medicine experts. Because many of the issues that are central to the transfusion care of adults are not always central to the care of neonates, infants, toddlers, and young adults and as textbooks covering this area are few, the *Handbook of Pediatric Transfusion Medicine* became a necessity in the arsenal of references for transfusion therapists in care of children. Thus, as the title states, this text is focused on *pediatric* transfusion medicine; this means that *all* chapters, even if similarly titled to those found in adult transfusion medicine texts (e.g. Component Preparation and Storage) are focused toward the special issues of *pediatric* transfusion patients.

The reader will find that a number of specialized pediatric transfusion situations are the focus of individual chapters in this book, including the technical considerations and mechanical devices required for the transfusion care of the pediatric patient; red blood cell transfusion in the neonate, infant, and child; transfusion in pediatric surgery, trauma, and the intensive care unit (ICU); exchange transfusion in the infant and child; and granulocyte and platelet transfusions—again, all focused on children and the special issues they raise. Extracorporeal membrane oxygenation and cardiac bypass surgery, hemolytic disease in the newborn, treatment of individuals with hemoglobinopathies, and the diagnosis and treatment of patients with autoimmune hemolysis are also given separate attention. A unique chapter addressing transfusion in developing countries is included, since more than three-quarters of individuals transfused in those countries are young children. Finally, full sections of the book are dedicated to hematopoietic stem cells and related cellular products, complications of transfusion, and pediatric therapeutic apheresis.

It is the editors' hope that this book will be valuable to pediatricians, many pediatric subspecialty practitioners, and certainly to the pediatric transfusion medicine specialist. I am privileged to have Drs. Strauss and Luban, internationally recognized experts in the field of pediatric transfusion medicine, as co-editors, superb colleagues, and rare friends. I hope you find this book to be helpful and enjoyable and look forward to any comments you might have.

C.D.H
Atlanta, GA

Acknowledgements

We, the editors, would like to acknowledge the outstanding technical and professional support of Sue Rollins, the expertise and guidance of Hilary Rowe and Mamata Reddy, and the exceptional organizational and editorial help of Cassandra Josephson. Each of these individuals played an instrumental role in the creation of this textbook, and we sincerely thank them. We would also like to thank our friends and families for their unconditional love and support, especially Krista, Whitney, Peter, Margot, and Jackson Hillyer; Norm, Matt, and Ben Luban; and all the members of the Strauss family. We wish to recognize our pediatric patients, for whom the provision of excellent clinical care is a requirement, not an option, and who continue to teach us new lessons every day.

Christopher D. Hillyer, Ronald G. Strauss,
and Naomi L.C. Luban, Co-Editors

About the Editors

Christopher D. Hillyer, MD is a tenured professor in the Department of Pathology and Laboratory Medicine at Emory University School of Medicine and serves as director of the Transfusion Medicine Program for Emory with oversight of blood banks at all affiliate institutions. Dr. Hillyer is directly responsible for the Emory University Hospital Blood Bank and the Blood Banks and Transfusion Services at the Children's Healthcare of Atlanta hospitals. He is the editor of three textbooks in transfusion medicine and author of more than 100 articles pertaining to transfusion, human immunodeficiency virus, cytokines, and herpes viruses (most notably cytomegalovirus), as well as more than 15 book chapters. Nationally recognized as an expert in hematology and blood transfusion, he is also the vice president of the American Association of Blood Banks (AABB), and a former Trustee of the National Blood Foundation. Dr. Hillyer has received wide recognition in transfusion medicine, haematology, and immunology and has been awarded more than $10 million in research funding from the National Institutes of Health (NIH), the Centers for Disease Control and Prevention (CDC), the National Blood Foundation, and other agencies. He has served as chairman, or been a member of, many NIH Study Sections or Scientific Review Panels since 1996. Dr. Hillyer is also a cofounder of a startup biotechnology venture, Transfusion & Transplantation Technologies, Inc (3Ti), which is involved in the design, production, and testing of fully automated immunohematology testing platforms. 3Ti is the recipient of NIH and Georgia Research Alliance grants, as well as the New Enterprise Grand Prize award for innovations in technology (2001). Dr. Hillyer is board certified in four specialty areas including transfusion medicine, hematology, medical oncology and internal medicine, and has served as Deputy Director of the Winship Cancer Institute at Emory University. He is highly regarded as a speaker and has been asked to present to committees of the Food and Drug Administration (FDA) and the Department of Health and Human Services. Dr. Hillyer is an associate editor of the journal *Transfusion* and serves on several other editorial boards including *Transfusion Medicine Reviews, Journal of Clinical Apheresis,* and *Blood Therapies in Medicine.* He received his B.S. from Trinity College (1980), and his M.D. from the University of Rochester School of Medicine (1984), with postgraduate training and fellowships in hematology-oncology, transfusion medicine, and bone marrow transplantation at Tufts-New England Medical Center in Boston.

Ronald G. Strauss, MD is a tenured professor in the Departments of Pathology and Pediatrics at the University of Iowa College of Medicine and serves as the medical director of the Elmer DeGowin Memorial Blood Center at the University of Iowa Hospitals and Clinics. He is the co-editor of six textbooks on transfusion medicine and author of more than 260 journal articles pertaining to transfusion medicine and pediatric hematology, 69 book chapters, and 242 abstracts. Dr. Strauss has served a variety of scientific and professional societies as a member of many committees and working groups, member of the board of directors for the AABB, trustee for the National Blood Foundation Research and Education Trust Fund, president and member of the board of directors for the American Society for Apheresis, president and member of the board of directors for the Midwest Society for Pediatric Research, and *ad hoc* member of several review groups for the NIH. He has been awarded more than $15 million in research funding from the NIH and other granting agencies. Dr. Strauss is board certified in pediatrics, pediatric hematology/oncology and transfusion medicine/blood banking. He has been invited to serve as a visiting professor or as speaker on hundreds of occasions at national and international sites. Dr. Strauss is an associate editor for the journal *Transfusion* and serves on the editorial boards for *Transfusion Medicine Reviews*, *Journal of Clinical Apheresis*, and *Transfusion and Apheresis Science*. He received his B.S. from Capital University (1961) and his M.D. from University of Cincinnati College of Medicine (1965), with postgraduate training and fellowships in pediatrics and pediatric hematology/oncology at the Boston City Hospital and the Cincinnati Children's Hospital and Research Foundation.

Naomi L.C. Luban, MD received her B.A. in zoology from Connecticut College for Women in 1968. After receiving her medical degree from Mount Sinai School of Medicine at the City University of New York in 1972, she completed an internship and residency in pediatrics at the Children's Hospital National Medical Center in Washington, D.C. She then completed fellowships in pediatric hematology/oncology at Cornell University School of Medicine and Rockefeller University and did post-fellowship rotations in pathology at Memorial Hospital-Sloan Kettering Cancer Institute and the New York Blood Center. She returned to Washington in 1976 to serve as Instructor of Child Health and Development at George Washington University School of Medicine and Children's National Medical Center (CNMC), where she has remained for her career. In 1989 she became a tenured professor with appointments in pedi-atrics and pathology. She was elected to the Society of Pediatric Research in 1990 and the American Pediatric Society in 2000, and was named an Associate Member of the American Board of Pediatrics in 2003. Dr. Luban has achieved international renown for her work in a variety of transfusion-related fields, including iron over-load, immune function in hemophilia and other hyper-transfused patients, modifications of blood bank techniques for neonatal and pediatric use, hazards of transfusions in newborns, irradiation of blood and blood components, and post-transfusion infections. She has been principal investigator, co-principal investigator, or co-investigator on grants and contracts since she began her academic career. She currently is principal investi-gator in a study of post-transfusion infections in pedi-atric patients, co-principal investigator of a Hemophilia Comprehensive Care grant and NIH Multicenter Hemophilia Cohort Study, and co-investigator and mentor for a K08 using probabilistic risk assessment to reduce error and improve transfusion safety. She has published more than 150 peer-reviewed and invited manuscripts on a variety of transfusion topics in jour-nals and has contributed to chapters in most textbooks on hematology and pediatric hematology and to several atlases of neonatology. Her book chapters number 50. She has edited several smaller texts for the AABB, as well as three textbooks in pediatric transfusion therapy, and reviews regularly for seven journals. She is on the Editorial Board of *Transfusion* and has been a member of several NIH study sections, scientific advisory com-mittees review panels, and Drug Safety and Monitoring Boards (DSMB). She was a member and chair of the Blood Diseases and Resources Advisory Committee of the NIH and member of the Blood Products Advisory of the Center for Biologicals, Evaluation and Research of the FDA. She currently serves as the chair of the DSMB for the Thalassemia Research Network and is a member of the Anti-infective Drugs Advisory Commit-tee of the FDA and the Blood and Blood Products Panel of the U.S. Pharmacopeia. She currently serves as interim executive director for Center for Cancer and Blood Disorders and chairwoman for Laboratory Med-icine and Pathology at the Children's National Medical Center. Additionally, she services as vice chairwoman for Academic Affairs in the Department of Pediatrics and professor of pediatrics and pathology at George Washington University Medical Center. Dr. Luban has recently assumed responsibilities for academic leader-ship of the Department of Pediatrics and the Divisions of Hematology-Oncology in Laboratory Medicine and Pathology at the George Washington University Medical Center in Washington, D.C. She is especially proud of her accomplishments in developing a mentor-ship program that has been successful in furthering the academic life of junior faculty at CNMC.

List of Contributors

Sheilagh Barclay, MT(ASCP)SBB, Department of Pathology and Laboratory Medicine, Emory University School of Medicine, Atlanta, GA

Laura C. Bowman, MD, Department of Pediatrics, Emory University School of Medicine, Atlanta, GA

Michael C. Briones, DO, Division of Pediatric Hematology, Oncology, and Bone Marrow Transplantation, Emory University School of Medicine, Atlanta, GA

David C. Burghardt, MT(ASCP)SBB, Transfusion Services, Children's Healthcare of Atlanta, Atlanta, GA

Mitchell S. Cairo, MD, Department of Medicine and Pathology, Columbia University, New York, NY

Kenneth A. Clark, MD, MPH, Division of HIV/AIDS Prevention-Surveillance and Epidemiology, Cneters for Disease Control and Prevention, Atlanta, GA

Jorge A. Di Paola, MD, Division of Hematology Oncology, University of Iowa College of Medicine, Iowa City, IA

Anne F. Eder, MD, PhD, Department of Pathology and Laboratory Medicine, University of Pennsylvania School of Medicine, Philadelphia, PA

David F. Friedman, MD, Department of Pediatrics, Children's Hospital of Philadelphia, Philadelphia, PA

Jed B. Gorlin, MD, Department of Laboratory Medicine, University of Minnesota, Minneapolis, MN

Alfred J. Grindon, MD, Department of Pathology, Emory University School of Medicine, Atlanta, GA

Ann E. Haight, MD, Department of Pediatrics, Emory University School of Medicine, Atlanta, GA

Lauren Harrison, RN, BSN, Department of Pediatrics, Columbia University, New York, NY

Christopher D. Hillyer, MD, Department of Pathology and Laboratory Medicine, Emory University School of Medicine, Atlanta, GA

Krista L. Hillyer, MD, American Red Cross Blood Services, Southern Region, and Emory University School of Medicine, Atlanta, GA

Heather A. Hume, MD, Department of Transfusion Medicine, Canadian Blood Service, Ottawa, Ontario, Canada

Faranak Jamali, MD, Transfusion Medicine Division, Johns Hopkins Medical Institutions, Baltimore, MD

Cassandra D. Josephson, MD, Department of Pathology, Emory University School of Medicine, Atlanta, GA

Susan T. Johnson, MSTM, MT(ASCP)SBB, Immunohematology Services, The Blood Center of Southeastern Wisconsin, Milwaukee, WI

Grace S. Kao, MD, Dana-Farber Cancer Institute, Boston, MA

Haewon C. Kim, MD, The Children's Hospital of Philadelphia, Philadelphia, PA

Karen E. King, MD, Departments of Pathology and Oncology, Johns Hopkins University School of Medicine, Baltimore, MD

Lennart E. Lögberg, MD, PhD, Department of Pathology and Laboratory Medicine, Emory University School of Medicine, Atlanta, GA

Bruce C. McLeod, MD, Rush-Presbyterian St. Luke's Medical Center, Chicago, IL

Lisa M. Montenegro, MD, Department of Anesthesia and Critical Care Medicine, The Children's Hospital of Philadelphia, Philadelphia, PA

Paul M. Ness, MD, Transfusion Medicine Division, Johns Hopkins Medical Institutions, Baltimore, MD

Anne-Monique Nuyt, MD, Division of Neonatology, Department of Paediatrics, Hospital Sainte-Justine, University of Montreal, Montreal, Quebec, Canada

Matthew Saxonhouse, MD, Department of Pediatrics, University of Florida, Gainsville, FL.

William Slayton, MD, Department of Pediatrics, University of Florida, Gainsville, FL.

Steven Sloan, MD, PhD, Joint Program in Transfusion Medicine, Boston Children's Hospital, Boston, MA

Martha C. Sola, MD, Department of Pediatrics, University of Florida, Gainsville, FL.

Ronald G. Strauss, MD, Department of Pathology and Pediatrics, University of Iowa, Iowa City, IA

Maria Luisa Sulis, MD, Department of Pediatrics, Columbia University, New York, NY

Brian M. Wicklund, MD, MPH, Department of Pediatric Hematology/Oncology, Children's Mercy Hospital and Clinics, Kansas City, MO

Edward C.C. Wong, MD, Department of Pediatrics and Pathology, Children's Hospital National Medical Center, George Washington University School of Medicine and Health Sciences, Washington, DC

Alexandros Panagopoulos, MD, Department of Paediatrics, Hospital Sainte-Justine, University of Montreal, Montreal, Quebec, Canada

Patricia T. Pisciotto, MD, Department of Laboratory Medicine, University of Connecticut School of Medicine, Farmington, CT

Tina M. Pugh, MT(ASCP), Transfusion Service, Children's Hospital of Wisconsin, Blood Center of Southeastern Wisconsin, Milwaukee, WI

Thomas J. Raife, MD, Department of Pathology, University of Iowa College of Medicine, Iowa City, IA

Jayashree Ramasethu, MD, FAAP, Division of Neonatology, Georgetown University Hospital, Washington, DC

John D. Roback, MD, PhD, Department of Pathology and Laboratory Medicine, Emory University School of Medicine, Atlanta, GA

Nancy Robitaille, MD, FRCP, Department of Paediatrics, Hospital Sainte-Justine, University of Montreal, Montreal, Quebec, Canada

S. Gerald Sandler, MD, Clinical Laboratories, Georgetown University Hospital, Washington, DC

Blood Donation

ALFRED J. GRINDON, MD

OVERVIEW

During a typical year, about 14 million units of whole blood are collected in the United States. Most of these come from volunteer donors who give an average 1.6 units per year. The remainder of the collected units comes from donors designating their units for themselves (autologous donors, 4% of the total) or for another specific recipient (directed donors, less than 2% of the total) (Sullivan 2003). Perhaps 60% of the population is eligible to donate blood. Therefore, about 5% of eligible volunteer donors give annually. This amount of blood is usually sufficient to meet the needs of the recipient population, since, with the use of measures such as postponement of elective surgery, significant shortages have been found only in some parts of the country at specific times of the year (for instance, late summer and during and after the winter holiday season). However, the margin between units collected and units transfused has become substantially smaller over the last few years, causing concern about the elasticity of the blood supply (Simon 2003).

In addition to these units of whole blood, cytapheresis collections provide an increasing number of red blood cell collections (perhaps now an additional 300,000 units). Also, a significant majority of platelet doses transfused are platelets collected by apheresis.

The whole blood collected is usually separated into red cells and plasma for transfusion, or red cells, platelets, and "recovered" plasma (used for fractionation into plasma derivatives), at a rate of more than 2 components per unit collected. Of the more than 14 million units of collected units of red cells, about 13.5 million are transfused. Approximately 1.7% is discarded as a result of abnormal screening tests, and the remainder of the blood losses (as much as 50% of autologous units) occurs as a result of outdating.

Between 15% to 25% of the 8 million donors in the donor pool are lost annually and must be replaced with first-time donors (or the remaining donors must donate more frequently) to keep the number of donations constant. Further, the number of potential first-time donors must be greater than the number of units actually collected and tested, since first-time donors are approximately twice as likely to have disqualifying medical conditions or screening tests as repeat donors. Increasing concerns about the losses of potential donors from addition of new tests or new health history requirements (for example, travel-related deferrals for variant Creutzfeld Jacob Disease [vCJD]) make improvements in donor recruitment or donor retention increasingly important.

DONOR RECRUITMENT

Many studies have focused upon donor motivation in order to improve donor recruitment (Gillespie and Hillyer 2002). It has been shown, however, that most individuals who are able to donate conveniently and who are asked directly to do so, will give blood (Drake, Finkelstein, and Sapolsky 1982). The challenges are to find a way to ask directly and make it convenient to donate.

The classical successful blood drive, designed to address these concerns, is organized and held at a business or industry. The chief executive is asked to offer support in the provision of a top management leader, who develops a committee of employees (typically one committee member for each of 10 employees). These committee members ask their co-workers directly to donate at an upcoming blood drive, provide a specific appointment time, and remind them of their pledge on the day before or the day of the drive. The physical collection venue is located at the business, in the lunchroom, conference room, or other convenient on-site location. Such classical blood drives are usually successful.

Another approach is telephone recruitment by blood center employees to a conveniently located collection facility. The added costs of recruitment and the permanent facility are offset by the more efficient use of collection staff, who do not need to travel to and set up at a distant site. A small number of collections come from self-contained bloodmobiles (typically of four- to six-bed capacity), but this approach is generally ineffective unless supported by a strong telerecruitment effort.

Telephone recruitment is also an important tool for recruitment of those donors needed in greater frequency than found in the general population. Group O donors generally, and O Rh negative specifically, are always in short supply. For this reason, those using red cell cytapheresis techniques will preferentially recruit Group O donors for this procedure. Similarly, Group AB plasma is generally in short supply, and facilities obtaining plasma for transfusion consider telerecruitment of Group AB donors for plasmapheresis. Blood from donors known to be cytomegalovirus (CMV) seronegative may be needed for transfusion to specific recipients, such as transplant recipients or neonates. Telerecruitment is useful here also; for the latter group, the donor blood may be placed in a bag containing an anticoagulant-preservative solution suitable for neonatal recipients.

Some regions of the country have supported these approaches with an additional incentive: a blood "insurance" plan whereby annual donations from a certain percentage of a given group ensures that all the members of the group (and their families) are "covered." Many centers use the provision of a personal health test, nonrelated to blood donation, such as cholesterol, prostate-specific antigen, blood sugar, or a test for hemochromatosis, as an incentive.

From time to time in the past, blood centers have used direct financial incentives or material incentives that have financial equivalents (event tickets, raffles). The increasing awareness that financial incentives are associated with a higher frequency of transfusion-transmitted disease (Eastlund 1998), along with current labeling requirements of the Food and Drug Administration (FDA) that such units be labeled as from a "paid donor" (but "Benefits, such as time off from work, membership in blood assurance programs, and cancellation of nonreplacement fees that are not readily convertible to cash, do not constitute payment within the meaning of this paragraph") (21 CFR 606.121 (c)(5)), have generally eliminated such usage. Two exceptions exist. Some smaller centers that have demonstrated that their infectious disease marker frequency is less than that of volunteer donors, or some facilities that focus on more difficult-to-recruit plateletapheresis donors may still use financial motivation. In addition, the "source plasma" industry, whose plasma products are collected by plasmapheresis and are subsequently pooled and subjected to viral inactivation steps during the fractionation manufacturing process (and therefore carry a greatly reduced risk of transfusion-transmitted diseases), has traditionally used financial incentives.

AUTOLOGOUS DONATION

Patients may give blood for themselves before elective surgery (preoperative) or have their blood returned during the operative procedure (intraoperative). Preoperative autologous donation requires a donor who is healthy enough to give blood, but who is undergoing an operative procedure in the next 6 weeks that usually requires blood. Autologous donors can be drawn with lower hemoglobin standards ($\geq$11 g/dL) and drawn weekly, but blood should not be drawn for a few days immediately before surgery to allow blood volume to be restored. Intraoperative salvage may be useful for procedures with large anticipated blood losses retrievable from a body cavity, such as for cardiovascular surgery.

The percentage of units of autologous blood collected preoperatively reached as high as 10% in some areas during the 1980s as a result of a general fear of transfusion-transmitted disease, but with the development of good testing for these diseases this percentage has subsequently fallen to about 4% of the total collected. While it is possible to collect, freeze, and store red blood cell units for up to 10 years, the cost and time required to prepare these units for transfusion make this practice impractical except in rare circumstances. These include patients whose red-cell antibodies make finding compatible allogeneic blood difficult, or for patients with several units of autologous blood already available but whose surgeries have been postponed.

Autologous donation for the pediatric population represents more of a challenge. To begin with, informed consent (or "assent") is difficult, since the concern is usually not of the patient but of the parents. Because of the child's size, collection of half units from older (larger) children may be all that is possible. Nevertheless, autologous blood may be used in the pediatric population for surgery where blood is usually required, such as for orthopedic spinal surgery.

DIRECTED DONATION

Directed donation is the donation of blood that is then specified for the use of a particular patient. Currently, directed donation represents only about 2% of the total blood collected. The collection of this blood represents a logistical challenge, since the blood must be handled separately and at the hospital, kept for as long a time as the patient might reasonably need it. Since donors of such blood must meet the same requirements as regular donors, directed donations are used for other patients when the need for transfusion of the intended recipient has passed. In a pediatric setting, directed donation is done at the desire of the parents and often with the intent of using parents as donors. One must be certain, however, that harm is not done. Maternal blood used for neonates may contain antibody to red cells, white cells, or platelets, and such antibody may not be detectable by regular antibody screening. Paternal blood may contain antigens to which the child has passive immunity, or may lead to transfusion-transmitted graft versus host disease. It is probably best not to use parental blood in the neonatal period, or if maternal cells are needed, to wash them before transfusion (Strauss, et al 1990). All blood from blood relatives must be irradiated to prevent graft versus host disease.

LIMITED DONOR EXPOSURE

It is at least theoretically desirable to limit the number of donor exposures to a given recipient. This can be accomplished for small children by using multiple aliquots from the same unit of blood. It is possible to provide repeated small transfusions from one donor (often the father) for transfusion of small children, by allowing an exception to the 8-week donor deferral period, and collecting smaller amounts of blood at shorter intervals. However, such exceptions may require examination of the donor by a physician before each donation.

BLOOD COLLECTION

Regulation

The collection of blood is regulated by the FDA and considered to be a manufacturing process. The govern-

ing regulations are found in several volumes of the Code of Federal Regulations (CFR), but primarily in sections describing the blood "current Good Manufacturing Practices" (21 CFR, sections 211 and 606). All facilities collecting blood (or "manufacturing" blood components) must register with the FDA, and those shipping blood in interstate commerce must be licensed by this agency. In either case, the process must meet the requirements of the CFR, as well as of guidelines published from time to time by the FDA (www.fda.gov/cber). In addition to these requirements, the American Association of Blood Banks (AABB), while a voluntary association, regularly publishes *Standards for Blood Banks and Transfusion Services (Standards)* (Friday 2003), which set a standard of practice and which are used by accrediting agencies.

Collection Site

Because all stages of the manufacturing process (donor selection, collection, testing, component preparation, storage, and distribution) are so highly regulated, it has become increasingly difficult for the average hospital to maintain a blood donation operation. For this reason most blood is collected by blood centers, rather than hospitals. Blood centers collect both at permanent locations ("fixed sites") and mobile units, both those set up for one or several days in businesses or schools and the smaller "self-contained" bloodmobiles. In all cases, the site must meet the CFR requirements for cleanliness and space (for instance, adequate to ensure donor privacy during the screening interview).

Donor Screening

Those potential donors who present at a site to give blood are screened by comparison to a registry of previously deferred donors, a focused health history, and a limited physical examination, on the day of donation. This process is designed to ensure, to the extent possible, that the donor is healthy enough to donate, and that the resulting blood will be as safe and effective for transfusion as possible. The historical and physical requirements change regularly (almost always becoming more restrictive). Within the past few years, requirements and guidelines have been added, for instance, to exclude donors with European (and particularly United Kingdom) residence because of fear of transfusion-transmitted vCJD, to address concerns regarding the transmission of West Nile virus disease, and to deal with potential donors who have been vaccinated recently for smallpox.

The most current directives from the FDA can be found at their Web site (www.fda.gov/cber). The

Standards is published every two years and also is supplemented by widely promulgated changes between editions. AABB donor screening criteria, current at the time of publication of this book, are given in Table 1.1. But one must always be certain that the latest edition of the *Standards* and the most recent directives of the FDA are those used.

Donor Registration

Potential donors need to provide some form of authentic identification, ideally photographic, along with numerical identifiers such as date of birth and social security number. This information is used to compare the potential donor to a registry of deferred donors, to ensure that the donor is eligible to donate on the day of presentation, has not donated within the last 8 weeks, and has not had a positive test result for a marker of transfusion-transmitted disease.

In some facilities, this comparison is done electronically, with hand-held computers. In others, the checking is done after collection at a central location. There is no national donor deferral registry, but the larger collection organizations (such as the American Red Cross) have a registry that includes donors from all their locations. As a part of the registration process, donors are given written information that explains the donation process, complications of donation, signs and symptoms of human immunodeficiency virus (HIV) infection, what constitutes high-risk behavior, confidentiality, and the fact that they will be notified of positive test results. This material is provided as part of the informed consent process.

Donor History

Donors are asked questions about their health to ensure both the well-being of the donor and the safety of the recipient. (See Table 1.1; reproduced with permission from Friday 2003.) Some facilities use systems that allow donors to answer the questions themselves, either manually or electronically. Those questions directed toward determining the well-being of the donor are typically open ended: "Are you feeling well and healthy today?" or "Have you been under a doctor's care?"

Those questions designed to protect the recipient can be quite specific: "In the past 12 months, have you had sex, even once, with anyone who has used a needle to take drugs not prescribed by a doctor?" Questions directed toward recipient safety are designed to detect current acute infection, infection with chronic blood-borne agents (or a history of a positive test marker for such an agent), or specific blood-borne diseases for which no test is currently available (babesiosis, Chagas' disease, malaria). Additional questions are asked about behaviors that place a donor at greater risk for HIV or hepatitis, such as intravenous drug abuse or exposure to prostitutes or parenteral blood (in order to detect those few donors who might be in the window between infectivity and the presence of detectable test marker).

Donors are asked to provide a medication history both to protect their health and that of the recipient. From a donor health perspective, medication provides an indication of a potentially serious underlying medical condition in the donor, such as antianginal medication suggesting significant heart disease. From the perspective of the recipient, antibiotics may indicate a greater risk of bacterial sepsis, and some drugs may be teratogenic or mutagenic in very small amounts. Finally, there are questions that are directed toward component efficacy: the presence of a clotting disorder or the use of aspirin-like drugs that might adversely affect platelet function.

Donor history requirements change regularly; recent additions (mentioned previously) include addressing concerns about transfusion transmission of vCJD, West Nile virus, and smallpox from recent vaccination. A set of questions that has been designed by members of the blood banking community and approved by the FDA, known as the Uniform Donor History Questions, may be found in the *Technical Manual of the AABB* (Brecher 2002). These questions are not the only way of meeting donor history requirements of the FDA and *Standards*, but they are at least one acceptable way.

Donor Physical Examination

Potential donors are screened with a limited physical examination consisting of temperature, blood pressure, pulse, and arm examination (the latter for evidence of intravenous drug abuse [IVDA] or dermatologic disease in the antecubital fossa limiting adequate cleansing). In general, these values must be within normal limits for the general population. While weight is recorded, the value is usually obtained historically, rather than with use of a scale.

Hemoglobin Determination

As part of the examination, a sample of blood (usually from a fingerstick) is tested for hemoglobin content. Here, an arbitrary standard of 12.5 g/dL (or a hematocrit of 38%) is used, even though the normal values for women are somewhat lower than those for men. A common system used is a determination of the ability of a drop of blood to sink in $CuSO_4$ of a specific gravity of 1.053, which generally indicates a hemoglobin level of

TABLE 1.1 Requirements for Allogeneic Donor Qualification

Category	Criteria
1 Age	≥17 years or applicable state law
2 Whole blood volume collected	Maximum of 10.5 mL per kilogram of donor weight, including samples, and blood collection container shall be cleared for volume collected
3 Donation interval	8 weeks after whole blood donation (Standard 5.6.7.1 applies)
	16 weeks after two-unit red cell collection
	4 weeks after infrequent apheresis
	≤2 days after plasma-, platelet-, or leukapheresis
	(See exceptions in Standard 5.5)
4 Blood pressure	≤180 mm Hg systolic
	≤100 mm Hg diastolic
5 Pulse	50–100 beats per minute, without pathologic irregularities
	<50 acceptable if an otherwise healthy athlete
6 Temperature	≤37.5°C (99.5°F) if measured orally, or equivalent if measured by another method
7 Hemoglobin/hematocrit	≥12.5 g/dL/≥38%; blood obtained by earlobe puncture shall not be used for this determination
8 Drug therapy	Medication evaluation:
	— Finasteride (Proscar, Propecia), isotretinoin (Accutane)—Defer 1 month after last dose
	— Dutasteride (Avodart)—Defer 6 months after last dose
	— Acitretin (Soriatane)—Defer 3 years after last dose
	— Etretinate (Tegison)—Defer indefinitely
	— Ingestion of medications that irreversibly inhibit plated function (e.g., aspirin) within 36 hours of donation precludes use of donor as sole source of platelets
9a Medical history General health	The donor shall be free of major organ disease (e.g., heart, liver, lungs), cancer, or abnormal bleeding tendency, unless determined suitable by blood bank medical director
9b Pregnancy	Defer if routine donation
9c Receipt of blood, component, or other human tissue	— Family history of CJD[1] or receipt of dura mater or pituitary growth hormone of human origin—Defer indefinitely
	— Receipt of blood, components, human tissue, or clotting factor concentrates—Defer for 12 months
9d Immunizations and vaccinations	— Receipt of toxoids, or synthetic or killed viral, bacterial, or rickettsial vaccines if donor is symptom-free and afebrile—No deferral
	[Anthrax, Cholera, Diptheria, Hepatitis A, Hepatitis B, Influenza, Lyme disease, Paratyphoid, Pertussis, Plague, Pneumococcal polysaccharide, Polio (injection), Rabies, Rocky Mountain spotted fever, Tetanus, Typhoid (by injection)]
	— Receipt of liver attenuated viral and bacterial vaccines—2-week deferral
	[Measles (rubeola), Mumps, Polio (oral), Typhoid (oral), Yellow fever]
	— Receipt of live attenuated viral and bacterial vaccines—4-week deferral
	[German measles (rubella), Chicken pox (varicella zoster)]
	— Small pox (refer to FDA Guidance)
	— Receipt of other vaccines—12-month deferral
	[Hepatitis B Immune Globulin (HBIG), unlicensed vaccines]
9e Infectious diseases	*Defer indefinitely:*
	— History of viral hepatitis after eleventh birthday
	— Confirmed positive test for HBsAg
	— Repeatedly reactive test for anti-HBc on more than one occasion
	— Present or past clinical or laboratory evidence of infection with HCV, HTLV, or HIV or as excluded by current FDA regulations and recommendations for the prevention of HIV transmission by blood and components
	— Donated the only unit of blood or component that resulted in the apparent transmission of hepatitis, HIV, or HTLV
	— A history of babesiosis or Chagas' disease
	— Evidence or obvious stigmata of parenteral drug use
	— Use of a needle to administer nonprescription drugs
	— Donors recommended for indefinite deferral for risk of vCJD, as defined in most recent FDA guidance
	12-month deferral from the time of:
	— Mucous membrane exposure to blood
	— Nonsterile skin penetration with instruments or equipment contaminated with blood or body fluids other than the donor's own. Includes tattoos or permanent make-up unless applied by a state regulated entity with sterile needles and non re-used ink.
	— Sexual contact with an individual with a confirmed positive test for HBsAg
	— Sexual contact with an individual who is symptomatic (clinical evidence or diagnosis) for any viral hepatitis

TABLE 1.1—cont'd

Category	Criteria
	— Sexual contact with an HCV positive individual who has had clinically apparent hepatitis within the past 12 months
	— Sexual contact with an individual with HIV infection or at high risk of HIV infection[2,3]
	— Incarceration in a correctional institution (including jails and prisons) for more than 72 consecutive hours
	— Completion of therapy for treatment of syphilis or gonorrhea or a reactive screening test for syphilis in the absence of a negative confirmatory test
	— History of syphilis or gonorrhea
	Other
	West Nile virus—defer in accordance with FDA Guidance
9f Malaria	— Prospective donors who have had a diagnosis of malaria shall be deferred for 3 years after becoming asymptomatic
	— Individuals who have lived in areas in which malaria is considered endemic by the Malarial Branch, Centers for Disease Control and Prevention, and U.S. Department of Health and Human Services may be accepted 3 years after departure from any malaria-endemic area if they have been free from unexplained symptoms suggestive of malaria during that length of time.
	— Residents of areas in which malaria is not endemic but who have traveled to an area in which malaria is considered to be endemic may be accepted 12 months after departing that area. However, they shall have been free of unexplained symptoms suggestive of malaria during the time period, irrespective of the receipt of antimalarial prophylaxis.
10 At risk	— Evaluation: Lesions on the skin at the venipuncture site

[1] FDA Guidance for Industry dated January 9, 2002 "Revised Preventive Measures to Reduce the Possible Risk of Transmission of Creutzfeldt-Jakob Disease (CJD) and Variant Creutzfeldt-Jakob Disease (vCJD) by Blood and Blood Products."

[2] FDA Memorandum dated April 23, 1992 "Revised Recommendation for the Prevention of Human Immunodeficiency Virus (HIV) Transmission by Blood and Blood Products."

[3] FDA Memorandum dated December 11, 1996 "Interim Recommendations for Deferral of Donors at Increased Risk for HIV-1 Group O Infection."

[4] The Department of Defense has recommended a 24-month deferral. Department of Defense Memorandum dated October 14, 1999 "Deferral of Service Members Stationed in Possible Malaria Areas in the Republic of Korea," and February 28, 2001, update.

[5] www.cdc.gov/travel.

greater than or equal to 12.5 g/dL. Often, those potential donors whose blood fails this test have a microhematocrit determination; perhaps 50% of those initially deferred pass this second test and are accepted as donors. These systems are used because they are relatively inexpensive and portable, yet have accuracy sufficient to protect the donor (Cable 1995). These systems do not protect against development of low levels of stored iron; however, iron deficiency that has not progressed to anemia has little clinical significance of itself and is generally not felt to be of major concern. In addition, hemoglobin levels are returned to normal in 3 to 4 weeks in donors with adequate iron stores. Eight weeks are required between donations as a safety measure, to allow additional dietary iron absorption for those with subnormal iron stores. Some centers have provided temporary iron supplementation for menstruating women.

Confidential Exclusion/Call Back

At the conclusion of the screening process, the donor is often given the opportunity to self-exclude, by placing a "use" or "do not use" bar-coded sticker on the donor information card. The donor completes this process out of sight of the collection staff, who cannot determine from the sticker the decision made by the donor. While initially helpful to exclude donors with a higher frequency of positive tests for transfusion-transmitted disease (who were perhaps under peer pressure to donate), donors who self-exclude have only a minimal impact on transfusion safety (Peterson et al. 1994). Nevertheless, it is still often used as an additional safeguard.

Whether confidential exclusion is used or not, facilities provide a telephone number with a copy of the donor's blood unit number, so that a donor who becomes ill with an infectious disease in the next day or two (or develops hepatitis or HIV infection in the next 12 months) can call the facility and provide the information keyed to the unit number.

Phlebotomy

Following the registration, health history, physical examination, and hemoglobin determination, the donor is typically given a collection set for his or her use, appropriate for his or her blood group and the center's daily needs. With collection set and donor registration card, he or she is directed to the blood drawing area. Phlebotomy is performed in a supine or semisitting position. One of the most important steps in the collection process is the verbal re-identification of the donor and the validation of that identification with the information on the collection set, donor card, and pilot tubes before phlebotomy in order to prevent mislabeling of the unit of blood. The next step is the cleansing of the antecubital area of the arm, usually with soap followed by an iodine-containing solution to obtain as aseptic a surface as possible, followed by the phlebotomy itself. Newer collection sets have a diversion system so that the pilot tubes for testing are filled first, before the filling of the primary collection bag begins, in order to reduce the possibility of bacterial contamination of the unit from a skin contaminant or in the skin plug, more likely to be found in the first few milliliters of blood. During the collection, it is important to mix the blood with the anticoagulant-preservative solution, in order to prevent activation of coagulation factors and platelet aggregation. This mixing can be performed either manually, several times during the collection, or using a mechanical agitation device.

The collection typically takes 6 to 10 minutes; if the collection occurs much more rapidly, the phlebotomist must suspect an arterial puncture and use special post-phlebotomy arm care. If the phlebotomy takes significantly longer, the potential for initiation of coagulation exists, and the blood is less desirable for preparation of components containing coagulation factor-rich plasma or platelets.

Following phlebotomy, mild pressure followed by a simple dressing is applied to the wound, and the donor is carefully conducted to an area where liquid refreshments are offered under observation. Donors are encouraged to drink liquids to replace missing blood volume (substantially repleted in the first 4 to 12 hours after phlebotomy) and to wait under observation for at least 10 minutes. These are measures taken to prevent/recognize donor reactions, most likely to occur immediately following phlebotomy.

Amount of Blood Withdrawn

Standards allows blood to be withdrawn to a maximum of 10.5 mL/kg (including pilot tubes), with the recognition that too large a reduction of donor blood volume may cause severe reactions, and that blood volume is proportional to weight. This means that a 50 kg (110 lb) donor can have a maximum of 525 mL removed. Since samples needed for testing may exceed

25 to 30 mL, care must be taken for donors of lower weight using 500 mL collection sets. Alternatively, sets designed for 450 mL can be used for lower weight donors. The 10.5 mL/kg standard is useful also for calculating the amount of blood that can be drawn from a larger child.

Donor Reactions

The most common untoward donation event is the subsequent development of a bruise or hematoma at the venipuncture site. The most common relatively serious event is the vasovagal reaction, seen in 2% to 3% of donors (Newman 1997). These reactions occur during or after blood donation. They begin as a simple faint, with lightheadedness, weakness, pallor, nausea, diaphoresis, and hypotension. About 5% of these reactions become more severe, with loss of consciousness, associated with a significant drop in systolic blood pressure (sometimes as low as 50 mm/Hg) and a pulse rate ranging from 40 to 60 beats per minute. Severe reactions may include convulsive movements, vomiting, and (rarely) fecal or urinary incontinence. These reactions are frightening but inconsequential in otherwise healthy donors.

While the primary cause of this vagal stimulation is felt to be psychological, it is associated with the volume of blood removed, the age and donation experience of the donor, and the ambience of the collection facility. Prevention includes ensuring that lower weight donors are not overbled and that additional attention is paid to young and first-time donors or donors otherwise apprehensive, because early reactions often can be reversed. It is important to ensure that donors are under observation in the immediate postdonation period so that signs and symptoms are noted before fainting and subsequent injury occur.

Treatment involves keeping the donor supine, with legs elevated, to enhance blood flow to the head. Recovery usually takes 5 to 15 minutes. While some facilities are prepared to provide intravenous fluids, oxygen, or vagal-blocking medications, these interventions are almost never necessary. The donor who has had such a reaction should be advised that these reactions may recur within the next several hours, and extra caution should be used in driving or operating heavy machinery. Donors with severe reactions should be discouraged from attempting to give blood again in the near future. Another occasionally seen reaction is hyperventilation-induced hypocapnea with consequent respiratory alkalosis and symptoms of hypocalcemia. With early recognition this reaction is readily reversed by rebreathing expired air.

APHERESIS

Apheresis Platelets

Apheresis techniques allow collection of a specific desired component in large amounts with the return to the donor of the remainder of the blood. These procedures use continuous flow (or less commonly, intermittent flow) techniques, with one or two venipunctures. Generally, such a collection takes place over 1 to 2 hours, and may provide one or more adult doses. For platelets, the collection process allows simultaneous leukoreduction, so that postcollection filtration is unnecessary. Since recipient exposure of a dose of such platelets is to only one adult, there is less risk of bacterial sepsis, less exposure to transfusion-transmitted viruses, and less HLA sensitization than with pooled whole blood derived platelets. The process also allows for the availability of a panel of HLA-typed individuals to provide a specific product for an alloimmunized refractory recipient. These advantages have led to provision by apheresis of most doses of platelets given today.

Apheresis platelet donors are telerecruited and must meet all requirements of whole blood donors. Since the end component is the only platelet product given to the recipient as one dose, the donor must not have taken aspirin-containing compounds for 36 hours before donation. In addition, there are frequency limits: platelets cannot be collected more than 24 times a year, or more frequently than twice weekly. If the collection interval is less than one month, the donor platelet count must be shown to be greater than 150,000/µL. Red cell and plasma losses must also be monitored and kept within specific limits.

The adverse effects of apheresis donation are similar to whole blood donation; however, vasovagal reactions are less frequent (since those giving are usually older and repeat donors), and symptoms of hypocalcemia are more frequent (McLeod et al. 1998), in large part because of the infusion of citrate during the procedure. These symptoms are relieved by the ingestion of oral calcium-containing tablets or by slowing the reinfusion rate. A decrease in platelet count of 30,000–50,000/µL is regularly seen. Poor recovery to normal levels may require a cessation of apheresis collections.

Apheresis Red Cells

Using apheresis techniques, it is possible to collect two units of red blood cells from a donor at one sitting. This facilitates recruitment of donors of blood groups always in short supply, such as O negative, or donors with rare phenotype (Shi and Ness 1999). Since two units of red cells are removed at once, the donor criteria are more stringent. There must be a 16-week interval between donations, and the hemoglobin must be 13.3 g/dL (or hematocrit 40), with additional minimal weight requirements.

Apheresis Granulocytes

The in vivo recovery of transfused granulocytes has been shown to be dose related; to obtain the large doses thought to be necessary for efficacy in adults, the donor must be given corticosteroid and subcutaneous granulocyte colony stimulating factor (G-CSF), usually 12 hours before the collection, as well as a sedimenting agent such as 6% hetastarch during the procedure. The adverse effects are similar to those of plateletpheresis, as well as those related to the administration of the necessary drugs (for instance, those receiving repeated doses of corticosteroid should be questioned about peptic ulcer disease and diabetes). Reactions to these drugs are regularly seen, but usually are of minimal severity and consequence (Price et al. 2000), except when given repeatedly over a short time, such as for family members recruited to give granulocyte support for a patient or for the procurement of stem cells for transplantation, facilitated by 5 days of G-CSF.

PLASMAPHERESIS

Manufacturers of blood plasma derivatives have collected plasma by apheresis for subsequent pooling and fractionation ("source plasma") for years, typically from paid donors. Plasma of blood groups in short supply (such as Group AB) can also be collected from volunteer donors by apheresis and used as a component for transfusion. For the latter group of donors, if collected at intervals of 4 weeks or greater, up to a total of 12 L may be collected annually.

References

Brecher ME, ed. 2002. *Technical manual.* 14th ed. Bethesda, MD: American Association of Blood Banks.

Cable RG. 1995. Hemoglobin determination in blood donors. *Transfus Med Reviews* 9:131–144.

Code of Federal Regulations, 21 CFR. 2002 (revised annually). Washington, DC: U.S. Government Printing Office.

Code of Federal Regulations, 21 CFR 606.121(c)(5). 2002. Washington, DC: U.S. Government Printing Office.

Drake AW, Finkelstein SN, and Sapolsky HM. 1982. *The American blood supply.* Cambridge, MA: MIT Press.

Eastlund T. 1998. Monetary blood donation incentives and the risk of transfusion-transmitted infection. *Transfusion* 39:874–882.

Friday JL, ed. 2003. *Standards for blood banks and transfusion ser-*

vices. 22nd ed. Bethesda, MD: American Association of Blood Banks.

Gillespie TW and Hillyer CD. 2002. Blood donors and factors impacting the blood donation decision. *Transfus Med Reviews* 16:115–130.

McLeod BC, et al. 1998. Frequency of immediate adverse effects associated with apheresis donation. *Transfusion* 38:350–358.

Newman BH. 1997. Donor reactions and injuries from whole blood donation. *Transfus Med Reviews* 11:64–75.

Peterson LR, et al. 1994. The effectiveness of the confidential unit exclusion option. *Transfusion* 34:865–869.

Price TH, et al. 2000. Phase I/II trial of neutrophil transfusions from donors stimulated with G-CSF and dexamethasone for treatment of patients with infections in hematopoetic stem cell transplantation. *Blood* 95:3302–3309.

Shi PA and Ness PM. 1999. Two-unit red cell apheresis and its potential advantages over traditional whole-blood donation. *Transfusion* 39:218–225.

Simon TL. 2003. Where have all the donors gone? A personal reflection on the crisis in America's volunteer blood program. *Transfusion* 43:273–279.

Strauss RG, et al. 1990. Directed and limited-exposure blood donations for infants and children. *Transfusion* 30:68–72.

Sullivan MT. 2003. *Executive summary of the report on blood collection and transfusion in the United States in 2001*. Bethesda, MD: National Blood Data Reference Center.

C H A P T E R

2

Component Preparation and Storage

DAVID C. BURGHARDT, MT(ASCP)SBB

INTRODUCTION

A transfusion service will usually provide a full spectrum of blood and blood products for inpatients, outpatients, and emergency room patients. The components supplied by the transfusion service are usually designed to meet the specific requirements of the patient at that time. Preparation of blood components for the pediatric patient, although similar in some respects to that which is provided for the adult oncology or surgical patient, is more challenging in many respects. The differences go well beyond the issue of modifying the adult dose to fit the smaller pediatric patient.

Pediatric transfusion practice is usually divided into two periods: (1) neonates from birth through 4 months and (2) older infants (>4 months) and children. An important aspect of pediatric transfusion practice is that blood product volume requirements vary with the age and size of the pediatric patient. An integral part of the many tasks of the staff technologists in a transfusion service is to select and prepare blood components in volumes compatible with patient needs.

To this end, the goal of this chapter is to provide a resource that will help ensure that children, large or small, receive the safest, most effective transfusion therapy.

UNIQUE TRANSFUSION NEEDS OF THE NEONATAL AND PEDIATRIC PATIENT

Just as the adult oncology and surgery patient have both common and distinct blood component needs for their transfusion support, so too, the pediatric patient is a definable subgroup with unique transfusion needs (Chambers and Issitt 1994; Rossi et al. 1995). Each of the items discussed below has implications on the efficacy of the various transfusion options for the neonatal and older pediatric patient.

Blood components are usually requested and prepared based on the individual need of a patient. Several "special" product requirements that may be requested for the neonatal or older pediatric patient include: (1) "fresh" red blood cells (RBCs) or whole blood, (2) cytomegalovirus (CMV) "safe" components, (3) leukocyte-reduced blood components, (4) irradiated cellular components, (5) units that are negative for hemoglobin S, (6) volume-reduced platelet products, and (7) components that are packaged into small volume containers. This chapter will briefly discuss these options only because they may be essential to the preparation of each requested blood component. The rationale for use and a more in-depth discussion of these components is presented in other chapters of this book.

It is important to recognize that it may not always be possible to meet all of the needs of a patient as requested by his or her physician. As the number of "special" requirements for a component increases, so does the likelihood that one or more of the requirements may result in a compromise to ensure timely provision of the component. Issues related to testing, directed donation, and specialized needs all will impact the timely availability of a blood product. When the ideal component is not available, the methods used to prepare the component and the properties of the resultant component must be assessed to determine the best alternatives. Each transfusion service is therefore obligated to have a mechanism to ensure that the components issued for the patient meet the specific request of the ordering physician. One effective method to accomplish this objective is to communicate the special needs request for a specific patient to a physician who is trained and knowledgeable in transfusion medicine. The patient's transfusion needs may then be evaluated with the patient's physician.

Some institutions recognize that groups of patients may have similar blood product needs and therefore develop standard protocols for ensuring that these needs are met (for example, extracorporeal membrane oxygenation [ECMO] patients, organ transplant patients, patients requiring exchange transfusion). Once a patient is identified as a member of this specific protocol group, blood components meeting these "special" requirements are provided routinely.

Fresh Blood

One of the more contentious issues in pediatric transfusion medicine is the request by a physician for components containing "fresh" RBCs. Special requests for fresh whole blood or RBCs may be received for selected patients such as the neonate or for the cardiovascular surgery patient. Although justification for fresh blood is often difficult, it may be ordered in a belief that fresh blood provides the greatest oxygen-carrying capacity because it has the maximum level of 2,3-diphosphoglycerate (2,3-DPG) and minimal amounts of metabolic waste products such as potassium when compared to older blood. Many scientific studies have established that levels of 2,3-DPG have been found to return to normal within 12 to 24 hours after transfusion in most recipients (Strauss 2000). This should not present a problem for routine transfusions for most patients. Concerns about 2,3-DPG and potassium levels may be important in selected patient groups such as the premature neonate, neonatal or pediatric patients with impaired cardiac function, or patients with severe impaired pulmonary function such as a patient on ECMO (Rudman 1995; Strauss 2000).

Newborns, especially the premature newborn, are sometimes candidates for fresh blood because a newborn may have a high percentage of fetal hemoglobin, which does not release oxygen to the tissues as well as adult hemoglobin A. Fresh blood less than 7 days old with higher levels of 2,3-DPG could be of clinical significance in the seriously ill premature infant, especially the patient with clinical conditions such as respiratory distress syndrome or pulmonary atresia. Neonates requiring red cell exchange may likewise benefit from blood less than 7 days old.

When it comes to the provision of specialized blood components, a request for "fresh" blood will involve two challenges: the first involves the definition of "fresh," and the second involves the process that begins with the collection of the donor blood and ends with the distribution of the requested product to the hospital transfusion service. In our current environment of transfusion recipient safety (with the emphasis on the testing used to reduce transfusion-transmitted diseases), the challenge is to complete the required testing in a timely manner. Blood that falls into the range of 3 to 7 days after collection will likely be readily available and meet the needs of the few special patients who require this product.

In some situations, the time frame required to complete the processing of a donated unit might in fact be at odds with the physician concept of "fresh." For example, pediatric intensivists insist on utilizing "fresh" RBCs for ECMO patients because of their concerns about elevated potassium (K+) in the ECMO circuit. Although relatively stable premature infants appear to be at no substantial risk from transfusion of a small volume of old, stored RBCs, rapid transfusion or transfusion of relatively large volumes of RBCs can cause severe problems in the recipient (Strauss 1990; Strauss 2000). Mortality and cardiac arrest related to the rapid transfusion of older blood has been reported and may be related to high plasma K+ levels. For those patients who are infused via a catheter directly into the heart, the importance of plasma K+ levels cannot be ignored. Therefore it is recommended that fresh (3 to 5 days old) blood be utilized in those patients who are most at risk for potentially elevated K+ levels in blood products (Hall et al. 1993).

Cytomegalovirus Safe Components

Donor collection centers and hospital transfusion services are sometimes faced with requests for CMV-reduced-risk blood products. Some donor centers offer RBCs and platelet products that are leukocyte-reduced as a "CMV-reduced-risk" alternative. Leukocyte reduction with high-efficiency filters may help to reduce posttransfusion CMV in high-risk neonates and transplant recipients. It is estimated that less than 2% of healthy donors are able to transmit CMV infection. However, the passive acquisition of CMV antibody from plasma units, which is estimated to occur in 40% to 70% of donors, can make it falsely appear that a patient has been infected. Posttransfusion CMV infection is generally of no clinical consequence in immunocompetent recipients, and intentional selection of CMV-reduced-risk blood is not warranted for all patients. Some transfusion medicine experts consider effectively leukocyte-reduced cellular components as equivalent to serologically screened components, although this is a controversial issue (Hillyer et al. 1994; Smith and Shoos-Lipton 1997; Preiksaitis 2000). But for the high-risk immunocompromised organ transplant recipient who is CMV–negative or the low birth weight infant, there is a high morbidity and mortality rate associated with CMV infection (Blajchman et al. 2001).

In infants, CMV infection may be acquired in utero, during the birth process, during breast-feeding or by close contact with mothers or nursery personnel. CMV can also be transmitted by transfusion, although it is estimated that the risk from the current blood supply is small. Infections in newborns are extremely variable in their manifestations, ranging from asymptomatic seroconversion to death. Studies of CMV infection in neonatal transfusion recipients (Strauss 1990; Strauss 1993; Brecher 2002) reveal the following observations:

1. The overall risk of symptomatic posttransfusion CMV infection seems to be inversely related to the seropositivity rate in the community. Where many adults are positive for CMV antibodies, the rate of symptomatic CMV infection in newborns is low.
2. Symptomatic CMV infection during the neonatal period is uncommon in children born to seropositive mothers.
3. The risk of symptomatic posttransfusion infection is high in multitransfused premature infants weighing less than 1200 g who are born to seronegative mothers.
4. The risk of acquiring CMV infection is directly proportional to the cumulative number of donor exposures incurred during transfusion.
5. CMV in blood is associated with leukocytes. The risk of virus transmission can be reduced by transfusing CMV-reduced-risk blood from seronegative donors or by using leukocyte-reduced components. Although deglycerolized and washed red cells also have a reduced risk of CMV infection, leukocyte reduction by filtration is the technique of choice.

CMV infection, which is a serious and often fatal complication in previously CMV seronegative organ transplant recipients, is related to the presence of CMV in the organ donor and recipient and the degree to which the recipient is immunosuppressed. It is important to ascertain the CMV status of the organ donor and the recipient. A recipient may develop primary CMV infection if he or she is CMV negative and receives a CMV-positive organ. Additionally, recipients who are CMV positive may experience reinfection or reactivation during the myeloablative regimen (Blajchman et al. 2001).

Leukocyte-Reduced Blood Components

Leukocyte reduction of RBC and platelet products has been used for many years in selected groups of patients. It is important to recognize that leukocytes in blood products can induce adverse effects in the recipient. Published data (TRAP 1997; Vamvakas et al. 1999; Hillyer et al. 2001; Ratko et al. 2001; Brecher 2002) suggest that leukocyte reduction of red cells and platelet products may reduce the risk of:

1. Febrile nonhemolytic transfusion reactions
2. Human leukocyte antigen (HLA) alloimmunization that may lead to patients becoming refractory to platelet transfusions or graft rejections
3. CMV transmission, Epstein-Barr virus (EBV), and human T-cell lymphotrophic virus type I (HTLV-I), which are transmitted by the cellular components of the blood (primarily the leukocytes)
4. Transfusion related immunomodulation (TRIM)

There are several methods for preparation of leukocyte-reduced products: (1) centrifugation, (2) saline washing, (3) freezing and deglycerolizing, (4) spin-cool filtration, (5) bedside filtration, and (6) prestorage filtration. Many of these methods will not always produce a final product that will meet the requirements for effective removal of leukocytes. Leukocyte-reduction by prestorage filtration is the method of choice for consistent quality of the final product (Food and Drug Administration 2001).

The benefit of leukocyte reduction of components transfused to infants and children remains controversial among experts in transfusion medicine. Neonates are rarely alloimmunized because of the immaturity of their immune system. It is difficult to identify transfusion reactions in neonates. Some studies have shown that the neonate who requires frequent transfusion support of red cell and platelet products may derive greater benefit from receiving leukocyte-reduced components. The studies showed that the risk of transfusion-transmitted CMV can be substantially reduced through CMV-safe blood products. However, the conclusion that leukocyte-filtered blood components are as effective as CMV seronegative blood components remains controversial (Hillyer et al. 1994; Bowden et al. 1993; Smith 1997). Leukocyte-reduced blood components may be of particular benefit for those older pediatric patients who require chronic transfusion. One recognized benefit would be to diminish the development of alloimmunization to HLA antigens in those pediatric patients who are likely to become candidates for marrow transplantation. Clinical studies suggest that leukocyte-reduced products may induce an immunosuppressive effect in transplant patients.

Irradiated Cellular Components

Transfusion-associated graft-versus-host disease (TA-GVHD) is a usually fatal immunologic transfusion complication that results when immunocompetent viable lymphocytes in a donor's blood product engraft in the recipient host. The engrafted lymphocytes mount an immunologic attack against recipient tissues, including hematopoietic cells, leading to refractory pancytopenia with bleeding and infectious complications, which can result in a 90% to 100% mortality rate. Although TA-GVHD can occur with any product containing lymphocytes, recipients at greatest risk are those who are severely immunosuppressed or immunocompromised, such as bone marrow transplant patients, low birth weight infants, infants who receive intrauterine transfusions and exchange transfusions, patients who are on ECMO, and patients who are recipients of designated donations from any blood relative (Petz et al. 1996; Kickler and Herman 1999; Strauss 2000; Hillyer et al. 2003).

Irradiation of cellular blood components is the only acceptable method to prevent TA-GVHD. Leukocyte-reduction of blood components by itself is not sufficient to avoid TA-GVHD, which has been reported after transfusion of leukocyte-reduced components. Irradiation of all cellular products with a minimum of 25 Gy (2500 cGy) inactivates the lymphocytes, leaving the platelets, red cells, and granulocytes undamaged (Food and Drug Administration 2000; Brecher 2002). Irradiation should never be performed on bone marrow or peripheral blood progenitor cells prior to their infusion.

There are currently no data to support a mandate for universal irradiation of blood products transfused to infants or children (Strauss 2000). Irradiated products may be transfused to patients who are immunologically normal with no adverse effects to the patient. Scientific studies have documented that irradiation of RBCs will produce some damage to the red cells causing them to leak intracellular fluids, especially potassium, and reduce their overall viability (24-hour-recovery). As a result, red cell components that have been irradiated require shortened expiration dating. Irradiated red cell products must be labeled to expire on their originally assigned expiration outdate or 28 days from the date of irradiation, whichever is shorter (Moroff and Luban 1997; Gorlin 2002). The use of older irradiated red cells is contraindicated in those patients who are most at risk for potentially elevated K+ levels in blood products (Hall et al. 1993). Platelets sustain minimal damage from irradiation, so their expiration date does not change. Granulocytes contain large numbers of lymphocytes and therefore should be irradiated to prevent TA-GVHD.

Hemoglobin S-Negative Red Cell Products

There are specific childhood conditions characterized by deficient or abnormal hemoglobin structures and anemia (for example, sickle cell disease, thalassemia) for which chronic red cell transfusions are used to treat tissue hypoxia and also to suppress erythropoiesis. The goal of transfusion in patients with sickle cell disease is to reduce the risk of stroke by decreasing the percentage of circulating red cells capable of sickling, while simultaneously avoiding an increase in blood viscosity. Maintaining a hemoglobin level of 8 to 9 g/dL with a hemoglobin S level less than 30% appears to be of benefit to children with sickle cell disease. Therapy may include routine transfusion every 3 to 4 weeks to maintain these levels or erythrocytopheresis to treat acute complications such as splenic sequestration, aplastic crisis acute chest syndrome, and priapism. Donor blood for transfusion to a patient with sickle cell disease should be screened for hemoglobin S (Capon and Chambers 1996; Rosse et al. 1998; Simon et al. 1998).

The goal of transfusion in children with thalassemia and severe anemia is to ameliorate the complications associated with the disease (Simon et al. 1998). Hypertransfusion, in which the hemoglobin is kept between 8 and 9 g/dL, allows for normal growth and development and normal levels of activity for these patients. Screening of donor blood for hemoglobin S offers no benefit for this group of patients.

Volume-Reduced Platelet Products

Routine centrifugation of platelets to reduce the volume of transfusion is not necessary. Patients who require fluid restriction, such as a low birth weight infant, may benefit from volume-reduced platelet products. It is important to recognize that if platelets have been volume-reduced and also placed in a syringe, the pH will decline rapidly. This is a potential problem for an already ill, acidotic patient. There is also an increased risk of bacterial contamination when volume-reducing a platelet product (Kickler and Herman 1999).

Centrifugation of a platelet product and subsequent removal of most of the plasma can be used to prepare a volume-reduced platelet component. However, centrifugation of the platelet product will result in unwanted clumping of the platelets, inadvertent platelet loss, and possible damage to the platelets. The quality and efficacy of the final product is questionable.

The product can be centrifuged for 10 minutes at $2000 \times g$ at 20° to 24°C followed by removal of all but 15 to 20 mL of the platelet-reduced plasma. The platelets should be allowed to "rest" undisturbed for 15 to 60 minutes followed by gentle manual kneading and mechanical rotation for 15 to 30 minutes. If leukocyte reduction and volume reduction are both required, the component should be leukocyte-reduced before centrifugation to prevent loss of platelets.

Components That Are Packaged into Small Volume Containers

The goal of transfusion therapy is to provide the most appropriate product for the patient. When components are prepared based on the individual needs of the patient, this greatly improves treatment outcomes while maximizing donor resources (Simon et al. 1998). The ability to provide transfusion products "packaged" for each patient's specific needs is an essential part of component therapy in the pediatric setting. In this setting, it is sometimes desirable to make components or smaller aliquots of a primary donor product (Hume and Bard 1995; Strauss 2000). For example, a request may be made to provide a 35-mL aliquot of red cells for transfusion to a neonatal patient. The product may be prepared utilizing one of a variety of methods without wasting the remaining original red cells. Effective component preparation results in better management of the donor product.

Component preparation must incorporate the principles of aseptic techniques using sterile, pyrogen-free equipment and solutions (Gorlin 2002). Most donor collection centers are equipped to provide transfusion services with red cells that have attached satellite bags that can be used to split off smaller volumes of red cells. The blood bags used for phlebotomy are designed to have integral, sterile bags known as "satellite bags" attached to the main collection bag. After collection and processing the donor whole blood product into red cells, the remaining attached satellite bags can be used to transfer smaller volumes of red cells through the closed system of tubing. The desired volume of red cells may be transferred into one of the satellite bags and then the tubing may be heat-sealed and separated from the primary bag. Preparing the product in this type of a closed system maintains sterility and allows the primary and separated products to keep the original unit outdate. Therefore, the primary bag may be split into many doses over time without shortening the outdate of the original product. Additionally, the chances of accidental contamination of the product are minimized.

When preparing smaller volume aliquots from the primary donor red cell component, each subsequent aliquot of the product must be labeled to mirror the original bag label and each piece must have a unique donor identification number. The usual practice is to label the product with the same original donor number followed by a predetermined nomenclature of succes-

sive alpha or numeric characters. For example, if the original unique donor identification number was 12LW12345, the first split product could be labeled as 12LW12345/1 or 12LW12345A; the second split could be labeled as 12LW12345/2 or 12LW12345B, and so on for each subsequent split piece.

If no satellite bags are attached to the primary red cell unit, it is possible to connect additional satellite bags using one of two methods, one of which opens the system and the other method which maintains the closed sterility of the system. The first method involves using one of the portals on the top of the primary bag as the point at which an extra satellite bag is attached. Because this method involves potential contamination of the original product as well as the separated piece, these products must be labeled with a significantly shortened expiration dating. During preparation of components, if the method of preparation compromises the sterility of the system or if there is a concern that it may have, the outdate of the products involved must be changed to a shorter date/time. In general, if a component is stored at $1°$ to $6°C$, the outdate of the product is 24 hours from preparation. If the product is stored at $20°$ to $24°C$, the product must be used within 4 hours of the preparation.

The second separation method allows for the attachment of a syringe or transfer bag(s) to the original product bag by means of a device known as a sterile connection device (sterile docking device). This device (Terumo Medical Corporation; Somerset, NJ) provides a method of joining two separate tubings together in a sterile manner so that neither item is "opened" to the air. The device welds two pieces of the same size blood tubing by melting the plastic tubing with a superheated metal wafer, moving the two ends to be joined together, removing the wafer, and allowing the two hot ends of the tubing to fuse, creating a sterile attachment of items. Tubing may be welded together from transfer bags, special syringe sets, needles, filters, and collection or apheresis sets as long as the diameter and materials of the tubing to be connected are similar. Because each part that is connected/welded together maintains its internal sterility during the process, the original outdating of each product or bag is maintained. When using the sterile connection device, each weld must be carefully inspected for completeness, integrity, leakage, and air bubbles. If the weld quality is in doubt or unsatisfactory, the outdating of the product should change as indicated in the previous paragraph (Rudmann 1995).

A variety of alternatives exists for selection of the bag or syringe use for packaging the separated blood component. Table 2.1 summarizes a variety of commercially available transfer bags that are approved for the separation of components. Table 2.2 summarizes a

TABLE 2.1   Blood Transfer Bags

Transfer Bag	Baxter	Terumo	Charter Medical	NHS
75 mL Pedi-Pak Quad with filter				402–04F
75 mL Pedi-Pak Quad without filter				402–04
75 mL Pedi-Pak Quad needleless Y site				402–01
75 mL Pedi-Pak Single				402–024
100 mL triple bag aliquot system with piercing pin			T2182	
100 mL triple bag aliquot system with piercing pin & injection site			T2183	
100 mL quadruple bag aliquot system with piercing pin			T3007	
100 mL sextuplet bag aliquot system with piercing pin & injection site			T2186	
100 mL eight bag aliquot system with piercing pin & injection site			T2188	
150 mL single with piercing pin			T3101	
150 mL single	4R2001	4BT015CB70		
150 mL double bag aliquot system with piercing pin			T3002	
150 mL quadruple	4R2004			
150 mL quadruple bag aliquot system with piercing pin & injection site			T3000	
150 mL quadruple bag aliquot system with 150 micron filter, piercing pin & injection site			T3001	
150 quadruple bag aliquot system with attached 600 mL transfer bag			T3605	
150 mL sextuplet bag aliquot system with attached 600 mL transfer bag			T3607	
150 mL sextuplet bag aliquot system with piercing pin & needleless syringe adaptor			T2286	
150 mL eight bag aliquot system with piercing pin & needleless syringe adaptor			T2288	
150 mL eight bag aliquot system with piercing pin & inject site			T2008	
300 mL single with coupler	4R2014	4BT030CB71		
400 mL single with piercing pin			T3104	
600 mL single with piercing pin			T3106	
600 mL single with coupler	4R2023	4BT060CB71		
600 mL single with needle adapter	4R2024			
600 mL single with slip lock adapter			T3135	
800 mL single with 2 couplers without outlet ports	4R2054			
800 mL single with 2 couplers & outlet ports	4R2055	4BT080BB71		
1000 mL single with piercing pin			T3107	
1000 mL single with coupler	4R2032	4BT100BB71		
1000 mL single with slip lock adapter			T3137	
2000 mL single with coupler	4R2041	4BT200BB71		
2000 mL single with piercing pin			T3108	
Platelet pooling unit 600 mL/8 couplers	4R2027			
400 mL bag with attached four-lead harness			T4004	
400 mL bag with attached six-lead harness			T4006	
600 mL bag with attached ten-lead harness			T6010	

NOTE: This is not a complete list of all products available. Please consult manufacturer for additional products.

variety of commercially available syringe devices that are approved for the separation of components. The transfer bags or syringes are designed to be attached by either method of attachment described above. The decision to use a transfer bag or syringe will depend on the method of separation, the number of connections/separations that is anticipated for the splitting of the product, and the cost of the materials for each option.

In the pediatric setting, because of the availability of syringe infusion pumps and the need to monitor fluids infused into neonates or smaller pediatric patients, patient care personnel may request that blood components be issued in syringes when appropriate (Chambers 1995; Hume and Bard 1995). Each component, whether packaged in a blood bag or syringe, must be labeled. It is recommended that the FDA uniform blood labeling guidelines be followed. When the units are subdivided, a standard labeling convention should be used to uniquely identify each successive component prepared from the original product (see previous discussion). In all cases, the donor identification must be traceable. Additionally, records must be kept of any modifications made to a product. The record must include the date, time, component used, component made, expiration date of the newly created product, identification of the person performing the modifica-

TABLE 2.2 Pediatric Syringe Sets

Pediatric Syringe Sets	Charter Medical
Blood administration set with 30 cc BD syringe and attached 100 mL bag with 150 micron filter	03-BDP-30
Blood administration set with 60 cc BD syringe and attached 100 mL bag with 150 micron filter	03-BDP-60
Blood administration set with 60 cc Monoject syringe and attached 100 mL bag with 150 micron filter	03-MJP-60
Blood administration set with 150 micron filter and 30 cc BD syringe	03-960-00
Blood administration set with 150 micron filter and 60 cc BD syringe	03-960-10
Blood administration set with 150 micron filter and 30 cc Monoject syringe	03-960-35
Blood administration set with 150 micron filter and 60 cc Monoject syringe	03-960-37
TRUFLOW NEO-40 Blood administration set with 40 micron filter and 30 cc BD syringe	N40-960-00
TRUFLOW NEO-40 Blood administration set with 40 micron filter and 60 cc BD syringe	N40-960-10
TRUFLOW NEO-40 Blood administration set with 40 micron filter and 35 cc Monoject syringe	N40-960-35
TRUFLOW NEO-40 Blood administration set with 40 micron filter and 60 cc Monoject syringe	N40-960-37
TRUFLOW NEO-40 Blood administration set with 30 cc BD syringe and attached 150 mL bag with 40 micron filter	N40-BDP-30
TRUFLOW NEO-40 Blood administration set with 60 cc BD syringe and attached 150 mL bag with 40 micron filter	N40-BDP-60
TRUFLOW NEO-40 Blood administration set with 60 cc Monoject syringe and attached 150 mL bag with 40 micron filter	N40-MJP-60

NOTE: This is not a complete list of all products available. Please consult manufacturer for additional products.

tion, and the lot numbers of the materials used to prepare the components. The purpose of this record keeping is to assist in identifying and resolving component quality and production problems.

BLOOD DONATION ISSUES

Although the intent of this chapter is to primarily discuss component preparation and storage, part of the decision related to the selection of blood products for the neonatal and older pediatric patient may ultimately involve consideration of donor exposure. Many parents of the neonatal or older pediatric patient will arrive at the decision to seek alternatives for the use of allogeneic donor blood products. For most parents, this decision results from a protective desire to provide the "safest" blood products possible for their child. This decision is sufficiently important that they will, given enough time, educate themselves about the alternatives to or substitutions for allogeneic blood. Pediatric health care providers are in a unique position to be supportive of the alternative sources for provision of blood components.

Autologous Donation

Pediatric autologous donation is most successful when the likelihood of transfusion is high. For the adolescent and preteen patient, preoperative donation of autologous red cells and frozen plasma components may offer the best choice in many situations. Autologous donation is a recognized alternative in planned orthopedic procedures. One or more donations of whole blood may be adequate to provide the anticipated prod-

ucts for surgical procedures. Younger children, however, are usually not suitable candidates for autologous donation because of their small blood volumes, limited or difficult venous access, and concerns regarding their cooperation during collection. Greater success will be achieved when careful evaluation of the pediatric patient occurs prior to autologous donation. Intraoperative salvage and isovolemic hemodilution in the adolescent and preteen may also represent an alternative source for autologous blood.

Directed Donation

A directed donation is one in which a person "directs" his or her donated blood to be used by a given patient. For pediatric patients, the parents or legal guardian will designate donors on behalf of the child. Directed donor blood is suitable for use in planned surgical procedures and routine transfusions. Concern about the overall safety of blood products has generated expectations by parents that they will be best suited to select suitable donors to supply the blood components that their child may need. Because of logistic and philosophical problems associated with these directed donations and scientific evidence that suggests that directed donor blood may not be as safe as allogeneic donor blood, most blood centers and hospitals will provide this service only upon request.

Biological parents often *want* to be the directed donor for their child or infant. As with all transfusions from biological relatives, certain risks are associated with transfusion of blood components donated by biological parents. TA-GVHD is a potential risk. Therefore, irradiation of all blood components that may contain viable lymphocytes is mandatory. Additionally, there may be other circumstances in which there is inherent risk. In cases of suspected or serologically confirmed neonatal alloimmune thrombocytopenia (NAIT), the mother may be the best candidate to donate platelet products (Petz et al. 1996; Kickler and Herman 1999). At the time of suspicion or diagnosis of NAIT, which usually occurs in the first few days of the infant's life, the mother may have a low-grade infection or anemia. This may compromise her ability to be a "good" donor. Maternal plasma alloantibodies to either white cell or platelet antigens are frequently present but remain undetected because routine pretransfusion compatibility testing does not screen for these antibodies. If the decision is made to use the mother as the platelet or RBC component donor, consideration must be given to the removal of maternal plasma antibodies that have the potential to cause hemolysis or immune cytopenia in the infant (Kickler and Herman 1999; Brecher 2002). Additionally, the mother may also be used as a donor if she possessed an antibody to a high incidence red cell antigen that the infant may have genetically inherited from the father. Likewise, cellular blood components (red cells, white cells, or platelets) from the father may contain the same antigen as the infant's red cells. These components may be susceptible to any maternally derived alloantibodies. Because of these and other concerns, it is recommended by transfusion medicine specialists that biological parents not be donors of blood for their infant during the first few weeks of life.

Limited Exposure Donor or "Dedicated" Donors

Given the relatively long life expectancy of the pediatric patient, serious consideration to reducing the risk of transfusion-associated infection as well as allogeneic antigen exposure may be appropriate for those pediatric patients who will require long-term transfusion support. One such consideration is in limiting the number of allogeneic donor exposures. A limited exposure blood donor program is an alternative form of allogeneic directed donation in which a single dedicated donor or a small number of dedicated donors provide all of the blood components necessary for a given patient's transfusion needs (Hare et al. 1994). In theory, fewer donor exposures results in both reduced risk of transfusion-associated disease transmission as well as decreased alloantibody production. Limited donor programs for the donation of red cells, platelets, or plasma products may be especially suited to meet the ongoing blood product needs of the pediatric patient. The number of donors needed is largely dependent on patient size and ongoing need for specific blood products. The logistics of having the appropriate blood

component available can be complicated. Coordination of the donation process with the blood center and the hospital may require dedicated time and resources on the part of a representative of the patient. The goal of a well-coordinated limited donor program is to efficiently utilize available donated blood components and limit wastage. Effective use of the limited donor resources may result in the provision of quality blood components at reasonable cost and bring about improved patient care.

ANTICOAGULANT-PRESERVATIVE SOLUTIONS

Anticoagulant-preservative solutions are used in blood collection/storage to: (1) maintain blood products in an anticoagulated state, (2) support the metabolic activity of red cells and maintain function of cellular constituents while in storage, and (3) minimize the effects of cellular degradation during storage. There are several anticoagulant preservative solutions, including CPDA-1, CPD, and ACD-A, that are approved by the FDA for storage of blood and blood components (Code of Federal Regulations 2001). Most anticoagulant-preservative solutions contain citrate, which maintains the anticoagulated state through chelation of calcium; dextrose and/or adenine, which support ATP synthesis in the stored cells; and sodium phosphate which provides phosphate for ATP synthesis and serves as a buffer to minimize the effects of decreasing pH in the stored products.

Additive systems, such as AS-1, AS-3, and AS-5, extend the shelf life of red cell products via an adenine additive solution. The additive solution must be added to the anticoagulated red cells according to the manufacturer requirements, which is usually within 72 hours of collection. The volume of additive solution is 100 mL, which increases the total volume of the red cell component and also decreases the hematocrit of the red cell component.

The selection of anticoagulant-preservative solution for red cells for pediatric transfusion is largely dependent on the size of the patient and to a lesser degree the clinical condition of the patient. In general, RBCs stored in CPDA-1 are the preferred product used for pediatric transfusion. The rationale for this selection is in the delivery of a higher percentage of red cells to total volume of product infused. The decision to use CPDA-1 versus additive solutions (AS) may also be dependent on availability of the product from the blood center. It would not be surprising to point out that blood centers have a different perspective on the distribution of blood components and their potential use by all of their customers. Transfusion services that treat neonatal and older pediatric patients must make the blood center aware of the unique needs of their patients.

There is some concern regarding potential side effects of the adenine, dextrose, and mannitol in AS solutions. Large amounts of adenine and mannitol have been reported to be associated with renal toxicity. Mannitol is also a potent diuretic and due to its effect on fluid dynamics of premature infants, may cause unacceptable fluctuations in cerebral blood flow. Published studies demonstrate that red cells preserved in extended storage media present no substantive risks when used for small volume transfusions in neonates (Luban et al. 1991; Hall et al. 1993; Strauss 2000). For premature infants with severe hepatic or renal insufficiency, however, use of red cells stored in CPDA-1 may be the preferred product (Luban et al. 1991). If RBCs stored in additive solutions must be used, removal of the preservative by washing or centrifugation and resuspending the RBCs in another appropriate fluid is recommended. The safety of red cells stored in all additive solutions and used in cardiathoracic surgery and massive transfusions has not been confirmed (Brecher 2000).

INDIVIDUAL COMPONENTS FOR TRANSFUSION

Whole Blood

Whole blood contains 450 to 550 mL +/−10% of donor blood plus the anticoagulant. Whole blood is routinely collected into collection bags with transfer bags attached for component preparation. Separation of whole blood into specific components allows selective transfusion therapy based on specific patient need and also maximizes the number of recipients who may benefit from any single donation. The most commonly used configuration is intended for the production of RBCs, fresh frozen plasma (FFP), and platelets (Circular of Information 2002; Triulzi 2002).

Whole blood is not routinely available for allogeneic transfusion. This is because whole blood is routinely collected and processed into RBCs, plasma, and other components by the blood center. The routine preparation of these components is necessary to meet the needs of most patients as well as manage the overall cost of the blood collection/testing/component preparation process. Whole blood may be useful for patients with concomitant RBC and volume deficits, such as actively bleeding patients. It will provide oxygen-carrying capacity, blood volume expansion, and stable coagulation factors. The standard of care in most hospitals for cases of active bleeding in trauma and surgery patients is to provide RBCs, plasma, and cryoprecipitate or platelets for the specific coagulation products needed. In these situations, transfusing individual components rather than whole blood better serves the patient's specific needs. One minor advantage of using whole blood is that there will be less donor exposures for the comparative volume of RBCs and plasma transfused. One downside of using whole blood is that the amount of RBCs delivered per milliliter of transfused blood product is significantly less than that attributed to use of red blood cell components.

Reconstituted whole blood may be prepared by combining FFP with RBCs in the hospital blood bank. 5% albumin or saline may also be substituted for FFP. This product is frequently used for exchange transfusions.

Red Blood Cells

RBCs (packed RBCs) are prepared by removing most of the plasma from a unit of centrifuged or sedimented whole blood. During initial component preparation, the plasma portion can be removed and then may be used for further component separation (Circular of Information 2002; Triulzi 2002).

RBCs have the same red cell volume and therefore the same oxygen-carrying capacity as whole blood in a significantly reduced volume. Red cells are the preferred product for patients with symptomatic anemia. Patients with deficient or abnormal hemoglobin structures may require chronic transfusion with red cells to ameliorate the complications associated with the disease. Red cells may also be used for exchange transfusion in newborns or patients with sickle cell disease in crisis (Simon et al. 1998).

A variety of modifications can be used when RBCs are provided for transfusion. They include leukocyte-reduced red cells, washed red cells, and frozen/deglycerolized red cells. There are a number of factors, such as cost, time requirements, need for special equipment, and ease of manipulation, to consider when making the decision to provide any of these RBC modifications. RBCs may also require irradiation in addition to the above modifications.

Leukocyte-reduced RBCs are generally the product of choice for transfusion to pediatric patients. Prestorage leukocyte-reduction of RBCs has now made it possible to provide this product with consistent quality and with little manipulation, though the expense is slightly increased. The advantages and disadvantages of routine use of leukocyte-reduced RBCs were previously discussed in this chapter.

Washed RBCs are preferable to use in the following situations: (1) to remove metabolic components such as potassium or plasma hemoglobin, (2) to remove additive solutions, (3) to remove plasma proteins and microaggregates that can cause febrile and urticarial reactions in some patients, and (4) for those patients who have developed antibodies to plasma proteins (for example, the IgA-deficient patient who may experience severe and even fatal anaphylaxis when receiving products carrying IgA). Use of leukocyte-reduction filters will provide a much better leukocyte-reduced product than washing of RBCs (Hillyer et al. 2001).

Frozen/deglycerolized RBCs are prepared for a variety of reasons, which may include the storage of blood from a rare donor, autologous units stored for a future scheduled surgery, or a means to conserve or stockpile inventory that may be needed at a future date. The deglycerolization process used for frozen RBCs will provide the same advantages as washed RBCs, but at an increased cost.

Undoubtedly, the easiest method of providing a RBC component for transfusion to pediatric patients is the use of unmodified RBCs. There are advantages and disadvantages that must be taken into consideration when making the decision of which of these types of RBC preparations will meet the needs of your patients.

Platelets

Platelets play a significant role in hemostasis by repairing breaks in small blood vessel walls and releasing phospholipids and other products that are required for in vivo hemostasis. Platelets are essential to the formation of the primary hemostatic plug and provide the hemostatic surface upon which fibrin formation occurs. Deficiencies in platelet number and/or function can have unpredictable effects that range from clinically insignificant prolongation of the patient's bleeding time to major life-threatening hemorrhage. Decreased platelet numbers result from many conditions that decrease platelet production, increase destruction, or occur due to dilutional thrombocytopenia during massive transfusion. Platelet function may be adversely affected by such factors as drugs, sepsis, liver or kidney disease, increased fibrin degradation products, cardiopulmonary bypass, and primary marrow disorders (Blanchette et al. 1995; Kickler and Herman 1999).

Before platelet therapy is initiated, it is important to determine if platelets are necessary to prevent or correct the bleeding episode. Many factors, such as the number and functional ability of the patient's own platelets and the cause of the thrombocytopenia, must be evaluated.

Platelets used for transfusion may come from two different sources: (1) those prepared by separating platelet rich plasma from routine whole blood donation ("platelet concentrates" or whole blood derived platelets), or (2) those collected from a single donor using the cytapheresis process ("single donor platelets" or apheresis derived platelets).

Platelet concentrates, which are the platelets that are prepared from a single unit of whole blood, are prepared by a two-step process. The first step uses low-speed centrifugation to separate the red cells from the plasma. The resultant plasma product is termed *platelet-rich plasma*. Then the platelet-rich plasma is subjected to high-speed centrifugation. The platelet-poor plasma on the top is expressed into another bag, and the residual platelet pellet is resuspended in a small volume of plasma. The resultant platelet bag must then be allowed to rest undisturbed at room temperature before resuspending the platelets, which undergo aggregation during the centrifugation process.

Platelet Concentrates

Platelet concentrates usually contain a minimum of 5.5×10^{10} platelets (Circular of Information 2002; Triulzi 2002). A single unit of platelet concentrate may be adequate for transfusion to a neonate or infant. The infusion of 10 mL/kg of platelet concentrate should be able to raise the blood platelet count to $>100 \times 10^9$/L in an infant and 1 U/10 kg for children and smaller adults (Kickler and Herman 1999). Individual platelet concentrates may be issued in the original container or transferred to a syringe. All platelets issued in a syringe must be transfused within 4 hours of preparation. Platelet concentrates may also be pooled together in a single bag to meet the transfusion dose required. If pooling of individual platelet concentrates is performed, all of the units in the pool should be the same ABO group. There are a number of different methods for pooling platelet concentrates, but all involve either the use of an open system or a closed system by use of the sterile connecting device. A platelet pooling set with a bag and eight couplers is available (Baxter Healthcare Corporation; Deerfield, IL). Alternately, a plasma transfer set of a size adequate to meet the final volume of the combined product and multiple couplers may be used. The drawback of either preparation process is the potential for contamination of the product in the bag. Pooled platelets in an open system outdate 4 hours after pooling. The final container should be labeled according to the FDA uniform blood labeling guidelines, must be labeled as "Platelets Pooled," must be assigned a single unique identification number, and this identification number must reflect all platelet concentrates included in the pool. A record must be kept of pooling. The record must include the date, time, component used, component made, expiration date of the newly created product, identification of the person performing the modification, and the lot numbers of the materials used to prepare the components. The purpose of this record keeping is to assist in identifying and resolving component quality and production problems.

Single Donor Platelets

Platelets collected by apheresis contain a minimum of 3×10^{11} platelets, which is the equivalent of five to eight random donor platelet concentrates (Circular of Information 2002; Triulzi 2002). The benefit of single-donor platelets in the multiply transfused recipient is the decrease in donor exposures. Apheresis platelets may be separated into smaller aliquots for transfusion to the same patient over the 5-day dating period. Although the separation process may be performed using an open system with a plasma transfer bag, which will then have an expiration of 4 hours from preparation of both pieces of the split product, the preferred method of separation is to utilize the sterile connecting device. A plasma transfer bag or syringe may be attached to the original apheresis bag using the sterile connecting device. The desired volume of product can then be transferred using a closed tubing system. Using this process will not change the original expiration date of the final products. As a cautionary note, unless the transfer bag is made of plastic suitable for platelet storage (bags made of PL-732 plastic are suitable), the aliquot should be used as soon as possible. Additionally, there is no method to accurately split the total number of platelets in the apheresis product when transferring the bag into smaller aliquots. Although adequate mixing of the primary apheresis bag is essential before separating the product, the accuracy of the final transferred product will be in volume only. A platelet count and calculation on the transferred product may be necessary to assess the total number of platelets in the final product.

Published clinical studies have indicated that a single donor plateletpheresis is no more effective than the equivalent number of random donor platelet concentrates in raising the platelet count in patients who have yet become alloimmunized (TRAP 1997; Kickler and Herman 1999). Although apheresis platelets or random donor platelet concentrates may be used interchangeably, the drawback of using multiple random donor platelet concentrates is the risk of increased donor exposure from the perspective of the potential for alloimmunization as well as infectious disease transmission.

Platelet components may require further modification to meet special patient needs. They can be irradiated, leukocyte-reduced, volume-reduced, washed, and split into smaller aliquots.

Fresh Frozen Plasma

FFP is prepared by removing plasma from a unit of donated whole blood and placing it at $-18°C$ or lower within 8 hours of collection (Circular of Information 2002; Triulzi 2002). FFP contains plasma proteins plus all coagulation factors. The component is usually packaged in a volume of 200 to 250 mL, but some blood centers will split a unit of FFP into multiple pieces of approximate volume of 50 to 60 mL. This split product is called "pedi-FFP" and is suitable for transfusion for patients who require smaller volumes of FFP. Blood centers may preferably collect "pedi-FFP" from group AB donors so that the product will be ABO-compatible with all recipients. "Pedi-FFP" may also be prepared by splitting plasma collected by apheresis donor procedure.

Clinical need for FFP in infants and children is relatively limited but similar to the indications for adults. Transfusion of FFP is appropriate for: (1) replacement of specific coagulation factors not available in concentrate form, (2) management of multiple factor deficiency, which is usually a result of liver disease, and (3) treatment of antithrombin III (ATIII) deficiency. FFP should not be used as a general volume expander because it carries the risk of disease transmission and safer, less expensive products are readily available. Nor should it be used when a more appropriate and specific therapy is available. For example, using FFP to replace factor VIII in a factor VIII deficient hemophiliac is inappropriate. Not only will more volume be infused than is necessary, but specific factor VIII concentrates, such as recombinant factor VIII, are readily available and are much safer from the perspective of disease transmission (Development Task Force of the College of American Pathologists 1994; Petz et al. 1996).

FFP may be issued in syringes. The syringed FFP outdates in 24 hours if prepared in a closed system; otherwise, the FFP outdates in 4 hours.

Cryoprecipitate-Poor Plasma

Cryoprecipitate-poor plasma is a by-product of cryoprecipitate preparation (Circular of Information 2002; Triulzi 2002). The product lacks labile clotting factors V and VIII, but contains adequate levels of the clotting factors II, VII, IX, and X. Cryoprecipitate-poor plasma lacks fibrinogen because it was concentrated in the cryoprecipitate during preparation. It also lacks von Willebrand's factor (vWF) and its multimers. Thus, cryoprecipitate-poor plasma may be suitable for transfusion into some patients with deficiencies of the stable clotting factors and is used in the treatment of thrombotic thrombocytopenic purpura (TTP).

Cryoprecipitated Antihemophiliac Factor

Cryoprecipitate is the cold-insoluble portion of plasma that remains after FFP has been allowed to thaw at 1° to 6°C (Circular of Information 2002; Triulzi 2002). The product is separated from the thawed plasma and refrozen as cryoprecipitate. The remaining plasma can be labeled as cryoprecipitate-poor plasma (as above) or processed into single donor plasma. Cryoprecipitiate is rich in coagulation factor VIII (both factor VIII-C and factor VIII-vWF), factor XIII and fibrinogen. On average, one unit of cryoprecipitate contains approximately 250 mg of fibrinogen. The product also contains variable amounts of fibronectin, a protein that participates in phagocytosis.

The primary clinical use of cryoprecipitate is for intravenous supplementation of fibrinogen, particularly in disseminated intravascular coagulation (DIC) or in hypofibrinogenemia, which is a rare congenital deficiency of fibrinogen. Secondarily, cryoprecipitate can be used in factor XIII deficiency, although a more suitable factor concentrate is currently available. It is also used in conjunction with thrombin as a topical agent to stop bleeding when applied directly as fibrin glue. Cryoprecipitate may also be useful in dysfibrinogenemias that are congenital or acquired. Patients with severe liver disease frequently exhibit a dysfribrinogenemia. Although cryoprecipitate had previously been used for patients with factor VIII-deficient hemophilia, it is not currently recommended for use because safer factor VIII products are available.

Granulocytes

Granulocytes are prepared upon request by cytapheresis and contain greater than 1.0×10^{10} granulocytes suspended in 200 to 300 mL of plasma (Circular of Information 2002; Triulzi 2002). This product also contains large numbers of red cells, platelets, and other leukocytes. In special circumstances, a granulocyte concentrate can be prepared from the buffy coat of a fresh unit of single donor whole blood. This is primarily used for neonates who may require only small volumes of granulocytes. Granulocytes prepared by cytapheresis are preferred over buffy coat preparation. In either case they should be transfused as quickly as possible following collection (≤24 hours).

Granulocytes are used as supportive therapy for patients with severe neutropenia and documented sepsis unresponsive to antibiotic therapy (Hillyer et al. 2003). Because granulocytes are only a temporary form of therapy, the patient should be evaluated to ensure that there is a reasonable chance for recovery of marrow function. Granulocytes have also been used in patients with chronic granulomatous disease that may have normal numbers of granulocytes, but the white cells are dysfunctional.

Neonates are more susceptible than older children to severe bacterial infections. Controversy surrounds several issues in granulocyte transfusion for neonates, including dose, neutrophil level at which to transfuse, type of component to use, and efficacy as compared to other forms of therapy.

Based on the fact that granulocyte concentrates contain large numbers of lymphocytes, there is consensus that all granulocyte transfusions should be irradiated to prevent TA-GVHD. Additionally, for granulocyte transfusions to neonates, donors are usually selected to be CMV seronegative and must be ABO-group compatible with the infant. Additional manipulation of granulocytes, such as splitting the product, may in fact be detrimental to the constituents of the product.

RECORD KEEPING FOR COMPONENT PREPARATION

Current good manufacturing practice (cGMP) regulations require that records be maintained for all aspects of component preparation (Brecher 2002; Gorlin 2002). Documentation is required to ensure that all blood and blood components were processed and/or modified under controlled conditions of temperature and other requirements of each component. It must be possible to trace, from its source to its final disposition, any unit of blood and every component prepared from each unit. The purpose of this record keeping is to assist in identifying and resolving component quality and production problems.

Records of any modifications performed to any blood component must include the date, time, component used, component made, expiration date of the newly created product, identification of the person performing each significant step of the process, and the records must document the lot numbers of the materials used to prepare the components. Records of temperature storage and product inspection must be kept indefinitely. Records of personnel involved in component preparation must document training and competency evaluation. Their inclusive dates of employment must also be kept.

Procedure manuals should include accurate procedures on how to prepare each component. These manuals must also include any form(s) used to document information related to component preparation. Each procedure must have annual review and there also must be documentation that personnel regularly reviewed each procedure. Superceded procedures must be saved, along with documentation of when and why the procedure was replaced.

If the transfusion service uses a computer system to facilitate record keeping, documentation of program development, installation, validation, modification, and integrity of data must be performed. An alternative or back-up system must be available at all times in case the computer is unavailable for any reason. Personnel must be familiar with both the uses of the computer system as well as the back-up system. Procedures must exist for the use of the system.

The purpose of record keeping in the component preparation process is to ensure that the highest quality products are prepared and manufactured in the safest and most effective manner. Components prepared in this manner provide to the final user, the patient, the most beneficial treatment available.

STORAGE OF BLOOD AND BLOOD PRODUCTS

Storage of blood and blood products should be designed to maintain maximum viability of each component in a strictly controlled and secure environment. Physical storage conditions best suited for maximum product survival are as important as the anticoagulant/preservative solution used. Storage conditions of blood and blood components are regulated by the FDA in the Code of Federal Regulations. The American Association of Blood Banks (AABB), the College of American Pathologists (CAP), and the Joint Commission on Accreditation of Healthcare Organizations (JCAHO) all provide voluntary and required standards of practice (Brecher 2002; Gorlin 2002).

Each blood product has a designated storage area (refrigerators, freezers, platelet environmental chambers, rotators), which may contain blood, blood components, derivatives, donor or patient samples, tissues for transplantation, and reagents, or laboratory supplies. The areas for all products should be clearly and distinctly labeled. Each storage device should have a separate area for untested, tested, and quarantined products. No food should be stored in these refrigerators, freezers, or environmental chambers. Each piece of storage equipment must have adequate capacity and air circulation to ensure optimal storage conditions for the particular blood products.

All storage equipment, whether a refrigerator, a freezer, or a platelet environmental chamber, must have a system for continuous internal temperature monitoring and for recording of temperature at least every 4 hours. If components are stored at room temperature, outside of a controlled storage device, the ambient room

temperature should be recorded every 4 hours. Any deviations from the expected temperatures should be documented, corrected, and reported. Temperature recording charts should be changed regularly. Policies and procedures must be in place for each of these devices, which should minimally outline the storage requirements, permissible temperatures, and include instructions to be followed and corrective action to be taken in the event of failure of the storage device or failure of power to the storage device. The goal of temperature monitoring is to ensure controlled temperatures in all compartments in which blood and components are stored.

All storage devices should have an audible alarm system, which is set to activate at a temperature that will allow removal of the blood products before they reach unacceptable temperatures. The audible alarm must be able to be heard in a staffed area that will ensure an immediate response by personnel 24 hours a day. Each storage device must have a backup power source.

Blood may be stored outside of the transfusion service in a remote area of the hospital (for example, operating room, nursery, acute patient care area, or trauma center). Blood stored in these areas is subject to the same storage requirements as blood stored in the transfusion service or blood center. These refrigerators should be monitored with recording thermometers and audible alarms. It is often most efficient if the monitoring of these refrigerators is the responsibility of the blood bank or transfusion service.

Appropriate Storage Temperatures

Appropriate storage temperatures for product viability from the AABB Standards (Gorlin 2002), include the following:

1. Whole blood and liquid RBCs, $1°$ to $6°C$
2. FFP and cryoprecipitate, $\leq-18°C$
3. Platelets, $20°$ to $24°C$
4. Granulocytes, $20°$ to $24°C$
5. RBCs frozen in 40% glycerol, $\leq-65°C$
6. RBCs frozen in 20% glycerol, $\leq-120°C$

Inspecting Blood Components Before Issuing

All blood and blood products should be inspected before issuing for patient use or shipping to another facility (Brecher 2002; Gorlin 2002). This inspection should take place immediately before the product is shipped or released to the patient care area. Inspection of the blood component should ensure the following: (1) that the container is intact, (2) that the product is normal in appearance, (3) that it is labeled appropriately, and (4) that the correct product is issued for the identified patient. If a product fails to meet any of these criteria, it should be quarantined in a segregated storage area and the nonconformity should be investigated and documented.

SUMMARY

A challenge to blood providers is created when neonates or older pediatric patients require transfusion. The preparation of blood components for the pediatric population should be based on the individual needs of the patient. Advances in the treatment of pediatric patients require a periodic review of transfusion practice. Members of the patient care team and transfusion service will need to work collaboratively so that they will successfully identify ongoing transfusions needs, solve problems, and modify policies and procedures. Effective preparation of each blood component is essential to the provision of quality blood components that will ensure that children, large or small, receive the safest, most effective transfusion therapy we can provide.

References

American Association of Blood Banks, America's Blood Centers, American Red Cross. 2002. Circular of information for the use of human blood and blood components. Bethesda, MD: AABB.

Blajchman MA, Goldman M, et al. 2001. Proceedings of a consensus conference: prevention of post-transfusion CMV in the era of universal leukoreduction. *Transfus Med Rev* 15:1–20.

Blanchette VS, Kuhne J, Hume HA, et al. 1995. Platelet transfusion therapy in newborn infants. *Transfus Med Rev* 9:215–230.

Bowden RA, Slichter SJ, Sayers M, et al. 1995. A comparison of filtered leukocyte-reduced and cytomegalovirus (CMV) seronegative blood products for the prevention of transfusion-associated CMV infection after marrow transplant. *Blood* 86:3598–3603.

Brecher ME, ed. 2000. Collected questions and answers. 6th ed. Bethesda, MD: AABB.

Brecher ME, ed. 2002. Technical manual. 14th ed. Bethesda, MD: AABB.

Capon SM and Chambers LA. 1996. New directions in pediatric hematology. Bethesda, MD: AABB.

Chambers L. 1995. Evaluation of a filter-syringe set for preparation of packed red cell aliquots for neonatal transfusion. *Am J Clin Pathol* 104:253–257.

Chambers LA and Issitt LA, eds. 1994. Supporting the pediatric transfusion recipient. Bethesda, MD: AABB.

Code of Federal Regulations, 21 CFR 610,53. 2001, rev. annually. Washington, DC: U.S. Government Printing Office.

Development Task Force of the College of American Pathologists. 1994. Practice parameter for the use of fresh frozen plasma, cryoprecipitate and platelets. *JAMA* 271:777–781.

Food and Drug Administration. January 23, 2001. Draft guidance for industry: prestorage leukocyte reduction with whole blood and blood components intended for transfusion. Rockwell, MD: CBER Office of Communications, Training and Manufacturers Assistance.

Food and Drug Administration. 2000. Guidance for industry: gamma irradiation of blood and blood components: a pilot program for licensing. Rockwell, MD: CBER Office of Communications, Training and Manufacturers Assistance.

Gorlin JE, ed. 2002. Standards for blood banks and transfusion services. 21st ed. Bethesda, MD: AABB.

Hall TL, Barnes A, Miller JR, et al. 1993. Neonatal mortality following transfusion of red cells with high plasma potassium levels. *Transfusion* 33:606–609.

Hare VW, Liles BA, Crandall LW, and Lufer CN. 1994. "Partners for life"—a safer therapy for chronically transfused children [abstract]. *Transfusion* 34 (suppl):925.

Hillyer CD, Emmens RK, Zago-Novaretti M, et al. 1994. Methods for the reduction of transfusion-transmitted cytomegalovirus infection: filtration versus the use of seronegative donor units. *Transfusion* 34:929–934.

Hillyer CD, Hillyer KL, Strobl FJ, et al., eds. 2001. Handbook of transfusion medicine. London: Academic Press.

Hillyer CD, Silberstein LE, Ness PM, et al., eds. 2003. Blood banking and transfusion medicine: basic principles and practice. 2nd ed. Philadelphia, PA: Churchill Livingstone.

Hume H and Bard H. 1995. Small volume red blood cell transfusion for neonatal patients. *Transfus Med Rev* 9:187–199.

Kickler TS and Herman JH, eds. 1999. Current issues in platelet transfusion therapy and platelet alloimmunity. Bethesda, MD: AABB Press.

Luban NLC, Strauss RG, and Hume HA. 1991. Commentary on the safety of red cells preserved in extended storage media for neonatal transfusions. *Transfusion* 31:229–235.

Mollison PC, Englefriet CP, and Contreras M. 1998. Blood transfusion in clinical medicine. Oxford: Blackwell Scientific.

Moroff G and Luban NL. 1997. The irradiation of blood and blood components to prevent graft-versus-host disease: technical issues and guidelines. *Transfus Med Rev* 11:15–22.

Petz LD, Swisher SN, Kleinman S, et al., eds. 1996. Clinical practice of transfusion medicine. New York: Churchill Livingstone.

Preiksaitis J. 2000. The cytomegalovirus "safe" blood product: is leukoreduction equivalent to antibody screening? *Transfus Med Rev* 14:112–136.

Ratko TA, Cummings JP, et al. 2001. Evidence-based recommendations for the use of WBC-reduced cellular blood components. *Transfusion* 41:1310–1319.

Rosse WF, Tilen M, and Ware RE. 1998. Transfusion support for patients with sickle cell disease. Bethesda, MD: AABB.

Rossi EC, Simon TL, Moss GS, and Gould SA, eds. 1995. Principles of transfusion medicine. 2nd ed. Baltimore, MD: Williams and Wilkins.

Rudmann SV. 1995. Textbook of blood banking and transfusion medicine. Philadelphia, PA: WB Saunders Company.

Simon TL, Alverson DC, AuBuchon J, et al. 1998. Practice parameter for the use of red blood cell transfusions. Developed by the Red Blood Cell Administration Practice Guidelines Task Force of the College of American Pathologists. *Arch Pathol Lab Med* 122:130–138.

Smith DM and Shoos-Lipton K. 1997. Leukocyte-reduction for the prevention of transfusion-transmitted cytomegalovirus. American Association of Blood Banks 97.2. Bethesda, MD: AABB Press.

Strauss RG. 1993. Selection of white cell-reduced blood components for transfusions during early infancy. *Transfusion* 33:352–357.

Strauss RG. 2000. Data-driven blood banking practices for neonatal RBC transfusions. *Transfusion* 40:1528–1540.

Strauss RG, Sacher RA, Blazina JF, et al. 1990. Commentary on small volume red cell transfusions for neonatal patients. *Transfusion* 30:565–570.

Trial to Reduce Alloimmunization to Platelets (TRAP) Study Group. 1997. Leukocyte reduction and ultraviolet B irradiation of platelets to prevent alloimmunization and refractoriness to platelet transfusion. *N Engl J Med* 337:1861–1869.

Triulzi D, ed. 2002. Blood transfusion therapy. A physician's handbook. 6th ed. Bethesda, MD: AABB.

Vamvakas EC, Dzik WH, Blajchman MA. 1999. Deleterious effects of transfusion-associated immunomodulation: appraisal of the evidence and recommendations for prevention. In: Vamkakas EC and Blajchman MA, eds. Immunomodulatory effects of blood transfusion. Bethesda, MD: AABB Press.

3

Blood Components

CASSANDRA D. JOSEPHSON, MD, AND CHRISTOPHER D. HILLYER, MD

INTRODUCTION

Transfusion of blood components is an integral part of treating various pediatric disorders. As such, a specific knowledge of the components is critical to providing excellent patient care. This chapter predominantly addresses the composition, volume, ordering, and dosing for neonates, children, and adolescents, as well as some unique aspects of whole blood (WB), packed red blood cells (PRBCs), platelets, plasma products, cryoprecipitate, granulocytes, coagulation factor preparations, albumin, and gamma globulins.

WHOLE BLOOD AND PACKED RED BLOOD CELLS

Description

The point of origin for the manufacture of most of the components used in transfusion is WB. WB contains RBCs, plasma, clotting factors, platelets, and approximately 10^9 white blood cell (WBCs). However, the most commonly transfused blood component is PRBCs. PRBCs are produced from WB collections by either centrifugation or less frequently directly from the donor, by apheresis techniques. In the United States over 12 million units of PRBCs are transfused each year. Table 3.1 provides important information on WB and PRBC products including approximate volumes, composition, dosing, and storage periods.

Indications

In general, WB and PRBCs are transfused to increase the oxygen carrying capacity of the patient's blood to maintain adequate tissue oxygenation by replenishing blood volume and red cell mass. WB availability is dependent upon each institution and blood center. WB use is discouraged in most situations, as the use of specific blood components is more appropriately tailored to the unique needs of each patient. Once WB has been stored for 24 hours at 4°C, platelet function is lost. Furthermore, by day 21 of storage, factors V and VIII lose 5% to 30% of their activity. Autologous donation, however, is one situation where WB is commonly used.

It is important to note that normal hemoglobin levels in *children* are lower than in *adults*, and that most children do not have underlying cardiopulmonary or vascular pathology. Thus, a more restrictive criterion for transfusion may be applied in many cases. Guidelines on specific indications for WB and PRBCs are given in Section IV, Chapters 12, 13, 15, 17, 18, 19, and 21.

Ordering

Informed consent is a vital part of the transfusion process. It is required for all blood product transfusions except during an emergency. The physician must explain to the patient and his or her parents the treatment plan, risks, benefits, and alternative approach to treatment. A witness, along with a parent and/or guardian of the patient, must sign a standardized transfusion consent form along side of the physician acknowledging consent

TABLE 3.1   Whole Blood and Packed Red Blood Cell Products

Component	Approximate Volume (mL)	Composition	Neonatal/Pediatric Dosage	Hematocrit	Comments
Whole blood (WB)	500	250 mL red cells 250 mL plasma 63 mL anticoagulant	10 mL/kg body weight transfused over 2–4 hours	<35%–40%	• Storage 4°C • Platelets, granulocytes, labile factors V and VIII are not reliable
Packed red blood cells (PRBCs)	250	CPD or CPDA* 200 mL red cells 50 mL plasma CPD or CPDA*	10 mL/kg body weight transfused over 2–4 hours	<50%–80%	• 21–35-day storage period • Made from WB • Storage 4°C • Contains 10^8 WBCs • Cannot be infused as rapidly as WB due to increased viscosity
PRBCs (additive solution) AS-1 AS-3 AS-5	350	200 mL red cells 50 mL plasma 100 mL adenine saline solution†	10 mL/kg body weight transfused over 2–4 hours	50%–60%	• 35-day storage period • Made from WB • Storage 4°C • Contains 10^8 WBCs • Red cell product most commonly available
Prestorage Leukoreduced (LR-PRBCs)	250–350	Depends on additive solution and anticoagulant	10 mL/kg body weight transfused over 2–4 hours	50%–80%	• 42-day storage period • Made from WB • $<5 \times 10^6$ WBCs • ≥85% of original red cell mass • Does *not* prevent transfusion-associated graft-versus-host disease (TA-GVHD)
Washed PRBCs	200	180 mL red cells 20 mL isotonic saline (0.9%)	10 mL/kg body weight transfused over 2–4 hours		• 21–42-day storage period • Washing removes most of plasma and approxiamtely 80% of leukocytes • 24-hour storage period after washing
Irradiated PRBCs	250–350	Depends on additive solution and anticoagulant	10 mL/kg body weight transfused over 2–4 hours		• Reduces storage time due to potassium leak after irradiation • Prevents TA-GVHD • Storage 28 days after irradiation, or by original expiration date, whichever comes first
Frozen deglycerolized PRBCs	200	180 mL red cells 20 mL isotonic saline/dextrose solution	10 mL/kg body weight transfused over 2–4 hours		• Usually reserved for rare blood phenotypes • May be stored frozen for 10 years • Plasma reduced • Approximately 90% leukocyte reduced after deglycerolization and washing • 24-hour storage period after deglycerization

*Citrate, phosphate buffer, dextrose, adenine.
†Adenine, dextrose, saline, mannitol (most plasma removed and replaced with additive solution).

to the transfusion. If the child is old enough to understand what a transfusion is and he or she can sign his or her name, the child should sign an "assent" form. Furthermore, prior to a red cell transfusion, ABO and Rh compatibility testing should be performed and compatible blood administered. A *type and screen* order should be placed when a nonurgent transfusion is anticipated. At that time the patient's ABO and Rh antigen types are determined, and an indirect antiglobulin test (IAT) or "screen" is performed on the patient's serum to detect unexpected alloantibodies to RBCs. Alloantibody formation typically occurs subsequent to foreign RBC exposure. Type and screen testing takes approximately 30 to 45 minutes. If the screen is negative and no unexpected antibodies are discovered, it should take 15 to 30 minutes to provide crossmatched PRBC units after an order is submitted (see below). However, if the antibody screen is positive and alloantibodies have been detected, further investigation must be done to determine if the antibody detected is clinically significant. Once specificity is determined, antigen-negative PRBCs can be provided by the blood bank, though this can significantly delay a transfusion.

A *type and crossmatch* order should be submitted if blood transfusion is required within the subsequent 12 hours. Patients are typed for ABO and Rh antigens, and an antibody screen is performed. However, with this order, unlike the type and screen, the appropriate blood product is crossmatched with the patient's serum prior to being issued (Brecher 2002). All blood banks perform a crossmatch procedure before issuing any WB or PRBC component, except in an emergency. All of these issues are discussed in greater detail in Section II, Chapters 4–6.

Emergency Transfusions or Emergency Release

When immediate transfusion of PRBCs is required there is often not enough time for ABO or Rh typing of the recipient. In that situation Group O, Rh negative, uncrossmatched PRBCs may be administered to all recipients. An emergency order for such units requires a signature from a physician to ensure that it is clear the units are not crossmatched. As Group O, Rh negative PRBCs are a scarce and valuable resource, the quantities required may be unavailable. The next appropriate choice is Group O, Rh positive PRBC. Pediatric transfusion medicine services try to limit this substitution only to males in order to avoid the risk of immunizing Rh negative females who may later become pregnant with an Rh positive fetus, predisposing to hemolytic disease of the newborn. In semi-emergent situations, where there is a limited amount of time available before transfusion, the blood bank can only provide type-specific blood (A

to A, B to B). In this case, screening and crossmatching are not performed before the PRBCs leave the blood bank. Ideally, if a patient's sample can be obtained quickly and delivered immediately to the blood bank, or if a current, prior sample is available, an emergency crossmatch can be performed within 15 to 20 minutes.

Expected Response

A patient who has received approximately 10 mL/kg of PRBCs or WB and is not actively bleeding, should experience a rise in hemoglobin by 1 to 2 g/dL or in hematocrit by 3% to 6%. If this does not occur, suspicion should be aroused regarding peripheral destruction or loss of blood as a cause for the unexpectedly low outcome.

Component and Recipient Identification

Labeling of samples is critical in the blood bank. The recipient's name and hospital identification number must be on the sample tube, which in turn must exactly match the information on the recipient's hospital identification band. A technologist within the blood bank then labels the donor PRBC units and matches them with the recipient's identification number before the units are issued. Before transfusion, the healthcare provider administering the blood product must check the identity of the recipient and match it with the PRBC unit before transfusing the patient. Every one of these steps is crucial, since clerical errors remain the leading cause of fatal hemolytic transfusion reactions (Myhre and McRher 2000).

Dosing

For infants younger than 4 months of age, PRBCs preserved in extended storage solutions such as AS-1, AS-3, and AS-5 transfused in small volumes (approximately 10 mL/kg) over 2 to 4 hours should not present a risk when transfused. This may be done safely without further processing, such as washing or resuspension in other solutions. However, for those premature infants with severe hepatic or renal compromise, removal of the extended storage solution is recommended and resuspension with saline or albumin should be performed prior to transfusion. For large volume transfusions such as exchange transfusions, cardiac surgery, extracorporeal membrane oxygenation (ECMO), and massive transfusion extended storage media is not recommended (Luban et al. 1991).

For those children older than 4 months of age who have good cardiac and vascular function, but require red cell transfusion, dosing can range between 10 to

20 mL/kg with a maximum of approximately 2 units to be infused over 2 to 4 hours depending on the patient's disease process and healthcare constraints. The choice of PRBCs (preservative solution) for these ages is much less critical due to the child's increased blood and plasma volume.

Adverse Reactions

RBC transfusion can elicit transfusion reactions of many types. These reactions include transfusion-transmitted diseases, which are extensively covered in Section IV, Chapters 26–28. However, toxicities from additives that extend PRBC storage is a specialized potential complication applicable mostly to infants younger than 4 months of age. As was discussed in the previous chapter, anticoagulant preservative solution formulations have different compositions of NaCl, dextrose, adenine, mannitol, trisodium citrate, citric acid, and sodium phosphate. Regardless of the preservative solution used, the volume of additive solution itself is small, especially in the setting of infants who are administered small-volume transfusions over a 2 to 4 hour period. Table 12.3 in Chapter 12 (adapted from Luban et al. 1991) can be used as a guide for clinicians involved in transfusion of infants younger than 4 months of age. It lists the theoretical toxicity calculations for the individual components stored in CPDA, AS-1, and AS-3 storage solutions.

Special Processing to Prevent Complications

Leukoreduction, gamma-irradiation, and washing of cellular products such as PRBCs are specialized processes performed to prevent specific complications. Leukoreduction removes most WBCs in PRBCs and platelets. The process is accomplished most often with a leukoreduction filter and is performed to prevent or delay febrile nonhemolytic transfusion reactions (FNHTR), human leukocyte antigen (HLA) alloimmunization, transfusion-transmitted cytomegalovirus (CMV) infection, and transfusion-related immunomodulation (TRIM). However, leukoreduction does not abrogate transfusion-associated graft-versus-host disease (TA-GVHD), which has a 90% mortality rate.

For immunocompromised patients at risk for TA-GVHD *gamma-irradiation*, another specialized procedure, is indicated. Irradiation stops donor T-cell replication and engraftment in the host preventing TA-GVHD.

Finally, *washing* of cellular products is done to reduce the recurrence of severe allergic or anaphylactic transfusion reactions. Washed PRBCs are not adequately leukoreduced to prevent WBC-associated complica-

tions, nor do they protect against the development of TA-GVHD (Hume and Preiksaitis 1999). All three of these special processes are discussed in greater detail in Section III, Chapters 7, 9, and 10.

Alternatives to Allogeneic PRBC Transfusion

Autologous Donation

Stable patients undergoing elective surgical procedures potentially requiring blood transfusion, such as orthopedics, are candidates for preoperative WB collection. Autologous donation can significantly reduce patient exposure to allogeneic red cell antigens and also infectious elements (Henry 2002). To safely donate autologous blood, a patient's hemoglobin should be at least 11 g/dL. Blood donation ≥4 weeks before the surgical procedure is recommended to allow time for adequate compensatory erythropoiesis. This time lag reduces the risk of anemia at the time of surgery by allowing the patient to produce adequate volume recovery. Weekly collection is most common and dietary supplementation with iron is recommended before initiation of autologous blood collections. The WB product is stored at 4°C for up to 35 days, after which it must be frozen or discarded. If the product is not used, it cannot be crossed over for allogeneic use, because autologous donors do not meet the strict criteria required of the general blood donor population. Absolute contraindications to autologous donation are as follows: infection or risk of bacteremia, aortic stenosis, unstable angina, active seizure disorder, myocardial infarction or cerebrovascular accident during previous 6 months, high-grade left main coronary artery disease, cyanotic heart disease, uncontrolled hypertension, and significant pulmonary or cardiac disease.

Intraoperative Blood Collection ("Cell Saver")

"Cell savers" are instruments that collect blood lost during surgery. The RBCs are washed with normal saline and concentrated to make an approximate 225 mL unit with a hematocrit of ~55%. RBC units can be either directly transfused into the patient or washed again and stored (Goodnough et al. 1996). If the unit is stored, it must be properly labeled and can only be stored for 6 hours at room temperature, or for 24 hours at 1° to 6°C, if it is chilled within 6 hours of beginning the collection. Patients are excluded from this procedure if they have malignant neoplasms, infections, or otherwise contaminated operative fields. The disadvantage of this procedure is that a lower percentage of RBCs are recovered than in preoperative autologous donation.

Acute Normovolemic Hemodilution (ANH)

This technique involves WB collection from patients immediately prior to a procedure in which blood loss is anticipated. Rapid replacement of the removed blood volume with crystalloid or colloid solution is done prior to surgery. Re-infusion of the collected blood typically occurs toward the end of the procedure, or as soon as major bleeding has stopped (Goodnough et al. 1992). The reduction of RBC loss during surgery is the purpose of this technique and is sometimes preferred to the cell saver WB collection, which ends up with lower hematocrits than ANH blood products.

Postoperative Blood Collection

This procedure involves recovery of blood from surgical drains and is usually filtered but not always washed before reinfusion. The salvaged product may be hemolyzed and dilute. The product must be transfused within 6 hours or it must be discarded. The primary indications for postoperative blood collection are cardiac and orthopedic surgery cases.

PLATELETS

Description

Two types of platelet components are available to most hospitals in the United States: pooled platelet concentrates (also called "random donor platelets") and apheresis platelets (also called "single-donor platelets"). Platelet concentrates are derived from WB donations from a single donor. Apheresis platelets are collected via an apheresis device, returning the other WB components to the patient. In addition to the difference in product production, the amount of platelets/unit is also quite distinct. It takes 5 to 8 pooled platelet concentrates ($\sim 7 \times 10^{10}$ platelets/concentrate) to achieve the same dose of platelets as a single apheresis platelet unit (3 to 6×10^{11} platelets). As a result, the recipient of pooled platelet concentrates is exposed to 5 to 8 times more blood donors per transfusion than a single apheresis platelet recipient. Additionally, a platelet concentrate unit must undergo leukofiltration to be rendered leukoreduced (WBC $<5 \times 10^6$) while an apheresis platelet unit is already "process" leukoreduced (WBC $<10^4$ to 10^6). Finally, RBC contamination is often less in the apheresis product than in WB-derived platelet concentrates; therefore, apheresis platelets may elicit less Rh sensitization. Table 3.2 lists the types of platelet products, with their approximate volumes, compositions, dosing, and storage periods.

Indications

The normal peripheral blood platelet count is 150,000 to 450,000/μL in premature infants, neonates, children, adolescents, and adults. In premature neonates, the threshold to transfuse is higher than in

TABLE 3.2 Platelet Products

Component	Approximate Volume (mL)	Composition	Neonatal/Pediatric Dosage	Storage Period	Comments
Platelet, apheresis (Single donor)	300	$\geq 3 \times 10^{11}$ platelets; $<10^4$–10^6 WBCs and plasma	Can be dosed at 10 mL/kg body weight, but most times is dosed by $^1/_4$, $^1/_2$, and whole pheresis units. Dose extrapolated back from adult dose of 1 pheresis for adult BSA 1.7 m²–70 kg adult.	4 hours if system opened (i.e., volume reduction or washing) 5 days (closed system)	• Storage 22°–26°C (room temp) with constant horizontal agitation • Equivalent to 5–8 units of platelet concentrates • Decreased number donor exposures to patient • Fewer lymphocytes than equivalent dose of platelet concentrates
Platelet concentrate (Random donor)	50	$\geq 5.5 \times 10^{10}$ platelets; variable numbers RBC, WBCs, and plasma	Transfused by gravity, pump, or IV push. 10 mL/kg body weight transfused by gravity, pump, or IV push.	4 hours if system opened (i.e., volume reduction or washing) 5 days (closed system)	• HLA-matched products may be provided • Cost equivalent to 6–8 units of concentrate • Storage 22°–26°C (room temp) with constant horizontal agitation • Average adult dose is 5–8 units, which are pooled for infusion

other age groups. When the platelet count drops below 10,000/µL there is a clinically significant risk of intracranial hemorrhage, especially in those <1.5 kgs at birth (Andrew et al. 1987). In contrast, most clinically stable, nonbleeding neonates, children, and adolescent patients tolerate platelet counts as low as 5 to 10,000/µL without experiencing major bleeding. Prophylactic transfusions for the prevention of future bleeding remain the most common reason for platelet transfusions (Pisciotto et al. 1995). Hanson and Slichter showed that approximately 7000 platelets/µL/day are required to maintain endothelial integrity in normal individuals (Hanson and Slichter 1985).

Two recent prospective clinical trials in adults support that the platelet transfusion trigger should be 10,000/µL instead of 20,000/µL in stable patients receiving prophylactic transfusions without coexisting conditions (Rebulla et al. 1997; Wandt et al. 1998). However, for patients with fever, active bleeding, or coexisting coagulation defects, a level of 20,000/µL is commonly selected. More detailed indications for platelet transfusions, specifically in children, will be covered in Section IV, Chapters 12, 17, 20, and 22.

Ordering

Informed consent must be obtained before transfusion. See the ordering PRBC section on p. 27 for more details. Platelets should be ABO and Rh matched, when possible, in order to attain the best response from the platelet transfusion and decrease the potential for RBC hemolysis. Therefore, the blood bank requires an order to ABO and Rh type the patient before transfusion. This has usually been performed with the type and screen/crossmatch order since PRBCs are often given before, or around the same time as, platelets are administered. However, ABO and Rh matching are not absolutely necessary, and platelet transfusion should not be denied if type-specific platelets are not available. The outcome from giving ABO/Rh incompatible platelets does not have as great a potential to yield a fatal outcome as does ABO/Rh mismatched red cells. Rh immunoglobulin should be administered (estimate 1 mL of PRBC transfused, per platelet concentrate) if the platelet Rh type is mismatched. When ABO-mismatched platelet transfusions occur, they may contribute to an eventual platelet refractory state (Carr et al. 1990). Thus, in an attempt to prevent platelet and HLA-alloimmunization, leukoreduced, ABO matched units are recommended. Additionally, hemolysis of RBCs has been reported when patients have received either large volumes of ABO-incompatible plasma or plasma with high-titer isohemagglutinins both of which are more likely to occur with an apheresis platelet product rather than with pooled platelet concentrates (Pierce et al. 1985). Therefore, it is generally recommended in the neonate and small child that the platelet products, regardless of type (apheresis or platelet concentrate), be volume-reduced, to eliminate most of the incompatible plasma, before transfusion. However, since volume-reduction practices have been shown to decrease the number and possibly the function of some of the platelets (as well as reducing the storage time to four hours), the procedure is not routinely recommended for older children or adult patients receiving ABO-mismatched platelet products. Further discussion of volume-reduction of platelet products can be found in Section IV, Chapter 22.

Dosing

Transfusion of 10 mL/kg of a platelet concentrate should provide approximately 10×10^9 platelets. Platelets dosed from an apheresis unit at 10 mL/kg may yield a slightly lower dose if more plasma than platelets are pulled into the syringe at the time of making smaller components from the apheresis unit. More often however, platelet apheresis products are ordered as "quarters" or "halves." Different institutions have defined patient subgroups' weights for the different portions of apheresis platelets. Alternatively, if one estimates that an adult has a body surface area of 1.7 m² and is 70 kgs then one can extrapolate to a child and neonate's body surface area and dose accordingly.

Expected Response

One way to assess the expected response is to calculate the corrected count increment with a 15 minute to 1 hour post platelet count. The corrected count increment (CCI) formula can help the physician determine if his or her patient is platelet refractory or is getting an adequate rise in platelet count based on dose and body surface area.

$$CCI = \frac{(1 \text{ hr Post PC} - \text{Pre PC} \times \text{BSA (m}^2))}{\text{Number of Platelets Transfused} \times 10^{-11}}$$

If it is less than 5000 to 7500/µL on 2 successive days, the patient is considered to be refractory. When this situation arises, the blood bank should be notified so they can help with the next steps in providing either crossmatched platelets or HLA-matched platelets. Both specialized products may require hours to days for the blood center to obtain and prepare. Platelet refractory states are discussed in greater detail in Section IV, Chapter 22.

Contraindications

Platelet transfusions have several caveats and/or contraindications. (1) Surgical or local measures should be pursued first to achieve hemostasis when a single anatomic site is thought responsible for the bleeding. Platelet transfusions are indicated in this situation only if the patient is thrombocytopenic. (2) Surgical intervention rather than platelet transfusion is likely needed if hemorrhage of >5 mL/kg/hour is occurring. (3) Thrombotic thrombocytopenic purpura (TTP) and heparin induced thrombocytopenia (HIT) patients should generally not be transfused with platelets, as the addition of platelets may worsen the thrombotic complications. (4) Although not absolutely contraindicated, ITP patients are unlikely to benefit from platelet transfusion due to rapid immune-mediated peripheral platelet destruction. (5) Bleeding uremic patients are usually unresponsive to platelet transfusions alone. However, if administered in conjunction with DDAVP, PRBCs to keep hematocrit >30 g/dL, and/or concurrent dialysis, bleeding uremic patients may respond well to platelet transfusion.

Adverse Reactions

There are three main adverse reactions that are more specific to platelet transfusion: (1) hypotension, (2) human leukocyte antigen (HLA) and/or human platelet antigen (HPA) alloimmunization, and (3) posttransfusion purpura. These reactions will be further detailed in Section VI, Chapters 26–28.

Special Processing

Leukoreduction, gamma-irradiation, washing, and volume reduction are all special processes relevant to platelets. The reader is referred to Section III, Chapters 7, 9, and 10, and Section IV, Chapter 22.

GRANULOCYTES

Description

Granulocyte collections are mainly performed via automated leukapheresis. The final product is approximately 300 mL in volume and contains, in addition to granulocytes, other elements such as RBCs (6 to 7 g/dL of hemoglobin per granulocyte product), platelets, and citrated plasma. The product is collected from volunteer apheresis donors who receive either corticosteriods

(dexamethasone) and/or growth factors such as G-CSF. Oral dexamethasone has been demonstrated to increase baseline peripheral blood granulocytes two- to threefold (1.7×10^9), whereas G-CSF stimulated donors have been shown to have a seven- to tenfold increase from baseline (4 to 5×10^{10}). The combination of dexamethasone and G-CSF has been deemed superior with a 9 to 12 fold increase in circulating granulocytes from baseline. Usually collections are daily for 4 to 5 days. The final granulocyte yield per collection depends upon the total volume of blood processed as well as the starting peripheral blood neutrophil count of the donor. Seven to 12 liters of blood are usually processed through a continuous flow blood cell separator over 2 to 4 hours (Price 1995).

Indications

Clinical indications for granulocyte transfusions include severe neutropenia ($<0.5 \times 10^9$ polymorphonuclear cells [PMNs]/μL), and the following: (1) progressive, nonresponsive, documented bacterial, yeast, or fungal infection nonresponsive to therapy after 48 hours of antimicrobial treatment, (2) a protracted period of neutropenia in stem cell transplant recipients, (3) congenital granulocyte dysfunction, and (4) bacterial infection in neonates (Klein et al. 1996). These indications will be discussed in greater detail in Chapter 16. It is important to note that the use of prophylactic granulocyte transfusion is not recommended (Vamvakas and Pineda 1997).

Ordering

Granulocytes constitute an unlicensed product and therefore have no official FDA product specifications. However, the American Association of Blood Banks (AABB) standards require the leukapheresis product to contain at least 1×10^{10} granulocytes ≥75% of units tested (AABB 2003). Ideally, the ordering physician should notify the hospital blood bank who will in turn notify the blood center that a granulocyte transfusion is necessary. The blood center will call potential donors, usually on a known registry, who have the same blood type as the patient. ABO compatibility is required because the granulocyte product has a large volume of RBC contamination. A crossmatch is also required prior to administration. Also, as granulocyte products contain a significant number of T-lymphocytes capable of causing TA-GVHD in these immunocompromised recipients, irradiation of all granulocyte products are recommended. While likely obvious, the product cannot be leukocyte depleted and should not be infused through a leukocyte reduction filter (Chanock and Gorlin 1996). It is not necessary to HLA match granulocytes, unless the patient is known to be HLA-alloimmunized. Finally, in most centers an emergency release needs to be signed by the ordering physician as the product needs to be infused soon after collection, a time when the blood supplier has yet to perform all of the infectious disease testing.

Dosing

For children, the average granulocyte dosage is 1×10^9/kg/day. The neonatal dose average is 1 to 2×10^9/kg (Vamvakas and Pineda 1996). The product is recommended to be given for a period of 4 to 7 days to increase the granulocyte count to combat nonantibiotic treatment-responsive infections in severely neutropenic patients.

Expected Response

It is difficult to accurately predict the posttransfusion increment of granulocytes. A measurement can be made; however, the increment has not been shown to correlate with granulocyte dose given, thus clinical significance is difficult to assess. The goal is to achieve a sustained granulocyte count above 500 PMN/μL (0.5×10^9/L) after transfusion. This is increasingly possible as the ability to collect large numbers of granulocytes improves.

Contraindications

Amphotericin B administration concurrent with granulocyte transfusions has been reported to be associated with pulmonary toxicity. Therefore, granulocyte transfusion is recommended to be separated by at least 4 hours from amphotericin B infusion (Chanock and Gorlin 1996).

Adverse Reactions

Transfusion reactions, such as fever, dyspnea, rigors, and hypotension, may occur with granulocyte infusions. Reduction of the infusion rate, antihistamines, corticosteroids, and meperidine may help control these symptoms (see Chapter 26).

PLASMA PRODUCTS

Description

Plasma, the aqueous, acellular portion of WB, consists of proteins, colloids, nutrients, crystalloids, hor-

mones, and vitamins. Albumin, the most abundant of the plasma proteins, is discussed on p. 38. Other plasma proteins include complement (C3 predominantly), enzymes, transport molecules, immunoglobulins (gamma-globulins), and coagulation factors. The latter two are also discussed later in this chapter. Coagulation factors in plasma include fibrinogen (2 to 3 mg/mL); factor XIII (60 μg/mL); von Willebrand factor (5 to 10 μg/mL); factor VIII, primarily bound to its carrier protein vWF (approximately 100 ng/mL); and vitamin K-dependent coagulation factors II, VII, IX, X (1 unit of activity/mL for each factor).

WB or plasmapheresis collections give rise to several types of plasma products. Single donor plasma or source plasma is produced by plasmapheresis and is stored at $-20°$C. All other plasma products are derived from WB, and the "time after collection to time of freezing" determines its designation. FFP must be frozen within 6 to 8 hours of collection and stored at $-18°$C or colder (Brecher 2002). F24 plasma must be frozen within 24 hours of collection and frozen at $-18°$C or colder. FFP and F24 are considered as essentially equivalent products, though factor VIII levels are slightly lower in F24. However, due to factor VIII's acute phase reactant property, its levels are quickly replenished in recipients without hemophilia A. Furthermore, specific factor VIII concentrates and recombinant factor VIII are available for use in patients with congenital factor VIII deficiency. Thus, FFP and F24 may be used interchangeably in patients without hemophilia A. Another FDA-approved plasma product is cryoreduced plasma (CRP) also, known as cryosupernatant. This product is depleted of its cryoprecipitate fraction; the cryosupernatant is then refrozen at the above temperature. Table 3.3 lists the plasma-derived products, appropriate volumes, composition, and storage periods.

Indications

The primary use of frozen plasma products (FFP and F24) is for the treatment of coagulation factor deficiencies in which specific factor concentrates are not available or when immediate hemostasis is critical. Specific indications include: bleeding diatheses associated with acquired coagulation factor deficits, such as end stage liver disease, massive transfusion (Crosson 1996), and disseminated intravascular coagulation (DIC); the rapid reversal of warfarin effect; plasma infusion or exchange for TTP; congenital coagulation defects (except when specific factor therapy is available); and C1-esterase inhibitor deficiency. A more detailed discussion of the indicated uses are addressed in Section IV, Chapters 13, 15, 18, and 20, and Section VII, Chapter 31.

TABLE 3.3 Plasma-Derived Products

Component	Approximate Volume (mL)	Composition	Neonatal/Pediatric Dosage	Storage Period	Comments
Source plasma (Single donor plasma)	180–300	• Plasma proteins • Immunoglobulins • Complement • Coagulation factors (II, VII, IX, X, VIII, XIII, vWF, fibrinogen) • Albumin	10–15 mL/kg body weight transfused over 1 hour or IV push	• One year if frozen • 24 hours if maintained at 1°–6°C	• Obtained through single donor plasmapheresis • Stored at –20°C after collection • Not for volume expansion or fibrinogen replacement
Recovered plasma	180–300	Same as above	10–15 mL/kg body weight transfused over 1 hour or IV push	• One year if frozen • 24 hours if maintained at 1°–6°C	• Plasma obtained from WB of regular donor • Not for volume expansion or fibrinogen replacement
Fresh frozen plasma (FFP)	180–300	Same as above	10–15 mL/kg body weight transfused over 1 hour or IV push	• One year if frozen • 1–5 days after thawing	• Separated from WB within 6–8 hours of collection • Stored frozen at –18°C • Not for volume expansion or fibrinogen replacement
Plasma frozen within 24 hrs (F24)	180–300	Same as above	10–15 mL/kg body weight transfused over 1 hour or IV push	• One year if frozen • 1–5 days after thawing	• Separated from WB and frozen within 24 hours of collection • Stored frozen at –18°C • Not for volume expansion or fibrinogen replacement
Cryoreduced plasma (CRP)	180–300	Same as above except depleted levels of factors VIII, XIII, fibrinogen, and vWF	• 1 bag/10 kg body weight • Bags pooled in blood bank before transfusion	• One year if frozen • 24 hours after thawing	• Depleted of its cryoprecipitate fraction

Ordering

No specific compatibility testing is performed prior to infusion of plasma products. However, the blood bank needs to have an order to ABO type the patient because plasma products must be ABO compatible despite the lack of formal compatibility testing. This requirement exists because plasma contains isohemagglutinins, which must be compatible with the recipient's blood type, otherwise hemolysis will ensue. However, if the recipient's ABO type is unknown prior to plasma infusion, AB plasma may be administered to all recipients, due to its lack of isohemagglutinins. Rh alloimmunization rarely occurs due to Rh mismatch of plasma products, as there are few RBCs in the plasma component. Therefore, Rh compatibility is not as essential as is ABO type when transfusing plasma.

Dosing

In children and adults, 10 to 20 mL/kg of plasma will usually yield a coagulation factor concentration of approximately 30% of normal. Multiple doses are usually required to correct a clinically significant coagulopathy. The infusion can be rapid, if the patient's cardiovascular status is stable. Timing of repeat doses depends upon the half-life of each factor deficiency being addressed.

Contraindications

The use of FFP or F24 is not without risk to the recipient and should not be used to expand plasma volume, increase plasma albumin concentration, or bolster the nutritional status of malnourished patients. Antithrombin (ATIII) or Activated Protein C concentrates may offer an advantage over FFP or F24 use when considering treatment of burns, meningococcal sepsis (Churchwell et al. 1995), or acute renal failure.

Adverse Effects

Anaphylactic allergic reactions have been attributed to antibodies in the donor's plasma that react with the recipient's WBCs, although the reactions are uncommon. Furthermore, isohemagglutinins may cause mild to severe hemolytic reactions or result in a positive direct antiglobulin test (Coombs' test) if "out-of-group" plasma is administered to the patient. Lastly, to avoid life-threatening anaphylaxis, IgA-deficient patients who have formed anti-IgA antibodies must receive IgA-deficient plasma from a national rare donor registry. However, the presence of absolute IgA deficiency with anti-IgA antibodies is an extremely rare occurence and should be confirmed by demonstration of 0% IgA levels using sensitive measures and presence of anti-IgA antibodies before requesting these rare plasma components.

CRYOPRECIPITATE

Description

Cryoprecipitate contains the highest concentrations of factor VIII (80 to 150 U/unit), vWF (100 to 150 U/unit), fibrinogen (150 to 250 U/unit), factor XIII, and fibronectin. Upon thawing FFP (1° and 6°C) an insoluble precipitate is formed, isolated, and is refrozen in 10 to 15 mL of plasma within 1 hour and is termed cryoprecipitate. Storage ($\leq$–18°C) is up to 1 year. Before the 1980s, cryoprecipitate was primarily used for the treatment of von Willebrand's disease and hemophilia A. However, with the development of recombinant factor products and improved viral inactivation procedures, cryoprecipitate's therapeutic role in treating these diseases has diminished. Presently, fibrinogen replacement is its primary use due to the high fibrinogen content.

Indications

Cryoprecipitate has a narrow range of indications, due to the development of safer, more specific factor concentrates. Congenital or acquired fibrinogen deficiencies, factor XIII deficiency, DIC, orthotopic liver transplantation, and poststreptokinase therapy (hyperfibrinogenolysis) are a few of its indicated uses. A more detailed discussion of these uses can be found in Section IV, Chapters 13, 17, and 20.

Ordering

Cryoprecipitate units have a small volume compared with other plasma products, PRBCs, and apheresis platelets. Thus, anti-A and anti-B isohemagglutinins are present only in small quantities. While the AABB *Standards* recommend (AABB 2003) ABO compatibility for cryoprecipitate transfusions, especially in pediatric patients, compatibility testing is not required. Furthermore, since cryoprecipitate does not contain red cells, Rh matching is not necessary.

Dosing

Dosing of cryoprecipitate is dependent upon the clinical condition being treated. For fibrinogen replacement, the most common condition treated with this product, 1 bag/10 kg will increase the fibrinogen level by 60 to 100 mg/dL. However, in a neonate 1 unit will increase fibrinogen by >100 mg/dL. The dosing frequency may vary from every 8 to 12 hours to days depending on the cause of hypofibrinogenemia. In von Willebrand's disease, cryoprecipitate is a second line therapy, and in children 1 unit/6 kg every 12 hours should be administered. In hemophilia A, cryoprecipitate is also a second line therapy. If an assumption is made that 1 unit of cryoprecipitate has 100 U factor VIII, then 1 unit/6 kgs will give an approximate factor VIII level of 35% if the patient has <1% at initiation of therapy. The interval for this dosing is discussed in greater detail in Chapter 20. For factor XIII deficiency, due to the low level necessary to achieve hemostasis (2% to 3%), only 1 unit/10 kg every 7 to 14 days is necessary.

Contraindications

The availability of recombinant factor VIII products, which go through viral inactivation steps unlike cryoprecipitate, have made the use of this product in that disease a relative contraindication. It should only be given if recombinant products are unavailable.

Adverse Reactions

Refer to FFP adverse reactions.

COAGULATION FACTORS

Description

Before the 1960s, plasma infusion was the only way to treat bleeding disorders. As was described above, plasma contains fibrinogen (I), factor XIII, von Willebrand factor (vWF), factor VIII, and all of the vitamin-K dependent coagulation factors: II (prothrombin), VII, IX, and X. However, Pool's discovery in 1964 of high concentrations of factor VIII in cryoprecipitate revolutionized the treatment of hemophilia A and eventually vWD (Pool et al. 1964). Subsequently, investigators, with the use of chromatography and monoclonal antibody immunoaffinity technology, were able to produce progressively more purified forms of factor VIII concentrates. However, the purification techniques did not change the fact that the pooled plasma source could and did transmit viral infection, such as HIV, hepatitis C, hepatitis B, nonenveloped viruses, and other pathogens. In order to create a product free of infectious disease transmission, recombinant DNA technology allowed the production of recombinant coagulation factor products via cloning of a desired factor gene and optimization of an expression system.

There are many products used to treat various clotting disorders (congenital and acquired) that are either plasma derived or recombinantly produced. Tables 3.4 and 3.5 summariz these products, their manufacturers, and unique characteristics.

Indications

Various indications for factor replacement exist for each type of factor deficiency. The specifics of therapeutic indications for various congenital and acquired disorders are addressed in Chapter 20.

Ordering

The specifics of ordering each product will not be addressed here but can be found in Chapter 20. Generally, however, the ordering physician should be aware of whether he or she desires a plasma-derived product or a recombinantly-derived product. The units and interval of dosing is critical for each factor deficiency because the therapy is necessary to achieve hemostasis, and if underdosed or overdosed the consequences could be fatal. Furthermore, these products are expensive, so

more than others, and should not be administered unless absolutely deemed necessary in consultation with either a hemophilia or transfusion medicine specialist.

Dosing

Dosing of any factor preparation is not only dependent on the product being infused but the type of insult being managed. When dosing factor VIII in general the calculation should be based on body weight in kilograms and desired factor VIII level to be achieved. This level will vary according to prophylaxis or treatment regimen being employed. Each FVIII unit per kilogram of body weight will increase the plasma FVIII level by approximately 2%. The half-life is 8 to 12 hours; therefore the interval of IV dosing can vary from 8 to 24 hours depending upon initial biodistribution and the desired FVIII level to be maintained.

$$\text{Bolus dose (U)} = \text{weight (kg)} \times (\% \text{ desired FVIII level}) \times 0.5$$
$$\text{Continuous infusion dose (U)} = \text{expected level } 100\%$$
$$= 4 - 5 \text{ U/kg/hr}$$

(individualize dose depending on postinfusion FVIII level)

When dosing factor IX products for hemophilia B disease, the ordering physician must know that Benefix, the only recombinant product available, has a 28% lower recovery in vivo than the more highly purified plasma derived FIX products, Mononine and Alpha Nine SD. There is no significant difference in half-life, approximately 24 hours, between the two products. Each FIX unit (plasma derived) per kilogram of body weight will increase the plasma FIX level by approximately 1%. However, when dosing Benefix one should use the following calculations:

$$(\text{FIX units required}) = \text{body weight (kg)} \times \text{desired FIX increase (\%)} \times 1.2 \text{ U/kg (Abshire et al. 1998)}$$

The dosing and specific uses of recombinant FVIIa, Humate-P, and aPCCs will be specifically addressed in Chapter 20.

Contraindications/Adverse Reactions

Generally, any allergic or anaphylactic type of reaction to infusion of any of these preparations would make a second dose contraindicated. The specifics

TABLE 3.4 Plasma-Derived Factor Products

Factor	Manufacturer	Virus Inactivation	Purification	Purity	Comments
Factor VIII *Cryoprecipitate AHF*	Blood Center	None		Low	Only used for FVIII deficiency when other factors unavailable
Factor VIII *Humate-P*	Aventis Behring	Pasteurization		Intermediate	Licensed for vWD treatment
Factor VIII *Alphanate*	Alpha Therapeutic	Solvent detergent (S/D), heat treated, filtered	Gel chromatography	High	Contains vWF
Factor VIII *Koate-DVI*	Baxter	S/D, polysorb 80, heat treated		High	Stabilized with human albumin
Factor VIII *Hemofil M*	Baxter	S/D	Immunoaffinity chromatography	Ultra-high	Mouse protein, trace
Factor VIII *Monoclate-P*	Aventis Behring	Pasteurization	Immunoaffinity chromatography	Ultra-high	Stabilized with human albumin
Factor IX *Konyne 80*	Bayer	Heat treated		Low	PCC, high content of FII, VII, IX, X
Bebulin VH	Immuno	Heat treated		Low	
Factor IX *Proplex T*	Baxter	Heat treated			PCC, high content of FII, VII, IX, X
Factor IX *Alpha Nine* *Mononine*	Alpha Therapuetic Aventis Behring	S/D Non-S/D	Immunoaffinity Chromatography		Contains factor IX only. Recovery after infusion is normal compared to recombinant FIX product (see text).
Factor IX *Autoplex-T*	Nabi	Heat treated			aPCC, high content FVIIa, IXa, Xa
Factor IX *FEIBA VH*	Bayer	Heat treated			aPCC, high content FVIIa, IXa, Xa
Factor XIII *Fibrogammin P*	Aventis Behring	Pasteurization			Administered every 4–6 weeks

surrounding each product will be addressed in Chapter 20.

Another problem is inhibitor formation with antibodies directed against an infused factor or protein contained in the preparation. Inhibitors render the product ineffective, blocking the product's ability to aid in hemostasis. This type of adverse reaction, technically the most severe, would make the product in question contraindicated. Further discussion of inhibitor formation and treatment can be found in Chapter 20.

ALBUMIN

Description

Albumin is the most abundant of the plasma proteins (3500 to 5000 mg/dL) and has multiple functions. Its main purpose is to maintain plasma colloid oncotic pressure. Synthesis of albumin occurs in the liver, and there are small body stores, which undergo rapid catabolism. Each molecule remains intact for approximately 15 to 20 days. Albumin produced specifically for trans-

TABLE 3.5 Recombinant Factor Products

Products	Factor/Generation	Manufacturer	Protein Additives	Comments
Kogenate	**Factor VIII**	Bayer	Human albumin	• Half-life 8–12 hours
Bioclate	First	Aventis Behring	Human albumin	• Dosing varies from continuous 4–5 U/
Helixate		Aventis Behring	Human albumin	kg/hour to every 24 hours depending
Recombinate		Baxter	Human albumin	upon hemostatic injury
Helixate FS	**Factor VIII**	Aventis Behring	Human albumin	• Kogenate and ReFacto formulated with
Kogenate FS	Second	Bayer	None	sucrose
ReFacto		Genetics Institute/Wyeth	None	• ReFacto is B-domain deleted FVIII
				• Half-life and dosing same as first generation products
Benefix	**Factor IX** First	Genetics Institute/Wyeth	None	• 28% lower recovery rate then plasma-derived FIX products, thus must dose 20% higher to achieve same level of hemostasis
NovoSeven	**Factor VIIa** First	NovoNordisk	None	• Half-life 2 hours • Dosing for factor VIII and IX inhibitor patients—only FDA-approved indication 90–300 µg/kg for first 48 hours of bleeding episode, then every 2–6 hours based on clinical hemostasis assessment

fusion purposes is separated from human plasma through a cold ethanol fractionation procedure. Commercially available human albumin preparations include a 5% solution, a 25% solution, and a plasma protein fraction 5% solution (PPF). All preparations are from pooled plasma and have a balanced physiological pH, contain 145 mEq of sodium, and contain less than 2 mEq of potassium per liter. The products contain no preservatives or coagulation factors.

Indications

Albumin has a wide variety of uses (Table 3.6). It is indicated after large-volume paracentesis, for nephrotic syndrome resistant to diuretics, and for volume/fluid replacement in plasmapheresis. Relative indications include adult respiratory distress syndrome (ARDS); cardiopulmonary bypass pump priming; fluid resuscitation in shock, sepsis, and burns; neonatal kernicterus; and enteral feeding intolerance. A further detailed discussion can be found in Section IV, Chapters 13 and 17, as well as in Section VII.

Ordering

Albumin is an acellular product virtually devoid of blood group isohemagglutinins. Therefore, neither serologic testing nor ABO or Rh compatibility is necessary prior to administration. It is important to specify the percent solution preparation of albumin when ordering because the volume infused will vary accordingly.

Dosing

In children with hypoproteinemia 0.5 to 1 g/kg/dose is recommended and may be repeated one to two times in a 24-hour period. No more than 250 grams should be administered within 48 hours and the infusion should run over 2 to 4 hours.

Contraindications

Albumin use is contraindicated in the following situations: correction of nutritional hypoalbuminemia or hypoproteinemia, nutritional deficiency requiring total parenteral nutrition, preeclampsia, and wound healing. Albumin should not be used for resuspending RBCs or simple volume expansion (for example, in surgical or burn patients). Furthermore, it should not be administered to those patients with severe anemia or cardiac failure or with a known hypersensitivity.

Adverse Reactions

These include hypertension due to fluid overload, hypotension due to hypersensitivity reaction, as well as fever, chills, nausea, vomiting, and rash.

GAMMA-GLOBULINS

Description

Immune Globulin Intravenous (Human) is the FDA-approved name for IVIG. The product was first licensed in the United States in 1981 and is currently the most widely used plasma product in the world. IVIG is prepared by fractionation of large pools of human plasma and has a half-life between 21 to 25 days, similar to native immunoglobulins. However, increased clearance of immunoglobulins has been seen in states of increased metabolism such as fever, infection, hyperthyroidism, or burns. There are numerous preparations available, each prepared in a slightly different manufacturing process. There are theoretical disadvantages and advantages linked to each licensed product. An ideally composed product should contain each IgG subclass; retain Fc receptor activity; have a physiologic half-life; demonstrate virus neutralization, opsonization, and intracellular killing; and possess antibacterial capsular polysaccharide antibodies. In addition, the product should be devoid of transmissible infectious agents and vasoactive substances. In reality, although each company strives for this composition, certain brands have better profiles than others regarding the treatment of different disease states. For example, Polygam S/D and Gammagard S/D, both produced by Baxter, have <3.7 µg/mL IgA content and are therefore the most suitable IVIG product for IgA-deficient patients.

IVIG's immunomodulatory effects are not well understood. There are several postulated mechanisms of action, such as autoantibodies inhibition, increased IgG clearance, complement activation modulation, macrophage-mediated phagocytosis inhibition, cytokine suppression, superantigen neutralization, and B and T cell function modulation. The wide range of potential effects explains the vast array of on- and off-label IVIG indications.

Indications

There are six FDA-approved uses for IVIG, four of which are directly applicable to children (Table 3.7). The efficacy of IVIG in the following four indications has been well substantiated in controlled clinical trials (Buckley et al. 1991; Anonymous 1999; Cines and Blanchette 2002). The approved uses are idiopathic thrombocytopenic purpura (ITP), congenital (that is, severe combined immunodeficiency syndrome [SCIDS]) and acquired immunodeficiences (that is, pediatric human immunodeficiency virus [HIV]), and Kawasaki syndrome (mucocutaneous lymph node syndrome) (Burns et al. 1998). Interestingly, greater than half of the IVIG produced yearly is used for off-label indications Table 3.8 (Nydegger et al. 2000; Anonymous 1999). More in-depth discussion of the on- and off-label uses of IVIG are covered in various chapters throughout Section IV.

TABLE 3.6   Albumin

Indicated

Nephrotic syndrome resistant to potent diuretics
Volume/fluid replacement in plasmapheresis

Possibly Indicated

Adult respiratory distress syndrome
Cardiopulmonary bypass pump priming
Fluid resuscitation in shock/sepsis/burns
Neonatal kernicterus
To reduce enteral feeding intolerance

Not Indicated

Correction of measured hypoalbuminemia or hypoproteinemia
Nutritional deficiency, total parenteral nutrition
Red blood cell suspension
Simple volume expansion (surgery, burns)
Wound healing

Investigational

Cadaveric renal transplantation
Cerebral ischemia
Stroke

Common Usages

Serum albumin <20 g/dL
Nephrotic syndrome, proteinuria, and hypoalbuminemia
Labile pulmonary, cardiovascular status
Cardiopulmonary bypass pump priming
Extensive burns
Plasma exchange
Hypotension
Liver disease, hypoalbuminemia, diuresis
Protein-losing enteropathy, hypoalbuminemia
Resuscitation
Premature infant undergoing major surgery

Ordering

When ordering an IVIG product it is good to know that most hospitals will use whatever immunoglobulin preparation they have available at the time unless the physician specifies otherwise. In most situations that substitution is appropriate. However, in certain disease states such as renal insufficiency and IgA-deficiency, a specific knowledge of the product is important. Table 3.9 (modified from Knezevic-Maramica and Kruskall 2003) lists seven licensed products with some of their specifications.

Dosing

The dose of IVIG used is dependent upon the disease being treated—not the type of product being adminis-

TABLE 3.7   FDA-Approved Pediatric Uses for Intravenous Immunoglobulin

Primary Immunodeficiency Syndromes

Common variable immunodeficiency
X-linked agammaglobulinemia
Severe combined immunodeficiency
Ataxia-telangiectasia
Wiskott-Aldrich syndrome
IgG subclass deficiency

Acquired Immunodeficiency

HIV

Infectious Disorders

Mucocutaneous lymph node syndrome

Immune-Mediated Disorders

Idiopathic thrombocytopenic purpura

TABLE 3.8   Off-Label Uses of Intravenous Immunoglobulin

Neurologic Disorders

Chronic inflammatory demyelinating polyneuropathy (CIDP)
Guillian-Barre syndrome
Multifocal motor neuropathy

Inflammatory Disorders

Dermatomyositis, polymyositis refractory
Inflammatory bowel disease (Crohn's disease; ulcerative colitis)

Infectious Disorders

Transplantation: CMV-negative recipients of CMV-positive organs
Infection prophylaxis in high-risk neonates
Parvovirus B19-associated anemia
Sepsis; toxic shock

Immune-mediated Disorders

Abortions, recurrent spontaneous
Diabetes mellitus
Hematologic coagulation disorders: acquired factor VIII inhibitors;
    acquired von Willebrand's disease
Hematologic immune-mediated cellular disorders: autoimmune
    hemolytic anemia, autoimmune neutropenia, fetal-neonatal
    alloimmune
Thrombocytopenia, HLA-alloimmune thrombocytopenia
Rheumatoid diseases
Myasthenia gravis
Multiple sclerosis
Posttransfusion purpura
Systemic lupus erythematosus
Toxic epidermal necrolysis (Lyell's syndrome)
Transplantation: renal graft rejection
Transplantation: solid organ (alloimmunization and
    hypogammaglobulinemia)

Miscellaneous

Asthma
Autism

---

terd. Dosing can range from daily 400 mg/kg/day times 5 days to 1 g/kg/day times 1 to 2 days for ITP to 2 g/kg times one dose for Kawasaki syndrome. Dosing for pediatric HIV is 200 to 400 mg/kg every 2 to 4 weeks and for congenital immunodeficiencies 300 to 400 mg/kg monthly, adjusting for trough IgG of 400 to 500 mg/dL. An extrapolation from the adult bone marrow transplant experience would suggest 500 to 1000 mg/kg weekly to prevent GVHD and infection.

Contraindications

The disease process drives contraindications of these products. If a patient has renal insufficiency, or IgA deficiency, then the type of product chosen should be monitored or the therapy should be changed. Furthermore, if volume overload or pulmonary compromise is a concern, IVIG treatment should be carefully considered.

Adverse Reactions

Most of the adverse effects experienced by patients are related to IgG aggregates and dimer formation in combination with complement activation. The symptoms include headache, fever, flushing, and hypotension, which are all usually mild and transient. Amelioration of symptoms may be accomplished by slowing down the IVIG infusion rate or changing brands of IVIG. Renal failure, aseptic meningitis, and thromboembolic events have also been described with IVIG infusion. Renal failure has been directly correlated to sucrose load and

aseptic meningitis to dose and patient history of migraines. Furthermore, in IgA-deficient recipients who receive IgA-containing products, severe anaphylactic reactions have been described. Other reactions include pulmonary edema, fluid overload, eczema, arthritis, and transfusion-related acute lung injury (TRALI).

Passive transfer of blood group antibodies such as anti-A, anti-B (IgG class), in addition to non-ABO antibodies such as anti-Kell, -C, and -Lewis[b], can occur. This transfer can result in positive antibody screens and positive direct antiglobulin tests in many instances. Therefore, cautious interpretation of results postinfusion must be performed. Furthermore, there have been rare instances of hemolytic anemias secondary to anti-D or

TABLE 3.9   Intravenous Immuoglobulin Preparations

Product	Manufacturer	Viral Inactivation	Stabilizing Agent: Sucrose	Stabilizing Agent: Other	IgA Content (µg/mL)
Sandoglobulin (now called *Carimune*) and Panglobulin (lyophilized)	ZLB Bioplasma AG	Pepsin (pH 4)	1.67 g/g Ig	0	<2400
Polygam S/D (lyophilized) and *Gammagard S/D* (lyophilized)	Baxter	S/D treatment	0	Albumin Glycine Glucose PEG	<3.7
Iveegam EN (lyophilized)	Baxter	Immobilized trypsin, PEG precipitation, DEAE Sephadex	0	Glucose (5 g/100 mL)	25
Gamimmune N (5% or 10% liquid)	Bayer	Filtration, pH 4.25 and low salt, S/D treatment	0	5% solution-maltose (10%) 10% solution-glycine (0.16–0.24 M)	120
Gammar-P I.V. (lyophilized)	Aventis-Behring	Heat treatment (10 h at 60°C)	1 g/g Ig	albumin	<50
Venoglobulin-S (5% or 10% liquid)	Alpha Therapuetic	PEG/bentonite precipitation, S/D treatment	0	albumin D-sorbitol (50 mg/mL, isoosmolar)	5%–15% 10%–50%

Modified from Knezevic-Maramica and Kruskell, 2003.

anti-A. Thus, close observation of the patient's hemoglobin after treatment is advised.

During the mid 1990s there were over 200 cases of hepatitis C transmission related to IVIG. It was temporarily removed from the market and since that time all manufacturers of IVIG have had to put in additional safeguarding steps for protection against hepatitis C virus such as pasteurization or solvent/detergent treatment. Donor screen has also intensified. Otherwise, IVIG has always had a good safety record.

References

AABB. 2003. Standards for Blood Banks and transfusion services, 22nd ed. Bethesda, MD, American Association of Blood Banks Press.

Abshire T, Shapiro A, Gill J, et al. 1998. Recombinant FIX (rFIX) in the treatment of previously untreated patients with severe or moderately severe hemophilia B. Presented at XXIII International Congress of the World Federation of Haemophilia, May 17–22, The Hague, The Netherlands.

Andrew M, Castle V, Saigal S, et al. 1987. Clinical impact of neonatal thrombocytopenia. *J Pediatr* 110:457–464.

Anonymous. 1999. Availability of immune globulin intravenous for treatment of immune deficient patients—United States, 1997–1998. *MMWR* 48:159–162.

Anonymous. 1991. National Institute of Child Health and Human Development Intravenous Immunoglobulin Study Group: intra-venous immune globulin for the prevention of bacterial infections in children with symptomatic human immunodeficency virus infection. *N Engl J Med* 325:73–80.

Brecher M. 2002. Technical Manual. 14th ed. Bethesda, MD: American Association of Blood Banks Press.

Buckley RH and Schiff RI. 1991. The use of intravenous immune globulin in immunodeficiency diseases. *N Engl J Med* 325:110–117.

Burns J, Capparelli E, Brown J, Newburger J, et al. 1998. Intravenous gamma-globulin treatment and retreatment in Kawasaki disease. *Pediatr Infect Disease J* 17:1144–1148.

Carr R, Hutton JL, Jenkins JA, et al. 1990. Transfusion of ABO mismatched platelets leads to early platelet refractoriness. *Br J Haematol* 75:408–413.

Chanock SJ and Gorlin JB. 1996. Granulocyte transfusions. Time for a second look. *Infect Dis Clin N Am* 10:327–343.

Churchwell KB, McManus ML, Kent P, et al. 1995. Intensive blood and plasma exchange for treatment of coagulopathy in meningococcemia. *J Clin Apher* 10:171–177.

Cines DB and Blanchette VS. 2002. Immune thrombocytopenic purpura. *N Engl J Med* 346:995–1008.

Crosson JT. 1996. Massive transfusion. *Clin Lab Med* 16:873–882.

Goodnough LT, Brecher ME, and Monk TG. 1992. Acute normovolemic hemodilution in surgery. *Hematology* 2:413–420.

Goodnough LT, Monk TG, Sicard G, et al. 1996. Intraoperative salvage in patients undergoing elective abdominal aortic aneurysm repair: an analysis of cost and benefit. *J Vasc Surg* 24:213–218.

Hanson SR and Slichter SJ. 1985. Platelet kinetics in patients with bone marrow hypoplasia: evidence for a fixed platelet requirement. *Blood* 66:1105–1109.

Henry DA, Carless PA, Moxey AJ, O'Connell D, et al. 2003. Preoperative autologous donation for minimising perioperative allogeneic blood transfusion [Review]. *The Cochrane Database of Systematic Reviews 3.*

Hume H and Preiksaitis JB. 1999. Transfusion associated graft-versus-host disease, cytomegalovirus infection and HLA alloimmunization in neonatal and pediatric patients. *Transfusion Science* 21:73–97.

Klein HG, Strauss RG, and Schiffer CA. 1996. Granulocyte transfusion therapy. *Semin Hematol* 33:722–728.

Knezevic-Maramica I and Kruskall MS. 2003. Intravenous immune globulin—an update for clinicians. *Transfusion* 43:1460–1480.

Luban NLC, Strauss RG, and Hume HA. 1991. Commentary on the safety of red cells preserved in extended-storage media for neonatal transfusions. *Transfusion* 31:229–235.

Myhre BA and McRuer D. 2000. Human error—a significant cause of transfusion mortality. *Transfusion* 40:879–885.

Nydegger UE, Mohacsi PJ, Escher R, and Morell A. 2000. Clinical use of intravenous immunoglobulins. *Vox Sang* 78:191–195.

Pierce RN, Reich LM, and Mayer K. 1985. Hemolysis following platelet transfusions from ABO-incompatible donors. *Transfusion* 25:60–62.

Pisciotto PT, Benson K, Hume H, et al. 1995. Prophylactic versus therapeutic platelet transfusion practices in hematology and/or oncology patients. *Transfusion* 35:498–502.

Pool JG, Hershgold EJ, and Pappenhagen AR. 1964. High potency antihaemophilic factor concentrate prepared from cryoglobulin in precipitate. *Nature* 203:312.

Price TH. 1995. Blood center perspective of granulocyte transfusions: future applications. *J Clin Apher* 10:119–123.

Rebulla P, Finazzi G, Marangoni F, et al. 1997. The threshold for prophylactic platelet transfusions in adults with acute myeloid leukemia. Gruppo Italiano Malattie Ematologiche Maligne dell'Adulto. *N Engl J Med* 337:1870–1875.

Vamvakas EC and Pineda AA. 1997. Determinants of the efficacy of prophylactic granulocyte transfusions: a meta-analysis. *J Clin Apher* 12:74–81.

Vamvakas EC and Pineda AA. 1996. Meta-analysis of clinical studies of the efficacy of granulocyte transfusion in the treatment of bacterial sepsis. *J Clin Apher* 11:1–9.

Wandt H, Frank M, Ehninger G, et al. 1998. Safety and cost effectiveness of a $10 \times 10 (9)$/L trigger for prophylactic platelet transfusions compared with the traditional $20 \times 10(9)$/L trigger: a prospective comparative trial of 105 patients with acute myeloid leukemia. *Blood* 91:3601–3606.

Red Blood Cell Antigens and Human Blood Groups

SHEILAGH BARCLAY, MT(ASCP)SBB

INTRODUCTION

Since Landsteiner's discovery of the ABO system in 1900, there has been tremendous growth in the understanding of human blood groups. More than 250 red blood cell (RBC) antigens have been described and categorized by the Working Party of the International Society of Blood Transfusion (ISBT) into 26 major systems (Daniels et al. 1995, 1996, 1999). Those RBC antigens not assigned to major systems have been grouped into five collections, a series of low-prevalence antigens and a series of high-prevalence antigens. This chapter describes the RBC antigens that are most commonly encountered in the clinical practice of transfusion medicine (Table 4.1). A detailed description of all the known blood group systems can be found in the text *Applied Blood Group Serology* (Issitt and Anstee 1998).

Blood group antigens are determined by either carbohydrate moieties linked to proteins or lipids, or by amino acid (protein) sequences. Specificity of the carbohydrate-defined RBC antigens is determined by terminal sugars; genes code for the production of enzymes that transfer these sugar molecules onto a protein or lipid. Specificity of the protein-defined RBC antigens is determined by amino acid sequences that are directly determined by genes. The proteins that carry blood group antigens are inserted into the RBC membrane in one of three ways: single-pass, multipass, or linked to phosphatidylinositol.

Many factors influence the clinical significance of alloantibodies formed against RBC antigens. The prevalence of different RBC antibodies depends on both the prevalence of the corresponding RBC antigen in the population and the relative immunogenicity of the antigen. The clinical importance of an RBC antibody depends on both its prevalence in a population and whether it is likely to cause RBC destruction (hemolytic transfusion reactions) or hemolytic disease of the newborn (HDN). The type and degree of transfusion reactions and the degree of clinical HDN caused by antibodies to each blood group antigen system will be reviewed in this chapter. The overall clinical significance of antibodies to each of the major blood group antigens is summarized in Table 4.2.

ABO BLOOD GROUP SYSTEM

Antigens

Three genes control the expression of the ABO antigens: *ABO*, *Hh*, and *Se*. The *H* gene codes for the production of an enzyme transferase that attaches L-fucose to the RBC membrane-anchored polypeptide or lipid chain. In the presence of the *A* gene-encoded transferase, N-acetyl-galactosamine is attached, which confers "A" specificity. In the presence of the *B* gene-encoded transferase, galactose is added and confers "B" specificity. If no A or B gene/enzyme is present, the H specificity remains, and the individual is of group O. If *A* and *B* genes/transferases are both present, "AB" specificity is defined. If the *H* gene is absent, L-fucose is not added to the precursor substance, and even if the *A* and/or *B* genes and their respective enzymes are present, the A and B antigens cannot be constructed. The secretor gene (*Se*) controls the individual's ability to secrete soluble A, B, and H antigens into body fluids and secretions. There are about 800,000 to 1,000,000

TABLE 4.1   Antibody Prevalence in U.S. Population

Antigens	Systems	IgM	IgG	Transfusion Reactions	HDN*	Caucasians	African-Americans
A	ABO	X	X	Mild-severe	None-moderate	40%	27%
B	ABO	X	X	Mild-severe	None-moderate	11%	20%
D	Rh	X	X	Mild-severe	Mild-severe	85%	92%
C	Rh		X	Mild-severe	Mild	68%	27%
E	Rh	X	X	Mild-moderate	Mild	29%	22%
c	Rh		X	Mild-severe	Mild-severe	80%	96%
e	Rh		X	Mild-moderate	Rare	98%	98%
K	Kell	X	X	Mild-severe	Mild-severe	9%	2%
k	Kell		X	Mild-moderate	Mild-severe	99.8%	>99%
Kp^a	Kell		X	Mild-moderate	Mild-moderate	2%	<1%
Kp^b	Kell		X	None-moderate	Mild-moderate	>99%	>99%
Js^a	Kell		X	None-moderate	Mild-moderate	<1%	20%
Js^b	Kell		X	Mild-moderate	Mild-moderate	>99%	99%
Fy^a	Duffy		X	Mild-severe	Mild-severe	66%	10%
Fy^b	Duffy		X	Mild-severe	Mild	83%	23%
Jk^a	Kidd		X	None-severe	Mild-moderate	77%	92%
Jk^b	Kidd		X	None-severe	None-mild	74%	49%
M	MNS	X	X	None	None-mild	78%	74%
N	MNS	X		None	None	70%	75%
S	MNS		X	None-moderate	None-severe	52%	31%
s	MNS		X	None-mild	None-severe	89%	94%
U	MNS		X	Mild-severe	Mild-severe	100%	>99%
Le^a	Lewis	X		Few	None	22%	23%
Le^b	Lewis	X		None	None	72%	55%
Lu^a	Lutheran	X	X	None	None-mild	8%	5%
Lu^b	Lutheran	X	X	Mild-moderate	Mild	>99%	>99%
Do^a	Dombrock		X	Rare	+DAT/ No HDN	67%	55%
Do^b	Dombrock		X	Rare	None	82%	89%
Co^a	Colton		X	None-moderate	Mild-severe	>99.9%	>99.9%
Co^b	Colton		X	None-moderate	Mild	10%	10%
P1	P	X		Rare	None	79%	94%

*HDN = hemolytic disease of the newborn.

TABLE 4.2   Clinical Significance of Antibodies to the Major Blood Group Antigens

Usually Clinically Significant	Sometimes Clinically Significant	Clinically Insignificant If Not Reactive at 37°C	Generally Clinically Insignificant
A and B	At^a	A_1	Bg
Diego	Colton	H	Chido/Rogers
Duffy	Cromer	Le^a	Cost
H in O_h	Dombrock	Lutheran	JMH
Kell	Gerbich	M and N	Knops
Kidd	Indian	P1	Le^b
P, $PP1P^k$	Jr^a	Sd^a	Xg^a
Rh	Lan		
S, s, and U	LW		
Vel	Scianna		
	Yt		

TABLE 4.3 ABO Blood Group Phenotypes and Prevalence

	Prevalence	
Phenotypes	Caucasians	African-Americans
A	40%	27%
B	11%	20%
AB	4%	4%
O	45%	49%

copies of the A antigen per group A adult RBC; 600,000 to 800,000 copies of the B antigen per group B adult RBC; and 800,000 copies of the AB antigen per group AB adult RBC. The antigens of the ABO system are not fully developed at birth. In newborns there are about 250,000 to 300,000 copies of the A antigen and 200,000 to 320,000 copies of the B antigen. At birth RBCs have linear oligosaccharide structures, which can accommodate the addition of only single sugars. Complex branching oligosaccharides, which permit the addition of multiple sugars, appear at about 2 to 4 years of age.

Phenotypes

The prevalence in the United States of the four phenotypes associated with the ABO blood group system is listed in Table 4.3.

Antibodies

The antibodies of the ABO system are "naturally occurring" in that they are formed as a result of exposure to ABH-like substances from the gastrointestinal tract, occurring in utero or immediately postpartum and peaking about 5 to 10 years of age. Thus, there is development of antibodies against whichever ABH antigens are absent on the person's own RBCs. ABO antibodies are mostly IgM but some IgG is present, and they efficiently fix complement. Antibodies present in cord blood are almost entirely of maternal origin. IgM is not transported across the placenta, but all four subclasses of IgG are.

Clinical Significance

Clinically, ABO is the most important RBC antigen system, as circulating A and B antibodies are complement-fixing and thus can cause intravascular hemolysis. *Transfusing a patient with the incorrect ABO group blood may have fatal consequences.* ABO incompatibility is the most common cause of HDN in the United States; however, the clinical significance of HDN caused by ABO incompatibility is typically none to moderate,

and only rarely severe, since placental transfer of ABO antibodies is limited to the IgG fraction found in maternal serum, and fetal ABO antigens are not fully developed. ABO-HDN is most often found in nongroup O infants of group O mothers because anti-A and anti-B from group O individuals often have a significant IgG component.

Rh BLOOD GROUP SYSTEM

Antigens

The Rh system contains at least 45 antigens of which the major antigens are D, C, E, c, and e. The Rh system is a complex system, and controversy over its genetics has resulted in the development over time of multiple nomenclature systems. In 1943 Wiener proposed the idea of a single Rh locus with multiple alleles. Fisher and Race later inferred the existence of reciprocal alleles on three different but closely linked Rh loci (1946). Later, Rosenfield proposed a numerical system for the Rh blood group system based on serological data (1962).

The isolation of the Rh antigen-containing components of the RBC membrane led to the definitive identification of several nonglycosylated fatty acid acylated Rh polypeptides. Genomic studies have identified two distinct Rh genes, *RHD* and *RHCE* (Cherif-Zahar et al. 1991; Le Van et al. 1992). The presence of *RHD* determines Rh(D) antigen activity. Rh(D)-negative individuals have no *RHD* gene, and thus have no Rh(D) antigen. The *RHCE* gene codes for both Cc and Ee polypeptides. There are four possible alleles: *RHCE*, *RHCe*, *RhcE*, and *Rhce*.

With the exception of the A and B antigens, Rh(D) is the most important RBC antigen in transfusion practice. The Rh(D) antigen has greater immunogenicity than virtually all other RBC antigens. Expression of Rh(D) antigen varies quantitatively and qualitatively among individuals. Weakened D reactivity can be caused by three different mechanisms: (1) If a C gene is on the chromosome opposite the D gene (trans position), the D antigen may be weakened. (2) There can be a qualitative difference in the D antigen in which an individual lacks a portion of the D antigen molecule (and if exposed to the D antigen may produce an antibody to the portion that they lack). This condition is called "partial D" and is defined in terms of the specific D epitopes possessed (Lomas et al. 1993; Cartron 1994). (3) The *RHD* gene in some individuals (primarily African-Americans) codes for an Rh(D) antigen that reacts more weakly. The Rh antigens are well developed on the RBCs of newborns.

Phenotypes

The prevalence of the various phenotypes associated with the Rh blood group system is listed in Table 4.4.

Antibodies

As was stated previously, the Rh(D) antigen has greater immunogenicity than virtually any other RBC antigen, followed by Rh(c) and Rh(E). Most Rh antibodies result from exposure to human RBCs through pregnancy or transfusion. Rh antibodies are almost always IgG and do not bind complement; thus, they lead to extravascular rather than intravascular RBC destruction.

Clinical Significance

The most common Rh antibody is anti-D; 15% of the U.S. Caucasian population lacks the Rh(D) antigen. Since it is a potent immunogen, the likelihood of an Rh(D)-negative person becoming immunized to Rh(D) following exposure to Rh(D)-positive RBCs is great. It is standard practice to type all donors and recipients for the Rh(D) antigen and to give Rh(D)-negative packed RBCs (PRBCs) to Rh(D)-negative recipients. The use of Rh(D)-positive blood for Rh(D)-negative recipients should be restricted to acute emergencies when Rh(D)-negative PRBCs are not available. Once formed, anti-D can cause severe and even fatal HDN. Anti-D is capable of causing mild to severe delayed transfusion reactions. Antibodies to other Rh antigens have also been implicated in both hemolytic transfusion reactions and HDN.

Percentage of Compatible Donors

The prevalence in the population of the antigen-negative phenotype determines the ease of, or difficulty in, providing compatible PRBCs for transfusion (Table 4.5).

KELL BLOOD GROUP SYSTEM

Antigens

The primary antigens of the Kell system are K and k. Other antithetical antigens are $Kp^a/Kp^b/Kp^c$, as well as Js^a/Js^b, K11/K17, and K14/K24. Three unpaired low-prevalence antigens and seven high-prevalence antigens complete the system. The antigens are carried on a single-pass membrane glycoprotein (type II). Kell antigens are well developed on the red cells of newborns.

Phenotypes

The prevalence in the United States population of the various phenotypes associated with the Kell blood group system is listed in Table 4.6.

TABLE 4.5   Prevalence of Rh Antigen-Negative Phenotypes

Phenotypes	Prevalence		
	Caucasians	African-Americans	Asians
D-negative	15%	8%	1%
C-negative	32%	73%	7%
E-negative	71%	78%	61%
c-negative	20%	4%	53%
e-negative	2%	2%	4%

TABLE 4.4   Rh Blood Group Phenotypes and Prevalence

Antigens	Phenotypes	Prevalence		
		Caucasians	African-Americans	Asians
CcDe	R_1r	34.9%	21%	8.5%
CDe	R_1R_1	18.5%	2%	51.8%
CcDEe	R_1R_2	13.3%	4%	30%
cDe	R_0r	2.1%	45.8%	0.3%
cDEe	R_2r	11.8%	18.6%	2.5%
cDE	R_2R_2	2.3%	0.2%	4.4%
CDEe	R_1R_z	0.2%	Rare	1.4%
CcDE	R_2R_z	0.1%	Rare	0.4%
CDE	R_zR_z	0.01%	Rare	Rare
cde	rr	15.1%	6.8%	0.1%
Cce	r'r	0.8%	Rare	0.1%
cEe	r''r	0.9%	Rare	Rare
CcEe	r'r''	0.05%	Rare	Rare

TABLE 4.6   Kell Blood Group Phenotypes and Prevalence

Phenotypes	Prevalence	
	Caucasians	African-Americans
K−k+	91%	98%
K+k−	0.2%	Rare
K+k+	8.8%	2%
Kp(a+b−)	Rare	0%
Kp(a−b+)	97.7%	100%
Kp(a+b+)	2.3%	Rare
Js(a+b−)	0%	1%
Js(a−b+)	100%	80%
Js(a+b+)	Rare	19%
K°	Exceedingly Rare	

TABLE 4.7 Prevalence of Kell Antigen-Negative Phenotypes

Phenotypes	Prevalence	
	Caucasians	African-Americans
K-negative	91%	98%
k-negative	0.2%	<1.0%
Kp^a-negative	98%	>99%
Kp^b-negative	<1.0%	<1.0%
Js^a-negative	>99%	80%
Js^b-negative	<0.1%	1%

TABLE 4.8 Duffy Blood Group Phenotypes and Prevalence

Phenotypes	Caucasian	Prevalence			
		African-American	Asian-Chinese	Asian-Japanese	Asian-Thai
Fy(a+b−)	17%	9%	90.8%	81.5%	69%
Fy(a−b+)	34%	22%	0.3%	0.9%	3%
Fy(a+b+)	49%	1%	8.9%	17.6%	28%
Fy(a−b−)	Very rare	68%	0%	0%	

Antibodies

Kell system antibodies are generally of the IgG type, react best at body temperature, and rarely bind complement. Anti-K is strongly immunogenic and is frequently found in the serum of transfused K-negative patients. Anti-k, -Kp^a, -Kp^b, -Js^a, and -Js^b are less commonly observed in the United States. Anti-Ku is sometimes seen in immunized K_0 persons.

Clinical Significance

Anti-K may cause both severe HDN and immediate and delayed hemolytic transfusion reactions. Anti-k, -Kp^a, -Kp^b, -Js^a, and -Js^b occur less often than anti-K, but when present may cause HDN and hemolytic transfusion reactions. The other Kell system antibodies have the potential to cause HDN and hemolytic transfusion reactions, but due to their high (or low) frequencies, these reactions seldom occur.

Percentage of Compatible Donors

The prevalence in the population of the antigen-negative phenotype determines the ease of, or difficulty in, providing compatible PRBCs for transfusion, as shown in Table 4.7.

DUFFY BLOOD GROUP SYSTEM

Antigens

The Duffy system comprises five antigens (Fy^a, Fy^b, Fy3, Fy5, Fy6). The molecule containing the Duffy antigens is a multipass membrane glycoprotein, and there are approximately 13,000 Duffy antigen sites per RBC in persons homozygous for Fy^a or Fy^b. RBCs from heterozygotes have about 6000 antigen sites per RBC. These heterozygous RBCs show weaker agglutination than homozygous cells in serological tests, a phenome-

non called the "dosage" effect. The Duffy antigens are well developed at birth, and frequency varies significantly in different racial groups. Interestingly, the Fy3 glycoprotein is the receptor for the malarial parasites *Plasmodium vivax* and *P. knowlesi*; thus, Fy(a-b-) RBCs resist infection by certain malarial organisms.

Phenotypes

The U.S. prevalences of the four phenotypes associated with the Duffy blood group system are listed in Table 4.8.

Antibodies

Duffy antibodies are almost always IgG, though rarely IgM, and only rarely bind complement. In spite of the high percentage of the Fy: -3 phenotype in African-Americans, anti-Fy3 is rare.

Clinical Significance

Anti-Fy^a may cause mild to severe transfusion reactions and HDN. Fy^b is a poor immunogen, and anti-Fy^b antibodies are only infrequently implicated as the cause of transfusion reaction and HDN.

Percentage of Compatible Donors

The prevalence of the antigen-negative phenotype determines the ease of, or difficulty in, providing compatible PRBCs for transfusion, as listed in Table 4.9.

KIDD BLOOD GROUP SYSTEM

Antigens

Three antigens make up the Kidd system: Jk^a, Jk^b, and Jk3. The carrier molecule is a multipass membrane protein, and there are 11,000 to 14,000 Kidd antigens per RBC. Kidd antigens are well developed at birth.

TABLE 4.9   Prevalence of Duffy Antigen-Negative
Phenotypes

Phenotypes	Prevalence	
	Caucasians	African-Americans
Fy^a-negative	34%	90%
Fy^b-negative	17%	77%
Fy3-negative	0%	68%
Fy5-negative	0%	68%
Fy6-negative	0%	68%

TABLE 4.10   Kidd Blood Group Phenotypes and Prevalence

Phenotypes	Prevalence		
	Caucasians	African-Americans	Asians
Jk(a+b−)	26.3%	51.1%	23.2%
Jk(a−b+)	23.4%	8.1%	26.8%
Jk(a+b+)	50.3%	40.8%	49.1%
Jk(a−b−)	Rare	Rare	0.9%

Phenotypes

The prevalences in the United States of the four phenotypes associated with the Kidd blood group system are listed in Table 4.10.

Antibodies

Kidd antibodies are usually IgG, but may be a mixture of IgG and IgM. *They often bind complement and may cause intravascular hemolysis.* It is not uncommon for anti-Jk^a antibody titers to fall rapidly following initial elevation and become undetectable in future antibody screening procedures. If the patient is then exposed to Jk^a antigen-positive PRBCs, a rapid rise in anti-Jk^a titer (anamnestic or "rebound" phenomenon) is often observed, leading to hemolysis.

Clinical Significance

Because Kidd antibodies often bind complement, severe hemolytic transfusion reactions are possible. However, only mild HDN is generally seen. Because of the above-described characteristic, rapid decline in antibody levels, a delayed transfusion reaction, with marked hemolysis of transfused PRBCs within a few hours, can be seen in subsequent exposures to Jk^a-positive PRBCs.

TABLE 4.11   Prevalence of Kidd Antigen-Negative
Phenotypes

Phenotypes	Prevalence	
	Caucasians	African-Americans
Jk^a-negative	23%	8%
Jk^b-negative	26%	51%

Percentage of Compatible Donors

The prevalence of the antigen-negative phenotype determines the ease of, or difficulty in, providing compatible PRBCs for transfusion, as listed in Table 4.11.

MNS BLOOD GROUP SYSTEM

Antigens

Forty-three antigens make up the MNS system. The major antigens are M, N, S, s, and U. MNS antigens are carried on single-pass membrane sialoglycoproteins. The M and N antigens are located on glycophorin A, while S and s antigens are located on glycophorin B. Also included are a number of low-prevalence antigens whose reactivity is attributed to either one or more amino acid substitutions, a variation in the extent or type of glycoslylation, or the existence of a hybrid sialoglycoprotein. MNS antigens are expressed on the RBCs of newborns.

Phenotypes

The prevalences of the numerous phenotypes associated with the MNS blood group system are listed in Table 4.12.

Antibodies

Anti-M antibodies can be IgM or IgG (cold-reactive). Rare examples are active at 37°C. Anti-N is almost always IgM. Both may be present as seemingly "naturally occurring" antibodies. Anti-S, anti-s, and anti-U are usually IgG and occur following RBC stimulation. Antibodies to M and N may frequently show "dosage" effects, reacting more strongly with RBCs with homozygous expression of these antigens. Anti-U is rare, but should be considered when serum from a previously transfused or pregnant African-American person contains antibody to an unidentified high-prevalence antigen.

TABLE 4.12 MNS Blood Group Phenotypes and Prevalence

Phenotypes	Prevalence	
	Caucasians	African-American
M+N–S+s–	6%	2%
M+N–S+s+	14%	7%
M+N–S–s+	10%	16%
M+N+S+s–	4%	2%
M+N+S+s+	22%	13%
M+N+S–s+	23%	33%
M–N+S+s–	1%	2%
M–N+S+s+	6%	5%
M–N+S–s+	15%	19%
M+N–S–s–	0%	0.4%
M+N+S–s–	0%	0.4%
M–N+S–s–	0%	0.7%

TABLE 4.13 Prevalence of MNS Antigen-Negative Phenotypes

Phenotypes	Prevalence	
	Caucasians	African-Americans
M-negative	22%	26%
N-negative	28%	25%
S-negative	48%	69%
s-negative	11%	6%
U-negative	0%	<1%

Clinical Significance

Anti-M has been only implicated in transfusion reactions or HDN in rare cases. Anti-N has no known clinical significance. Antibodies to S and s are capable of causing hemolytic transfusion reactions and HDN. Anti-U has been implicated in mild to severe hemolytic reactions and HDN.

Percentage of Compatible Donors

The prevalence of the antigen-negative phenotype determines the ease of, or difficulty in, providing compatible PRBCs for transfusion, as listed in Table 4.13.

P BLOOD GROUP SYSTEM

Antigens

The sequential transcription of multiple genes is required for the expression of the P1 antigen; this occurs

TABLE 4.14 Prevalence of P1 Antigen-Negative Phenotypes

Phenotypes	Prevalence	
	Caucasians	African-Americans
P1-negative	21%	6%

upon the addition of a galactosyl residue to paragloboside. The P, Pk, and LKE antigens were previously included in the P system, but because a different locus and biochemical pathway has been found to be involved in their production, they have been moved to the globoside (glob) collection. The P1 antigen is more strongly expressed on fetal cells than on neonatal cells, P1 expression weakening as the fetus ages. Adult levels are not reached until 7 years of age.

Antibodies

Anti-P1 is IgM and is naturally occurring in many P1-negative individuals. Complement binding by anti-P1 is rare.

Clinical Significance

Antibodies to P1 rarely cause transfusion reactions and have not been implicated in HDN.

Percentage of Compatible Donors

The prevalence of the antigen-negative phenotype determines the ease of, or difficulty in, providing compatible PRBCs for transfusion, as listed in Table 4.14.

LUTHERAN BLOOD GROUP SYSTEM

Antigens

The Lutheran system comprises four antithetical pairs, Lu^a and Lu^b, Lu6 and Lu9, Lu8 and Lu14, and Au^a and Au^b, as well as 10 high-prevalence antigens. The antigens are carried on a single-pass membrane glycoprotein with 1500 to 4000 copies per RBC, depending on zygosity. The Lutheran antigens are expressed weakly on cord blood cells.

Phenotypes

The prevalences of the four phenotypes associated with the Lutheran blood group system are listed in Table 4.15.

TABLE 4.15   Lutheran Blood Group Phenotypes and Prevalence

Phenotypes	Prevalence in Most Populations
Lu(a+b−)	0.2%
Lu(a−b+)	92.4%
Lu(a+b+)	7.4%
Lu(a−b−)	Rare

TABLE 4.16   Prevalence of Lutheran Antigen-Negative Phenotypes

Phenotypes	Prevalence	
	Caucasians	African-Americans
Lu^a-negative	92%	95%
Lu^b-negative	<1%	<1%

Antibodies

The most common Lutheran antibodies are anti-Lu^a and Lu^b, but they are not often encountered. Lutheran antibodies may be IgG or IgM, are generally not reactive at body temperature, and rarely bind complement.

Clinical Significance

The Lutheran antigens are not well developed at birth, and anti-Lu^a has not been reported as a cause of HDN. Anti-Lu^a has not been associated with hemolytic transfusion reactions. Anti-Lu^b may cause accelerated destruction of transfused PRBCs, but only causes mild (if at all) HDN.

Percentage of Compatible Donors

The prevalence of the antigen-negative phenotype determines the ease of, or difficulty in, providing compatible PRBCs for transfusion, as listed in Table 4.16.

LEWIS BLOOD GROUP SYSTEM

Antigens

Le^a, Le^b, and Le^x make up the antigens of the Lewis system. The Lewis antigens are not intrinsic to RBCs, but are located on type 1 glycosphingolipids that are adsorbed onto the red cells from the plasma. The biosynthesis of Lewis antigens results from the interac-

TABLE 4.17   Lewis Blood Group Phenotypes and Prevalence

Phenotypes	Prevalence	
	Caucasians	African-Americans
Le(a+b−)	22%	23%
Le(a−b+)	72%	55%
Le(a−b−)	6%	22%
Le(a+b+)	Rare	Rare

tion of two independent genetic loci, *Le* and *Se*. The transferase that is the product of the *Le* gene attaches fucose to the subterminal GlcNAc of type 1 oligosaccharides. This confers Le^a activity. The *Se* gene determines a transferase that attaches a fucose to the terminal Gal, but only if the adjacent GlcNAc is already fucosylated. This configuration has Le^b activity. Le^b is adsorbed onto the RBC preferentially over Le^a. Individuals possessing both *Le* and *Se* genes will have RBCs that express Le^b, but not Le^a. RBCs from a person with *Le* but not *Se* genes will express Le^a. The Lewis antigens are not expressed on cord cells.

Phenotypes

The prevalence of phenotypes associated with this blood group system is listed in Table 4.17.

Antibodies

Anti-Le^a and anti-Le^b are frequently occurring antibodies made by Le(a-b-) individuals, usually in the absence of a foreign RBC stimulus. Lewis antibodies are predominantly IgM, and some bind complement.

Clinical Significance

Most Lewis antibodies react at colder temperatures than body temperature and are not clinically significant. Since Lewis antigens are poorly developed at birth, and IgM antibodies cannot cross the placenta, Lewis antibodies do not cause HDN. Anti-Le^a that has activity at 37°C has caused hemolytic transfusion reactions on rare occasions.

Percentage of Compatible Donors

The prevalence of the antigen-negative phenotype determines the ease of, or difficulty in, providing compatible PRBCs for transfusion, as listed in Table 4.18.

TABLE 4.18   Prevalence of Lewis Antigen-Negative Phenotypes

Phenotypes	Prevalence	
	Caucasians	African-Americans
Le^a-negative	78%	77%
Le^b-negative	28%	45%

TABLE 4.19   Diego Blood Group Phenotypes and Prevalence

Phenotypes	Prevalence		
	Caucasians	African-Americans	Asians
Di(a+b−)	<0.01%	<0.01%	<0.01%
Di(a−b+)	>99.9%	>99.9%	90%
Di(a+b+)	<0.1%	<0.1%	10%

DIEGO BLOOD GROUP SYSTEM

Antigens

The Diego system includes two pairs of antithetical antigens, Di^a/Di^b and Wr^a/Wr^b, and the low-prevalence antigens, Wd^a, Rb^a, WARR, ELO, Wu, Bp^a, Mo^a, Hg^a, Vg^a, Sw^a, BOW, NFLD, Jn^a, KREP, Tr^a, Fr^a, and SW1. The antigens are carried on a multipass membrane protein, having 1,000,000 antigen copies per RBC. Diego antigens are expressed on the RBCs from newborns.

Phenotypes

The prevalences of phenotypes associated with the Diego blood group system are listed in Table 4.19.

Antibodies

Anti-Di^a and anti-Di^b are of the IgG class and do not bind complement. Anti-Wr^a and Wr^b may be IgM or IgG and do not bind complement.

Clinical Significance

Anti-Di^a is not common because of the low prevalence of the antigen in the United States, *but anti-Di^a can cause RBC destruction and should be considered clinically significant when present.* Anti-Wr^a has been implicated in mild to severe transfusion reactions and in HDN.

CARTWRIGHT BLOOD GROUP SYSTEM

Antigens

The high-prevalence antigen Yt^a and the low-prevalence antigen Yt^b make up the Cartwright system. The antigens are carried on the PI-linked glycoprotein, acetylcholinesterase. There are 10,000 antigen copies per RBC. Cartwright antigens are expressed weakly on the RBCs of newborns.

Antibodies

The Cartwright antibodies are IgG and do not bind complement. Because of the high prevalence in the United States of the Yt^a antigen, anti-Yt^a is not common. Because Yt^b is a poor immunogen, anti-Yt^b is also uncommon.

Clinical Significance

Many examples of anti-Yt^a have been shown to be clinically benign in vivo, while other examples have shown increased RBC destruction in in vivo survival studies. Anti-Yt^a is not known to cause HDN. Anti-Yt^b has not been implicated in hemolytic transfusion reactions or HDN.

Xg BLOOD GROUP SYSTEM

Antigens

Xg was the first blood group system to be assigned to the X chromosome. Xg^a is located on a single-pass glycoprotein (type 1). Xg^a is expressed weakly on cord blood cells. CD99 has been assigned to the Xg system. The expression of CD99 on RBCs is correlated with the expression of Xg^a. All Xg(a+) individuals have high RBC expression of CD99, all Xg(a−) females have low RBC expression of CD99 and the RBCs of Xg(a−) males may have either high or low expression of CD99.

Phenotypes

The prevalence of phenotypes associated with the Xg blood group system is listed in Table 4.20.

Antibodies

Xg^a is a poor immunogen. IgG-type antibodies to Xg^a are more common than IgM. Some examples of anti-Xg^a are naturally occurring.

TABLE 4.20   Xg Blood Group Phenotypes and Prevalence

| Phenotypes | Prevalence | |
	Females	Males
Xg(a+)	88.7%	65.6%
Xg(a−)	11.3%	34.4%

TABLE 4.21   Prevalence of Xg Antigen-Negative Phenotypes

Phenotypes	Prevalence
Xg^a-negative	11% (females)
Xg^a-negative	34% (males)

TABLE 4.22   Scianna Blood Group Phenotypes and Prevalence

Phenotypes	Caucasians
Sc:1,−2	99.7%
Sc:1,2	0.3%
Sc:−1,2	Very rare
Sc:−1,−2	Very rare

Clinical Significance

Antibodies to Xg^a have not been shown to cause transfusion reactions or HDN.

Percentage of Compatible Donors

The prevalence of the antigen-negative phenotype determines the ease of, or difficulty in, providing compatible PRBCs for transfusion, as listed in Table 4.21.

SCIANNA BLOOD GROUP SYSTEM

Antigens

There are three antigens associated with the Scianna system: the high-prevalence Sc:1, the low-prevalence Sc:2, and Sc:3 (which is present if either Sc:1 or Sc:2 is present). These antigens are carried on a membrane glycoprotein and are expressed on the RBCs of newborns.

Phenotypes

The prevalences of phenotypes associated with the system are listed in Table 4.22.

TABLE 4.23   Prevalence of Scianna Antigen-Negative Phenotypes

Phenotypes	Prevalence in Caucasians
Sc1-negative	Very rare
Sc2-negative	99.7%

Antibodies

Antibodies to Scianna RBC antigens are rare, but if present, are generally of the IgG type.

Clinical Significance

Antibodies directed against Scianna antigens have not been reported to cause transfusion reactions. Anti-Sc1 may cause a positive direct antiglobulin test (DAT), but it does not cause clinically significant HDN.

Percentage of Compatible Donors

The prevalence of the antigen-negative phenotype determines the ease of, or difficulty in, providing compatible PRBCs for transfusion, as listed in Table 4.23.

DOMBROCK BLOOD GROUP SYSTEM

Antigens

The five antigens that make up the Dombrock system are Do^a, Do^b, and the high-prevalence antigens Hy, Gy^a, and Jo^a. The antigens are carried on a GPI-linked glycoprotein and are expressed on the RBCs of newborns.

Phenotypes

The prevalences of phenotypes associated with the Dombrock system are listed in Table 4.24.

Antibodies

Dombrock system antibodies are IgG and do not bind complement. Do^a and Do^b are poor immunogens, and thus, antibodies to Do^a and Do^b are not common. Gy^a is highly immunogenic, but because of its high prevalence in the United States, anti-Gy^a is rarely observed.

Clinical Significance

Anti-Do^a may cause increased RBC destruction, but anti-Do^b is considered to be clinically insignificant.

TABLE 4.24 Dombrock Blood Group Phenotypes and Prevalence

| Phenotypes | Prevalence | | | | | | |
	Do^a	Do^b	Gy^a	Hy	Jo^a	Caucasians	African-Americans
Do(a+b−)	+	0	+	+	+	18%	11%
Do(a+b+)	+	+	+	+	+	49%	44%
Do(a−b+)	0	+	+	+	+	33%	45%
Gy(a−)	0	0	0	0	0	Rare	0%
Hy−	0	Weak	Weak	0	0	0%	Rare
Jo(a−)	Weak	0/Weak	+	Weak	0	0%	Rare

TABLE 4.25 Prevalence of Dombrock Antigen-Negative Phenotypes

| Phenotypes | Prevalence | |
	Caucasians	African-Americans
Do^a-negative	33%	45%
Do^b-negative	18%	11%
Hy-negative	0%	Rare
Gy^a-negative	Rare	0%
Jo^a-negative	0%	Rare

Antibodies to Do^a, Do^b, Gy^a, Hy, and Jo^a have not been observed to cause HDN.

Percentage of Compatible Donors

The prevalence of the antigen-negative phenotype determines the ease of, or difficulty in, providing compatible PRBCs for transfusion, as listed in Table 4.25.

COLTON BLOOD GROUP

Antigens

Three antigens have been assigned to the Colton system: Co^a, Co^b, and Co3. The antigens are carried on a multipass membrane glycoprotein that is part of the water transport protein CHIP-1. Colton antigens are expressed on the RBCs of newborns.

Phenotypes

The prevalences of the four phenotypes associated with the Colton blood group system are listed in Table 4.26.

TABLE 4.26 Colton Blood Group Phenotypes and Prevalence

Phenotypes	Prevalence in Most Populations
Co(a+b−)	90%
Co(a−b+)	0.5%
Co(a+b+)	9.5%
Co(a−b−)	<0.01%

TABLE 4.27 Prevalence of Colton Antigen-Negative Phenotypes

Phenotypes	Prevalence in All Populations
Co^a-negative	0.1%
Co^b-negative	90%
Co3-negative	<0.01%

Antibodies

Antibodies to Colton system antigens are IgG and rarely bind complement (with the exception of Co3).

Clinical Significance

Colton antibodies are rare, but anti-Co^a has been implicated in both HDN and in vivo RBC destruction. Mild HDN caused by Co^b has been reported. Anti-Co3 has been known to cause severe HDN.

Percentage of Compatible Donors

The prevalence of the antigen-negative phenotype determines the ease of, or difficulty in, providing compatible PRBC for transfusion, as listed in Table 4.27.

TABLE 4.28 LW Blood Group Phenotypes and Prevalence

Phenotypes	Prevalence	
	Most Populations	Finnish
LW(a+b−)	97%	93.9%
LW(a+b+)	3%	6%
LW(a−b+)	Rare	0.1%
LW(a−b−)	Rare	

TABLE 4.29 Prevalence of Chido-Rogers Antigen-Negative Phenotypes

Phenotypes	Prevalence in All Populations
Ch-negative	4%
Rg-negative	2%
Wh-negative	85%

LW BLOOD GROUP SYSTEM

Antigens

The LW system is composed of the antigens LW^a, LW^b, and LW^{ab}. The antigens are located on a single-pass membrane glycoprotein (type 1). The original anti-Rh described by Landsteiner and Weiner (1940) was in fact directed against the LW antigen. There is a phenotypic relationship between LW and the Rh(D) antigen. In adults, Rh(D)-negative RBCs have less expression of LW than do Rh(D)-positive RBCS. The expression of LW antigens is stronger on cord cells than on the RBCs from adults, and it is expressed in equal amounts regardless of the Rh(D) type.

Phenotypes

The prevalences of phenotypes associated with the LW blood group system are listed in Table 4.28.

Antibodies

LW antibodies may be IgM or IgG and do not bind complement. Autoanti-LW antibodies have been reported upon immunosuppression of LW antigens, observed primarily during pregnancy and in patients with certain malignancies.

Clinical Significance

While Rh(D)-negative, LW+ PRBCs have successfully been transfused to persons with anti-LW in their sera, LW antibodies may cause accelerated destruction of LW+ RBCs. Significant HDN has not been reported.

CHIDO-ROGERS BLOOD GROUP SYSTEM

Antigens

The Chido-Rogers (Ch/Rg) system is made up of nine antigens (Ch1, Ch2, Ch3, Ch4, Ch5, Ch6, Rg1, Rg2, and WH) that reside on the complement component C4. Chido antigens are located in the C4d region of C4B. Rogers antigens are located in the C4d region of C4A. C4A and C4B are glycoproteins that are adsorbed onto the RBC membrane from the serum. These antigens are absent or weakly expressed on the RBC of newborns.

Antibodies

Ch/Rg antibodies are IgG, do not bind complement, and may be neutralized (in serological tests) with serum or plasma from Ch/Rg antigen-positive individuals.

Clinical Significance

Antibodies to Ch/Rg antigens are usually clinically insignificant, but they may create difficulties in interpretations of serological investigations. None have been known to cause HDN or increased RBC destruction.

Percentage of Compatible Donors

The prevalence of the antigen-negative phenotype determines the ease of, or difficulty in, providing compatible PRBCs for transfusion, as listed in Table 4.29.

Hh BLOOD GROUP SYSTEM

Antigens

The H antigen is the precursor molecule on which the A and B antigens are built. On group O RBCs there are no A or B antigens, and the RBC membrane expresses abundant H antigen. The amount of H antigen present on RBCs by blood group is, in order of diminishing quantity, $O > A_2 > B > A_2B > A_1 > A_1B$. The red cells of newborns have weak H antigen. In the absence of the H gene, no L-fucose is added to the protein or lipid, and thus no H antigen is expressed (the Bombay phenotype). In the Bombay phenotype, even in the presence of functional A and B genes, no A or B antigen will be formed; the RBCs will type as group O, and hence the notation O_h.

Antibodies

A few group A_1, A_1B, and B individuals have so little unconverted H antigen that they produce an anti-H that is a weak IgM antibody reactive at colder than room temperatures. The anti-H formed by an O_h individual is potent, reacts over a thermal range of 4° to 37°C, and rapidly destroys RBCs with A, B, and/or H antigens.

Clinical Significance

The benign anti-H formed by group A_1, A_1B, and B individuals is not considered clinically significant. However, *the anti-H formed by O_h individuals can cause hemolysis in vivo and subsequent severe hemolytic transfusion reactions.* HDN may occur in O_h mothers.

Percentage of Compatible Donors

The only suitable PRBCs for transfusion to a Bombay patient are those from another Bombay (O_h) individual.

XK BLOOD GROUP SYSTEM

Antigens

Only one antigen is included in the XK system, Kx. It is carried on a multipass membrane protein. The antigen is present on virtually all RBCs, but its expression is weak on RBCs of common Kell phenotypes. Elevated levels of Kx are found on the null (K_o) RBC. Absence of the Kx protein is associated with McLeod syndrome, a spectrum of abnormalities including weakened Kell system antigens, bizarre RBC shapes, chronic anemia, and abnormalities of the nervous system, of the cardiac system, and of granulocytes. The McLeod phenotype has been identified in patients with chronic granulomatous disease (CGD) and appears to result from deletion of a portion of the X chromosome that includes the *XK* locus (as well as the *X-CGD*).

Antibodies

Anti-Kx is IgG and does not bind complement. Males with both McLeod syndrome and CGD may be immunized by PRBC transfusion and produce anti-KL (anti-Kx + anti-Km). Only Kx-negative PRBCs are compatible. Non-CGD males may make anti-Km; both Kx-negative and K_o PRBCs are compatible.

Clinical Significance

Mild delayed transfusion reactions have been reported with anti-Kx. As the antibody is only made by males, the possibility of HDN does not arise.

GERBICH BLOOD GROUP SYSTEM

Antigens

The Gerbich system contains seven antigens: Ge2, Ge3, and Ge4 (all high prevalence), and Wb, Ls^a, An^a and Dh^a (all low prevalence). The antigens are carried on single-pass membrane sialoglycoproteins (type 1). Ge3 and Ge4 are located on glycophorin C (GPC), while Ge2, Ge3, and An^a are located on glycophorin D (GPD). Dh^a and Wb are located on altered forms of GPC. Ls^a is found on an altered form of GPC and GPD. Rare phenotypes that lack one or more of the high-prevalence antigens are Leach (Ge: -2, -3, -4), Gerbich (Ge: -2, -3, 4), and Yus (Ge: -2, 3, 4). Gerbich antigens are expressed on the RBCs of newborns.

Antibodies

Antibodies to Gerbich antigens are usually IgG, but may also have an IgM component. Some examples of anti-Ge2 and anti-Ge3 are hemolytic in vitro.

Clinical Significance

Gerbich system antibodies are of variable clinical significance. Some have been known to cause mild HDN and slightly accelerated RBC clearance.

CROMER BLOOD GROUP SYSTEM

Antigens

Ten antigens have been assigned to the Cromer system: Cr^a, Tc^a, Dr^a, Es^a, IFC, UMC, and Wes^b (all high prevalence), and Tc^b, Tc^c, and Wes^a (all low prevalence). These antigens are located on the complement regulatory protein, decay accelerating factor (DAF). These antigens are expressed on the RBCs of newborns.

Antibodies

Antibodies to Cromer system antigens are extremely rare. Most examples of anti-Cr^a, anti-Wes^b, and anti-Tc^a have been found in serum from African-American individuals.

Clinical Significance

The clinical significance of Cromer antibodies is variable. Some have reportedly caused decreased survival of transfused PRBCs in vivo. Antibodies to Cromer system antigens do not cause HDN. Cromer antigens are present on trophoblasts of the placenta, which may adsorb maternal antibody.

KNOPS BLOOD GROUP SYSTEM

Antigens

Kn^a, Kn^b, McC^a, Sl^a, McC^b, Yk^a, and Vil make up the Knops system. These antigens are located on the single-pass membrane glycoprotein (type 1) CR1, the primary complement receptor on RBCs. Knops system antigens are only weakly expressed on the RBCs of newborns.

Antibodies

The antibodies directed against Knops antigens are IgG and do not bind complement. They were previously grouped among the "high-titer, low-avidity (HTLA) antibodies." These antibodies commonly show weak, variable reactivity in the antiglobulin phase of testing, but may continue to react in vitro, even at high dilutions. This variable reactivity has been shown to be a direct reflection of the number of CR1 sites on RBCs.

Clinical Significance

The Knops-directed antibodies are of no known clinical significance, since they do not cause transfusion reactions, increased RBC destruction, or HDN.

INDIAN BLOOD GROUP SYSTEM

Antigens

The Indian system is comprised of two antigens, the low-prevalence In^a and the high-prevalence In^b. The antigens are located on a single-pass membrane glycoprotein (CD44). CD44 has wide tissue distribution and is a cellular adhesion molecule involved in immune stimulation and signaling between cells. These antigens were discovered in persons from the Indian subcontinent (hence the name "Indian"). These antigens are weakly expressed on cord cells.

Antibodies

Indian antibodies are of the IgG class and do not bind complement.

TABLE 4.30 Prevalence of Indian Antigen-Negative Phenotypes

Phenotypes	Prevalence			
	Caucasians	African-Americans and Asians	Indians	Iranians
In^a-negative	99.9%	99.9%	96%	89.6%
In^b-negative	1%		4%	

Clinical Significance

Both anti-In^a and anti-In^b have been known to cause transfusion reactions and decreased RBC survival in vivo. Both antibodies may cause a positive DAT, but neither has been known to cause HDN.

Percentage of Compatible Donors

The prevalence of the antigen-negative phenotype determines the ease of, or difficulty in, providing compatible PRBCs for transfusion, as listed in Table 4.30.

Ok BLOOD GROUP SYSTEM

Antigens

The antigen of the Ok system, Ok^a, is also known by several names (Basigin [BSG], EMMPRIN, and M6 leukocyte activation antigen). The high prevalence antigen is carried on CD17, a single-pass (type 1) glycoprotein. The null phenotype (Ok^a-) has only been described in the Japanese. The antigen is well developed at birth.

Antibodies

The original anti-Ok^a is IgG and does not bind complement.

Clinical Significance

Anti-Ok^a has caused reduced RBC survival, but not HDN.

RAPH BLOOD GROUP SYSTEM

Antigens

The RAPH system consists of only one antigen, MER2 (Raph). The molecular basis of the antigen is not known, but expression is variable. The antigen is expressed on the RBCs of newborns.

Antibodies

Anti-Raph has, to date, only been identified in three Jews of Indian origin. The antibodies have been IgG and some have bound complement.

Clinical Significance

All antibody makers have been on renal dialysis because of kidney disease. Transfusion reactions have not been reported after Raph-positive transfusions. It is not known whether the antibody is capable of causing HDN.

JMH BLOOD GROUP SYSTEM

Antigens

The high prevalence JMH is carried on a GPI-linked glycoprotein. The antigen is expressed on cord cells.

Antibodies

JMH antibodies are IgG and do not bind complement. Autoantibodies to the protein are relatively common and are often associated with loss of JMH expression by circulating RBCs.

Clinical Significance

Neither transfusion reactions nor HDN has been reported.

COST BLOOD GROUP COLLECTION

Antigens

The antigens in this collection are Cs^a and Cs^b. They demonstrate variable expression on RBCs.

Antibodies

Antibodies to Cost antigens are IgG and do not bind complement. These antibodies were formerly grouped with the HTLA antibodies.

Clinical Significance

The Cost antibodies are not considered clinically significant.

Er BLOOD GROUP COLLECTION

Antigens

The Er collection consists of the high-prevalence Er^a and the low-prevalence Er^b. The Er antigens are expressed on the RBCs of newborns.

Antibodies

The Er antibodies are IgG and do not bind complement.

Clinical Significance

Anti-Er^a has not been implicated in transfusion reaction, but has been reported to cause a positive DAT without evidence of clinical HDN.

Ii BLOOD GROUP COLLECTION

Antigens

The antigens included in this collection, I and i, are not the products of alleles, but instead arise from the sequential actions of multiple glycosyltransferases. The antigens are located on precursor A, B, and H active oligosaccharide chains (I on branched type 2 chains and i on linear type 2 chains). Fetal RBCs carry few branched oligosaccharide chains and therefore are rich in i and poor in I. With age, the linear chains are modified by the addition of branched structures. This branching configuration confers I specificity. So-called compound antigens have been described: IA, IB, IAB, IH, IP, ILe^{bH}, iH, iP1, and iHLe^b.

Phenotypes

The prevalences of phenotypes associated with the blood group system are listed in Table 4.31.

TABLE 4.31 Ii Blood Group Phenotypes and Prevalence

| Phenotypes | Prevalence | | |
	Adult	Cord	i adult
I	Strong	Weak	Trace
i	Weak	Strong	Strong

Antibodies

Anti-I is an IgM antibody that reacts preferentially at room temperature or colder and binds complement. Only rare examples of I-negative adult RBCs exist; thus, alloanti-I is rare. Since all RBCs have trace amounts of I, anti-I is considered an autoantibody.

Clinical Significance

Anti-I is a low-titer, cold-reactive, common autoantibody that reacts within a narrow thermal range. *Anti-I assumes pathological significance in cold agglutinin disease and mixed-type autoimmune hemolytic anemia,* in which it behaves as a complement-binding hemolytic antibody with a high titer and wide thermal range. Although rare, accelerated RBC destruction has been observed in the adult I phenotype in the presence of alloanti-I. HDN has not been associated with Ii antibodies.

GLOBOSIDE COLLECTION

Antigens

Three antigens are included in this collection: P, P^k, and LKE. The sequential action of multiple gene products is required for expression of these antigens. Lactosyl ceramide is the precursor substance.

Phenotypes

The prevalences of phenotypes associated with the collection are listed in Table 4.32.

Antibodies

The antibodies directed against globoside antigens may be IgG or IgM. They may bind complement, and some are hemolytic in vitro.

Clinical Significance

Anti-P has been implicated in rare cases of severe transfusion reactions, but has been known to cause only mild HDN. The biphasic autohemolysin in paroxysmal cold hemoglobinuria (PCH) has anti-P specificity. Anti-P, P1, P^k (formerly anti-Tj^a) is a potent IgM antibody that has caused hemolytic transfusion reactions and occasional cases of HDN.

LOW-PREVALENCE ANTIGENS

Antigens

Low-prevalence antigens occur in <1% of most populations, have no known alleles, and cannot be placed in

TABLE 4.32　Globoside Collection Phenotypes and Prevalence

Phenotypes	Antigens	Prevalence	Antibodies
P_1	P,P1,P^k	79%	None
P_2	P,P^k	21%	Anti-P1
P_1^k	P1P^k	Rare	Anti-P
P_2^k	P^k	Rare	Anti-P
P	None	Rare	Anti-P,P1,P^k

TABLE 4.33　High-Prevalence Antigens and Their Characteristics

Antigen	IgM/IgG	Complement—Binding	HTR	HDN
Vel	IgM or IgG	Yes	None-severe	+DAT, no HDN
Lan	IgG	Some	None-severe	None-mild
At^a	IgG	No	None-moderate	+DAT, no HDN
Jr^a	IgG	Some	↓RBC survival	+DAT, no HDN
EMM	IgG	Some	Unknown	Unknown
AnWj	IgG	Rare	Severe-1 case	None
Sd^a	IgG	No	Rare	None
Duclos	IgG	Unknown	Unknown	Unknown
PEL	IgG	Unknown	↓RBC survival	None
ABTI	IgG			
MAM	IgG			

a particular blood group system or collection. They include By, Bi[a], Bx[a], Chr[a], HJK, HOFM, JFV, JONES, Je[a], Kg, Li[a], LOCR, Milne, Ol[a], Pt[a], Rd, RASM, REIT, Re[a], SARA, and To[a].

Antibodies

These antibodies may be IgM or IgG. Antibodies to low-prevalence antigens may cause HDN. The relative prevalence of antigens and antibodies in the population make incompatibilities rare.

HIGH-PREVALENCE ANTIGENS

Antigens

High-prevalence antigens occur in >90% of the U.S. population, but have no known alleles and cannot be placed in a specific blood group system or collection.

Antibodies

Since the individual who makes high-prevalence antibody lacks the respective antigen, antibodies directed against high-prevalence antigens are rarely encountered. When these antibodies do occur, it may be exceedingly difficult to find compatible blood for the patient (as most of the population is positive for the antigen; Table 4.33).

References

Cartron JP. 1994. Defining the Rh blood group antigens. Biochemistry and molecular genetics. *Blood Rev* 8:199–212.

Cherif-Zahar B, Mattei M, Le Van KC, Bailly P, Cartron JP, and Colin Y. 1991. Localization of the human Rh blood group gene structure to chromosome 1p34.3–1p36.1 region by *in situ* hybridization. *Hum Genet* 86:98.

Daniels GL, Anstee DJ, Cartron JP, et al. 1999. ISBT Working Party on terminology for red cell surface antigens. Oslo Report. *Vox Sang* 77:52–57.

Daniels GL, Anstee DJ, Cartron JP, et al. 1995. Blood group terminology 1995: from the Working Party on Terminology for Red Cell Surface Antigens. *Vox Sang* 69:265–279.

Daniels GL, et al. 1996. Report of ISBT Working Party. Makuhari, Japan.

Fisher RA and Race RR. 1946. Rh gene frequencies in Britain. *Nature* 157:48–49.

Issitt PD and Anstee DJ. 1998. Applied blood group serology. 4th ed. Durham, NC: Montgomery Scientific Publication.

Landsteiner K and Wiener AS. 1940. An aggutinable factor in human blood recognized by immune sera for rhesus blood. *Proc Soc Exp Biol NY* 43:223.

Le Van KC, Mouro I, Cherif-Zahar B, et al. 1992. Molecular cloning and primary structure of the human blood group Rh D polypeptide. *Proc Natl Acad Sci USA* 89:10925.

Lomas C, McColl K, and Tippett P. 1993. Further complexities of the Rh antigen D disclosed by testing category D[II] cells with monoclonal anti-D. *Transfusion Med* 3:67–69.

Lomas C, Tippett P, Thompson KM, et al. 1984. Demonstration of seven epitopes of the Rh antigen D using monoclonal anti-D antibodies and red cells from D categories. *Vox Sang* 57:261–264.

Mollison PL, Englefriet CP, and Contreras M. 1987. Blood transfusion in clinical medicine. 8th ed. Cambridge, MA: Blackwell Science.

Rosenfield RE, Allen FH, Jr., Swisher SN, and Kochwa S. 1962. A review of Rh serology and presentation of a new terminology. *Transfusion* 2:187.

Wiener AS. 1943. Genetic theory of the Rh blood types. *Proc Soc Exp Biol Med* 54:316–319.

5

Pretransfusion Compatibility Testing

SUSAN T. JOHNSON, MSTM, MT(ASCP)SBB, AND TINA M. PUGH, MT (ASCP)

INTRODUCTION

The goal of pretransfusion compatibility testing is to provide an individual with blood that is ABO, Rh-compatible and antigen-negative for any clinically significant antibodies the recipient possesses to decrease the possibility of a hemolytic transfusion reaction.

These results are first accomplished with proper specimen collection. Positive identification of the patient at the time of specimen collection and transfusion is critical as most errors are caused by mislabeled specimens or transfusion of the wrong unit of blood to a patient—that is, clerical mistakes. According to the *American Association of Blood Banks (AABB) Standards for Blood Banks and Transfusion Services,* patient samples must be labeled with two unique identifiers, the date and a way to identify the individual who drew the sample before leaving the bedside. Many pediatric institutions also use a special blood bank number to aid in correctly identifying the patient.

Once the sample is received in the transfusion service a review of previous records is performed to check for special requirements and past transfusion history. The specimen is typed for ABO antigens and antibodies and Rh(D) antigen, screened for unexpected antibodies, and and an identification of any alloantibodies detected. Neonates also require a direct antiglobulin test (DAT) to check for passive acquisition of IgG antibodies (Figure 5.1).

ABO(H)

Antigens and Antibodies

ABO antigens are constructed of a series of carbohydrates attached to glycoproteins or glycolipids. These antigens are also in soluble forms in other body fluids, such as saliva, plasma, and tears, and are expressed by tissues. Glycosyltransferases attach sugars to the end of these carbohydrate chains to determine the ABO type. The H transferase attaches fucose (Fuc) to a terminal galactose residue creating the H antigen, or the Group O blood type. The A and B transferases then add N-acetylgalactosamine (GalNAc) and D-galactosamine (Gal) respectively to create A and B antigens. Group AB individuals contain all glycosyltransferases. ABO antigens are not fully developed in neonates (younger than 4 months old) and may be suppressed in patients who have hematopoietic disorders. Newborn infants have decreased levels of a branching enzyme that allows multiple copies of an antigen to be present on one chain. This sometimes results in weaker reactions with antisera, especially Group A. This decreased level of antigen could result in mixed field agglutination (two cell populations) being observed even though all of the red cells are Group A.

A "naturally occurring" immune response is produced to the A and B antigens. Anti-A and anti-B are stimulated not by transfusion of incompatible blood but by exposure to environmental flora ("gut-associated" bacteria, plants, and other exogenous material) and are usually IgM, containing some IgG antibodies. The exception to this is Group O anti-A,B, which is predominately IgG. IgG antibodies, unlike IgM types, are

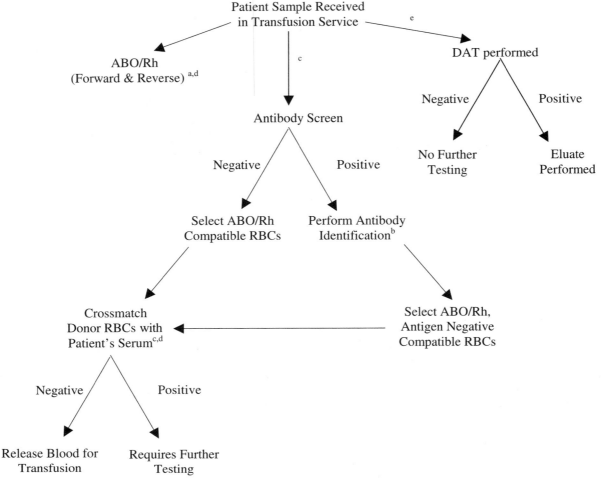

FIGURE 5.1   Pretransfusion testing flowchart.
[a] Neonates do not require testing for anti-A or anti-B unless non-Group O blood is required.
[b] In neonates, the patient's or maternal serum may be used to identify an antibody.
[c] In neonates, if the initial antibody screen is negative and the patient has not left the hospital, Group O packed red blood cells (PRBCs) are able to be transfused without repeat testing of the antibody screen for the first 4 months.
[d] If a non-Group O neonate needs type-specific RBCs, a new sample must be drawn and the serum tested for passive anti-A or anti-B.
[e] DATs with anti-IgG are performed on neonates to detect RBC sensitization via placental transfer.

able to cross the placenta and cause hemolytic disease of the newborn (HDN). In most cases, infants with HDN due to ABO incompatibility have Group O mothers. Newborn infants cannot produce IgG or IgM antibodies efficiently until they are 3 to 6 months old because of their decreased immune system. Many times it takes up to a year before ABO antibodies are detected. Adult levels are produced between 2 to 4 years, and maximum levels found at ages 5 to 10. Individuals do not make antibodies that react with the ABO antigens that they possess (Table 5.1).

Transfusion of ABO incompatible red blood cells (RBCs) or plasma leads to complement activation and immediate intravascular hemolysis. Transfusing as little as 1 to 20 mL of ABO-incompatible RBCs may produce

serious side effects. This hemolytic transfusion reaction is swift and severe, producing hemoglobinuria and hemoglobinemia and possibly resulting in renal failure, disseminated intravascular coagulation (DIC), shock, or death.

Transfusion

The selection of RBCs and plasma is based on the ABO type of the recipient. Every effort should be made to transfuse type-identical products. If this is not readily feasible, type-compatible products may be used. For example, Group A individuals should receive A RBCs and plasma. If these products cannot be supplied, type-compatible products can be safely transfused. For

TABLE 5.1 Forward and Reverse Typing for ABO and Blood Component Compatibility

Forward Type Patient RBCs +			Reverse Type[a] Patient Serum +		ABO Results	Component Compatibility (in order of priority)	
Anti-A	Anti-B	Control[b]	A₁ Cells	B Cells		RBCs	FFP/PLTS
0	0	N/A	+	+	O	O	O, A, B, AB
+	0	N/A	0	+	A	A, O	A, AB
0	+	N/A	+	0	B	B, O	B, AB
+	+	0	0	0	AB	AB, A, B, O	AB

N/A = Test not applicable.

0 = Negative (no agglutination).

+ = Positive (agglutination present).

[a] Reverse typing is not performed on infants <6 months. Antibody production may not be detectable until after 1 year.

[b] An ABO control (saline or albumin) must be performed on Group AB individuals to detect polyagglutination.

example, a group A individual can be given RBCs that are Group A or O. "Universal donor" Group O RBCs and AB plasma are usually reserved for patients who can only receive Group O red cells or AB plasma, who are neonates, or individuals who need emergency uncrossmatched products. Massive transfusion of type-compatible red cells could cause mixed field results or a complete exchange of the patient's red cells, that is, the individual who is really type A, typing as type O. If this occurs, a patient's ABO type cannot be determined and Group O packed red cells should be transfused until the patient's type can be resulted. Passive transfer of incompatible RBC antibodies also occurs in massive transfusion, especially in neonates. This passive transfer of anti-A or anti-B may make the patient incompatible with type-identical blood. When type-identical blood is incompatible, no whole blood may be safely transfused. Only type-compatible RBCs may be transfused.

Platelets possess ABO antigens, and some studies have shown that there may be decreased survival with transfusions of platelets that are incompatible with the recipient's serum antibodies. This phenomenon is generally not considered clinically significant, especially in neonates who do not possess ABO antibodies. The incompatible plasma component of platelets may react with the recipient's RBCs. If incompatible plasma is to be transfused, the platelets may be volume reduced to remove as much plasma as possible if multiple transfusions are anticipated. If incompatible platelet transfusions are transfused without volume reduction, a positive DAT could result from anti-A or anti-B binding to the recipient's RBCs leading to hemolysis. In neonates, this phenomenon could be similar to an ABO HDN.

Testing

In determining a patient's ABO type, two tests are performed. One detects antigens on the red cell surface (forward type) and the other screens for expected antibodies in the serum (reverse type). The forward type is performed by adding patient RBCs with anti-A and anti-B and looking for agglutination. To confirm these results, a reverse typing is performed. Patient serum is combined with known A and B cells and inspected for agglutination (see Table 5.1). If there is a discrepancy between the forward and reverse typings, the patient's ABO type cannot accurately be determined and group O RBCs and AB plasma must be given until the discrepancy is resolved. Some individuals, such as patients receiving chemotherapy, patients with hypogammoglobulemia diseases, and prior bone marrow transplants may not have the expected reverse typing because of a decrease in their immune system or a change of blood type. Neonates also have diminished antibody production and do not produce serum antibodies, and for this reason, do not have to be screened for IgM anti-A or anti-B, that is, only a forward typing is performed. If, however, type-specific, non-Group O blood needs to be transfused, the neonate's serum must be screened for the presence of passively transferred anti-A or anti-B demonstrable with an indirect antiglobulin test.

Rh(D)

Antigens and Antibodies

Of the over 600 red cell antigens besides ABO, the only one for which typing is routinely performed is the

Rh(D) antigen. The Rh system is the largest and most complex blood group system comprising over 50 antigens. For the purpose of this section, the term Rh will refer only to the D antigen. It would be nearly impossible and not cost efficient to type for every antigen and provide completely phenotyped blood. Since most people who are transfused do not make antibodies to foreign RBC antigens this extensive matching is not required. But, it can be expected that 80% of immunocompetent Rh-negative individuals transfused with Rh-positive blood will develop an anti-D that will remain for the rest of the individual's life. Since anti-D in an IgG form is very efficient at crossing the placenta, causing HDN in an Rh-positive fetus, every effort should be made to reserve Rh-negative blood for Rh-negative female children who may become pregnant in the future.

Transfusion

For transfusion purposes, an Rh-positive patient may receive either Rh-positive or Rh-negative blood. Rh-negative patients must receive only Rh-negative blood. Since only 15% of the population is Rh-negative and Group O Rh-negative individuals are "universal" donors, every effort should be made to conserve Rh-negative blood for only those Rh-negative patients who require transfusions. In the case where there is transfusion of Rh-positive red cells to an Rh-negative individual, a prophylactic Rh immunglobulin (RhIg) should be given to bind the Rh-positive red cells before the immune system can mount an antibody response to the foreign antigen.

Treatment with RhIg should be started within 72 hours after exposure. If the Rh-positive transfused cells comprise ≥20% of the total blood volume of the patient, an exchange transfusion with Rh-negative RBCs may be performed. RhIg can either be transfused intravascularly (IV) or intramuscularly (IM). If Rh-positive whole blood was transfused, one recommended dose is 9 ug (45 IU)/mL blood (IV) and 12 ug (60 IU)/mL blood (IM). Transfusion of Rh-positive RBCs requires 18 ug (90 IU)/mL blood (IV) and 24 ug (120 IU)/mL blood (IM). Consult the manufacturer's package insert for correct dosing.

Platelets do not possess the D antigen, but are still labeled as to the Rh status of the donor. Pooled, whole blood-derived platelet units could contain up to 5 mL of red cells, which could lead to immunization. One possible way to minimize exposure to the D antigen is to transfuse apheresis products, which contain little or no red cells. If Rh-positive platelets are transfused to an Rh-negative recipient, RhIg should be administered to avoid immunization. A 300 ug dose of RhIg will suppress about 17 mLs of RBCs.

Plasma does not contain intact red cells and therefore the Rh factor is considered not to be of consequence.

Testing

D antisera come in a variety of types (IgM, IgG, monoclonal, polyclonal, or blends of these) and it is the discretion of an institution as to which anti-D they use. The principle of the test to detect the D antigen is the same regardless of what antisera are used.

Patient's red cells are mixed with anti-D and observed for agglutination. This test may also be carried through to an indirect antiglobulin phase to detect the presence of a weak D antigen (the so-called D^u test or weak D test) (Table 5.2). If a patient types as weak D-positive, he or she either has a decreased expression of the D antigen or a partial D antigen. There is some controversy over the transfusion of these patients since a person who has a partial D antigen could make an antibody directed against that part of the D antigen they are missing. The frequency of the weak D antigen is less than 1% and so this test is not routinely performed on recipients. If someone is mistyped as Rh-negative when they are truly weak D-positive, it is of no consequence since the transfusion of Rh-negative blood will not precipitate any problems. The weak D test is performed on all Rh-negative donors to confirm the Rh status as negative and not weakly positive. This testing ensures that a weak D-positive unit of blood is not mislabeled as Rh negative. If weak D-positive RBCs are transfused to a Rh-negative individual, anti-D could be made.

OTHER RED CELL ANTIGENS

As mentioned previously, red cell antigens other than A, B, and Rh are not typed for on a routine basis. However, if an antibody is encountered, there are two reasons why antigen typing is performed. First, the patient's red cells are typed for the antigen corresponding to the antibody he or she produced. This is used as a "confirmatory" test because individuals generally do not make antibodies to antigens that they possess, unless it is an autoantibody. Secondly, donor RBCs are typed because RBCs negative for the antigen to which the clinically significant antibody (CSA) is directed against need to be found and crossmatched through an antiglobulin test.

In neonates, some red cell antigens are not fully developed, for example, the Lewis system, I, P_1, the Lutheran system, and Xg^a. Typing for these antigens is not beneficial until after the age of 1 year.

Some institutions provide partially phenotyped blood for patients with sickle cell disease because of

TABLE 5.2 Rh Typing of Patient's Red Cells for the D Antigen and Compatibility

Immediate Spin Patient RBCs +		Indirect Antiglobulin Test Patient RBCs +			
Anti-D	Control[a]	Anti-D	Control[a]	Rh Results	Rh Compability
0	NA/0	NA	NA	Negative[b]	Rh-negative
+	NA/0	NA	NA	Positive	Rh-positive, Rh-negative
0/ +W	0	+	0	Positive (Weak D positive)	Rh-positive[c], Rh-negative

N/A = Test not required.

0 = Negative (no agglutination).

+ = Positive (agglutination present).

+W = Positive (weak agglutination present).

[a] Controls for anti-D (patient RBCs mixed with albumin or saline) need only be performed if testing for the weak D antigen or if the antisera used contain a high concentration of protein (20% to 24%).

[b] A weak D test does not have to be performed on Rh-negative recipients.

[c] If the weak D positive result is due to a partial D antigen, only Rh-negative blood may be chosen to be transfused.

the extensive transfusion requirements these patients have and their increased antibody response to foreign antigens. A complete phenotyping is performed and red cells matched to the individual's Rh (D, C, c, E, e), Kell (K), and possibly Duffy (Fy^a, Fy^b) and/or Kidd (Jk^a, Jk^b) antigens are transfused to deter antibody production.

DIRECT ANTIGLOBULIN TEST

Patients who possess antibodies, IgG, or complement components (C3d, C3b,d), passive or immune stimulated, to red cell antigens that they possess on their own or transfused red cells can experience a positive DAT. The antibody will bind to the red cell antigen and possibly cause intra- or extravascular red cell destruction in vivo. In vitro, an eluate is prepared from DAT positive samples and the causative antibody identified. Eluates are prepared via various chemical or physical processes to remove antibody from the RBC surface and to make it available for testing and identification.

DATs are routinely performed on neonates using anti-IgG only, since any antibody found in the neonate is passively transferred from the mother. Passive transfer of IgG antibodies that result in HDN may only have a positive DAT, that is, the antibody is not strong enough to be detected in the serum. Antigen-negative or antiglobulin crossmatch-compatible blood must be transfused if an antibody is eluted from the surface of the red cell.

DATs performed using polyspecific AHG (IgG and C3d) reagents are routinely used when investigating autoimmune hemolytic anemia, cold hemagglutinin disease (CHD), paroxsymal cold hemoglobinuria (PCH), and drug-induced hemolytic anemia. Some antibodies, especially those that are newly forming, "cold-reacting antibodies," which are mostly IgM, or complement-dependent antibodies like some of those to the Kidd blood group may only be detected via detection of complement components on the RBC surface.

Testing

In testing, polyspecific AHG is combined with patient red cells and observed for agglutination immediately to detect mostly IgG antibodies. The test is then incubated at room temperature to allow for full complement binding. If a positive result is detected, follow-up testing requires differentiation between IgG and/or complement components expressed on the RBC surface. Patient red cells are combined individually with anti-IgG, anti-C3d, and anti-C3b,d to detect which proteins are bound to the red cell surface.

ANTIBODY DETECTION TEST (SCREEN)

The "immune"-mediated alloantibodies are produced in response to foreign red cells via transfusion or pregnancy, unlike the ABO system's anti-A and anti-B, which are "naturally" occurring. The antibody detection test or screen is used to detect alloantibodies and help identify a specificity. Once antibody specificity is known, antigen-negative cells can be crossmatched using an antiglobulin technique and transfused.

Neonates, even though they rarely produce RBC antibodies, must be initially screened for the presence of passively acquired alloantibodies. Maternal serum may be used for the work up of a positive screening test since any RBC antibody that the neonate possesses is presumed to have come from the mother. Antigen testing for any antibody found must be performed using the neonate's red cells to determine if a potential HDN condition exists.

Patient and, in the case of neonates, maternal transfusion history, including where and when last transfusion occurred, maternal antibody production, or the administration of RhIg, are all important information that the transfusion service uses to determine the clinical significance of an antibody detected.

CSAs are reactive at body temperature and have the ability to cause hemolysis. These antibodies are usually detected at 37°C incubation or the indirect antiglobulin test (IAT). Antibodies that react only at room temperature or are "prewarmable" are generally not clinically significant. The "prewarm" test warms RBCs and serum to 37°C before testing so cold-reacting IgM and subsequent complement components do not bind to the RBC and carry over to the antigloblin phase of testing.

Once a CSA is detected, antigen-negative blood must be transfused for the rest of the patient's life even if the screen becomes negative in the future. If antibody levels fall below the level of detection and antigen-positive blood is transfused, a delayed hemolytic transfusion reaction may occur due to renewed production of the antibody. This secondary response usually occurs within

48 hours and is primarily IgG. In emergency situations the need for blood is weighed against the possibility of a transfusion reaction. If a transfusion is needed to sustain life, it must be given. The patient may suffer red cell removal over a few days time, but this removal is usually through the reticuloendothelial system and is often extravascular. With time, compatible RBCs may be located.

Testing

In the laboratory, patient's serum is mixed separately with at least two screening RBCs of known antigen phenotypes. These cells are typed only for the more common antigens as an antibody could be missed if it is directed against an antigen of low frequency. The set of screening RBCs used by an institution is up to the discretion of the institution, but all common antigens must be present in a single (heterozygous) or double (homozygous) expression on at least one cell. Screening cells should also exhibit a homozygous expression for certain antigens that exhibit dosage. Antibodies that exhibit dosage react only with or have stronger reactions with RBCs that possess a double expression of antigen. Ideally, but often not possible, screening RBCs should have homozygous expression of C, c, E, e, Jk^a, Jk^b, Fy^a, and Fy^b to ensure adequate binding of antibodies to these red cells antigens (Table 5.3).

There are different test methods available, but the one thing that they all have in common is an incubation at 37°C, an indirect antiglobulin phase and Coomb's "check cells" (RBCs sensitized with anti-IgG), or another appropriate control to detect false-negative results. Each different technology has its pros and cons, and no one test system will be able to identify every CSA.

TABLE 5.3   Red Cell Antigen Screening Cells Anagram

	Screening Cell Phenotype																						Patient Results	
	D	C	c	E	e	M	N	S	s	P_1	Le^a	Le^b	Lu^a	Lu^b	K	k	Kp^a	Js^a	Fy^a	Fy^b	Jk^a	Jk^b	I.S.	37°C	IAT
SC I	+	+	0	0	+	+	+	0	+	+	0	+	0	+	+	+	0	0	0	+	+	0	0	0	+
SC II	+	0	+	+	0	+	0	+	+	+	+	0	0	+	0	+	0	0	+	0	0	+	0	0	0

SC I = Screening cell 1 of 2.
SC II = Screening cell 2 of 2.
Screening Cell 0 = RBC is negative for the antigen.
Screening Cell + = RBC is positive for the antigen.
I.S. = Immediate spin results.
37°C = 30–60 min. incubation at 37°C results.
IAT = Indirect antiglobulin test results.
Patient Results 0 = Negative (no agglutination).
Patient Results + = Positive (agglutination present).

Test Tubes

Antibody screening with test tubes can be performed with enhancement media such as low ionic strength saline (LISS) or polyethylene glycol (PeG), or without (saline). There are three phases with tube testing: immediate spin (IS), 37°C incubation, and the IAT. The red cells are incubated at 37°C, washed to remove unbound antibody, and antihuman globulin (AHG) is added and centrifuged. The tubes are then observed for agglutination or hemolysis. Check cells are added to all antiglobulin-negative tests to ensure that the AHG was functional. IS testing is sometimes disregarded if no enhancements are used, since some cold-reacting antibodies will be able to bind complement to the red cells when centrifuged at room temperature and will carry over to the antiglobulin phase of testing. These antibodies are generally not clinically significant and usually have M, N, P_1, I or Lewis antigen specificity.

Column Affinity Technology/Gel

If column affinity testing or "gel" technology is used, there is no IS test. The screening cells and patient plasma are mixed and incubated at 37°C with anti-IgG. If there is a clinically significant antibody present, it will bind to the red cells and with anti-IgG to create agglutinates. The RBCs are loaded onto the gel and centrifuged so that the agglutinates that form in the test will have to travel down the gel column. The size of the agglutinates (strength of antibody) determines the position of the red cells at the end of the centrifugation and will determine a positive or negative reaction. A negative test results when all of the RBCs pass to the bottom of the column with centrifugation. If the agglutinates get trapped within the gel column a positive result is recorded.

Solid Phase Technology

Solid phase technology uses solid-phase red cell adherence (SPRCA) in its test system. Microwells are coated with red cells, and then patient serum and LISS are added and incubated at 37°C. If an antibody is present it will bind to the RBCs fixed over the surface of the wells. Indicator cells (RBCs coated with AHG) are added and centrifuged. A positive test will have a uniform layer of red cells coating the entire bottom surface of the well due to indicator cells binding to the patient's antibody attached to antigens expressed by the fixed layer of red cells. A negative test results when antibody on the indicator cells does not adhere to the antigen-antibody complex bound to the test well and, consequently, is centrifuged to form a button at the center of the well.

CROSSMATCH (XM)

A crossmatch is performed before a blood transfusion to ensure compatibility, mainly ABO, with the patient. As mentioned earlier, ABO antibodies cause intravascular hemolysis and may be fatal. There are five ways of crossmatching RBCs. The method chosen is based on the need of a transfusion versus time, the age of the patient, previous history, and the current antibody screen.

Uncrossmatched

"Uncrossmatched" emergency units are the first means of providing RBCs. Because these RBCs are transfused usually without any pretransfusion testing, they should be given only if time does not permit an antibody screen to be performed. In pediatrics, because most children are not multiply transfused and have not been pregnant, their chances of making a clinically significant antibody are small. Even so, the possibility exists and the risk of a delayed hemolytic transfusion reaction versus the need for red cells should always be considered. Emergency units are usually O-negative and should not cause an immediate transfusion reaction. When a sample is eventually received by the transfusion service, ABO/Rh typing and an antibody screen will be performed, and any emergency units already given will be crossmatched posttransfusion. If more blood is required before compatibility testing is completed, type-identical crossmatch-incomplete (antiglobulin phase of testing not resulted) red cells may be given.

Immediate Spin/Abbreviated

The next crossmatch is the "immediate spin" or "abbreviated" crossmatch. If a patient has a negative antibody screen and a negative history of alloantibodies, the only requirement for transfusing RBCs is to check for ABO compatibility. Patient serum is mixed with donor red cells at room temperature, immediately centrifuged, and observed for agglutination or hemolysis. If this reaction is positive, further investigation must be performed to distinguish between an ABO incompatibility or a cold-reacting antibody. An antibody that reacts at the antiglobulin phase to a low incidence antigen, not on the phenotypically known RBCs used to do the initial antibody screen, may be missed. If by chance, the patient is transfused with an antigen-positive unit, a transfusion reaction may occur.

Complete/Antiglobulin

If a patient possesses, or has ever demonstrated a clinically significant antibody, then a "complete" or "antiglobulin" crossmatch must be performed. Red cells negative for the antigen for which the antibody is directed against is combined with the patient's serum, incubated at 37°C and an antiglobulin test is performed to ensure no antigen-antibody reactivity occurs. If there is no reactivity at 37°C or in the antiglobulin phase of testing, the unit may be transfused. If the crossmatch is positive at 37°C or IAT, it is not suitable for transfusion. Further testing needs to be performed to identify the reason for the incompatibility.

Not Required

The fourth crossmatch is the "not required" crossmatch and is used when transfusing neonates. According to the *AABB Standards*, infants less than 4 months old do not have to be crossmatched for any current admission as long as Group O red cells that are negative for antigens corresponding to any passive antibody present in the infant's serum are transfused. If a non-Group O infant is to be given type-identical RBCs, a blood sample must be tested to see if the neonate possesses passively acquired anti-A or anti-B. Patient serum and A or B cells are tested through an antiglobulin phase to ensure red cell survival. If anti-A or anti-B is detected, the neonate may only receive RBCs lacking that A or B antigen until the antibody test is negative. The neonate's serum should also be tested for passive anti-A or anti-B and/or an eluate prepared from the RBCs if transfused with plasma-incompatible platelets and there are signs or symptoms of hemolysis.

Electronic

The last form of crossmatch is the "electronic" crossmatch. If the transfusion service uses a validated computer system, one could rely on it to select units based only on ABO/Rh and crossmatch blood without testing patient serum and donor cells. There are certain criteria and systems that must be implemented before this process may be utilized. These requirements include: two patient ABO typings on file; the patient must have pretransfusion testing performed; the donor unit has on file ABO typing, unit number, and component type; there is verification of data entry; and there is a computer system that alerts the user to discrepan-

cies in ABO/Rh. For further details, see the *AABB Standards*.

TURNAROUND TIMES

Turnaround time is the time it takes to perform a certain procedure. Each institution should do studies and evaluations of its own turnaround time to let the physicians know about how long a procedure takes. An estimate of times is given below. If additional processing needs to be performed, such as irradiation, washing blood, aliquoting blood, or a positive antibody screen, workup turnaround times may be extended (Table 5.4).

Type and Screen

A type and screen should be ordered if the likelihood of a transfusion is unlikely. Units of blood may be added on with this sample if there is a need to transfuse. An ABO/Rh is performed along with an antibody screen. An ABO/Rh requires 10 to 15 minutes. If the screen is positive, any unexpected antibody will be identified. If saline techniques are used the incubation time is 30 to 60 minutes. When enhancements are added in the test system, the incubation time decreases to 10 to 30 minutes. Enhancement media allow the red cells to come into closer contact with each other or increase antibody uptake. A positive antibody screen will require testing patient serum against a panel of red cells to iden-

TABLE 5.4   Turnaround Times[a]

Test	Time (Minutes)
ABO/Rh	10–15
Antibody Screen	
Negative	25–45 (with enhancements[b])
	45–75 (without enhancements)
Positive	50–90 (with enhancements)
	90–150 (without enhancements)
Crossmatch	
Uncrossmatched	1–2
Not required	2–5
Electronic	2–5
Immediate spin	5–10
Complete	25–45 (with enhancements)
	45–75 (without enhancements)

[a] Turnaround times are estimates of the time of receiving the sample in the blood bank to notification of the nurse that blood is available. Special circumstances such as the workup of a positive antibody screen, washing units, irradiating units, or aliquoting blood may require more time.

[b] Enhancements are added to the test system to shorten the reaction time via various methods.

tify a specifity. An identification panel may take up to 45 minutes or longer if multiple antibodies or antibodies to high frequency antigens are present. A negative type and screen normally requires 45 to 75 minutes. If antibody identification is required, the time may be up to two hours or more.

Crossmatch

A crossmatch needs to be ordered when there is a good probability the patient will need to be transfused. If two units are to be transfused, two units should be ordered. Extra units of blood can be crossmatched from the sample if more is needed. Many hospitals also have defined C/T (crossmatch/transfusion) ratios for surgical orders. These can be used as a good guide to correctly order units of blood to be available for a procedure. A crossmatch consists of a type and screen with allocating units of blood to a specific patient. A "not-required crossmatch" does not involve testing of patient serum and donor cells. This crossmatch takes 2 to 5 minutes and only involves allocating the unit to the patient.

An IS crossmatch is performed if the antibody screen is negative and there is no history of CSA. A check for ABO compatibility is the only requirement. The IS crossmatch takes 5 to 10 minutes. If there is a CSA, a complete crossmatch must be performed with antigen-negative blood. The first step is to antigen screen units of blood, which can take up to 30 minutes. Once antigen-negative blood is obtained, an antiglobulin crossmatch needs to be performed. If enhancements are used, the incubation time is 10 to 30 minutes; saline requires a 30- to 60-minute incubation. Crossmatching units with a negative antibody screen requires a total time of 55 to 85 minutes. If the screen is positive, it may take up to four hours or longer before blood is available.

Special Circumstances

Multiple antibodies or antibodies to high frequency antigens may require testing with rare phenotyped cells found in immunohematology reference laboratories. Patient samples must be sent out of the hospital to specialized laboratories where identification of the antibody(ies) will occur. In this instance, depending on the feasibility of finding antigen-negative blood, crossmatch-compatible blood may not be available for two to three days. If a transfusion is required before compatible blood is located, one might try to give the least incompatible crossmatched blood, starting out with 5 to 20 mL. If the patient can tolerate the transfusion, more blood may be administered. The patient may have a transfusion reaction, and in this case it must be evaluated if more blood should be given, depending on the prognosis of the patient.

Autoantibodies react with antigens that are common to all red cells. All cells are positive at the antiglobulin phase of testing. Underlying alloantibodies need to be screened for, and this is achieved with adsorptions. A series of adsorptions are routinely performed in reference laboratories to detect underlying alloantibodies and crossmatch blood with serum that has had the autoantibody reactivity removed. This process may take up to 12 to 24 hours. The autoantibody is allowed to bind to a series of RBCs of a known phenotype in order to remove all of the autoantibody from the serum. Once that procedure is performed, only alloantibodies should be left in the serum. If there are no alloantibodies present or the patient requires a transfusion before all testing is completed, one could transfuse the least incompatible crossmatched blood to the patient and watch for in vivo signs of hemolysis or a transfusion reaction if blood is needed, a "biological crossmatch."

Suggested Reading

Brecher ME, ed. 2002. Technical manual. 14th ed. Bethesda, MD: American Association of Blood Banks.
DePalma L. 1992. Red cell alloantibody formation in the neonate and infant: considerations for current immunohematologic practice. *Immunohematology* 8:33–7.
Floss AM, Strauss, RG et al. 1986. Multiple transfusions fail to provoke antibodies against blood cell antigens in human infants. *Transfusion* 26:419–22.
Gorlin JB, ed. 2002. Standards for blood banks and transfusion services. 21st ed. Bethesda, MD: American Association of Blood Banks.
Herman JH and Manno CS. 2002. Pediatric transfusion therapy. Bethesda, MD: American Association of Blood Banks.
Issitt PD and Anstee DJ. 1999. Applied blood group serology. Durham, NC: Montgomery Scientific Publications.
Petz LD, Swisher SN, et al. 1996. Clinical practice of transfusion medicine. New York, NY: Churchill Livingstone Inc.
Rudmann SV 1995. Textbook of blood banking and transfusion medicine. Philadelphia, PA: W.B. Saunders Co.

References

AABB. 2003. Standards for Blood Banks and Transfusion services 22nd ed. Bethesda, MD: American Association of Blood Banks Press.

6

Serologic Investigation of Unexpected Antibodies

SUSAN T. JOHNSON, MSTM, MT(ASCP)SBB, AND TINA M. PUGH, MT(ASCP)

INTRODUCTION

Investigation of unexpected antibody problems and blood group incompatibilities occurs less frequently in children than in adults. Infants less than 4 months of age usually do not make antibodies, and if antibodies are detected they are generally passively acquired from the mother. The incidence of alloimmunization in the general pediatric population is 1.2% (Pugh et al. 2001) as compared to the adult population whose alloimmunization rate is reported as 1.5% to 35% depending on the patient population and number of transfusions (Issitt and Anstee 1998). Sample size available for testing is often limited, and the types of antibodies seen are more often clinically insignificant. These differences provide unique challenges when evaluating these patients.

COMMON ANTIBODY IDENTIFICATION METHODS

Antibody Detection Test (Screen)

Several methods (Table 6.1) are utilized to detect unexpected antibodies in patient's serum/plasma. As reviewed in the previous chapter, test tube methods include most commonly low ionic strength solution (LISS) and polyethylene glycol (PEG). These methods require two drops or approximately 100 uL of serum/ plasma. Newer methods include gel testing or solid phase, which utilize as little as 10 uL of sample. This is advantageous in working with the pediatric population given its smaller blood volume and difficulty in obtaining samples.

Antibody Identification Panels/Selected Cell Panels and Evaluation

Several approaches may be taken in identification once an antibody is detected. A variety of special techniques may be performed to resolve serologic problems (Box 6.1). To start, the best approach is to use the same method that detected the antibody. If a positive reaction was seen when using a gel test, the initial antibody identification panel should be run in gel.

Each laboratory determines how testing will be performed and at what phase the patient serum will be tested (for example, immediate spin [IS], after a 37°C incubation, or in the indirect antiglobulin test [IAT]). The phase of testing at which the antibody reacted should also be evaluated to decide how to proceed. If using a test tube method with LISS or PEG and positive reactions are detected at IS, a saline, no additive screen can be run to rule out the presence of alloantibodies.

If the routine antibody detection method is gel or solid phase, an antibody identification panel may be run to determine if specificity can be determined. If not, an alternative test tube method may be chosen. Both gel and solid phase utilize LISS so a comparative tube method could be LISS or commercial polyethylene glycol that is normally prepared in a LISS diluent.

Regardless of the method used, patient serum/ plasma is tested against a panel of red cells of known phenotype. Results of testing are evaluated by performing a "cross-out" or "rule-out" technique. Negative reactions are used to eliminate the presence of alloantibodies. As shown in Table 6.2, the first negative reaction is against panel cell 1. The following antibodies

Box 6.1   Tests Performed for Antibody Identification

Antibody identification panels—Warm and cold
Selected cell panels
RBC phenotyping
Direct antiglobulin test
Elution
Neutralizations
Lectin typing
Prewarm indirect antiglobulin test
Thermal amplitude studies
Donath-Landsteiner test
Adsorptions—Autologous and allogeneic

TABLE 6.1   Methods for Detecting Antigen-Antibody Reactions

Test Tube Methods	Micromethods
Saline—no additive	Gel
LISS	Solid phase
PEG	
Albumin	

LISS = Low ionic strength solution.
PEG = Polyethylene glycol additive solution.

TABLE 6.2   Antibody Identification Panel

	Rh						MNS				Lu		P	Lewis		Kell		Duffy		Kidd		
	D	C	E	c	e	f	M	N	S	s	Lu^a	Lu^b	P_1	Le^a	Le^b	K	k	Fy^a	Fy^b	Jk^a	Jk^b	IAT
1	+	+	0	0	+	0	+	+	0	+	0	+	0	+	0	0	+	0	+	0	+	0√
2	0	0	0	+	+	+	+	0	+	+	0	+	+	0	+	+	0	+	+	+	0	3+
3	0	0	+	+	0	0	0	+	0	+	0	+	+	0	+	+	+	+	0	0	+	3+
4	+	+	0	0	+	0	+	0	+	+	0	+	0	0	0	0	+	+	+	+	+	0√
5	0	0	+	+	+	+	+	+	0	+	0	+	0	0	+	0	+	0	+	0	+	0√
6	0	+	0	0	+	0	+	+	0	+	0	+	+	0	+	+	+	+	+	+	+	3+
7	0	0	+	+	0	0	0	+	0	+	0	+	0	0	0	0	+	0	+	+	0	0√
8	0	0	0	+	+	+	+	0	+	0	+	+	+	0	+	0	+	+	0	0	+	0√
9	+	+	0	+	+	+	+	+	0	+	0	+	+	+	0	0	+	0	+	+	0	0√
10	+	0	+	+	0	0	0	+	+	+	0	+	+	+	0	0	+	+	0	+	0	0√
11	+	0	0	+	+	+	0	+	0	0	0	+	+	0	+	0	+	0	0	+	+	0√
A																						0√

0 = Negative (no agglutination or clumping of RBCs), √ = IgG-sensitized Coombs' control cells were positive, 3+ = Positive (several large agglutinates or clumps of RBCs).

can be ruled-out: -D, -C, -e, -s, -Lu^b, -Le^b, -k, -Fy^b, and -Jk^b. If these antibodies were present, the serum would have been positive with the panel cell expressing these antigens. This process is repeated for all negative cells on the panel. The rule-out results are than evaluated to determine if any specificity is clear-cut and if all other significant alloantibodies are eliminated. Dosage must be kept in mind when evaluating results. Some antibodies will show stronger positive reactions with red cells that have a double dose of the antigen versus red cells that have a single dose. This means a donor with a double dose has inherited two genes coding for the antigen versus one gene. Ideally it is preferable to rule-out the presence of an antibody that has a double dose of antigen ensuring that weak antibodies are detected. Additional selected cells are often required to complete this evaluation. In this example, anti-K is identified and the presence of all other alloantibodies is eliminated. Criteria for the number of antigen-positive and antigen-negative cells necessary to rule in and rule out antibodies vary among institutions.

Red Blood Cell Phenotyping

Typing (phenotyping) a patient's red cells with antisera of known specificity to determine if the individual possesses or lacks an antigen provides useful information. If a patient's red cells lack the antigen, they have the ability to make an alloantibody to that antigen. When a person's red cells possess an antigen, they will most likely not make an antibody to it. Unusual situations do occur, however, such as when an individual who types antigen-positive makes the corresponding antibody; for example, Rh(D)-positive people with a partial-D antigen can make anti-D.

Phenotyping is most often used to confirm antibody specificity identified in a patient serum. A patient who has made anti-K should be K-negative. It is also useful when there are multiple antibodies present and/or a small amount of sample is available to help determine what antibodies a patient could make. Caution must be taken, however, when using phenotype to aid in identification. Sometimes it is easy to presume antibody specificity when the investigator knows the person lacks the antigen. Just because an individual lacks an antigen does not mean he or she has the antibody. Once antibody specificity is given to a patient, it will remain in the history forever, and antigen-negative red blood cells (RBCs) will be required point forward. This makes it more difficult to find compatible blood and may add unnecessary expense.

Direct Antiglobulin Test

The direct antiglobulin test (DAT) detects IgG and/or complement (C3) coating of the patient's red cells in vivo. The following are clinical situations in which a positive DAT can occur:

- Hemolytic disease of the newborn (HDN)—Antibody from the mother, classically anti-D, crosses the placenta and coats antigen-positive fetal RBCs in utero. When born, the baby's red cells will show a positive DAT. The strength of DAT varies, usually depending on the antibody specificity. Rh and other IgG antibodies usually cause a strongly positive DAT while ABO antibodies classically exhibit weaker (1 to 2+) positive DATs.
- Hemolytic transfusion reaction—Antibody in the transfusion recipient coats antigen-positive transfused red cells. The DAT is generally weaker because there are far fewer antigen-positive red cells in the background of the patient's own antigen-negative red cells.
- Autoimmune hemolytic anemia (AIHA)—Autoantibody, either warm (IgG) or cold (IgM) reactive, binds to an individual's red cells and causes red cell destruction. The DAT in most cases is strongly positive (2 to 4+).
- Drug-induced immune hemolytic anemia—There are three proposed ways that drugs may induce a positive DAT in a patient. (1) Penicillin, penicillin-derivatives, or cephalosporins can bind covalently to red cell membrane proteins and stimulate hapten-dependent antibodies; (2) methyldopa or procainamide, through an unknown mechanism, can induce autoantibodies specific for red cell membrane components; and finally (3) quinine, quinidine, and some nonsteroidal antiinflammatory drugs (NSAIDS) induce antibodies that bind to RBCs only when the drug is present in a soluble form.
- Antibodies from passenger lymphocytes in transplanted organs—Lymphocytes from a transplanted organ can produce antibody against foreign antigens in the organ recipient. For example, a Group A patient transplanted with a Group O marrow (minor incompatibility) may have lymphocytes present that recognize A antigen as foreign and begin to produce anti-A. This anti-A will coat the transplant recipient's A red cells causing a positive DAT.
- Major ABO incompatible bone marrow transplant—Opposite of the example just described, if a Group O recipient receives a Group A bone marrow (major incompatibility), anti-A in the recipient may attach to any Group A red cells made by the new marrow.
- Passively acquired antibodies present in equine antihuman lymphocyte globulin (ALG) or high dose immune globulin (IVIG)—Antibodies with blood group specificity may be present in ALG and IVIG. Examples include anti-A, anti-B, anti-D, and others. If a patient's red cells are antigen-positive corresponding to a passive antibody present, the antibody will bind to the red cells causing the positive DAT.

Table 6.3 lists common serologic findings of immune hemolytic anemia.

Patient history, presence of antibody in the patient's serum/plasma, along with the strength of a positive DAT provide invaluable information in evaluating the cause of a positive DAT.

Eluates

Elutions are performed to determine specificity of antibody coating the patient's red cells. There are two common elution methods used. The one selected is based on the type of suspected antibody coating the patient's red cells.

An eluate that is simple to perform and requires no special reagents uses freezing, then thawing (historically called the Lui Freeze-Thaw method). Patient RBCs are washed free of unbound plasma antibody, diluted with a few drops of saline or albumin and frozen. Once the cells are completely frozen they are thawed under warm water. The red cells lyse leaving a hemoglobin-stained eluate. ABO antibodies are released from the red cell membrane and are left in the eluate. The eluate is then tested using a standard saline IAT. This method is ideal for detecting ABO antibodies. Table 6.4 lists clinical situations in which to perform this eluate.

The second most common elution method is the rapid acid method. Antibody will elute off the red cells after adding a glycine acid (approximately pH 3) to the coated red cells. The red cells are centrifuged, supernatant with antibody is removed, and then buffered

TABLE 6.3 Serologic Findings in Immune Hemolytic Anemias

	WAIHA	CAS	PCH	DIHA	HTR
Direct Antiglobulin Test	3-4+	2-4+	W-2+	W-4+	w-2+
Anti-IgG	3-4+	0	0-W	W-4+	w-2+
Anti-C3	0-3+	2-4+	W-2+	W-4+	0-2+
Eluate	Positive w/all RBCs	Not indicated	Not indicated	Negative	Antibody
Antibody Detection Test	0-4+	W-4+	W-2+	O-3+	0-4+
Reactivity	IAT	Immediate spin, room temperature (Reactive at 37°C &/or 30°C)	Immediate spin, room temperature biphasic hemolysin	IAT Variable, depending on drug and course of therapy	IS, 37C, IAT
Antibody Specificity	Broad Rh	–I, –i	–P		Any

Modified from Johnson ST. 2002. Immune hemolytic anemias. In *Clinical laboratory medicine*. 2nd ed. Edited by McClatchey KD. Philadelphia, PA: Lippincott Williams and Wilkins. With permission.

TABLE 6.4 Type of Elution Method for Clinical Situation

Lui Freeze-Thaw	Rapid Acid
ABO hemolytic disease of the newborn (HDN)	Passively acquired ABO and IgG antibodies
Passively acquired ABO antibodies	ABO and other antibodies from passenger lymphocytes in bone marrow and solid organ transplants
ABO antibodies from passenger lymphocytes in transplanted organs	Hemolytic transfusion reaction
Incompatible (major) bone marrow transplant	Autoimmune hemolytic anemia
	Drug-induced immune hemolytic anemia
	HDN other than ABO
	ABO HDN when sample is limited

back to normal pH for testing. Rapid acid eluates very efficiently remove IgG alloantibodies and autoantibodies. It will remove ABO antibodies but at times is not as effective as the freeze-thaw method. Rapid acid methods are ideal when working with small volumes. An eluate can be prepared with as little as four to six drops of red cells since equal volumes of eluting solution and buffering solution are added to the number of drops of red cells available. For example, if four drops of red cells are used, four drops of acid solution are added and at least four drops of buffering solution are added. The resulting volume for testing is 12 drops, enough to select cells and/or perform an antibody detection test.

Thinking about what cells to run against the eluate is prudent, especially when the sample is limited. Reviewing patient history and antibody identified in the serum may help limit the number of selected cells that need to be tested. If anti-Jk[a] is identified in the serum there is a good chance anti-Jk[a] will be in the eluate. There is no need to run an entire 11 cell panel with half the panels cells being Jk(a+). Depending on the laboratory's policy for identifying an antibody, two to three Jk(a+)-selected cells and enough Jk(a-) cells to rule out other antibodies are required. As few as six selected red cells can be run against 12 drops of eluate.

Neutralizations

Neutralizations can be useful in antibody investigations. They are used most often to confirm antibody specificity. Anti-Sd[a] is neutralized by urine and anti-Ch and anti-Rg are neutralized by plasma. These antibodies usually show variable, weak (weak to 2+) positive reactions in an indirect antiglobulin test. Anti-Sd[a] cause characteristic tight, shiny agglutinates. Anti-Ch and anti-Rg are not reactive when testing enzyme-treated red cells. Mixing urine or plasma with patient serum and incubating it will allow the substance in the urine (Sd[a]) or plasma (Ch/Rg) to bind with corresponding antibody, neutralizing its activity.

Lectin Typing

Polyagglutinated red cells, particularly T activation occurs most often in the pediatric population. Red cells may become polyagglutinated after a bacterial or viral infection. For example, in children this can occur in necrotizing enterocolitis. Bacteria produce an enzyme such as neuraminidase that works on red cells cleaving N-acetyl-neuraminic acid residues exposing a "new" antigen, T. Anti-T is a naturally occurring antibody in the majority of people. If these red cells are tested against serum from any normal individual they will agglutinate, thus the term polyagglutination. There are

TABLE 6.5 Lectin Reactivity with Polyagglutinable Red Cells

Lectin	T	Th	Tk	Tn
Arachis hypogaea	+	+	+	−
Salvia sclarea	−	−	−	+
Salvia horminum	−	−	−	+
Glycine soja	+	−	−	+

several forms of polyagglutination that can be characterized by lectin typing (Table 6.5).

Routine screening for polyagglutination in newborns is generally not performed. Red cells from normal, healthy infants can show T activation and even infants with necrotizing entercolitis can have T-activated red cells and show no signs of hemolysis. Most experts believe that testing for polyagglutination in neonates should be selectively performed when the neonate has received RBCs or plasma products and has demonstrated evidence of hemolysis or an unexplained lack of rise in posttransfusion hemoglobin.

If it is determined that a neonate has, for example, clinically significant T-activated red cells and requires transfusion, all RBCs should be washed. Platelets may also be washed. Plasma-containing products from low-titer anti-T donors may also be advisable. These products are generally not available from the local blood supplier but can be determined rather quickly in the transfusion service. A minor crossmatch using plasma from the product to be transfused is mixed with patient's T-activated red cells. If anti-T present in the donor is low in titer, the crossmatch will be negative or weak and the product may be transfused.

PROCESS FOR EVALUATION OF POSITIVE ANTIBODY DETECTION TESTS

Before beginning any serologic testing it is important to review patient history and initial reactivity found in the antibody detection test (Box 6.2).

Preliminary Considerations

Patient History

Patient history will provide information useful to the evaluation. If this is the first time a child is in the hospital, it is very unlikely he or she has had previous transfusions or pregnancy. Previous transfusion and pregnancy history will indicate if the patient has been exposed to foreign red cells. Passive anti-A and anti-B can be detected in a patient after transfusion with ABO types different than the patient's and even cryoprecipitate.

Diagnosis is important in the investigation. A patient with sickle cell disease will likely be multiply transfused. A diagnosis of anemia may indicate autoimmune hemolytic anemia.

Patient age will also provide valuable information. If the patient is younger than 4 months old, antibody detected in his or her serum is likely passively acquired from the mother. A 3-year-old child is more likely to possess a naturally occurring cold-reactive antibody such as anti-M or anti-P_1 than a 37°C, immune-stimulated antibody like anti-Jk^a.

Medication history is also important. A patient with idiopathic thrombocytopenia purpura (ITP) may be receiving Rh immune globulin as treatment. Any antibody screens performed posttreatment will possess anti-D. Some drugs induce drug-dependent antibodies and subsequent hemolysis. The presence of drug-dependent antibody is rare in adults and even rarer in children but should be considered.

Knowing which antibodies occur most frequently can give the investigator clues as to identification. Table 6.6 lists the frequency of common antibody specificities.

Factors Specific to Children

Age of Patient

Patient age should always be evaluated when a positive antibody detection test is found. If the patient is younger than 4 to 6 months of age, any antibody detected is most likely passively acquired from the mother. A check of the mother's history may quickly determine antibody specificity. Further testing will confirm that specificity.

A child older than 1-year is unlikely to have produced clinically significant alloantibodies if not transfused or minimally transfused. A patient transfused with several units of blood is more likely to have made antibody.

Antibody Frequency

Knowing which antibody specificity occurs most frequently in a given patient population can shorten an investigation. Antibodies found most often in the pediatric patient are cold-reactive and generally clinically insignificant. Specificities include anti-P_1, anti-Le^a, anti-Le^b, and autoanti-I or -IH (see Table 6.6). These antibodies generally show positive reactions at IS and often react in the IAT.

Anti-E and anti-K are the most commonly detected IgG antibodies in children just as in adults. These antibodies react best at 37°C and are detected in an IAT.

Passive Antibody

Passive antibody is found more often in the pediatric population than in adults. One reason for this is patient size. A child's blood volume is smaller than adults, therefore, antibody found in therapies such as intravenous immune globulin or ABO specific plasma products are more easily detected in a patient's plasma.

Infants younger than 6 months of age will have passive ABO antibodies from their mothers as described above. In addition, if the mother has received Rh immune globulin antenatally anti-D may be detected in the infant's plasma. It is extremely rare for these infants to produce alloantibodies of their own.

Rh immune globulin is a common therapy in ITP, causing anti-D to be detected.

A small patient transfused with multiple ABO-compatible cryoprecipitate or platelets may show extra ABO antibody. For example, a Group A patient may have detectable anti-A in his or her plasma from anti-A present in Group O platelets.

A careful check of the patient's diagnosis, medication, and transfusion history should alert the investigator to possible passive antibody being the cause of serologic problems. Table 6.7 summarizes possible scenarios. Passive antibodies are usually not clinically significant. They may cause a positive DAT and/or antibody screen and even an incompatible crossmatch if antigen-positive blood is selected for transfusion. In most cases RBCs chosen for transfusion lack the antigen corresponding to the passive antibody present in the patient allowing for a compatible (negative) crossmatch.

After considering patient age, common antibodies detected, and checking patient transfusion and treatment history, the next step in the process, testing, can occur.

TABLE 6.6   Frequency of Clinically Insignificant and Significant Antibodies Detected in Children

Clinically Insignificant Antibodies	% of Total Antibodies Detected
–I	30
–M	16.1
Warm autoantibody	7.9
–P_1	9.2
–Le^a	5.2
–Le^b	3.4
Others	1.8
Clinically Significant Antibodies	
–E	8.6
–K	4.5
–Jk^a	1.9
–Jk^b	1.9
Warm autoantibody	3.7
Others	6.8

Modified from Pugh et al. 2001.

TABLE 6.7   Passive Antibody Source and Specificity Found in Children

Patient History	Source of Passive Antibody	Antibody Specificity
HDN-ABO	Mother	Anti-A, -B, -A,B
Rh-HDN	Mother	Anti-D, -C, -E, -c, -e
HDN-Other	Mother	Kell, Kidd, Duffy, others
Pregnancy-Rh negative	Mother Rh immune globulin	Anti-D
ITP	Rh immune globulin	Anti-D
BMT	Passenger lymphocytes	Anti-A, -B, -A,B
BMT	Intraveneous immune globulin	Anti-A, -B, -D, others
Transfused	ABO specific cryo, platelets	Anti A, -B, -A,B

HDN = Hemolytic disease of the newborn, ITP = idiopathic thrombocytopenia purpura, BMT = bone marrow transplant.

Cold-Reactive Antibody Evaluations

Cold-Reactive Alloantibody

No History of Antibody

Cold-reactive antibody evaluations are the most common antibody workups performed in the pediatric setting. This is due to the fact that many patients have not had previous foreign red cell exposure. Few patients have had multiple transfusions and even fewer have had a previous pregnancy. Given this, antibodies most commonly detected are naturally occurring and cold-reactive (see Table 6.6).

IgM cold-reactive alloantibodies react optimally at IS, causing direct agglutination of RBCs at colder temperatures. They can agglutinate red cells after the 37°C incubation and, occasionally, positive reactions are seen in the IAT. IgM binding to reagent cells at IS may cause complement activation. If polyspecific AHG is used, positive reactions in the IAT may be due to anti-C3d binding to bound C3d on the red cells.

Cold Panel A cold antibody identification panel is performed by mixing patient's serum with selected red cells. Again, since serum is often limited when evaluating children, thorough evaluation of the antibody detection test is critical and should guide the choice of selected cells to be run. If a negative reaction is seen with one of the screening cells in the antibody detection test, the antibody may have alloantibody specificity such as anti-M or anti-P$_1$. If this is suspected a standard antibody identification panel (see Table 6.2) or cells selected that are M+, P$_1$+ and M– or P$_1$– to rule out other antibodies may be run. The serum and red cells are then incubated at cold temperatures (room temperature, 18°C, 4°C). This method is used to identify antibodies that are often IgM, reacting preferentially at room temperature and colder and is the method of choice if positive reactions are obtained when testing at IS.

Prewarm Occasionally a cold-reactive antibody may be so potent that it causes a positive reaction to occur in the IAT. Most cold-reactive antibodies are enhanced when using a LISS or PEG test tube method, and they can be detected using gel or solid phase as well. Often this reactivity can be avoided by simply removing the additive solution and performing a 30- to 60-minute incubation followed by an antiglobulin test using polyspecific antihuman globulin (AHG).

If reactivity still remains, a prewarm IAT can be used. It is extremely important to know the specificity of the antibody before performing this technique. Several studies have shown that weakly reactive, clinically sig-

nificant antibodies will be negative in a prewarm test causing these antibodies to be missed. Patient serum and reagent red cells are warmed to 37°C separately and then mixed. Warming the mixture avoids cold-reactive antibody binding. Although there are risks in performing a prewarm, its benefits allow confirmation of antibody specificity, rule out other antibodies, and provide compatible crossmatches. If positive reactions persist when using the prewarm test, most laboratories will provide crossmatch-compatible, antigen-negative blood for transfusion since it is presumed that this antibody is able to bind at body temperature, 37°C (Figure 6.1).

History of Antibody

Patients with a history of cold-reactive antibody should be approached differently. An initial antibody screen can be run to determine if similar reactivity is noted. If not, the next step is to run a method that will allow detection of warm-reactive antibodies while avoiding cold-reactive antibody. A saline IAT or prewarm IAT may be required.

Cold-Reactive Autoantibody

IgM cold-reactive autoantibodies, like alloantibodies, usually react optimally at IS, causing direct agglutination of RBCs at colder temperatures. If all screening cells are positive there is a good chance that the antibody is autoantibody with a -I or -IH specificity. Most IgM cold-reactive autoantibodies have I, IH, or i specificity. Autoanti-I is seen most often in chronic cold agglutinin syndrome (CAS) as well as following mycoplasma pneumonia. Autoanti-i is most often associated with infectious mononucleosis. Other specificities, such as anti-Pr, which is enzyme sensitive, have been reported but are rare. Cells selected for performing the cold panel should include screening cells, ABO compatible A$_1$, A$_2$ and/or B cells, cord cells or I- cells, and an autocontrol.

The majority of cold-reactive autoantibodies detected in children are nuisance antibodies. Most cold autoantibodies detected do not cause CAS. If the cold-reactive autoantibody is pathologic, the DAT is most often positive due to complement coating patient red cells. Cold-reactive autoantibodies are usually IgM and bind to red cells at lower body temperatures in the circulation. When red cells move to warmer areas IgM elutes off the cells, but complement remains attached, and red cells are subsequently destroyed by the activation of complement.

Another difference in benign versus pathologic cold autoantibodies is their thermal reactivity. Pathologic autoantibodies can attach to red cells in vivo at temperatures greater than 30°C. Determining true thermal

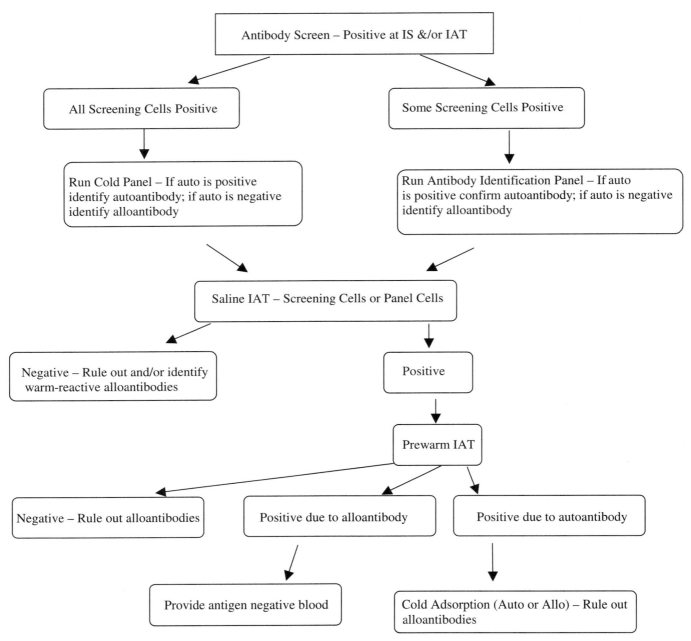

FIGURE 6.1 Evaluation of patients with cold-reactive antibodies
Auto = Autologous control, patient serum mixed with patient RBCs.

reactivity of an autoantibody must be performed on a sample that has been drawn and allowed to clot at 37°C. The tube must be spun immediately and the serum removed quickly. This prevents cold autoantibody from adsorbing onto the RBCs in vitro.

Thermal Amplitude

Thermal amplitude studies are performed to confirm a cold-reactive antibody is capable of binding to red cells at warmer temperatures (>30°C). This is important when a cold-reactive autoantibody is present but the clinician is unclear as to the cause of hemolysis. These studies, while not critical to perform, may help confirm a diagnosis.

Paroxysmal Cold Hemoglobinuria

Paroxysmal cold hemoglobinuria (PCH) is the rarest form of autoimmune hemolytic anemia. It is most often characterized as a sudden onset of hemolysis following viral infection in children. In the past it was associated with syphilis. The patient's DAT will usually be positive (weak to 2+) due to IgG and/or complement. Children will sometimes experience an anti-P_1-like antibody. Patient serum will react preferentially with P_1 positive red cells versus anti-P that reacts with all cells except rare P-negative red cells. PCH is caused by an autoantibody that is IgG in nature but is capable of binding to RBCs at cold temperatures, causing complement activation and subsequent hemolysis of patient RBCs.

Transfusion is rarely necessary for patients with PCH unless hemolysis is severe. Crossmatch-incompatible random donor blood is usually chosen for transfusion. P-negative blood is extremely rare, occurring in approximately 1 in 200,000 random individuals, and should be reserved only for patients experiencing severe hemolysis with random donor blood.

Donath-Landsteiner Test

The diagnostic test for PCH is the Donath-Landsteiner test. Patient serum is mixed with normal red cells and incubated at 4°C followed by 37°C incubation. Antibody binds at cold temperatures and hemolysis occurs during warming if the Donath-Landsteiner antibody is present.

Warm-Reactive Antibody Evaluation

Warm-Reactive Alloantibody

No History of Antibody

An antibody detection test positive in the IAT usually indicates the presence of warm-reactive antibody. Running a panel using the same method in which a positive reaction was detected is usually done. If the sample is limited, an entire antibody identification panel (11 to 16 cells) is not necessary as long as a representative selection of antigen-negative and antigen-positive red cells are tested to rule in and rule out antibody. Once an antibody has been identified and all other alloantibodies have been ruled out according to laboratory policy, phenotyping the patient's red cells is usually performed.

History of Antibody

Some believe repeat antibody investigation is not necessary for subsequent transfusion requests. Patients with previously identified antibodies are required by *Standards for Blood Banks and Transfusion Services* (2002) to be tested by methods that identify additional clinically significant antibodies. Clinically significant antibodies are defined as those that are IgG in nature and reactive at 37°C. Some IgG antibodies will fall below the level of routine detection and will not be reactive in a crossmatch. Ruling out other antibodies can be accomplished by repeating an entire antibody identification panel, but, as noted, is usually not practical in children. Again, since the sample is often limited, caution in the approach to ruling out new antibodies is often required. There is no need to run antigen-positive cells to confirm historical antibody. Antigen-negative blood will be provided for patients with a history of a clinically significant antibody whether the antibody is detectable or not. It is more important to select cells negative for the corresponding antibody to rule out the presence of new antibody. If a patient's phenotype is known even fewer cells may need to be tested.

The newest approach to determining if a patient has made additional antibodies is to only do the antibody screen. If reactivity on the antibody detection test fits the pattern of previously identified antibody(ies), no additional testing is required unless one of the following occurs:

- Serologic evidence of a new antibody because of unexplained reactivity in the antibody detection test or an unexplained incompatible crossmatch.
- Clinical indication of a new antibody suggested by a report of unexplained hemolysis.
- Rare RBCs are required.
- All screening cells are positive indicating multiple alloantibodies, autoantibody, or an antibody to a high incidence antigen.
- Previous workup was inconclusive.

Regardless of methodology chosen, the most important thing to remember is to rule out the presence of additional antibodies. An evaluation process is outlined in Figure 6.2.

Warm-Reactive Autoantibody Evaluation

The DAT is usually strongly reactive (3 to 4+) with polyspecific AHG reagent in patients with warm-reactive autoimmune hemolytic anemia (WAIHA). Tests with AHG specific for IgG and C3 show IgG alone or a combination of IgG and C3.

Approximately 50% of patients with WAIHA have detectable autoantibody in their serum when using standard, no additive, antibody screening methods. Autoantibody spills into the plasma in vivo when all antigen sites on the RBCs are occupied, and there are no antigen sites left for autoantibody to bind. Warm autoantibody is usually IgG and reacts optimally at 37°C.

Autoantibody is normally detected in the IAT and reacts with all reagent red cells tested including the patient's own cells. It is rarely necessary to attempt to identify autoantibody specificity. Most autoantibodies are directed to antigens in the Rh blood group. Some

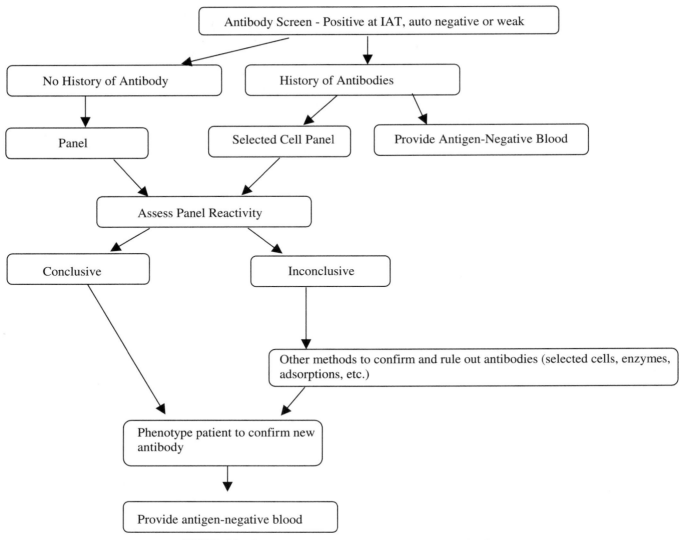

FIGURE 6.2 Evaluation of patients with warm-reactive antibodies.

have apparent, some relative, and some broad specificity within the Rh blood group. An autoantibody with apparent Rh specificity would react with common Rh antigens such as D, C, or more commonly c or e.

An autoantibody with relative specificity would react stronger with red cells possessing the corresponding antigen than if the cells lacked the antigen. For example, a relative autoanti-e may be 3+ with e+ RBCs but 1+ with e− red cells.

Most autoantibodies have broad specificity. The only way to confirm that these autoantibodies have Rh specificity would be to test them against rare Rh_{null} and/or D− red cells.

There are rare examples of autoantibodies that mimic alloantibodies. For example, a patient with anti-Jk^a in the serum is detected and his or her red cells type as Jk(a-).

These serologic results normally indicate alloanti-Jk^a. The unusual finding is anti-Jk^a is also found in the eluate. This most often occurs in individuals producing new alloantibody. Their immune system, now primed, makes autoantibody along with the alloantibody.

One of the biggest challenges in evaluating serum of patients with WAIHA is determining if there are any underlying alloantibodies, particularly if the patient should require transfusion. If the patient's serum reacts with all panel cells, it is difficult to determine if underlying alloantibodies are also present. It is reported that 23% of patients with autoantibodies also have alloantibodies.

To determine if alloantibodies are present, the autoantibody reactivity must be removed from the patient's serum. This is accomplished by performing adsorptions on the patient's serum. There are two types of adsorptions routinely performed. The first is an autologous adsorption.

Autologous Adsorption

If the patient has not been transfused within the last 3 months the patient's own RBCs are first treated to remove autoantibody to free antigen sites on these cells. Then the patient serum is mixed with these cells.

Several different methods can be used to free antigen sites. The most effective method is in-house prepared ZZAP, a combination of proteolytic enzyme and a sulfhydryl reagent.

Autoantibody is adsorbed onto the treated autologous cells and any alloantibody remains in the adsorbed serum since patient red cells lack antigen corresponding to antibody made. More than one adsorption is often required to fully remove autoantibody reactivity.

Allogeneic Adsorption

When a patient has been recently transfused, or when autologous cells are in short supply, as with pediatric and severely anemic patients, allogeneic adsorptions are necessary. Patient serum is mixed with aliquots of RBCs of known phenotype, usually a R_1R_1 (c negative, E negative), R_2R_2 (e negative, C negative) and rr (D negative, C negative, E negative), one of which is Jk(a-), Jk(b-), and s- and incubated at 37°C. Autoantibody will be adsorbed onto all RBCs regardless of phenotype. Alloantibody will also be removed if corresponding antigen is present on the adsorbing cells. Alloantibody will be left in the serum if adsorbing cells lack the antigen. Selecting cells from donors of various phenotypes is required to avoid missing alloantibody. It normally requires a minimum of 4 to 6 hours to determine if a patient has underlying alloantibodies. The extent of the workup needs to be considered when a request for blood is obtained for these patients.

The practice of repeating a workup to determine if a patient with warm autoantibody has an underlying alloantibody varies between institutions.

Repeating the workup every 3 days is consistent with AABB *Standards* requiring repeat antibody screens for unexpected clinically significant antibody. These patients are known to have an increased incidence of underlying alloantibodies so one could argue that repeating the screen for newly formed alloantibody(ies) is even more important in these patients than in patients with no history of antibody.

A moderate approach would be to repeat the work every 7 days. This would detect any new antibodies within 1 week of the patient making the antibody.

A more liberal approach would be to repeat the work once per month. This may be an especially suitable alternative in a patient that has been transfused for years, yet has never made alloantibody.

The presence or absence of alloantibodies, transfusion history, pregnancy history, and rate of transfusion should be evaluated when deciding how often to repeat workups.

Elutions

In patients with WAIHA, an elution should be performed the first time a patient is evaluated to confirm autoantibody and to rule out the presence of drug-dependant antibody. An eluate will typically be positive with all cells tested showing broad specificity.

It is not necessary to perform an eluate on every sample evaluated since it is likely that the serologic results will not change from sample to sample.

Some feel that it is not necessary to perform an elution even on the initial sample evaluated. The caution is that some cases of drug-induced immune hemolytic anemia will give identical serologic results.

DAT Negative Autoimmune Hemolytic Anemia

A negative DAT is seen in approximately 1% of patients even though they have signs and symptoms of immune hemolysis. A well-performed DAT detects approximately 100 to 500 molecules of IgG per red cell. A DAT-negative WAIHA can be due to low-affinity IgG autoantibodies that elute off the RBCs during routine washing of the RBCs before testing with AHG, RBCs sensitized with a small level of IgG molecules that are below the level of detection of the standard DAT, or IgA or IgM autoantibodies not detected by a DAT performed using monoclonal AHG. More sensitive DATs can be performed in specialized laboratories in an attempt to confirm the diagnosis of AIHA and to characterize the type of DAT-negative WAIHA a patient may be experiencing.

References

Issitt PD and Anstee DJ. 1998. *Applied blood group serology*. Durham, NC: Montgomery Scientific.

Johnson ST. 2002. Immune hemolytic anemias. In *Clinical laboratory medicine*. 2nd ed. Edited by McClatchey KD. Philadephia, PA: Lippincott Williams & Wilkins.

Pugh TM, Johnson ST, Baylerian DM, Gottschall JL. 2001. Red cell alloimmunization in transfused pediatric patients. *Transfusion* 41S:104S–105S.

Standards for blood banks and transfusion services. 21st ed. 2002. Bethesda, MD: American Association of Blood Banks.

Leukoreduced Products:
Prevention of Leukocyte-Related
Transfusion-Associated Adverse Effects

LENNART E. LÖGDBERG, MD, PhD

INTRODUCTION

The selective reduction of leukocytes, termed leuko-reduction, leukoreduced, or leukocyte-reduced (LR), in blood products intended for transfusion has become a widespread practice in recent years. LR products, whether making up a separate blood component inventory, used for targeted patient populations, or by preemptive LR of the entire blood supply ("universal leukoreduction" [ULR]), are used to prevent allogeneic leukocyte-related transfusion complications. This chapter will provide a basic understanding of the clinical indications for LR, a brief review of the technology used to achieve it, and the currently applied quality standards for LR. The ongoing debate regarding the implementation of national ULR will be discussed. Finally, the chapter will focus on the interface of LR and the practice of pediatric transfusion medicine.

RATIONALE FOR LEUKOREDUCTION OF BLOOD COMPONENTS

In current transfusion practice, the bulk of prescribed products are specific blood components, such as packed red blood cells (PRBCs), platelets (PLTs), or fresh frozen plasma (FFP), rather than whole blood (WB). During component preparation, the white blood cells (WBCs) (leukocytes) present in WB distribute primarily with all the cellular components. In recent decades, it has become apparent that these passenger leukocytes, far from being harmless, are implicated as important contributors to the majority of transfusion-associated

adverse reactions (Raife 1997; Walker 1987). This has led to the development and widespread application of methods to selectively reduce the leukocyte content in blood components before they are transfused, and many countries now exclusively use ULR products. In the United States, although a substantial portion (~75%) of the blood supply now undergoes LR, whether to institute ULR is still debated as of the writing of this chapter.

SPECIFIC INDICATIONS FOR LR PRODUCTS

Leukocytes in blood components have been implicated in a number of transfusion-related adverse events of clinical importance (Table 7.1) (Dzik and Szczepiorkowski 2002; Vamvakas and Blajchman 2001b). Some of these events are rooted in immunological/inflammatory mechanisms, such as recipient sensitization to the allogeneic major histocompatibility antigens (HLA) of the donor leukocytes and the release of inflammatory cytokines by either donor or recipient leukocytes. Other events result from the transmission of leukotropic viruses such as cytomegalovirus (CMV). The scientifically established indications and possible other clinical benefits of LR products are systematically discussed in the following section (see Table 7.1).

Prevention of Recurrent Febrile Nonhemolytic Transfusion Reactions (FNHTRs)

Febrile nonhemolytic transfusion reactions (FNHTRs) are classically the most frequent adverse

TABLE 7.1   Known and Potential Clinical Benefits from Using Leukoreduced Blood Products

Scientifically established indications for use of LR products

Prevention of recurrent febrile nonhemolytic transfusion reactions
Prevention of primary HLA-alloimmunization and resulting complications, including platelet refractoriness
Prevention of transfusion-transmitted cytomegalovirus infection in at-risk patients

Possible other clinical benefits (some support in randomized clinical trials)

Reduction of transfusion-associated immunomodulation related deleterious effects (postoperative infections and possibly other ailments)
Reduction of transfusion-associated postoperative mortality

Speculative clinical benefits **(some support by theoretical considerations, by animal experiments and/or by observational studies; not supported by randomized clinical trials)**

Reduction of prion (vCJD)-transmission (reason for introduction of ULR in the United Kingdom, Ireland, and Portugal)
Reduction of transmission of other leukotropic viruses (HTLV I–II and EBV)
Reduction of viral reactivation
Reduction of parasitic or bacterial infections
Reduction of transfusion-related acute lung injury

effects following non-LR RBC (0.5% to 6%) or PLT transfusions (20% to 30%) (Heddle and Kelton 2001). These reactions, typically occurring towards the end of or shortly after the transfusion, consist of nonspecific inflammatory-type symptoms (one or more of the following: fever, chills, cold, discomfort, rigors, headache, nausea/vomiting, and dyspnea) and usually resolve spontaneously posttransfusion. Most FNHTRs are due to the in vivo action of proinflammatory cytokines released from leukocytes (or PLTs) in the blood components, either in vivo, after transfusion, or in vitro, during storage. Thus, prevention of FNHTRs through LR of the blood components is an attractive strategy.

Mechanisms of FNHTRs

Many transfusion recipients have preformed alloantibodies to HLA or other leukocyte antigens, which can cause recurring FNHTRs by reacting with the transfused donor leukocytes. These reactions lead to the in vivo secretion of proinflammatory cytokines, such as interleukin-1 (IL-1), IL-6, IL-8, or tumor necrosis factor alpha (TNFα), produced either directly by the alloantibody-activated donor leukocytes or indirectly

by recipient leukocytes (probably reacting to released alloantibody-generated immune complexes) and appear to be a predominant cause of FNHTRs when transfusing non-LR RBCs. The leukocytes in stored PLT products also release the above cytokines into the storage medium over time, even in the absence of activation by alloantibodies, which explains much of the higher and more variable (different storage times) incidence of FNHTRs in recipients of PLT products as compared to RBCs (Heddle et al. 1994). Of note, PLTs also release biologically active factors during storage, such as sCD154 (Phipps, Kaufman, and Blumberg 2001), which provide another mechanism for FNHTRs, that would not be preventable through LR.

Evidence for the Effectiveness of LR as a Preventive Measure

An appreciation of the causal role of donor leukocytes in FNHTRs existed almost 40 years ago when it was shown that reducing the leukocyte content of the products to below 5×10^8/unit via washing prevented recurrent RBC-transfusion-induced FNHTRs (Perkins et al. 1966). This has been corroborated in subsequent studies (Heddle and Kelton 2001). Results concerning the prevention of PLT-induced FNHTRs are more complex to interpret, given the added mechanism of cytokine accumulation in stored PLT units and the varying degree of leukocyte content in PLTs produced from PLT-rich plasma or buffy coats, or through apheresis. A number of observational and experimental studies, including some randomized controlled trials, suggest that prestorage LR significantly reduces FNHTRs due to PLT transfusion (Heddle and Kelton 2001). However, residual severe LR-PLT transfusion-related FNHTRs of unknown cause remain (Heddle and Kelton 2001). FNHTRs caused by LR products deserve particular scrutiny, since they imply additional mechanisms for this type of transfusion reactions, such as released platelet factors (see previous paragraph).

Prevention of Primary HLA Alloimmunization

Patients who develop antibodies to human major histocompatibility antigens (HLA alloantibodies) are at increased risk of complications when receiving blood transfusions (platelet refractoriness) or tissue transplants (transplant rejection). While RBCs express negligible amounts of HLA-antigens, PLTs and most leukocytes express HLA class I but not class II antigens. However, a specialized subset of mononuclear blood leukocytes, the antigen presenting cells, serves as the main site of HLA class II antigen expression. Efficient alloimmunization to HLA antigens appears to depend on HLA class II antigen-expression and other properties of leukocyte subsets, such as the expression of co-stimulatory molecules. Accordingly, LR of blood products should and does reduce their alloimmunizing potential.

Evidence for the Effectiveness of LR as a Preventive Measure

In mice, repeated injections of pure platelet suspensions did not induce allo-HLA-antibodies, whereas contamination of the same suspensions with small numbers of leukocytes restored their alloimmunizing capacity (Claas et al. 1981). This finding led to a number of prospective, randomized clinical trials addressing the effect of LR on alloimmunization by PLT transfusions (Vamvakas 1998), notably the TRAP (Trial to Reduce Alloimmunization to Platelets Study Group) trial (1997). These trials primarily investigated patients with chronic thrombocytopenia, most often due to acute myelogenous leukemia, who received multiple allogeneic PLT transfusions. Overall, the results showed a clear reduction (~70%) in HLA alloimmunization in the groups receiving LR products versus controls, which translated to a corresponding reduction in the incidence of PLT refractoriness. LR pooled random donor PLT and LR apheresis PLT were equally efficient in conferring this beneficial effect.

Prevention of Transfusion-Transmitted CMV Infection in At-Risk Patients

In principle, LR would seem to be a promising strategy for the prevention of any predominantly leukotropic transfusion-transmitted virus. This possibility has been substantiated in practice and applied clinically with respect to transfusion-transmitted cytomegalovirus (TT-CMV) infection (Bowden et al. 1995; Hillyer et al. 1994; Nichols et al. 2003; Roback et al. 2003). TT-CMV can cause serious to lethal disease in immunocompromised patients and has historically been reported with an incidence ranging from 13% to 37% in at-risk groups (Roback et al. 2003). Latently infected peripheral blood mononuclear leukocytes appear to be the vector for TT-CMV (Roback et al. 2003). Thus, the two main strategies to ensure transfusion safety for TT-CMV have been the use of LR blood components or the use of blood components from CMV-seronegative donors (likely uninfected by CMV).

Evidence for the Effectiveness of LR as a Preventive Measure

The use of CMV-seronegative blood components to reduce TT-CMV ratios was shown to be effective and was established as a standard of care in the 1980s (Bowden et al. 1986; Yeager et al. 1981). Thus, the effectiveness of LR in this setting has usually been compared with results using CMV-seronegative blood components. Both strategies reduce the incidence of TT-CMV quite effectively (Bowden et al. 1995; Hillyer et al. 1994; Nichols et al. 2003). Many authors have interpreted the data to support the equivalent efficacy of the two strategies, favoring ULR in an effort to avoid dual inventories (separate LR products *and* seronegative products). Others, however, have pointed to residual differences in the TT-CMV incidence with LR versus CMV-seronegative components (for example, 2.4% and 1.4%, respectively [$p > 0.5$]) in a landmark study (Bowden et al. 1995), suggesting that seronegative units may yield a still higher degree of transfusion safety in the at-risk patient (Nichols et al. 2003). A recent study indicates that this may indeed be true at least for RBC products (Nichols et al. 2003). This finding is consistent with a recent consensus conference concerning the prevention of posttransfusion CMV in Canada, which, after the implementation of ULR, supported the maintenance of CMV-seronegative LR products for high-risk patients (congenital immunodeficiency; HIV+/CMV-; very low birth weight infants of CMV-mothers; CMV-hematopoietic stem cell transplant recipients; and so on) (Blajchman et al. 2001).

Prevention of Transfusion-Associated Immunosuppression/Immunomodulation

Allogeneic blood transfusions appear to induce an immunosuppressed state in the recipient that can have beneficial or deleterious effects. In 1973, Opelz demonstrated that these transfusions could prolong subsequent organ graft survival (Opelz et al. 1973). This clinically important immunomodulatory effect was used to enhance organ transplant outcome, particularly in the decade preceding the introduction of cyclosporine A and other modern antirejection drugs, and still appears to confer an independent, additional beneficial effect on graft rejection over and above HLA-matching and pharmaceutical immunosuppression (Opelz et al. 1997). Although this *transfusion-associated immunomodulation*, called "TRIM," can be clinically beneficial in this context, it has been postulated that TRIM could affect transfusion recipients adversely through an increased incidence of postoperative infections (Waymack, Robb, and Alexander 1987) and by reduced tumor immunity resulting in worse outcome of malignant disease (Gantt 1981). A large number of observational studies, and a more limited number of randomized clinical trials, have addressed the issue of such deleterious TRIM effects (Vamvakas and Blajchman 2001a). Due to issues such as the heterogeneity of outcome and trial design, the insufficient power of many trials, and the possibility of bias (that is, mostly not double-blinded studies) and other confounding factors, questions remain as to whether these TRIM effects are a clinical reality, and, if so, how quantitatively significant they are and through what mechanisms they may operate (Vamvakas and Blajchman 2001a).

Proposed Mechanisms of TRIM

One reason for the controversy surrounding the clinical relevance of TRIM-related adverse effects is the relatively poor and diffuse understanding of the possible underlying mechanisms. Although donor leukocytes appear to play a causal role in the etiology of most TRIM effects, soluble donor factors such as soluble HLA antigens and the lipocalin PP14 (produced by PLTs) have also been linked to TRIM. Moreover, mechanisms for leukocyte-induced TRIM are described very generally, such as a shift of Th1 $\rightarrow$ Th2 helper cell cytokine responses with a resulting down regulation of cell-mediated immunity, or the persistence of microchimerism, with roles for both partial HLA-match dependent and nonspecific mechanisms (Blajchman 1998; Vamvakas 2002b). Thus, a spectrum of mechanisms has been implicated in TRIM (Blajchman 1998; Vamvakas 2002b), likely with different subsets of mechanisms responsible for different clinical TRIM effects, which contributes to the heterogeneity and confounding effects reported in the clinical studies in this area.

Evidence for the Effectiveness of LR as a Preventive Measure

Clinical studies fail to consistently verify that allogeneic blood transfusions are associated with increased postoperative infections and tumor recurrence, and even if they were, such a correlation would not necessarily imply a causal relationship. Nonetheless, the transfusion medicine community has been intrigued by the latter possibility, and randomized clinical trials have examined whether LR products are less likely than their non-LR counterparts to be associated with the proposed TRIM-related postoperative infections (Vamvakas 2002a). A recent meta-analysis did reveal a statistically significant LR-related reduction of these complications (Vamvakas 2002a). In addition, in one of the clinical trials included in the meta-analyses, LR-reduction also was associated with a decreased mortality due to noncardiac causes, possibly related to nonrecognized postoperative infections (Vamvakas and Blajchman 2001b). More recent, post-ULR outcomes studies in Canada continue to suggest potential benefits of LR on deleterious TRIM effects (Corwin and AuBuchon 2003).

CONTRAINDICATIONS, SIDE EFFECTS, AND HAZARDS

Leukoreduction of granulocytes or of stem cells from bone marrow, peripheral blood, or umbilical cord would compromise the therapeutic efficacy of these products and is therefore contraindicated. As currently practiced clinically, LR is insufficient to ensure the prevention of transfusion-associated graft-versus-host disease due to residual leukocytes. Thus, LR products are contraindicated for that application, which is instead addressed by gamma irradiation of blood components (gamma-irradiated LR blood components may provide additional safety, but it has not been specifically studied). LR by leukofiltration can lead to the release of bradykinin, which can then provoke hypotensive reactions during bedside leukofiltration transfusion to patients receiving ACE inhibitor therapy (due to the reduced elimination of infused bradykinin). The trend toward prestorage LR (see the following section) has diminished the risk for this complication since bradykinin is unstable during storage.

LEUKOREDUCTION TECHNOLOGIES

Inspired by the recognition that leukocytes in blood components can cause adverse transfusion events, investigators have developed several technological advances to selectively remove them from the final products. Currently, specialized leukofiltration or apheresis collection (process LR) devices achieve a >10^4-fold reduction of leukocyte content in the final products, with some differential reduction efficacy in different leukocyte subsets (Roback, Bray, and Hillyer 2000). Several different versions of leukocyte filters are on the market, including those consisting of nonwoven filter media designed to have a large, hydrophilic surface area and a controlled functional pore size (usually ~4 μm) that fills an external housing, thereby preventing bypass during the filtration process. Leukocyte removal depends on several features including a combination of barrier filtration and cell adsorption to the filter material as well as the deformability of the RBCs and platelets. Different filters are used to produce LR-RBC and LR-platelet products. Toward the same end, several design principles have been added to modern apheresis devices to similarly allow efficient process LR during apheresis collection of PLTs.

With the exception of process LR apheresis PLTs, blood components usually undergo LR soon after blood collection (during component manufacturing or early storage [first 2 days]), typically at a blood center (*prestorage LR*), or in the context of usage (*poststorage LR*) either by the blood-bank (*in-laboratory LR*) before release or at the time of transfusion (*bedside LR*). The increased understanding of the role of cytokines released from leukocytes during storage in adverse transfusion effects has prompted a trend towards using prestorage rather than poststorage LR.

PRODUCTS, STANDARDS, AND QUALITY CONTROL

The major leukoreduced blood products are: (1) WB, LR; (2) RBCs, LR; (3) PLTs, LR; and (4) PLTs, Pheresis, LR. According to current standards from the American Association of Blood Banks (AABB) and guidelines from the U.S. Food and Drug Administration (FAD), products 1, 2, and 4 should contain $<5 \times 10^6$ residual leukocytes per unit while retaining $>85\%$ of the therapeutic component. The random donor PLT concentrates (product 3) typically used as pools of 6 units for transfusion should contain $<0.83 \times 10^6$ residual leukocytes per unit (which leads to $<5 \times 10^6$ residual leukocytes per standard therapeutic dose of pools of 6 random donor platelets). European countries maintain

a stricter standard of $<1 \times 10^6$ residual leukocytes per LR RBC or LR platelet (apheresis) unit, and the United States may follow suit. Quality control is mandated and blood manufacturers are to have procedures and processes to validate and monitor their adherence to these standards (for example, by manual counting, using a hematocytometer, or by flow cytometry). There are currently no standards for leukoreduction with respect to known leukocyte subsets.

UNIVERSAL LEUKOREDUCTION

Based on the proven or suspected deleterious effects of blood transfusion, attributed to the leukocyte content of the transfused products, Canada and many countries in Europe have implemented universal leukoreduction (the exclusive use of LR inventories of RBCs and PLTs), and many additional countries, including the United States, have moved closer towards the same objective (Dzik and Szczepiorkowski 2002; Vamvakas and Blajchman 2001b). Some controversy (Table 7.2) exists in the United States over this trend (Dzik 2002; Dzik and Szczepiorkowski 2002; Vamvakas and Blajchman 2001a; Vamvakas and Blajchman 2001b) even though committees of both the Department of Health and Human Services and the FDA have recommended ULR-implementation. Some authorities have suggested that the cost-to-benefit ratio of ULR is unacceptably high and favor LR for specific patient populations only.

Debate exists concerning the magnitude of clinical benefit resulting from ULR with differences in interpretation of the relevant clinical studies with respect to both the presence and magnitude of some of the

TABLE 7.2 Universal versus Selective Leukoreduction in the United States

For ULR	Against ULR
Large potential clinical benefit to be realized (see Table 7.1). Overall improved safety (initial processing step).	Much of the projected clinical benefit is controversial (see Table 7.1). Failure to show efficacy of LR in many randomized controlled trials.
In many cases, pivotal future randomized controlled trials may not be forthcoming. Decision about ULR should be made based on currently available data. Before/after studies can be used to evaluate effects of ULR-implementation.	Want to wait for results from additional pivotal randomized controlled trials as basis for making decision about implementation of ULR. Such implementation will make future RCTs harder to perform.
May simplify choices of blood component usage.	May remove medical decision from blood product selection process.
Streamlined blood supply with gains in inventory management.	Potential for increased vulnerability of blood supply (recall of LR devices; reduced sickle trait blood can diminish availalable rare African-American blood types).
Increase in cost represents a small fraction of the healthcare budget and will be offset due to realized health benefits. Marginal cost increase of blood components will be lowered by the fact that ~75% of the blood supply already is subjected to LR.	Significant cost of blood components.

proposed adverse effects due to leukocytes and the extent to which it is appropriate to generalize from the clinical study populations to the general patient population (Dzik and Szczepiorkowski 2002; Vamvakas and Blajchman 2001b). Furthermore, concerns have been raised that additional randomized clinical trials to resolve such controversies may not be conducted and that decisions with respect to ULR cannot await future clarification of these issues (Vamvakas and Blajchman 2001a; Vamvakas and Blajchman 2001b). Juxtaposed to these questions about the scientific basis for ULR are issues such as cost of its implementation—indeed, it has been argued that were it not for the magnitude of these costs, most of the transfusion medicine community would favor ULR (Vamvakas and Blajchman 2001b). Regardless of these controversies, the U.S. blood supply is approaching a state of ULR due to the political force of questions of blood safety and resultant management decisions at the blood centers.

LEUKOREDUCED BLOOD PRODUCTS AND PEDIATRIC TRANSFUSION MEDICINE

The clinical benefits of LR have been established by studies in adult patients, and there is a paucity of comparable trials in pediatric patients (Couban et al. 2002; Fergusson et al. 2002). The current usage guidelines for LR products in children and infants are thus, of necessity, derived mostly by extrapolations from experiences in adult patients. The few available clinical studies in pediatric patients (Couban et al. 2002; Fergusson et al. 2002) do suggest support for an extension of the major adult indications for LR products to pediatric populations, although further studies are needed to fully establish this and quantify the clinical benefit.

With the U.S. blood supply moving toward ULR, it is worth pointing out two possible changes in the future usage of LR blood applicable to the pediatric population. (1) In recent years, many transfusion services have switched to supplying LR products, instead of CMV-seronegative units, to prevent TT-CMV in at-risk patients, based on clinical trials suggesting equivalency between such components in terms of CMV-transmission risk (see previous section and Chapter 8). Recent data (Nichols et al. 2003) and alternative interpretations of previous trials (Blajchman et al. 2001) suggest that CMV-seronegative units may provide greater protection than LR products and that even for ULR products, the selection of seronegative LR products may still have a role. (2) Existence of a subset of patients who continue to experience FNHTRs when receiving LR products, particularly to PLT products, suggest that poststorage washing of LR PLTs to remove PLT-derived factors released into the supernatant may have a future role in selected subsets of patients. Both of these developments would suggest a trend toward combining LR with additional preventive measures to minimize the leukocyte-based deleterious effects of blood transfusion.

References

Blajchman MA. 1998. Immunomodulatory effects of allogeneic blood transfusions: clinical manifestations and mechanisms. *Vox Sang* 74 Suppl 2:315–319.

Blajchman MA, Goldman M, Freedman JJ, and Sher GD. 2001. Proceedings of a consensus conference: prevention of post-transfusion CMV in the era of universal leukoreduction. *Transfus Med Rev* 15(1):1–20.

Bowden RA, Sayers M, Flournoy N, Newton B, Banaji M, Thomas ED, and Meyers JD. 1986. Cytomegalovirus immune globulin and seronegative blood products to prevent primary cytomegalovirus infection after marrow transplantation. *N Engl J Med* 314(16):1006–1010.

Bowden RA, Slichter SJ, Sayers M, Weisdorf D, Cays M, Schoch G, Banaji M, Haake R, Welk K, Fisher L, et al. 1995. A comparison of filtered leukocyte-reduced and cytomegalovirus (CMV) seronegative blood products for the prevention of transfusion-associated CMV infection after marrow transplant. *Blood* 86(9):3598–3603.

Claas FH, Smeenk RJ, Schmidt R, van Steenbrugge GJ, and Eernisse JG. 1981. Alloimmunization against the MHC antigens after platelet transfusions is due to contaminating leukocytes in the platelet suspension. *Exp Hematol* 9(1):84–89.

Corwin HL and AuBuchon JP. 2003. Is leukoreduction of blood components for everyone? *JAMA* 289(15):1993–1995.

Couban S, Carruthers J, Andreou P, Klama LN, Barr R, Kelton JG, and Heddle NM. 2002. Platelet transfusions in children: results of a randomized, prospective, crossover trial of plasma removal and a prospective audit of WBC reduction. *Transfusion* 42(6):753–758.

Dzik WH. 2002. Leukoreduction of blood components. *Curr Opin Hematol* 9(6):521–526.

Dzik WH and Szczepiorkowski ZM. 2002. Leukocyte reduced products. In CD Hillyer, LE Silberstein, PM Ness, and KC Anderson, eds. *Blood banking and transfusion medicine.* Philadelphia: Churchill Livingstone.

Fergusson D, Hebert PC, Barrington KJ, and Shapiro SH. 2002. Effectiveness of WBC reduction in neonates: what is the evidence of benefit? *Transfusion* 42(2):159–165.

Gantt CL. 1981. Red blood cells for cancer patients. *Lancet* 2(8242):363.

Heddle NM and Kelton JG. 2001. Febrile nonhemolytic transfusion reactions. In MA Popovsky, ed. *Transfusion reactions.* Bethesda, MD: AABB Press.

Heddle NM, Klama L, Singer J, Richards C, Fedak P, Walker I, and Kelton JG. 1994. The role of the plasma from platelet concentrates in transfusion reactions. *N Engl J Med* 331(10):625–628.

Hillyer CD, Emmens RK, Zago-Novaretti M, and Berkman EM. 1994. Methods for the reduction of transfusion-transmitted cytomegalovirus infection: filtration versus the use of seronegative donor units. *Transfusion* 34(10):929–934.

Nichols WG, Price TH, Gooley T, Corey L, and Boeckh M. 2003. Transfusion-transmitted cytomegalovirus infection after receipt of leukoreduced blood products. *Blood* 101:4195–4200.

Opelz G, Sengar DP, Mickey MR, and Terasaki PI. 1973. Effect of blood transfusions on subsequent kidney transplants. *Transplant Proc* 5(1):253–259.

Opelz G, Vanrenterghem Y, Kirste G, Gray DW, Horsburgh T, Lachance JG, Largiader F, Lange H, Vujaklija-Stipanovic K, Alvarez-Grande J, Schott W, Hoyer J, Schnuelle P, Descoeudres C, Ruder H, Wujciak T, and Schwarz V. 1997. Prospective evaluation of pretransplant blood transfusions in cadaver kidney recipients. *Transplantation* 63(7):964–967.

Perkins HA, Payne R, Ferguson J, and Wood M. 1966. Nonhemolytic febrile transfusion reactions. Quantitative effects of blood components with emphasis on isoantigenic incompatibility of leukocytes. *Vox Sang* 11(5):578–600.

Phipps RP, Kaufman J, and Blumberg N. 2001. Platelet derived CD154 (CD40 ligand) and febrile responses to transfusion. *Lancet* 357(9273):2023–2024.

Raife TJ. 1997. Adverse effects of transfusions caused by leukocytes. *J Intraven Nurs* 20(5):238–244.

Roback JD, Bray RA, and Hillyer CD. 2000. Longitudinal monitoring of WBC subsets in packed RBC units after filtration: implications for transfusion transmission of infections. *Transfusion* 40(5):500–506.

Roback JD, Drew WL, Laycock ME, Todd D, Hillyer CD, and Busch MP. 2003. CMV DNA is rarely detected in healthy blood donors using validated PCR assays. *Transfusion* 43(3):314–321.

The Trial to Reduce Alloimmunization to Platelets Study Group. 1997. Leukocyte reduction and ultraviolet B irradiation of platelets to prevent alloimmunization and refractoriness to platelet transfusions. *N Engl J Med* 337(26) 1861–1869.

Vamvakas EC. 1998. Meta-analysis of randomized controlled trials of the efficacy of white cell reduction in preventing HLA-alloimmunization and refractoriness to random-donor platelet transfusions. *Transfus Med Rev* 12(4):258–270.

Vamvakas EC. 2002a. Meta-analysis of randomized controlled trials investigating the risk of postoperative infection in association with white blood cell-containing allogeneic blood transfusion: the effects of the type of transfused red blood cell product and surgical setting. *Transfus Med Rev* 16(4):304–314.

Vamvakas EC. 2002b. Possible mechanisms of allogeneic blood transfusion-associated postoperative infection. *Transfus Med Rev* 16(2):144–160.

Vamvakas EC and Blajchman MA. 2001a. Deleterious clinical effects of transfusion-associated immunomodulation: fact or fiction? *Blood* 97(5):1180–1195.

Vamvakas EC and Blajchman MA. 2001b. Universal WBC reduction: the case for and against. *Transfusion* 41(5):691–712.

Walker RH. 1987. Special report: transfusion risks. *Am J Clin Pathol* 88(3):374–378.

Waymack JP, Robb E, and Alexander JW. 1987. Effect of transfusion on immune function in a traumatized animal model. II. Effect on mortality rate following septic challenge. *Arch Surg* 122(8):935–939.

Yeager AS, Grumet FC, Hafleigh EB, Arvin AM, Bradley JS, and Prober CG. 1981. Prevention of transfusion-acquired cytomegalovirus infections in newborn infants. *J Pediatr* 98(2):281–287.

Preparation of Blood Components to Reduce Cytomegalovirus and Other Infectious Risks

JOHN D. ROBACK, MD, PhD

INTRODUCTION

Extensive work over the past decades has produced dramatic improvements in our ability to screen blood before transfusion. Serological assays are now used to detect human immunodeficiency virus (HIV), hepatitis B virus (HBV), hepatitis C virus (HCV), human T-cell lymphotropic viruses I and II (HTLV-I, -II), and syphilis, while nucleic acid testing (NAT) is applied to identify blood from donors recently infected with HIV and HCV. When these tests are used in combination with donor medical histories, only 1 in every 2 million blood components is potentially infectious for HIV or HCV. For perspective, complications resulting from general anesthesia occur about once in 10,000 procedures, or 200 times more frequently than a screened HIV- or HCV-infected blood unit is transfused. Routine screenings by donor history, viral serology, and NAT testing are discussed in Chapter 1.

The present chapter focuses on additional methods that can be used to reduce transfusion transmission of other viruses (for example, cytomegalovirus [CMV]), or to sterilize other products (for example, coagulation factor concentrates). These methodologies are typically applied only to specific blood components or blood that will be used for specific patient populations. Most attention is devoted to CMV, which can produce lethal infections in premature or newborn infants, children with congenital immune deficiencies, and hematopoietic stem cell transplant recipients. The relative advantages of CMV-seronegative and filtered units are discussed. Additional viruses that have specific implications for pediatric transfusion, such as Epstein-Barr virus (EBV), are also discussed. Finally, recently developed methods for inactivating viruses, both known and unknown, are nearing clinical use and may further improve the safety profile of blood products.

RED BLOOD CELL AND PLATELET COMPONENTS WITH REDUCED INFECTIOUS RISKS

The preparation and use of standard blood components and plasma derivatives are discussed in Chapter 2. Components that have undergone additional testing and processing to further improve their safety are discussed below.

CMV-Seronegative Units

While the human herpesvirus family is composed of eight viral species, only CMV routinely causes clinical sequelae following transmission by blood transfusion. CMV is leukocytotropic and can be transmitted by blood transfusion, hematopoietic stem cell transplantation, and solid organ transplantation. Primary CMV infection, or reactivation of latent CMV infection, can produce morbidity and death in immunocompromised patients, including children with immature or congenitally defective immune systems.

CMV is the largest of the herpesviruses, with a genome of approximately 230 kb in length that likely encodes at least 200 viral proteins. Mature virions can be up to 300 nm in diameter. CMV, like other herpesviruses, can replicate in infected cells eventually producing cell lysis (lytic growth) but can also persist in the cells indefinitely without replicating (latency). While lytic growth requires extensive coordinated viral gene expression, few if any viral genes are reproducibly expressed during latency. Rather, the latent genome can remain in an otherwise healthy cell indefinitely, with the capacity to reactivate and undergo lytic growth at a later time. Free virus as well as lytic or latently infected cells can transmit CMV infection to a recipient.

Community-acquired CMV infection is typically due to close contact with an infected person shedding CMV into body fluids. Shed virus can be identified in saliva and urine. CMV can also be detected in semen and the female genital tract, consistent with epidemiological evidence of sexual transmission. In the prenatal and neonatal settings, in utero transplacental transmission and shedding into breast milk are significant risk factors for CMV infection. Fetal infection occurs in up to 50% of pregnancies in which a seronegative mother acquires a primary CMV infection. In utero infection is a significant cause of mental retardation and other birth defects. In contrast to seronegative mothers, only about 2% of seropositive mothers show evidence of transplacental transmission. This is likely due to the rarity of CMV viremia in healthy seropositive donors as well as the protective effects of maternal IgG antibodies passively transferred to the fetus.

In the pediatric setting, transfusion-transmitted CMV infection (TT-CMV) is of most concern in low birth weight (<1200 to 1500 g) neonates born to seronegative mothers, as well as in seronegative stem cell transplant recipients. Evidence suggests that latently infected monocytes and their progenitors are the primary vectors for TT-CMV, although rare cases may be caused by blood products containing free infectious virions. The two most effective methods to prevent TT-CMV are through transfusion of either blood from donors that do not have anti-CMV antibodies and thus are unlikely to have been exposed to the virus (CMV-seronegative units) or the use of units filtered to remove white blood cells (WBCs) including those harboring latent CMV (Table 8.1).

In the early 1980s, Yeager and colleagues first demonstrated a marked reduction in the incidence of TT-CMV through exclusive use of CMV-seronegative blood for transfusion. Ten of 74 (13.5%) infants born to seronegative mothers contracted CMV infection when transfused with blood containing anti-CMV antibodies. When 50 mL of red blood cells (RBCs) or more was transfused, the incidence of TT-CMV increased to 24%. In comparison, none of 90 infants (also of seronegative mothers) transfused exclusively with CMV-seronegative blood developed TT-CMV (Yeager et al. 1981). In half of the infected infants, all weighing less than 1200 grams, TT-CMV produced significant morbidity or death. In other studies, seronegative neonates weighing less than 1500 grams also experienced a high incidence of TT-CMV (Preiksaitis 1991). Yeager's results did not show a difference in the incidence of TT-CMV when seronegative or seropositive units were transfused to infants of seropositive mothers (1981). However, in other investigations, low birth weight infants of seropositive mothers were at risk for lethal CMV infection, despite the transfer of humoral immunity (de Cates et al. 1988). Exclusive use of seronegative blood also significantly reduces the incidence of TT-CMV in immunocompromised adults (Hillyer et al. 1999; Miller et al. 1991). Seronegative RBCs, platelets, and granulocytes are appropriate for susceptible patients; fresh frozen plasma, cryoprecipitate, and clotting factors need not be from seronegative donors since these components do not transmit CMV primarily due to their acellular nature (Table 8.2).

Leukoreduced Units

RBC and platelet units prepared under standard conditions for transfusion contain large numbers of passenger donor WBCs (10^8 to 10^{10}). Passenger WBCs can cause a number of adverse effects in transfusion recipients including CMV transmission, HLA alloimmunization, and repetitive febrile nonhemolytic transfusion

TABLE 8.1   Appropriate Use of CMV-Safe Components

Possibility of CMV Disease Following TT-CMV	Patient Groups
Significant: CMV-seronegative components are preferred; filtered components are acceptable alternatives	• Low birth weight infants (<1500 grams) of seronegative mothers
Possible: CMV-safe components (seronegative or filtered) are preferred when available	• Other infants (>1500 grams) of seronegative mothers • Low birth weight infants (<1500 grams) of seropositive mothers • All neonates receiving extensive transfusion support • Children of HIV-infected mothers • Children with congenital immunodeficiencies or iatrogenic immunosuppression
Negligible: CMV-safe components not indicated	• Other nonimmunocompromised infants (>1500 grams) of seropositive mothers

TABLE 8.2   CMV-Safe Components

Component	TT-CMV Risk?	Processing to Reduce TT-CMV
Red blood cells	Yes	Serology, filtration
Platelets	Yes	Serology, filtration
Granulocytes	Yes	Serology
Fresh frozen plasma	No	N/a
Cryoprecipitate	No	N/a
Clotting factors	No	N/a

N/a = Not applicable.

reactions (see Chapters 22, 26, and 28). Specialized filtration devices can remove over 99.99% of WBCs (4 $\log_{10}$ leukoreduction) before transfusion, with a resulting decrease in the adverse effects. Typically, blood is filtered at the blood center shortly after collection and before storage and shipment to hospital transfusion services. Unlike countries such as Canada that mandate universal leukoreduction of blood products, there is currently no such requirement in the United States. Nonetheless, current estimates suggest that approximately 75% of red cell and platelet units transfused in the United States are leukoreduced.

Leukoreduction for CMV

A number of different cells can be infected by CMV and serve as long-term reservoirs of latent virus. With respect to transmission of CMV by transfusion and hematopoietic stem cell transplantation, latently infected monocytes and their progenitors, respectively, are the likely primary vectors. These cells may also play a role in transmission by solid organ transplantation, although donor endothelial and epithelial cells are also likely to be important.

Latent CMV DNA has been clearly identified in monocytes of healthy, seropositive donors (Bolovan-Fritts et al. 1999). The possibility that some seronegative donors may also carry latent CMV DNA (Larsson et al. 1998), while reported, requires further verification since the CMV polymerase chain reaction assays used in these studies may not be optimally sensitive and specific (Roback et al. 2001). PCR has detected CMV DNA in bone marrow CD34+ cells as well as early and more differentiated hematopoietic progenitor cells (HPCs) (Sindre et al. 1996). These results suggest that while peripheral blood monocytes contain latent CMV, the true reservoir for long-term CMV latency in seropositive individuals may be bone marrow HPCs.

Since filters can remove monocytes and other leukocytes from blood products (Roback et al. 2000), and as there are logistical complications associated with

transfusion of seronegative units (blood donations are not routinely tested for CMV antibodies, and 50% or more of most donor populations are CMV-seropositive), filtration leukoreduction was investigated as an alternative approach to provide "CMV-safe" blood to at-risk transfusion recipients. Multiple clinical trials have uniformly demonstrated significant reductions in the incidence of TT-CMV following filtration of unscreened components (Bowden et al. 1995; Luban et al. 1987). In a large, randomized, multicenter trial, the incidence of TT-CMV was similar in adult hematopoietic transplant recipients that received filtered (but not CMV-screened) blood components as compared to those receiving seronegative units, leading to the suggestion that seronegative and filtered units can be used interchangeably in at-risk patient populations (Bowden et al. 1995). However, neither serological screening nor leukoreduction provides absolute protection from TT-CMV. In fact, up to 2.5% of immunocompromised patients will develop CMV infection after receipt of seronegative or filtered blood (Bowden et al. 1995). RBC and platelet components can be filtered before transfusion, but granulocyte components cannot be since the filter would remove the therapeutic leukocytes (see Table 8.2).

The American Association of Blood Banks (AABB) has published guidelines affirming the equivalence of seronegative and filtered units for the purpose of preventing TT-CMV (1997). However, clinical practice panels from Switzerland and Canada have concluded that seronegative units remain the standard of care (Blajchman et al. 2001; Zwicky et al. 1999). The latter position is supported by a recent retrospective analysis. A large center involved in the trial discussed previously (Bowden et al. 1995), reviewed the occurrence of CMV infection in adult HPC transplant recipients and found a small, but statistically higher incidence of TT-CMV associated with the use of filtered units, particularly filtered RBCs (Nichols et al. 2003). Although this study focused on immunocompromised adults, it has implications for transfusion of seronegative low birth weight neonates who are particularly susceptible to TT-CMV. Studies demonstrating the incidence of TT-CMV following transfusion of seronegative and filtered units in this population are needed.

Leukoreduction for Other Herpesviruses

While other herpesviruses are also leukocytotropic, only EBV has been definitively shown to be transfusion-transmitted. Typically, transfusion-transmitted EBV infection is not a significant concern in adult patients because well over 90% have been previously infected and have residual immunity. However, since a large percentage of pediatric patients are EBV-seronegative, it becomes important to prevent primary EBV infection and potential sequelae, such as posttransplant lymphoproliferative disease (PTLD). In one report, a 16-year-old EBV-seronegative recipient of a seronegative liver transplant developed EBV-related PTLD after transfusion of EBV-containing blood components. Molecular analysis demonstrated that the EBV strain in the patient matched the strain in one of the implicated units of blood (Alfieri et al. 1996). While EBV transmission could theoretically be prevented by transfusing only seronegative blood, this approach is impractical since only a small fraction of donors are expected to be seronegative. In contrast, leukoreduction filtration removes 99.99% of B-cells, which harbor latent EBV (Roback et al. 2000). Thus, transfusion of leukodepleted blood should markedly decrease the risk of transfusion-transmitted EBV infection in at-risk pediatric patients, although this hypothesis has not been adequately tested.

The possibility that HHV-8, the etiologic herpesvirus of Kaposi's sarcoma and body-cavity based lymphomas, is transmissible by transfusion has been a concern (Hillyer et al. 1999). However, there have been no documented cases of transfusion transmitted HHV-8. Since this virus resides primarily in leukocytes, leukoreduction would be expected to provide some degree of protection.

Leukoreduction for Other Pathogens

Ehrlichia chaffeensis, the first identified human ehrlichial agent, infects monocytes, while the human granulocytic ehrlichiosis agent infects neutrophils (Dawson et al. 1991; Klein et al. 1997). HTLV-I and -II preferentially infect CD4+ and CD8+ T-cells, respectively. For each of these pathogens, filtration is likely to be somewhat effective at reducing the pathogen load before transfusion. The magnitude of this effect is not known.

The Effects of Leukoreduction on Prion Transmission

Prions, proteinacious infectious agents without detectable associated nucleic acids, cause variant Creutzfeldt-Jakob disease (vCJD) and other transmissible spongiform encephalopathies (TSEs). TSEs are incurable, rapidly progressive, and fatal diseases. Evidence for transfusion-transmission of prions is contradictory. The prion agents of bovine spongiform encephalopathy or natural scrapie can be transmitted by blood transfusion from experimentally infected sheep donors to naïve sheep recipients (Hunter et al. 2002). However, the distribution of prion proteins into

the WBC, RBC, platelet, and plasma compartments of blood varies from species to species, suggesting that the sheep studies may not serve as an accurate model for transfusion-transmission in humans. Furthermore, to date there have been no clinical epidemiological data suggesting that vCJD can be transmitted by transfusion in humans, as reviewed recently (Dodd and Busch 2002).

Although the data regarding prion transmission by transfusion are not conclusive, the association of prions with WBCs in human blood prompted a number of countries to institute universal leukoreduction as a method to mitigate the possibility of transfusion-transmitted prion disease. It will likely be many years before the efficacy of this measure can be determined.

Nucleic Acid-Based Pathogen Inactivation Strategies

There are limits to the protection against transfusion-transmitted infections provided by screened and leukoreduced blood components. For example, there is a "window phase" of 10 to 14 days following HIV infection where a blood donation would be potentially infectious but could not be identified as such using serology and NAT. Furthermore, there were no screening assays available at the time of the West Nile virus outbreak in 2002, which subsequently necessitated extensive quarantine of plasma components. These issues illustrate the need for alternative but complementary approaches to prevent infectious disease transmission by transfusion. The attractiveness of broad-spectrum sterilization methodologies derives from the fact that they should be equally effective against common, rarely occurring, and even currently unidentified pathogenic microorganisms that contaminate blood components.

With the exception of prions, all human pathogens contain DNA and/or RNA molecules that are critical to their replication and pathogenesis. Since the function of RBCs and platelets does not depend on nucleic acids, agents that target DNA/RNA represent approaches to inactivate both known and unknown pathogens in blood components.

Psoralens are a class of heterocyclic planar compounds that can intercalate into the DNA of viruses, bacteria, and other pathogens. After exposure to UV-A light (320 to 400 nm), the activated psoralens form photoadducts that crosslink DNA, blocking pathogen replication (Lin et al. 1997; Margolis-Nunno et al. 1997). Psoralens can also inactivate RNA through photoadduct formation. These agents are highly effective at blocking replication of HIV, HCV, and other pathogens in experimentally infected platelet units. Amotosalen HCl, a proprietary psoralen compound developed by Cerus Corporation and marketed as the Helinx

INTERCEPT Blood System, has received final European CE Mark approval for sterilization of platelet components and is in clinical use. This system is currently under regulatory review in the United States and not yet available for routine use. Cerus' INTERCEPT Blood System for RBCs is in Phase III clinical trials. INACTINE, an alternate compound developed by V.I. Technologies, is currently in Phase III clinical trials for sterilization of RBC components. While INACTINE was originally developed as a nucleic acid-targeting molecule, it also appears to reduce infectivity of prion proteins such as those that cause vCJD. While the mechanism of this effect is currently unclear, it may relate to the washing procedure employed following INACTINE treatment. Other agents for pathogen inactivation of blood components are also under investigation. For example, the porphyrin-like phthalocyanines are activated by long wavelength red light (650 to 700 nm), which can be efficiently delivered to red cell units without interference from hemoglobin. All of these agents inactivate viruses as well as parasites, including *Plasmodium falciparum* and *Trypanosoma cruzi* (Lustigman and Ben-Hur 1996). Based on the recent progress in developing reagents to sterilize cellular blood components, pathogen inactivation methodologies will likely soon be used as a complement to serological screening, NAT, and leukoreduction to reduce the incidence of transfusion-transmitted infections.

PLASMA COMPONENTS, DERIVATIVES, AND FACTORS PREPARED TO ELIMINATE PATHOGENS

The issues of screening discussed above also apply to plasma components and derivatives. However, since the therapeutic efficacy of plasma and derivatives depends on relatively stable soluble proteins and not labile cellular elements such as RBCs or platelets, there is more latitude in the use of methods to inactivate pathogens.

Solvent/Detergent (SD) Plasma and Coagulation Factor Concentrates

HIV, HCV, and many other transfusable pathogens are encapsulated by lipid envelopes that can be disrupted by detergents. Through combined use of detergents (Tween 80 or Triton X-100), solvents (ethyl ether or TNBP), and chromatography, lipid-enveloped pathogens can be inactivated in plasma while preserving significant protein function. Plasma is typically treated in large pools composed of 2500 units, followed by filtration to remove leukocytes, bacteria, and the largest von Willebrand factor multimers. Treated plasma is then dispensed into 200 mL units and frozen. Originally marketed under the name PLAS+ SD in 1998, the use of this product has been linked to a number of deaths, including six liver transplant patients in 2000. The FDA subsequently stated that PLAS+ SD should not be used in patients undergoing liver transplant or patients with liver disease or known clotting abnormalities. In addition, viruses such as hepatitis A virus (HAV) and parvovirus B19 do not contain lipid envelopes and are thus resistant to SD treatment. B19, which can be dangerous to infants particularly those with compromised immune systems, has been transmitted to at least two patients by SD-treated plasma and has necessitated the recall of 37 lots of plasma. PLAS+ SD is not currently being produced, and it is unclear whether manufacturing will be reinitiated.

The SD treatment technique is also effective at inactivating lipid-enveloped viruses in protein derivatives prepared from plasma, such as coagulation factors. For example, following treatment of coagulation factor concentrates, it is estimated that less than 1 in 10^{16}, 10^{13}, and 10^6 will contain infectious HIV, HBV, and HCV, respectively. Eleven million units of SD-treated plasma derivatives, such as antihemophilia factor concentrates, have been transfused through 1997 without reported transmission of lipid-enveloped viruses (Klein et al. 1998). However, HAV and B19 have been transmitted by SD-treated factor concentrates (Klein et al. 1998; Pehta 1996).

Nucleic Acid Targeted Inactivation Strategies

The Cerus Helinx System previously described can also be used to inactivate pathogens in plasma and is currently undergoing Phase III clinical trials for this use. Methylene blue represents an alternate reagent for sterilizing plasma. Methylene blue has been well accepted as a safe agent for intravascular infusion as treatment of methemoglobinemia. Following illumination, methylene blue, like other photosensitizing dyes, produces reactive oxygen species, which can then damage surrounding molecules, such as nucleic acids. In 2002, the United Kingdom Blood Services began to treat fresh frozen plasma (FFP) with methylene blue before infusion into neonates and young children. Among the cited indications were children undergoing heart surgery, liver transplant, or massive transfusion. Methylene blue-treated plasma has also been used in other European countries.

Heat-Treated Plasma Derivatives

Some contaminating viruses can be effectively inactivated by heat treatment. A variety of protocols have been used, including heating in solution (pasteur-ization; usually at 60°C for 10 hours), heating previously freeze-dried plasma derivatives (80°C), and heating in hot vapor at 60°C (Mannucci et al. 1992). Pasteurization produces a large decrease in viral load, with a minimum calculated reduction of 14.9-$\log_{10}$ for HIV and 6.47-$\log_{10}$ for a nonenveloped virus (Chandra et al. 1999). HBV and HCV are not as susceptible to heat treatment as HIV, and parvovirus B19 is even more resistant (Santagostino et al. 1997).

Recombinant Coagulation Factors

The molecular biology revolution has impacted transfusion safety through the production of recombi-nant coagulation factors in vitro, virtually eliminating concerns of infectious disease transmission. In multiple clinical trials in patients with hemophilia A, recombi-nant human factor VIII has been shown to be effective, with hemostatic activity profiles similar to those of plasma-derived factor VIII (VanAken 1997). Recombi-nant DNA technology also allows coagulation factors to be modified for greater clinical efficacy. For example, recombinant activated factor VII, marketed as Novo-Seven, is effective in patients with hemophilia A or B even in the presence of inhibitors (Lusher et al. 1998), as well as in patients with other hemostatic defects (Monroe et al. 2000). Coagulation factor concentrates are discussed in more detail in Chapter 3.

SUMMARY

While CMV-seronegative and filtered units are similar in their reduced risk of transmitting CMV infec-tions, recent studies suggest that seronegative compo-nents may be safer for susceptible patients. While these studies examined adult hematopoietic stem cell transplant recipients, cautious extrapolation of these results to neonates with immature or otherwise defi-cient immune systems appears warranted. In the pedi-atric population, use of filtered units should be effective at reducing the load of latent EBV that is transfused and the associated risk of sequelae such as posttransplant lymphoproliferative disease, although this has not been proven. Agents that disrupt pathogen replication, such as the INTERCEPT and INACTINE systems, may soon be approved for routine use and may markedly decrease the risks of transfusion-transmitted infections. Coagula-tion factors and other plasma derivatives currently have a very low risk of pathogen transmission, which in some cases had been reduced to negligible levels through in vitro expression of recombinant constructs.

References

Alfieri C, Tanner J, Carpentier L, Perpete C, Savoie A, Paradis K, Delage G, and Joncas J. 1996. Epstein-Barr virus transmission from a blood donor to an organ transplant recipient with recovery of the same virus strain from the recipient's blood and oropharynx. *Blood* 87:812–817.

American Association of Blood Banks: Association Bulletin #97-2; Leukocyte reduction for the prevention of transfusion-transmitted cytomegalovirus (TT-CMV); April 23, 1997.

Blajchman MA, Goldman M, Freedman JJ, and Sher GD. 2001. Pro-ceedings of a consensus conference: prevention of post-transfusion CMV in the era of universal leukoreduction. *Transfus Med Rev* 15:1–20.

Bolovan-Fritts CA, Mocarski ES, and Wiedeman JA. 1999. Peripheral blood CD14(+) cells from healthy subjects carry a circular con-formation of latent cytomegalovirus genome. *Blood* 93:394–398.

Bowden RA, Slichter SJ, Sayers M, Weisdorf D, Cays M, Schoch G, Banaji M, Haake R, Welk K, Fisher L, et al. 1995. A comparison of filtered leukocyte-reduced and cytomegalovirus (CMV) seronegative blood products for the prevention of transfusion-associated CMV infection after marrow transplant. *Blood* 86:3598–3603.

Chandra S, Cavanaugh JE, Lin CM, Pierre-Jerome C, Yerram N, Weeks R, Flanigan E, and Feldman F. 1999. Virus reduction in the preparation of intravenous immune globulin: in vitro experiments [In process citation]. *Transfusion* 39:249–257.

Dawson JE, Anderson BE, Fishbein DB, Sanchez JL, Goldsmith CS, Wilson KH, and Duntley CW. 1991. Isolation and characterization of an Ehrlichia sp. from a patient diagnosed with human ehrli-chiosis. *J Clin Microbiol* 29:2741–2745.

de Cates CR, Roberton NR, and Walker JR. 1988. Fatal acquired cytomegalovirus infection in a neonate with maternal antibody. *J Infect* 17:235–239.

Dodd RY and Busch MP. 2002. Animal models of bovine spongiform encephalopathy and vCJD infectivity in blood: two swallows do not a summer make [Comment]. *Transfusion* 42:509–512.

Hillyer CD, Lankford KV, Roback JD, Gillespie TW, and Silberstein LE. 1999. Transfusion of the HIV seropositive patient: immunomodulation, viral reactivation, and limiting exposure to EBV (HHV-4), CMV (HHV-5) and HHV-6, 7, and 8. *Transfusion Med Rev* 13:1–17.

Hunter N, Foster J, Chong A, McCutcheon S, Parnham D, Eaton S, MacKenzie C, and Houston F. 2002. Transmission of prion diseases by blood transfusion. *J General Virology* 83:2897–2905.

Klein HG, Dodd RY, Dzik WH, Luban NL, Ness PM, Pisciotto P, Schiff PD, and Snyder EL. 1998. Current status of solvent/detergent-treated frozen plasma. *Transfusion* 38:102–107.

Klein MB, Miller JS, Nelson CM, and Goodman JL. 1997. Primary bone marrow progenitors of both granulocytic and monocytic lineages are susceptible to infection with the agent of human granulocytic ehrlichiosis. *J Infect Dis* 176:1405–1409.

Larsson S, Soderberg-Naucler C, Wang FZ, and Moller E. 1998. Cytomegalovirus DNA can be detected in peripheral blood mononuclear cells from all seropositive and most seronegative healthy blood donors over time. *Transfusion* 38:271–278.

Lin L, Cook DN, Wiesehahn GP, Alfonso R, Behrman B, Cimino GD, Corten L, Damonte PB, Dikeman R, Dupuis K, et al. 1997. Photochemical inactivation of viruses and bacteria in platelet concentrates by use of a novel psoralen and long-wavelength ultraviolet light. *Transfusion* 37:423–435.

Luban NL, Williams AE, MacDonald MG, Mikesell GT, Williams KM, and Sacher RA. 1987. Low incidence of acquired cytomegalovirus infection in neonates transfused with washed red blood cells. *Am J Dis Child* 141:416–419.

Lusher JM, Roberts HR, Davignon G, Joist JH, Smith H, Shapiro A, Laurian Y, Kasper CK, and Mannucci PM. 1998. A randomized, double-blind comparison of two dosage levels of recombinant factor VIIa in the treatment of joint, muscle, and mucocutaneous haemorrhages in persons with haemophilia A and B with and without inhibitors. rFVIIa Study Group, *Haemophilia* 4:790–798.

Lustigman S and Ben-Hur E. 1996. Photosensitized inactivation of *Plasmodium falciparum* in human red cells by phthalocyanines. *Transfusion* 36:543–546.

Mannucci PM, Schimpf K, Abe T, Aledort LM, Anderle K, Brettler DB, Hilgartner MW, Kernoff PB, Kunschak M, McMillan CW, et al. 1992. Low risk of viral infection after administration of vapor-heated factor VIII concentrate. International Investigator Group. *Transfusion* 32:134–138.

Margolis-Nunno H, Bardossy L, Robinson R, Ben-Hur E, Horowitz B, and Blajchman MA. 1997. Psoralen-mediated photodecontamination of platelet concentrates: inactivation of cell-free and cell-associated forms of human immunodeficiency virus and assessment of platelet function in vivo. *Transfusion* 37:889–895.

Miller WJ, McCullough J, Balfour HH, Jr, Haake RJ, Ramsay NK, Goldman A, Bowman R, and Kersey J. 1991. Prevention of cytomegalovirus infection following bone marrow transplantation: a randomized trial of blood product screening. *Bone Marrow Transplant* 7:227–234.

Monroe DM, Hoffman M, Allen GA, and Roberts HR. 2000. The factor VII-platelet interplay: effectiveness of recombinant factor VIIa in the treatment of bleeding in severe thrombocytopathia. *Sem Thrombosis Hemostasis* 26:373–377.

Nichols WG, Price TH, Gooley T, Corey L, and Boeckh M. 2003. Trans-fusion-transmitted cytomegalovirus infection after receipt of leukoreduced blood products. *Blood* 101:4195–4200.

Pehta JC. 1996. Clinical studies with solvent detergent-treated products. *Transfus Med Rev* 10:303–311.

Preiksaitis JK. 1991. Indications for the use of cytomegalovirus-seronegative blood products. *Transfus Med Rev* 5:1–17.

Roback JD, Bray RA, and Hillyer CD. 2000. Longitudinal monitoring of WBC subsets in packed RBC units after filtration: implications for transfusion transmission of infections. *Transfusion* 40:500–506.

Roback JD, Hillyer CD, Drew WL, Laycock ME, Luka J, Mocarski ES, Slobedman B, Smith JW, Soderberg-Naucler C, Todd DS, et al. 2001. Multicenter evaluation of PCR methods for detecting CMV DNA in blood donors. *Transfusion* 41:1249–1257.

Santagostino E, Mannucci PM, Gringeri A, Azzi A, Morfini M, Musso R, Santoro R, and Schiavoni M. 1997. Transmission of parvovirus B19 by coagulation factor concentrates exposed to 100 degrees C heat after lyophilization. *Transfusion* 37:517–522.

Sindre H, Tjoonnfjord GE, Rollag H, Ranneberg-Nilsen T, Veiby OP, Beck S, Degre M, and Hestdal K. 1996. Human cytomegalovirus suppression of and latency in early hematopoietic progenitor cells. *Blood* 88:4526–4533.

VanAken WG. 1997. The potential impact of recombinant factor VIII on hemophilia care and the demand for blood and blood products. *Transfus Med Rev* 11:6–14.

Yeager AS, Grumet FC, Hafleigh EB, Arvin AM, Bradley JS, and Prober CG. 1981. Prevention of transfusion-acquired cytomegalovirus infections in newborn infants. *J Pediatr* 98:281–287.

Zwicky C, Tissot JD, Mazouni ZT, Schneider P, and Burnand B. 1999. Prevention of post-transfusion cytomegalovirus infection: recommendations for clinical practice. *Schweiz Med Wochenschr* 129:1061–1066.

9

Irradiated Products

EDWARD C.C. WONG, MD

HISTORY

Irradiation of blood products is necessary to prevent transfusion-associated graft-versus-host disease (TA-GVHD), an often fatal immunological complication. This complication of blood transfusion was first reported in the 1960s in individuals with hematological malignancies and infants with congenital immunodeficiencies who received blood and then developed what was called runt disease (von Fliedner, Higby, and Kim 1972).

Despite a better understanding of the pathogenesis of TA-GVHD and the advent of preventive strategies (Anderson and Weinstein 1990; Webb and Anderson 2000; Linden and Pisciotto 1992), there are still many unanswered questions. There are no adequate estimates of prevalence or incidence in the United States, as most at-risk patients receive irradiated products. However, in Japan, the Japanese Red Cross GVHD Study Group and registry has estimated approximately 40 cases per year based on the homozygosity for one-way human leukocyte antigen (HLA) haplotype sharing in Japan, the use of familial donors, and the use of fresh rather than stored red blood cells (RBCs) (Ohto and Anderson 1996).

PATHOGENESIS

Three prerequisites for the development of GVHD have been proposed in the transplant setting: (1) differences in histocompatibility between recipient and donor; (2) presence of immunocompetent cells in the graft; and (3) inability of the host to reject the immuno-

competent cells. In TA-GVHD, a similar set of circumstances can occur. Most immunocompetent recipients will destroy the donor-derived T cells through lymphocytolysis. However, transfusion from an HLA homozygous donor into an HLA heterozygous recipient who shares one HLA haplotype with the recipient may result in a failure of recognition of the donor cells as being foreign and engraftment ensues (McMilan and Johnson 1993; Petz et al. 1993). In both cases, the foreign major histocompatibility complex (MHC) antigens and/or minor histocompatibility antigens (minor HA) of the host stimulate clonal T-cell expansion and the induction of an inflammatory response with cytokine release that is ultimately responsible for the induction of the process and clinical manifestations of the disorder.

Ferrara and Krenger (1998) have described the Th1/Th2 paradigm, which delineates the cellular origins and differential biologic functions of T-cell–derived cytokines based on murine CD4 differentiation. Th1 CD4 T cells secrete IL-2 as their allostimulator. Th2 CD4 T cells produce IL-4, IL-5, IL-6, IL-10, IL-13, and lesser amounts of TNFα, while Th1 and Th2 both produce IL-3 and granulocyte-macrophage colony stimulating factor. The type 1 cell is pro-inflammatory and induces cell-mediated immunity, while the type 2 cell is considered anti-inflammatory. Differentiation toward type 1 or type 2 is a complicated process that involves early exposure to IL-4 or IL-12, the type of antigen-presenting cells, costimulating molecules, and the presence of macrophages and their unique cytokines. Based on their mouse work, Ferrera and Krenger (1998) propose a three-step process for the development of acute GVHD in the transplant setting that may be comparable to the development of TA-GVHD. In this model, host tissues are damaged through irradiation or chemotherapy and secrete TNFα and IL-1, which enhance recognition of host MHC and/or minor HLA antigens by donor T cells. Donor T-cell activation results in proliferation of Th1 T cells and secretion of IL-2 and TNFα, which in turn activate T cells further and induce cytotoxic T lymphocytes and natural killer (NK) responses. Whether there is a direct comparability to TA-GVHD remains a question. Subsequently, additional donor and residual host phagocytes are stimulated to produce IL-1 and TNFα. Stimulated macrophages release the free radical nitric oxide (NO), which has further deleterious effects on host tissues. In addition, NO upregulates alloreactivity as well as mediates the cytotoxic function of macrophages (Worrall et al. 1995). A secondary triggering signals lipopolysaccharides (LPS) to stimulate gut-associated macrophages, lymphocytes, keratocytes, and dermal fibroblasts and further promotes the inflammatory response and end organ damage that are classical hallmarks of the disorder.

The importance of CD4 and CD8 cells in the pathogenesis of TA-GVHD has been studied by Fast et al. (1995) in a mouse model and by Nishimura et al. (1997) in a patient with TA-GVHD and is further supported by clinical correlation with HIV/AIDS (Ammann 1993). In the mouse, depletion of CD4+ cells increased the number of donor cells needed to induce TA-GVHD, while depletion of CD8+ and/or NK cells decreased the number of donor cells needed to induce the disorder. In HIV/AIDS, there has been only one report of TA-GVHD (Klein et al. 1996), despite widespread use of supportive transfusion and profound immunosuppression. Early depletion of CD4 may well protect against establishment of GVHD. Alternatively, activation of CD8+ lymphocytes against HIV-infected CD4 T cells may limit the development of the GVHD process (Ammann 1993). Studies by Kruskal et al. (2001) failed to demonstrate microchimerism in HIV-infected subjects enrolled in the Viral Activation Transfusion Study, suggesting that HIV patients are not at risk for TA-GVHD.

CLINICAL MANIFESTATIONS

Fever, anorexia, nausea, vomiting, and diarrhea are seen. Skin manifestations are variably severe and begin as an erythematous maculopapular eruption that may proceed to erythroderma with bullae and frank desquamation. Gastrointestinal bleeding is commonly seen, most usually as bloody diarrhea. Hepatic dysfunction with hyperbilirubinemia, with a progressively increasing direct fraction, is also seen. TA-GVHD differs from acute GVHD in the setting of allogeneic transplanta-

tion in that the pancytopenia seen is severe and often results in the death of the patient. Diagnosis is often made postmortem and is based on pathognomonic histopathological findings of lymphocyte infiltration in skin, lymph nodes, liver, and the gastrointestinal track (Brubaker 1993).

DIAGNOSIS OF TA-GVHD

Clinical suspicion may warrant a skin biopsy. Skin biopsy often reveals vacuolization of the epidermal basal cell layer, dermal-epithelial layer separation, and formation of bullae. Other findings include mononuclear cell migration into the epidermis, hyperkeratosis, and dyskeratosis. Liver biopsies may reveal eosinophilic infiltration and degeneration of small bile ducts, peripheral inflammation, and lymphocyte infiltration. The bone marrow will demonstrate what is classically described as an "empty marrow" with pancytopenia, fibrosis, and some lymphocytic infiltration.

Definitive confirmation is more complicated. Several methods have been utilized to identify lymphocytes of foreign origin in the circulation of a suspected patient or in affected tissue. Serological HLA typing, DNA-based HLA class II typing, karyotype analysis, restriction length polymorphism analysis using probes from both HLA and non-HLA regions, and genetic fingerprinting using polymorphic microsatellite markers have all been used (Blundell et al. 1992; Capon et al. 1991; Drobyski et al. 1989; Kunstmann et al. 1992; DePalma et al. 1994). Fibroblast and buccal mucosal cells of the recipient are often needed as the lymphocytolysis accompanying the disorder prohibits standardized serological HLA typing. Parental or familial specimens may be necessary to deduce a recipient's HLA type (Wang et al. 1994). Donor lymphocytes obtained from suspected blood products present in attached remaining blood bag segments often require polymerase chain reaction (PCR) amplication and sequence specific oligonucleotide probe (SSOP) methodologies to provide confirmation of donor cell origin (Friedman et al. 1994).

TIME COURSE OF MICROCHIMERISM

In 1995, Busch and Lee demonstrated a thousandfold expansion of donor lymphocytes in the circulation of otherwise healthy recipients 3 to 5 days following transfusion for elective orthopedic procedures; within 2 weeks, the allogeneic cells were cleared (Lee et al. 1996). In another study of adult trauma victims receiving large numbers (4 to 18 units) of fresh packed red blood cells (PRBCs), eight of ten had confirmed microchimerism. Two of the eight still had persistence of microchimerism when studied as long as one and a half years posttransfusion (Lee et al. 1999). Kruskall et al. (1995) in a study of 93 HIV-infected women who had received blood from male donors found that five of 47 women randomized to receive nonleukoreduced RBCs had detectable circulating male donor lymphocytes 1 to 2 weeks after transfusion and undetectable male donor lymphocytes after 4 weeks. This was in contrast to 46 HIV-infected women who did not have detectable male donor lymphocytes within a month of transfusion (Kruskall et al. 1995). Wang-Rodriquez et al. (2000) studied posttransfusion immune modulation in 14 premature infants. Through collaboration with Busch and Lee, two of six female infants were identified, transfused with nonleukodepleted RBCs, who developed transient microchimerism detected by Y-chromosome PCR amplification; both cleared these cells by 2 weeks posttransfusion. An additional three infants who received leukodepleted RBCs also had transient microchimerism. Vietor et al. (2000) studied nine surviving recipients of intrauterine transfusion whose donors were still available for testing. Using fluorescent in situ hybridization (FISH), PCR of Y chromosome-specific sequences, and assays for the frequencies of cytotoxic T lymphocyte and T helper-lymphocyte precursors, they detected true microchimerism in six out of seven young adults studied 20 years posttransfusion. Recently, Reed et al. (2001) have developed sequence-specific amplification of DRB1, which permitted identification of minor chimeric populations at the 0.01% level. The establishment of stable microchimerism and identification of its biological consequences is critical for pediatric patients who are expected to live to adulthood and may well be stable transfusion-induced chimeras, an intriguing and, at the same time, worrying concept. The persistence of microchimerism may predispose to autoimmune disease (Nelson et al. 1998; Arlett, Smith, and Jimenez 1998; Evans et al. 1999), chronic GVHD, recurrent abortion (Daya, Gunby, and Clark 1998), and may serve as an allogeneic stimulate of latent viral reactivation in the recipient.

GROUPS AT RISK

Patients who may develop TA-GVHD have been described in a number of recent reviews (Anderson and Weinstein 1990; Webb and Anderson 2000; Linden and Pisciotto 1992; McMilan and Johnson 1993; Petz et al. 1993; Williamson and Warwick 1995), and there has been a growing concern as to the types of patients who may be affected. For example, based on two reports of

TA-GVHD development in severe combined immunodeficiency in older infants (Klein et al. 1996; Friedman et al. 1994), recommendations have extended the age range of irradiation well past the neonatal age group (Luban and De Palma 1996; Ohto and Anderson 1996). Given the lack of adequate animal models and definite laboratory tests describing individual TA-GVHD risk, many reports stratify the need for irradiation using such terms as "clearly indicated" or "probably indicated" (Webb and Anderson 2000; Williamson and Warwick 1995; Kanter 1992). In reality, the spectrum of individuals at risk will likely grow as intensive immunomodulatory therapies expand beyond oncological disease and transplantation (Table 9.1).

THE IRRADIATION PROCESS

Among methodologies (for example, photoinactivation, peglyation, ultraviolet light, and irradiation) that can be used to prevent TA-GVHD, only irradiation of whole blood and cellular components is currently accepted practice by the Food and Drug Administration (FDA). Irradiation of cellular components with ionizing radiation results in the inactivation of T lymphocytes by damaging nuclear DNA either directly or by generating ions and free radicals that have damaging biological actions. This prevents posttransfusion donor T-cell proliferation in response to host antigen-presenting cells,

TABLE 9.1   Clinical Indications for Irradiated Products

Fetus/Infant

Intrauterine transfusion
Premature infants
Congenital immunodeficiency
Those undergoing exchange transfusion for erythroblastosis

Child/Adult

Congenital immunodeficiency
Hematological malignancy or solid tumor (neuroblastoma, sarcoma, Hodgkin's disease receiving ablative chemo/radiotherapy)
Recipient of PBSC, marrow, or cord blood
Recipient of familial blood donation
Recipient of HLA-matched products
Lupus or any other condition requiring fludarabine

Potential Indications

Term infant
Recipient and donor pair from a genetically homogeneous population
Other patients with hematological malignancy or solid tumor receiving immunosuppressive agents

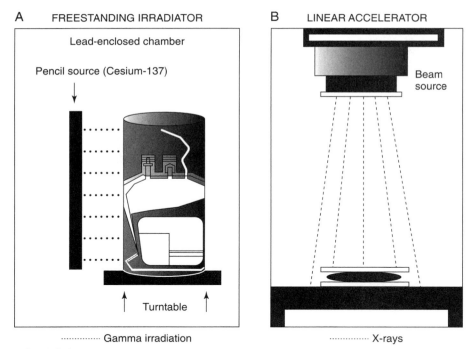

FIGURE 9.1 Diagrammatic views of two common types of instrumentation used for blood irradiation. *A*, Configuration of a freestanding irradiator using a cesium 137 source. *B*, Configuration of a linear accelerator.

which, in turn, abrogates GVHD (Shlomchik et al. 1995; Davey 1992; Fearon and Luban 1986). Two types of ionizing radiation, γ-rays and x-rays, are equivalent in inactivating T lymphocytes in blood components at a given absorbed dose. Gamma rays originate from the radioactive decay process within the atomic nucleus of cesium 137 (^{137}Cs) or cobalt 60 (^{60}Co). Freestanding blood bank gamma irradiators, which are the predominant instruments for blood component irradiation, use either of these two isotopes as an irradiation source. In contrast, x-rays are generated from the interaction of a beam of electrons with a metallic surface.

Linear accelerators that generate x-rays for patient therapy (teletherapy) may serve as an irradiation source for blood and blood components. Recently, the FDA has also approved the use of a freestanding x-ray machine (No-dion International, Ontario, Canada) for irradiation of blood components. This instrument does not require federal or Nuclear Regulatory Agency licenses.

INSTRUMENTATION FOR IRRADIATION

The basic operating principles and configurations of a freestanding irradiator with either a cesium (^{137}Cs) source or a linear accelerator are shown schematically

in Figure 9.1. With a freestanding ^{137}Cs irradiator, blood components are contained within a metal canister that is positioned on a rotating turntable. Continuous rotation allows for the γ-rays, originating from one to four closely positioned pencil sources, to penetrate all portions of the blood component. The number of sources and their placement depend on the instrument and model. The speed of rotation of the turntable also depends on the make or model of the instrument. A lead shield encloses the irradiation chamber. Freestanding irradiators employing ^{60}Co as the source of γ-rays are comparable except that the canister containing the blood component does not rotate during the irradiation process; rather, tubes of ^{60}Co are placed in a circular array around the entire canister within the lead chamber. When freestanding irradiators are used, the γ-rays are attenuated as they pass through air and blood but at different rates (Fearon and Luban 1986). The magnitude of attenuation is greater with ^{137}Cs than with ^{60}Co sources.

Linear accelerators generate a beam of x-rays over a field of given dimension. Routinely, the field is projected on a table-top structure. The blood component is placed (flat) between two sheets of biocompatible plastic several centimeters thick. The plastic on the top of the blood component (that is, nearer to the radiation source) generates electronic equilibrium of the secondary electrons at the point where they pass through the component container. The plastic sheet on the bottom of the blood component provides for irradiation backscattering that helps to ensure the homogenous delivery of the x-rays. The blood component is usually left stationary when the entire x-ray dose is being delivered. Alternatively it may be flipped over when half of the dose has been delivered; this process involves turning off and restarting the linear accelerator during the irradiation procedure. Although it seems as if the practice of flipping is not required, further data are needed.

COMPONENTS TO BE IRRADIATED

The single most important characteristic of a blood component's ability to induce TA-GVHD is its white blood cell (WBC) content and, specifically, its lymphocyte content. Based on animal models and estimates from the bone marrow transplant literature 5×10^4 to 1×10^5 T cells per kg in an ablated host will induce GVHD (Korngold 1993), while a greater number is likely needed in a nonablated host. The content of lymphocytes in each blood component differs based on the donor's initial lymphocyte count, the method of collection, and any postcollection manipulation and processing. In general, there are likely sufficient lymphocytes present in almost all blood products to induce TA-GVHD in a susceptible recipient (Table 9.2). Over storage, fewer lymphocytes can be isolated from both RBC and platelet concentrates. Detail on the nature of

lymphocyte subsets and the molecular changes occurring during storage in different blood products has not been well studied, making risk assessment between a specific class of product and a subcategory of at-risk patients impossible. Further, there could be a cumulative or synergistic effect from viable T cells present in the multiple transfusions received by a given patient whose own immunological status fluctuates with time from treatment and infectious disease state.

For patients at risk for GVHD, all components that might contain viable T lymphocytes should be irradiated. These components are listed in Table 9.3. All types of red cells suspended in citrated plasma or in an additive solution, postfreezing and -thawing (Suda, Leitman, and Davey 1993; Miraglia, Anderson, and Mintz 1994; Crowley, Skrabut, and Valeri 1974), and filtered products should be irradiated. Although extensive leukoreduction through filtration may decrease the potential for GVHD, it is not a substitute as there have been reports of TA-GVHD in patients who have received leukodepleted (filtered) red cells; however, the extent of leukoreduction of the components was not uniformly quantified in such reports (Akahoshi et al. 1992; Heim et al. 1991; Hayashi et al. 1993; Anderson 1995).

In addition, to the number and specific subtype of T lymphocytes present in a product, patient immunocompetence at the time of transfusion may also influence the development of TA-GVHD. It is likely that the greater the degree of immunosuppression, the fewer the viable

TABLE 9.2 White Blood Cell (WBC) Content of Different Blood Components

Component	Volume (mL)	Average WBC Content
Whole blood	450	$1-2 \times 10^9$
Red blood cells (RBCs)	250	$2-5 \times 10^9$
Washed RBCs	Variable	$<5 \times 10^8$
Deglycerolized RBCs	250	$\sim 10^7$
Platelet concentrate	50-75	4×10^7
Plateletpheresis unit	200-500	$3 \times 10^{8*}$
Cryoprecipitate	25	0
Fresh frozen plasma	125	0
Pediatric frozen plasma	Variable	0
Liquid plasma	125	1.5×10^5
Single-donor plasma	125	0
Granulocyte concentrate	200-500	1×10^{10}

Modified from Luban NLC. Basics of transfusion medicine. In Fuhrman BP, Zimmerman JJ, eds. 1992. *Pediatric critical care.* St. Louis: Mosby.

*Less with new modified chambers.

TABLE 9.3 Blood Components Requiring Irradiation for Patients at Risk of GVHD

All components that might contain viable T lymphocytes (including products collected by apheresis)
Whole Blood

Cellular components
  Red cells (regardless of anticoagulant or preservative)
  Leukofiltered red cells
  Platelet concentrates
  Leukofiltered platelet concentrates
  Previously frozen red blood cells
  Nonfrozen plasma (fresh plasma)
  Granulocyte concentrates

Questionable Components

  Fresh frozen plasma
  Frozen plasma

Unlikely to Contain Viable T Lymphocytes

  Cryoprecipitate
  PLAS+®SD—Plasma

T lymphocytes that will be required to produce GVHD in susceptible patients. In a recent review, it was suggested that cytotoxic T lymphocytes or interleukin-2 secreting precursors of helper T lymphocytes may be more predictive of GVHD than the number of proliferating T cells alone. Accordingly, this suggests that until further data are available to confirm adequate removal of these T-cell subtypes by leukoreduction, irradiation should be used for blood products destined for patients at risk for GVHD (Anderson 1995).

Blood components given to recipients, whether immunocompromised or immunocompetent, containing lymphocytes that are homozygous for an HLA haplotype shared with the recipient, pose a specific risk for TA-GVHD. This circumstance occurs when first- and second-degree relatives serve as directed donors (McMilan and Johnson 1993; Petz et al. 1993; Kanter 1992; Ohto and Anderson 1996) and when HLA-matched platelet components donated by related or unrelated individuals are transfused (Benson et al. 1994; Grishaber, Birney, and Strauss 1993). Irradiation of blood components must be performed in these situations.

Platelet components that have low levels of leukocytes because of their collection through the apheresis process and/or leukofiltration should also be irradiated if intended for transfusion to susceptible patients. This is because the minimum number of T lymphocytes that induces TA-GVHD has not yet been delineated.

In contrast, there is controversy over fresh frozen plasma. It is generally accepted that the freezing and thawing processes destroy the T lymphocytes that are present in such plasma. Two brief articles suggested that immunocompetent progenitor cells may be present in frozen-thawed plasma. These authors recommended that frozen-thawed plasma be irradiated (Wielding et al. 1994; Bernvill et al. 1994). Further studies are needed to validate these findings and to assess whether the number of immunocompetent cells, which may be present in thawed fresh frozen plasma, is sufficient to induce GVHD. In rare instances, when nonfrozen plasma (termed *fresh plasma*) is transfused, it should be irradiated because of the presence of a sizable number of viable lymphocytes, approximately 1×10^7 cells, in a component prepared from a unit of whole blood.

STORAGE OF RED CELLS AND PLATELETS AFTER IRRADIATION

Red Blood Cells

Irradiation of RBCs is not a benign process. The in vivo viability of irradiated red cells, evaluated as the 24-hour recovery, is reduced during storage when compared with nonirradiated red cells (Davey et al. 1992; Mintz and Anderson 1993; Moroff et al. 1992; Friedman, McDonough, and Cimino 1991; Moroff 1999). This has raised questions concerning the maximum storage time for red cells after irradiation. Davey et al. (1992) found that following irradiation with 3000 cGy on day 0, the mean ± SD 24-hour recovery for Adsol-preserved red cells after 42 days of storage was 68.5 ± 8.1% compared with 78.4 ± 7.1% for control, nontreated red cells. Subsequent studies employed total storage periods of between 21 and 35 days after day 0 or day 1 irradiation. After storage for 35 days, the mean (±SD) 24-hour recovery for irradiated (3000 cGy) and control Adsol red cells was 78.0 ± 6.8% and 81.8 ± 4.4%, respectively. In studies with a 28-day storage period, the values for irradiated (2500 cGy) and control Adsol red cells were 78.6 ± 5.9% and 84.2 ± 5.1%, respectively (Moroff et al. 1992). With Nutricel preserved red cells treated with 2000 cGy on day 1, mean 24-hour recovery for control and irradiated red cells were 90.4% and 82.7% after 21 days of storage and 85.0% and 80.7% after 28 days of storage (Friedman, McDonough, and Cimino 1991). Moroff et al. (1999) evaluated the effect of irradiation on Adsol red cells stored from day 1 to 28 (irradiated day 1, protocol 1), day 14 to 28 (irradiated day 14, protocol 2), day 14 to 42 (irradiated day 14, protocol 3), and day 26 to 28 (irradiated day 26, protocol 4), respectively. In comparison to previous investigations, this study was unique because red cells were studied after being irradiated for various times in storage and then studied after further storage. Protocol 1 mean ± SD recovery was 84.2 ± 5.1% for control RBCs and 78.6 ± 5.9% for irradiated RBCs (n = 16; p < 0.01). With protocol 3, the recoveries were 76.3 ± 7.0% for control RBCs and 69.5 ± 8.6% for irradiated RBCs (n = 16; p < 0.01). Protocols 2 and 4 demonstrated comparable 24-hour recoveries between control and irradiated RBCs. Long-term survival between control and irradiated RBCs were comparable in all protocols confirming previous data that the long-term survival of RBCs is minimally influenced by irradiation. Based on multiple linear regression analysis only the length of storage after irradiation had a significant effect on the 24-hour recovery. No effect was observed with day of irradiation or total storage time. In another study by Moroff et al. (1999) in vitro red cell properties such as adenosine triphosphate (ATP) levels and the amount of hemolysis were altered to only a small extent relative to control values with extended storage after irradiation, and it was found that potassium leakage from the red cells during storage is substantially enhanced by irradiation (Davey 1992; Miraglia, Anderson, and Mintz 1994; Ramirez et al. 1987; Rivet, Baxter, and Rock 1989). Despite elevated potassium levels, the association between irradiation induced changes in RBC viability and potassium leakage was not complimentary. Based on analysis of these studies, the FDA guidelines call for a 28-day maximum storage period for red cells after irradiation, irrespective of the day of storage on which the treatment was performed, with the proviso that the total storage time cannot exceed that for nonirradiated red cells.

Neonatal RBC Transfusion Concerns

Irradiated red cells undergo an enhanced efflux of potassium during storage at 1° to 6°C (Ramirez et al. 1987; Rivet, Baxter, and Rock 1989). Comparable levels of potassium leakage occur with or without prestorage leukoreduction (Swann and Williamson 1996). Washing of units of red cells before transfusion to reduce the supernatant potassium load does not seem to be warranted for most red cell transfusions because posttransfusion dilution prevents increases in plasma potassium (Strauss 1990). On the other hand, when irradiated red cells are used for neonatal exchange transfusion or the equivalent of a whole blood exchange is anticipated, red cell washing should be considered to prevent the possible adverse cardiotoxicity caused by hyperkalemia associated with irradiation and storage (Luban, Strauss, and Hume 1991).

Platelets

In contrast to red cells, platelets appear to be unaffected by irradiation. The storage period at 20° to 24°C for irradiated platelet components does not need to be modified. Both in vitro and in vivo platelet properties are not influenced to any extent by irradiation. Many studies have confirmed that platelet properties are retained immediately after conventional levels of irradiation and at the conclusion of a 5-day storage period, whether irradiation is performed prestorage or midstorage (Moroff et al. 1986; Espersen et al. 1988; Duguid et al. 1991; Read et al. 1988; Rock, Adams, and Labow 1988; Sweeney, Holme, and Moroff 1994; Seghatchian and Stivala 1995; Bessos et al. 1995). One recent report indicated some differences in selected in vitro parameters between irradiated and control platelets after storage (Seghatchian and Stivala 1995).

Granulocytes

Several studies have suggested that irradiation at doses recommended for irradiation of blood components does not affect granulocyte function. Wolber et al. (1987) found that there was no effect of irradiation

doses between 1500 and 2500 cGy on the ability of neutrophils to generate intracellular hydrogen peroxide in the conversion of the nonfluorescent compound 2'-7'-dichlorofluorescein into fluorescent 2'-7'-dichlorofluorescein. Wheeler et al. (1984) found that buffy coat-separated polymorphonuclear leukocytes when treated with 1500 cGy also did not demonstrate oxidative or migratory differences in chemiluminescence or chemotaxis in agarose as compared to nonirradiated controls. Patrone et al. (1979) found that granulocytes harvested by continuous flow centrifugation using the Aminco Celltrifuge and irradiated at 1500 cGy did not demonstrate any differences in chemotaxis compared to granulocytes in precollection venous samples. It is only when high irradiation doses such as 10,000 to 40,000 cGy are reached that functions such as granulocyte motility and bactericidal activity are significantly affected (Valerius et al. 1981).

SELECTION OF RADIATION DOSE

Studies utilizing a sensitive limiting dilution assay (LDA) indicate that 2500 cGy (measured at the internal midplane of a component) is the most appropriate dose (Pelszynski et al. 1994; Luban et al. 2000). In these experiments, red cell and platelet components were irradiated in their original plastic containers (blood bags) with increasing doses of radiation. After each irradiation dose, the LDA samples were removed and the clonogenic proliferation of T lymphocytes was measured in the system. With red cell units, 500 cGy had a minimal influence, whereas 1500 cGy inactivated T lymphocyte proliferation by approximately four logs; however, some growth was still observed in each experiment. Increasing the dose to 2000 cGy resulted in no T-lymphocyte proliferation in all but one experiment. No growth was observed after 2500 cGy (Pelszynski et al. 1994). In a subsequent study that used platelet apheresis components with sufficient T lymphocytes to perform the LDA, the influence of 1500 cGy and 2500 cGy was evaluated (Luban et al. 1994). With 1500 cGy, substantial inactivation was measured; however, some growth was still observed in all experiments. As noted with the red cell experiments, 2500 cGy resulted in complete abrogation of clonogenic T-lymphocyte proliferation. LDA measures the clonigenic potential of both CD4+ and CD8+ T cells in a functional assay. It provides a quantification at low T-cell numbers. It has been used to determine residual, functional T cells in bone marrows purged of T cells, thus providing a clinical correlate of prevention of GVHD (Quinones et al. 1993). Assays of T-cell proliferation using mixed lymphocyte culture or mitogens or detection of T cells by flow cytometry can detect up to a two-log reduction. PCR techniques are capable of detecting up to a six-log reduction but cannot distinguish viable versus nonviable cells and hence are noninformative if cells are inactivated. LDA assay, however, may fail to detect an as yet undescribed human T-cell subset that contributes to GVHD, but despite this limitation it is believed that the current studies provide experimental evidence for the selection of irradiation doses for plateletpheresis and red cell components to abrogate TA-GVHD. Currently, the FDA has recommended that the irradiation process should deliver 2500 cGy to the internal midplane of a freestanding irradiation instrument canister, with a minimum of 1500 cGy at any other point within the canister (Center for Biologics Evaluation and Research 1993).

QUALITY ASSURANCE MEASURES

One must document that the instrument being used for irradiation is operating appropriately and confirm that blood components have been irradiated. To ensure that the irradiation process is being conducted correctly, specific procedures are recommended for free-standing irradiators and linear accelerators, which are summarized in Table 9.4 and are detailed in a recent review (Moroff and Luban 1997).

Dose mapping measures the delivery of radiation within a simulated blood component or over an area in which a blood component is placed. This applies to an irradiation field when a linear accelerator is used or to the canister of a freestanding irradiator. Dose mapping is the primary means of ensuring that the irradiation process is being conducted correctly. It documents that the intended dose of irradiation is being delivered at a specific location (such as the central midplane of a canister), and it describes how the delivered irradiation dose varies within a simulated component or over a given area. This allows conclusions to be drawn about the maximum and minimum doses being delivered. Dose mapping should be performed with sensitive dosimetry techniques. A number of commercially available systems have been developed in recent years. Other quality assurance measures that need to be done include the routine confirmation that the turntable is operating correctly (for ^{137}Cs irradiators), measurements to ensure that the timing device is accurate, and the periodic lengthening of the irradiation time to correct for source decay. With linear accelerators, it is necessary to measure the characteristics of the x-ray beam to ensure consistency of delivery. Confirming that a blood component has, in actuality, been irradiated is also an important part of a quality assurance

TABLE 9.4   Quality Assurance Guidelines for Irradiating Blood Components

Dose

2500 cGy to the central midplane of a canister (freestanding irradiator) or to the center of an irradiation field (linear accelerator) with a minimum of 1500 cGy.

Dose Mapping (Freestanding Irradiators)

Routinely, once a year (^{137}Cs) or twice a year (^{60}Co) and after major repairs; using a fully filled canister (water/plastic) with a dosimetry system to map the distribution of the absorbed dose.

Dose Mapping (Linear Accelerators)

Recommend yearly dose mapping with an ionization chamber and a water phantom. More frequent evaluation of instrument conditions to ensure consistency of x-rays.

Correction for Radioisotopic Decay

With ^{137}Cs as the source, annually.
With ^{60}Co as the source, quarterly.

Turntable Rotation (Freestanding ^{137}Cs Irradiators)

Daily verification.

Storage Time for Red Cells after Irradiation

For up to 28 days; total storage time cannot exceed maximum storage time for unirradiated red cells.

Storage Time for Platelets after Irradiation

No change due to irradiation procedure.

Modified from Moroff G and Luban NL. 1997. The irradiation of blood and blood components to prevent GVHD: technical issues and guidelines. *Transfus Med Rev* 11:15–26.

program. Several indicator labels are available for this purpose.

CONFIRMING THAT IRRADIATION OCCURRED

It is important to have positive confirmation that the irradiation process has taken place in the case of operator failure and/or instrumentation malfunction. Several commercial products are available including a radiation-sensitive indicator label developed specifically for this purpose (International Specialty Products, Wayne, NJ). This label contains a radiation-sensitive film strip placed on the external surface of the blood component. Irradiation causes visually distinct changes and results in an appearance change from clear red

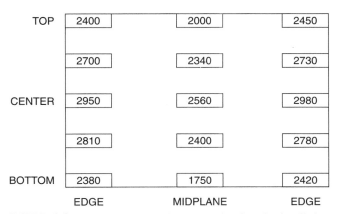

	EDGE	MIDPLANE	EDGE
TOP	2400	2000	2450
	2700	2340	2730
CENTER	2950	2560	2980
	2810	2400	2780
BOTTOM	2380	1750	2420

FIGURE 9.2 Two-dimensional dose map showing the irradiation dose distribution through a fully filled canister of a freestanding cesium 137 irradiator.

to opaque with obliteration of the word "NOT." Thus, when the label is placed on a blood component, there is a visual record that the irradiation process took place. The reliability of this type of indicator has been documented in a multisite study (Leitman et al. 1992).

Two versions of the indicator label have been manufactured. The difference is the range of radiation needed to cause a change in the radiation-sensitive film. The ratings for these indicators are 1500 cGy or 2500 cGy. The ratings serve as an approximate guideline for the amount of absorbed radiation that will be needed to completely change the window from reddish to opaque with complete obliteration of the word "NOT." Because the indicator labels are designed for and are used to confirm that the irradiation process has occurred, our laboratory utilizes the 1500 cGy label as the most appropriate tool to perform this quality control measure. This is based on the routinely observed pattern of dose distribution to a blood component in a canister of a freestanding irradiator. Despite a targeted central dose of 2500 cGy, there will be spots at which the dose will be less. If the theoretical dose map presented in Figure 9.2 is used as an example, there will be a spot that will receive only 1800 cGy. If the 2500 cGy-rated label were to be located on the external surface of a component, there may be minimal changes in the appearance of the radiation-sensitive film window. This would result in a judgment that the blood component was not irradiated, when in actuality it was treated satisfactorily.

NEW METHODS IN THE PREVENTION OF TA-GVHD

In the future, photochemical treatment (PCT) using psoralens and long wavelength ultraviolet irra-

diation (UVA) that have been developed to reduce the risk of bacterial and viral contaminants of platelet transfusions as well as other agents that disrupt or modify DNA replication of microorganisms may be used in the inactivation of leukocytes in blood products. Psoralens bind reversibly to nucleic acids by intercolation and, after UVA illumination, form covalent monoadducts and crosslinks with RNA and DNA. Among a broad group of compounds, the psoralen S-59 has been shown to be particularly effective in inactivating bacteria and viruses, without adversely affecting in vitro and in vivo platelet function (Lin et al. 1997).

Recently, S-59 and PCT has been studied for its possible inactivation of leukocytes in platelet concentrates (Hei et al. 1999; Grass et al. 1998). PCT inactivation of T cells was evaluated utilizing four assay systems. These assay systems included T-cell quantitation, inhibition of cytokine synthesis, modification of leukocyte genomic DNA by quantification of psoralen-DNA adducts, and inhibition replication of T cells using PCR amplification of genomic DNA sequences. These studies demonstrated significantly reduced or absent cytokine generation and greater DNA strand break induction in platelet concentrates treated with PCT as compared to those treated with irradiation (Hei et al. 1999; Grass et al. 1998). Furthermore, limiting dilution analysis was used to confirm inactivation of T cells in the platelet concentrates. To further support the efficacy of S-59 and PCT, a murine F1 hybrid-transfusion-induced GVHD model was tested (Grass et al. 1999). No GVHD was noted in mice receiving splenocytes treated with either 2500 cGy or 150 μmol/L S-5-9 and 2.1 J/cm² UVA. In another set of experiments, PCT to prevent GVHD in an immunocompromised mouse model was studied (Grass et al. 1997). These studies taken together suggest that PCT may well be an alternative to irradiation and further provide a mechanism to prevent increase in cytokine concentration in platelet concentrates. The limitation of PCT methodology is the need for UVA penetration, which is not currently possible with RBC products. Several strategies are currently being studied in an attempt to overcome this limitation. Other agents such as PEN110, a low-molecular weight electrophilic compound, have been shown to chemically modify DNA and result in inhibition of DNA replication. Such compounds do not require an activation step and have been demonstrated to inactivate mononuclear cells present in RBC units (Fast et al. 2002). Another method of inactivating WBCs uses ultraviolet B irradiation to induce immune tolerance in donor leukocytes (del Rosario, Zuclai, and Kao 1999). Further research in the safety of such approaches is warranted.

TREATMENT OF TA-GVHD

Despite the mortality (80% to 90%) associated with TA-GVHD, effective treatment remains elusive (Linden and Pisciotto 1992). Immunosuppressive therapy, the mainstay of treatment of TA-GVHD is of unproven value in TA-GVHD, and immunosuppressive agents such as steroids, methotrexate, antithymocyte globulin, Cyclosporin, and OKT3 have been associated with disappointing results (Linden and Pisciotto 1992; Lee et al. 1999). A recent report described the early detection and successful treatment of TA-GVHD using Solu-Medrol, Cyclosporin A, followed by high dose Cytoxan and antithymocyte globulin and autologous peripheral blood stem cell infusion (Hutchinson et al. 2003). Such cases are the exception—the best treatment for TA-GVHD remains prevention.

References

Akahoshi M, Takanashi M, Masuda M, et al. 1992. A case of transfusion-associated graft-versus-host disease not prevented by white cell-reduction filters. *Transfusion* 32:169–172.

Ammann AJ. 1993. Hypothesis: absence of graft-versus-host disease in AIDS is a consequence of HIV-1 infection of CD4+ T cells. *J Acquir Immune Defic Syndr* 6:1224.

Anderson KC. 1995. Leukodepleted cellular blood components for prevention of transfusion-associated graft-versus-host disease. *Transfus Sci* 16:265–268.

Anderson KC, Goodnough LT, Sayers M, et al. 1991. Variation in blood component irradiation practice: implications for prevention of transfusion-associated graft-versus-host disease. *Blood* 77:2096–2102.

Anderson KC and Weinstein HJ. 1990. Transfusion-associated graft-versus-host disease. *New Eng J Med* 323:315–321.

Arlett CM, Smith JB, and Jimenez SA. 1998. Identification of fetal DNA and cells in skin lesions from women with systemic sclerosis. *N Engl J Med* 333:1186–1191.

Benson K, Marks AR, Marshall MJ, et al. 1994. Fatal graft-versus-host disease associated with transfusions of HLA-matched, HLA-homozygous platelets from unrelated donors. *Transfusion* 34:432–437.

Bernvill SS, Abdulatiff M, Al-Sedairy S, et al. 1994. Fresh frozen plasma contains viable progenitor cells—should we irradiate. *Vox Sang* 67:405.

Bessos H, Atkinson A, Murphy WG, et al. 1995. A comparison of in vitro storage markers between gamma-irradiated and non-irradiated apheresis platelet concentrates. *Transfus Sci* 16:131–134.

Blundell EL, Pamphilon DH, Anderson NA, et al. 1992. Transfusion-associated graft-versus-host disease, monoclonal gammopathy and PCR. *Br J Haematol* 82:622–623.

Brubaker DB. 1993. Immunopathogenic mechanisms of post-transfusion graft versus host disease. *Proc Soc Exp Biol Med* 202:122–147.

Capon SM, DePond WD, Tyan DB, et al. 1991. Transfusion associated graft-versus-host disease in an immunocompetent patient. *Ann Intern Med* 114:1025–1026.

Center for Biologics Evaluation and Research, Food and Drug Administration. 1993. Recommendations regarding license amendments and procedures for gamma irradiation of blood products. Department of Health and Human Services.

Crowley JP, Skrabut EM, and Valeri CR. 1974. Immunocompetent lymphocytes in previously frozen washed red cells. *Vox Sang* 26:513–517.

Davey RJ. 1992. The effect of irradiation on blood components. In Baldwin ML and Jefferies LC, eds. *Irradiation of blood components*. Bethesda, MD: American Association of Blood Banks.

Davey RJ, McCoy NC, Yu M, et al. 1992. The effect of pre-storage irradiation on post-transfusion red cell survival. *Transfusion* 32:525–528.

Daya S, Gunby J, and Clark DA. 1998. Intravenous immunoglobulin therapy for recurrent spontaneous abortion: a meta-analysis. *Am J Reprod Immunol* 39:69–76.

del Rosario ML, Zuclai JR, and Kao KJ. 1999. Prevention of graft-versus-host disease by induciton of immune tolerance with ultra-violet B-irradiated leukocytes in H-2 diparate bone marrow donor. *Blood* 15:3558–3564.

DePalma L, Bahrami KR, Kapur S, et al. 1994. Amplified fragment length polymorphism analysis in the evaluation of post-transfusion graft versus host disease. *J Thoracic Cardiovasc Surg* 108:182–184.

Drobyski W, Thibodeau S, Truitt RL, Baxter LLA, Gorski J, Jenkins R, Gottschall J, and Ash RC. 1989. Third-party-mediated graft rejection and graft-versus-host disease after T-cell-depleted bone marrow transplantation, as demonstrated by hypervariable DNA probes and HLA-DR polymorphism. *Blood* 74:2285–2294.

Duguid JK, Carr R, Jenkins JA, et al. 1991. Clinical evaluation of the effects of storage time and irradiation on transfused platelets. *Vox Sang* 60:151–154.

Espersen GT, Ernst E, Christiansen OB, et al. 1988. Irradiated blood platelet concentrates stored for five days—evaluation by in vitro tests. *Vox Sang* 55:218–221.

Evans PC, Lambert N, Maloney S, et al. 1999. Long-term fetal microchimerism in peripheral blood mononuclear cell subsets in healthy women and women with scleroderma. *Blood* 93:2033–2037.

Fast LD, DiLeone G, Edson CM, and Purmal A. 2002. PEN110 treatment functionally inactivates the PBMNCs present in RBC units: comparison to the effects of exposure to gamma irradiation. *Transfusion* 42:1318–1325.

Fast LD, Valeri CR, and Crowley JP. 1995. Immune responses to major histocompatibility complex homozygous lymphoid cells in murine F1 hybrid recipients: implications for transfusion-associated graft-versus-host disease. *Blood* 86:3090.

Fearon TC and Luban NLC. 1986. Practical dosimetric aspects of blood and blood product irradiation. *Transfusion* 26:457–459.

Ferrera JM and Krenger W. 1998. Graft-versus-host disease. The influence of type 1 and type 2 T cell cytokines. *Transfus Med Rev* 12:1–17.

Friedman DF, Kwittken P, Cizman B, et al. 1994. Dan-based HLA typing of nonhematopoeitic tissue used to select the marrow transplant donor for successful treatment of transfusion-associated graft-versus-host disease. *Clin Diagn Lab Immunol* 1:590–596.

Friedman KD, McDonough WC, and Cimino DF. 1991. The effect of pre-storage gamma irradiation on post-transfusion red blood cell recovery. *Transfusion* 31:50S.

Grass J, Delmonte J, Wages D, et al. 1997. Prevention of transfusion-associated graft vs. host disease (TA-GVHD) in immunocompromised mice by photochemical treatment (PCT) of donor T cells. *Blood* 90(suppl 1):207a.

Grass JA, Hei DJ, Metchette K, et al. 1998. Inactivation of leukocytes in platelet concentrates by psoralen plus UVA. *Blood* 91:2180–2188.

Grass JA, Wafa T, Reames A, et al. 1999. Prevention of transfusion-associated graft-versus-host disease by photochemical treatment. *Blood* 93:3140–3147.

Grishaber JE, Birney SM, and Strauss RG. 1993. Potential for host-transfusion-associated graft-versus-host disease due to apheresis platelets matched for HLA class I antigens. *Transfusion* 33:910–914.

Hayashi H, Nishiuchi T, Tamura H, et al. 1993. Transfusion associated graft-versus-host disease caused by leukocyte filtered stored blood. *Anesthesiology* 79:1419–1421.

Hei DJ, Grass J, Lin L, Corash L, et al. 1999. Elimination of cytokine production in stored platelet concentrate aliquots by photochemical treatment with psoralen plus ultraviolet A light. *Transfusion* 39:239–248.

Heim MU, Munker R, Sauer H, et al. 1991. Graft-versus-host Krankheit(GVH mit letalem ausgang nach der gabe von gefilterten erythrozytenkonzentraten(Ek). *Infusionstherapie* 18:8–9.

Hutchinson K, Kopko PM, Muto KN, et al. 2003. Early diagnosis and successful treatment of a patient with transfusion-associated GVHD with autologous peripheral blood progenitor cell transplantation. *Transfusion* 42:1567–1572.

Kanter MH. 1992. Transfusion-associated graft-versus-host disease: do transfusions from second-degree relatives pose a greater risk than those from first-degree relatives? *Transfusion* 32: 323–327.

Klein C, Fraitag S, Foulon E, et al. 1996. Moderate and transient transfusion-associated cutaneous graft versus host disease in a child infected by human immunodeficiency virus. *AmJ Med* 101:445–446.

Korngold R. 1993. Biology of graft-versus-host disease. *Am J Pediatr Hematol Oncol* 15:18–37.

Kruskall MS, Lee TH, Assmann SF, et al. 2001. Survival of transfused donor white blood cells in HIV-infected recipients. *Blood* 98:272–279.

Kunstmann E, Bocker T, Roewer L, et al. 1992. Diagnosis of transfusion-associated graft versus-host disease by genetic fingerprinting and polymerase chain reaction. *Transfusion* 32:766–770.

Kutcher GJ, Coia L, Gillin M, et al. 1994. Comprehensive QA for radiation oncology: report of AAPM radiation therapy committee task group 40. *Med Phys* 21:581–618.

Lee TH, Donegan E, Slichter S, et al. 1996. Transient increase in circulating donor leukocytes after allogeneic transfusions in immunocompetent recipients compatible with donor cell proliferation. *Blood* 85:1207–1214.

Lee TH, Paglieroni T, Ohto H, et al. 1999. Survival of donor leukocyte subpopulations in immunocompetent transfusion recipients: frequent long-term microchimerism in severe trauma patients. *Blood* 93:3127–3139.

Leitman SF. 1993. Dose, dosimetry and quality improvements of irradiated blood components. *Transfusion* 33:447–449.

Leitman SF, Silberstein L, Fairman RM, et al. 1992. Use of a radiation-sensitive film label in the quality control of irradiated blood components. *Transfusion* 32:4S.

Lin L, Cook DN, Wiesehahn GP, et al. 1997. Photochemical inactivation of viruses and bacteria in human platelet concentrates using a novel psoralen and long wavelength UV light. *Transfusion* 37:423.

Linden JV and Pisciotto PT. 1992. Transfusion-associated graft-versus-host disease and blood irradiation. *Transfus Med Rev* 6:116–123.

Luban NL and DePalma L. 1996. Transfusion-associated graft-versus-host disease in the neonate-expanding the spectrum of disease. *Transfusion* 36:101–103.

Luban NLC, Drothler D, Moroff G, et al. 2000. Irradiation of platelet components: inhibition of lymphocyte proliferation assessed by limiting dilution analysis. *Transfusion* 40:348–352.

Luban NLC, Fearon T, Leitman SF, et al. 1995. Absorption of gamma irradiation in simulated blood components using cesium irradiators. *Transfusion* 35:63S.

Luban NLC, Strauss RG, and Hume HA. 1991. Commentary on the safety of red cells preserved in extended-storage media for neonatal transfusion. *Transfusion* 31:229–235.

Masterson ME and Febo R. 1992. Pre-transfusion blood irradiation: clinical rationale and dosimetric considerations. *Med Phys* 19:649–457.

McMiln KD and Johnson RL. 1993. HLA-homozygosity and the risk of related-donor transfusion-associated graft-versus-host disease. *Transfus Med Rev* 7:37–41.

Mintz PD and Anderson G. 1993. Effect of gamma irradiation on the in vivo recovery of stored red blood cells. *Ann Clin Lab Sci* 23:216–220.

Miraglia CC, Anderson G, and Mintz PD. 1994. Effect of freezing on the in vivo recovery of irradiated red cells. *Transfusion* 34:775–778.

Moroff G, George VM, Siegl AM, et al. 1986. The influence of irradiation on stored platelets. *Transfusion* 26:453–456.

Moroff G, Holme S, AuBuchon JP, et al. 1999. Viability and in vitro properties of AS-1 red cells after gamma irradiation. *Transfusion* 39:128–134.

Moroff G, Holme S, Heaton A, et al. 1992. Effect of gamma irradiation on viability of AS-1 red cells. *Transfusion* 32(suppl):70S.

Moroff G and Luban NL. 1997. The irradiation of blood and blood components to prevent graft-versus-host disease: technical issues and guidelines. *Transfus Med Rev* 11:15–26.

Moroff G, Luban NLC, Wolf L, et al. 1993. Dosimetry measurements after gamma irradiation with cesium-137 and linear acceleration sources. *Transfusion* 33:52S.

Nath R, Biggs PJ, Bova FJ, et al. 1994. AAPM code of practice for radiotherapy accelerators: report of AAPM radiation therapy task group no. 45. *Med Phys* 21:1093–1121.

Nelson J, Furst D, Maloney S, et al. 1998. Microchimerism and HLA compatible relationships of pregnancy in scleroderma. *Lancet.* 351:559–562.

Nishimura M, Uchida S, Mitsunaga S, et al. 1997. Characterization of T-cell clones derived from peripheral blood lymphocytes of a patient with transfusion-associated graft-versus-host disease: Fas-mediated killing by CD4+ and CD8+ cytotoxic T-cell clones and tumor necrosis factor beta production by CD4+ T-cell clones. *Blood* 89:1440.

Ohto H and Anderson KC. 1996a. Post-transfusion graft-versus-host disease in Japanese newborns. *Transfusion* 36:117–123.

Ohto H and Anderson KC. 1996b. Survey of transfusion-associated graft-versus-host disease in immunocompetent recipients. *Transfus Med Rev* 10:31–43.

Patrone F, Dallegri F, Brema F, et al. 1979. Effects of irradiation and storage on granulocytes harvested by continuous-flow centrifugation. *Exp Hematol* 7:131–136.

Pelszynski MM, Moroff G, Luban NLC, et al. 1994. Effect of γirradiation of red blood cell units on T-cell inactivation as assessed by limiting dilution analysis: implications for preventing transfusion-associated graft-versus-host disease. *Blood* 83:1683–1689.

Perkins JT and Papoulias SA. 1994. The effect of loading conditions on dose distribution within a blood irradiator. *Transfusion* 34: 75S.

Petz LD, Calhoun L, Yam P, et al. 1993. Transfusion-associated graft-versus-host disease in immunocompetent patients: Report of a fatal case associated with transfusion of blood from a second-degree relative and a survey of predisposing factors. *Transfusion* 33:742–750.

Quinones RR, Gutierrez RH, Dinndorf PA, et al. 1993. Extended cycle elutriation to adjust T-cell content in HLA-disparate bone marrow transplantation. *Blood* 82:307–317.

Ramirez AM, Woodfield DG, Scott R, et al. 1987. High potassium levels in stored irradiated blood. *Transfusion* 27:444–445.

Read EJ, Kodis C, Carter CS, et al. 1988. Viability of platelets following storage in the irradiated state. A paired-controlled study. *Transfusion* 28:446–450.

Reed WF, Lee TH, Trachlenberg E, et al. 2001. Detection of microchimerism by PCR as a function of amplification strategy. *Transfusion* 41:39–44.

Rivet C, Baxter A, and Rock G. 1989. Potassium levels in irradiated blood. *Transfusion* 29:185.

Rock G, Adams GA, and Labow RS. 1988. The effects of irradiation on platelet function. *Transfusion* 28:451–455.

Rosen NR, Weidner JG, Bold HD, et al. 1993. Prevention of transfusion-associated graft-versus-host disease: selection of an adequate dose of gamma irradiation. *Transfusion* 33:125–127.

Seghatchian MJ and Stivala JFA. 1995. Effect of 25 Gy gamma irradiation on storage stability of three types of platelet concentrates: a comparative analysis with paired controls and random preparation. *Transfus Sci* 16:121–129.

Shlomchik WD, Couzens MS, Tang CB, et al. 1995. Prevention of graft versus host disease by inactivation of host antigen-presenting cells. *Science* 285:412–415.

Sprent J, Anderson RE, and Miller JF. 1974. Radiosensitivity of T and B lymphocytes. II Effect of irradiation on response of T cells to alloantigens. *Eur J Immunol* 4:204–210.

Strauss RG. 1990. Routine washing of irradiated red cells before transfusion seems unwarranted. *Transfusion* 30:675–677.

Suda BA, Leitman SF, and Davey RJ. 1993. Characteristics of red cells irradiated and subsequently frozen for long term storage. *Transfusion* 33:389–392.

Swann ID and Williamson LM. 1996. Potassium loss from leukodepleted red cells following γ-irradiation. *Vox Sang* 70:117–118.

Sweeney JD, Holme S, and Moroff G. 1994. Storage of apheresis platelets after gamma irradiation. *Transfusion* 34:779–783.

Valerius NH, Johansen KS, Nielsen OS, et al. 1981. Effect of in vitro x-irradiation on lymphocyte and granulocyte function. *Scand J Hematol* 27:9–18.

Vietor HE, Hallensleben E, et al. 2000. Survival of donor cells 25 years after intrauterine transfusion. *Blood* 95:2709–2714.

von Fliedner V, Higby DJ, and Kim U. 1972. Graft-versus-host reaction following blood product transfusion. *Am J Med* 72:951.

Wang L, Juji T, Tokunaga K, et al. 1994. Polymorphic microsatellite markers for the diagnosis of graft-versus-host disease. *N Engl J Med* 330:398–401.

Wang-Rodriguez J, Fry E, Fiebig E, et al. 2000. Immune response to blood transfusion in very-low-birthweight infants. *Transfusion* 40:25–34.

Webb IJ and Anderson KC. 2000. Transfusion-associated graft-versus-host disease. In Anderson KC and Ness PM, eds. *Scientific basis of transfusion medicine*. Philadelphia, PA: WB Saunders.

Wheeler JG, Abramson JS, and Ekstrand K. 1984. Function of irradiated polymorphonuclear leukocytes obtained by buffy-coat centrifugation. *Transfusion* 24:238–239.

Wielding JU, Vehmeyer K, Dittman J, et al. 1994. Contamination of fresh-frozen plasma with viable white cells and proliferable stem cells. *Transfusion* 34:185–186.

Williamson LM and Warwick RM. 1995. Transfusion-associated graft-versus-host disease and its prevention. *Blood Reviews* 9:251–261.

Wolber RA, Duque RE, Robinson JP, and Oberman HA. 1987. Oxidative product formation in irradiated neutrophils. A flow cytometric analysis. *Transfusion* 27:167–170.

Worrall NK, Lazenby WD, Misko TP, et al. 1995. Modulation of in vivo alloreactivity by inhibition of inducible nitric oxide synthetase. *J Exp Med* 181:63–70.

Washed and/or Volume-Reduced Blood Components

S. GERALD SANDLER, MD, AND JAYASHREE RAMASETHU, MD, FAAP

INTRODUCTION

In special situations, saline-washed red blood cells (RBCs) may be needed to reduce the risk of adverse reactions to certain constituent(s) in plasma or in the preservative storage solution. For example, small infants may require saline-washed RBCs if freshly collected RBCs are not available for rapid or large-volume transfusions during exchange transfusions, extracorporeal membrane oxygenation (ECMO) procedures, cardiopulmonary bypass, or liver transplantation. Also, children with recurrent, severe allergic reactions, including IgA anaphylaxis, may require saline-washed RBCs to reduce exposure to allogeneic plasma proteins. RBCs salvaged from sterile surgical fields or traumatic injury sites should be saline-washed to remove tissue debris and other potentially biologically active material before return to the patient. In children with T activation and hemolysis, saline-washed RBCs avert the risk of further hemolysis induced by anti-T in (normal) donor plasma.

Washed platelet concentrates are rarely required. However, washed maternal platelets may be the only source of serologically compatible platelet transfusions for a fetus or newborn with neonatal alloimmune thrombocytopenia. Occasionally, group A, B, or AB patients may require volume-reduced pooled random-donor or single-donor (apheresis) platelets if "out-of-group" platelets are the only platelet components available to avoid ABO-related hemolysis. "Super-concentrated" or recentrifuged platelets to reduce volume and prevent circulatory overload are rarely necessary. Conventional doses of standard platelet components (5 to 10 mL/kg) should be tolerable for most children, including preterm infants and newborns, and are usually adequate for increasing the platelet count to a safe level. For the rare small infant who requires reduction of all intravenous (IV) fluids to avoid circulatory overload, protocols are available for preparing volume-reduced platelets. Blood banks providing this service must maintain strict quality control to ensure adequate number and function of platelets in the final component.

WASHED RED BLOOD CELLS

Indications

Large Volume Transfusions: Hyperkalemia

Small volume transfusions (<25 mL/kg) of RBCs stored as citrate-phosphate-dextrose-adenine (CPDA)-1-anticoagulated "packed" RBCs or in conventional extended-storage media (AS-1 [ADSOL], Baxter Healthcare Corporation, Deerfield, IL; AS-3 [Nutricel], Medsep Corporation, Covina, CA; AS-5 [Optisol], Terumo Corporation, Somerset, NJ) may be safely transfused to small children, including newborns and preterm infants (Strauss 2000a). However, larger volume transfusions (>25 mL/kg) of conventionally stored RBCs, particularly if transfused rapidly, may cause acute hyperkalemia, resulting in cardiac arrest and death (Hall et al. 1993). During storage, K^+ leaks from the intracellular fluid of RBCs, increasing the concentration of K^+ in the plasma or RBC preservation storage solution to levels that are potentially dangerous for rapid transfusions. The K^+ concentration in supernatant plasma of CPDA-1–stored RBCs increases continuously from 4.2 mmol/L to almost 80 mmol/L during the 35-day storage period (Strauss 2000a). For a newborn weighing 1 kg, a conventional transfusion (15 mL/kg) of 42-day stored RBCs prepared to a hematocrit of 80% may have a K^+ concentration of 50 mEq/L but contains a total of only 0.15 mEq of K^+. While that relatively high concentration of K^+ is tolerable and safe if transfused slowly (2 to 4 hours), the same RBC component could present a potentially fatal acute K^+ load if transfused rapidly. Situations when rapid transfusions of stored RBCs may increase recipient K^+ levels acutely and, therefore, require saline-washed or freshly-collected RBC components, include exchange transfusions, ECMO, cardiopulmonary bypass, and liver transplantation (Luban 1995; Scanlon and Krakaur 1980; Estrin et al. 1986). Saline-washed RBCs should be considered also for neonates and small children with renal failure, hyperkalemia, or severe acidosis who may require large volumes (>20 mL/kg) or rapid transfusion.

Some neonatologists are concerned that for large-volume transfusions, such as cardiopulmonary bypass, the doses of certain additives in extended-storage solutions may exceed known limits for safety in small children (Luban et al. 1991). Concerns derive from the possibility that, based on theoretical calculation of constituents of the storage media, such transfusions could result in hyperosmolality, hyperglycemia, hypernatremia, or hyperphosphatemia, in addition to hyperkalemia. For that reason, some transfusion services volume-reduce or wash RBCs stored in additive solutions for large-volume or rapid transfusions in neonates or small children.

An innovative alternative to saline-washed or freshly collected RBCs to reduce the risk of acute hyperkalemia from stored RBCs is transfusion via an investigational K^+-adsorption filter containing an ion exchange resin that has the capability to bind as much as 20 mEq of K^+ (Model 7R-001, Mitsubishi Chemical, Tokyo, Japan) (Inaba et al. 2000). While this filter was developed to reduce the risk of hyperkalemic reactions from stored gamma-irradiated RBCs, this approach could provide a practical solution to the problem of quickly preparing safe RBCs for small children needing rapid or large-volume transfusions of stored RBCs. Although the manufacturer has reported satisfactory safety and efficacy on 65 patients treated with this filter in a Phase III clinical trial, its status remains investigational.

Recurrent Allergic or Anaphylactic Reactions

Some chronically transfused children require saline-washed RBCs to prevent recurrent, acute allergic reactions, such as generalized urticaria, brochospasm, rash, flushing, nausea, vomiting, or diarrhea. Most first-time allergic-type reactions are *product*-related, rather than *patient (recipient)*-related and do not recur when additional plasma-containing components are transfused. However, some chronically transfused children have recurrent reactions that are *not* prevented by pretreatment with an antihistamine. In this situation, the only reliable way to avert an acute allergic reaction is transfusion of saline-washed RBCs.

Other children may require saline-washed RBCs because of the severity of their reactions to plasma-containing components. These severe generalized reactions may occur as anaphylaxis with severe brochospasm, hypotension, and life-threatening circulatory collapse (Sandler et al. 1995). In such cases, a diagnosis of an IgA anaphylactic reaction may be established on the basis of a conventional passive hemagglutination assay for IgA antibodies and a passive hemagglutination inhibition assay for IgA concentration (Sandler et al. 1994). The authors advise caution when interpreting results of these hemagglutination assays because they may yield nonspecific results. The same hemagglutination assay that is used to diagnose *patients* with IgA anaphylactic reactions, detected IgA deficiency and the presence of anti-IgA in 1:1200 asymptomatic healthy *blood donors* (Sandler et al. 1994). Clearly, this number of healthy persons at-risk for an IgA anaphylactic reaction greatly exceeds the incidence of IgA anaphylactic transfusion reactions in clinical practice.

Other kinds of severe generalized reactions to plasma-containing components have been identified recently; these include transfusion-related acute lung injury (TRALI), leukocyte-mediated cytokine, and antihaptoglobin (Sandler et al. 2003). It is likely that some of the early reports of IgA anaphylactic reactions, particularly those attributed to IgA subclass antibodies, represent misdiagnoses of more common generalized transfusion reactions that were erroneously "confirmed" by nonspecific hemagglutination assay results (Gilstad et al. 2002). It is important to carefully document both the clinical and laboratory diagnoses of IgA anaphylactic reactions before committing a child to a life-long requirement for saline-washed RBCs.

Intraoperative Red Cell Salvage

Intraoperative salvage of shed RBCs has been utilized effectively to reduce allogeneic blood transfusions in surgical procedures associated with large-volume blood loss (for example, correction of craniosynostosis, spinal surgery, or liver resection) (Jimenez and Barone 1995; Kruger and Colbert 1985; Estrin et al. 1986). Conventionally, shed RBCs are aspirated from the sterile surgical field, anticoagulated using heparin-saline, and washed with sterile saline in a rotating polycarbonate bowl. The effluent-containing clots, white blood cells (WBCs), platelets, cellular debris, and heparin are discarded, and saline-washed, concentrated RBCs are returned to the patient.

An alternative method of salvaging RBCs has been introduced whereby sanguinous drainage from the chest, large joints, or other sites is heparinized and returned directly (unwashed) from collection canisters. This method is controversial because of concerns that such drainage may contain procoagulant material, variable amounts of anticoagulant, and/or unsterile debris. In fact, even saline-washed, salvaged RBCs collected under optimal conditions may contain residual biologically active materials capable of causing increased vascular permeability, acute respiratory distress syndrome, or disseminated intravascular coagulation (DIC) (Bull and Bull 1990; Ramirez et al. 2002). This complication, described as the "salvaged blood syndrome," may be averted by standardizing aspiration techniques, controlling saline-wash volumes, and monitoring the rotating bowl for an abnormal accumulation of debris on the inner wall (Bull and Bull 1990).

A similar process for salvaging and saline-washing RBCs has been adapted for pediatric bone marrow donors to reduce the risk of symptomatic anemia following bone marrow harvest (Kletzel et al. 1999). These investigators used a semi-automated closed system (Stericel, Terumo Somerset, NJ) to process bone marrow hematopoietic cells from healthy, human leukocyte antigen (HLA)-matched, pediatric bone marrow donors (mean age 8 years). The marrow was aspirated under general anesthesia, filtered in the operating room, processed by density gradient separation with Ficoll-Hypaque, and washed. The aliquot containing mononuclear cells was infused in the recipient. A second aliquot containing the washed RBCs was returned to the donor. This method reduces the risk of post-bone marrow harvest anemia in pediatric donors, enriches the mononuclear and CD34+ cell population to be infused in the recipient, and does not appear have a negative effect on hematopoietic reconstitution.

T Activation

Immune-mediated hemolysis has been reported following transfusion of plasma-containing blood components to patients whose RBC T cryptantigen has been exposed by bacterial infection, resulting in T activation (van Loghem et al. 1955; Williams et al. 1989; Ramasethu and Luban 2001). In this situation, T activation occurs because bacterial neuraminidase removes *N*-acetyl neuraminic acid residues and exposes T cryptantigens on RBC membranes. Exposure of the T antigen allows binding with IgM anti-T, a normal constituent of adult plasma, which results in RBC agglutination and hemolysis. T activation has been reported in neonates with necrotizing enterocolitis, especially in those with severe disease requiring surgical intervention. It has also been reported in septicemic infants and children with other surgical problems, in children with hemolytic uremic syndrome associated with *Streptococcus pneumoniae* infection, and in adult surgical intensive care patients.

T activation should be suspected in at-risk children who have evidence of intravascular hemolysis following transfusion of plasma-containing blood components. Routine compatibility testing may not detect polyagglutination due to T activation if routine ABO grouping of patients is performed using (non–plasma-derived) monoclonal reagents. If human plasma-derived ABO grouping reagents are used, discrepancies in forward and reverse typing results and/or in vitro hemolysis may indicate T-antigen activation. The diagnosis is confirmed by specific agglutination tests using peanut lectin *Arachis hypogea* and *Glycine soja*. Further hemolysis following component transfusion may be prevented using saline-washed RBCs and platelets. Exchange transfusion with plasma-reduced components may be necessary for infants with ongoing severe hemolysis (Ramasethu and Luban 2001). A dilemma develops when a patient with T activation requires an urgent transfusion of fresh frozen plasma or platelets to control hemostasis. Some physicians recommend completely avoiding transfusion of plasma-containing components, while others question the clinical relevance of T activation and recommend that indicated plasma-containing components not be withheld (Eder and Manno 2001). Recommendations, which are based on case reports, are conflicting and there are neither evidenced-based guidelines nor results of a randomized, controlled clinical trial to direct practice. Pending additional information on this controversial matter, the preponderance of clinical evidence favors not withholding plasma-containing blood components when they are indicated to control hemostasis (Eder and Manno 2001). For those blood banks that routinely use murine or other monoclonal blood group typing reagents, the authors do not recommend adding human plasma-derived typing reagents as a screening procedure to detect T activation.

Paroxysmal Nocturnal Hemoglobinuria

Paroxysmal nocturnal hemoglobinuria (PNH), an acquired disorder associated with a somatic mutation in totipotent hematopoietic stem cells, is uncommon in children (Hillmen et al. 1995). However, PNH may be associated with bone marrow failure (Wynn et al. 1998; Rizk et al. 2002) or Fanconi anemia in children

(Wainwright et al. 2003), and it is therefore important that pediatricians are aware of issues related to blood transfusions in PNH.

Whole blood transfusions have been reported to increase hemoglobinuria and hemoglobinemia in patients with PNH as the consequence of an exacerbation of complement-mediated intravascular hemolysis. Thus, some hematologists request saline-washed RBCs for transfusion to patients with PNH. While posttransfusion hemolysis is rare in PNH when saline-washed RBCs are transfused (Dacie 1948), it is observed (Rosse 1990) (Barnett et al. 1951; Hirsch et al. 1964). However, most PNH patients do not have hemolysis after RBC transfusions, even when whole blood is transfused (Manchester 1945). A retrospective review of 13 patients with PNH at the Mayo Clinic who were transfused on 82 occasions with 138 units of blood (15 units of whole blood, 100 units of RBCs, 22 units of saline-washed RBC, and 1 unit of FFP) showed minor febrile reactions occurring with the same frequency after transfusions of whole blood or RBCs (11%) or saline-washed RBCs (10%); only one hemolytic transfusion reaction occurred among the 82 transfusion events when a unit of nongroup-specific blood was transfused. This was the same situation that occurred in the patient originally reported by Dacie in 1943 and 1948 and was responsible for the origin of the practice of transfusion saline-washed red blood cells to patients with PNH (Sherman and Taswell 1977). Based on this information, as well as our experience of transfusing patients with PNH for more than 10 years with conventional AS-1 RBCs without reactions, the authors recommend that use of saline-washed RBCs is not necessary for patients with PNH.

Methods for Preparing Saline-Washed Red Blood Cells

RBC components may be washed manually using a refrigerated centrifuge or, more typically, using one of several commercially marketed automated cell washers. Washing a unit of RBCs using 1 to 2 L of 0.9% sodium chloride will remove approximately 99% of plasma proteins, electrolytes, and antibodies but may result in loss of up to 20% of the red cell mass, depending on the protocol used (Brecher 2002). In small children who receive syringe transfusions, loss of 20% of the original component's red cell mass is unlikely to result in an increased donor exposure. However, in older children, that loss of red cell mass could increase donor exposure if the blood bank needed an additional component to compensate for the loss.

Rarely, a recipient who is highly reactive to small quantities of residual plasma may require RBCs that have been washed using 4 to 6 L of saline. Conventional washing devices use an "open" system, which requires use of the washed RBC component within 24 hours.

Methods for Preparing Volume-Reduced Red Blood Cells

AS-3 Red Blood Cells

Conventionally, RBCs are prepared by removing supernatant plasma from whole blood following a "heavy spin" (5000 × g, 5 minutes) (Brecher 2002). Removing 225 to 250 mL of plasma will result in an RBC component with a hematocrit of 70% to 80%. The hematocrit of the final component may be decreased by removing proportionally less plasma (Brecher 2002). Strauss et al. developed a method to prepare multiple aliquots of RBCs for neonatal transfusions with hematocrits >90% using RBC components stored at 4°C for as long as 42 days (Strauss et al. 1995). For this method, donor blood is collected in a primary bag containing CP2D anticoagulant (Leukotrap RC System, Miles Inc., Elkhart, IN), which is centrifuged at 5000 × g for 5 minutes. Platelet-rich plasma is transferred into a second bag, which is disconnected. Extended storage media (100 mL) (Nutricel, AS-3, Miles Inc., Elkhart, IN) is added to the storage bag and, after mixing, the RBCs are transferred via the leukocyte-reduction filter to another storage bag. A cluster of small-volume bags is attached to the storage bag using a sterile connecting device. When a transfusion is ordered, the storage bag is centrifuged in an inverted position to pack the AS-3 RBCs to a hematocrit of approximately 90%. The original method has been modified to centrifuge at 4000 × g *for 4 minutes, not faster,* (Personal correspondence, R.G. Strauss to S.G. Sandler, May 15, 2003). The volume of RBCs requested flows from the bottom of the storage bag through the outlet tubing into one of the small-volume bags. The aliquot is disconnected, and the residual contents of the storage bag are mixed thoroughly and returned to storage until the next aliquot is required. During storage, AS-3 RBCs are mixed and resuspended weekly. In the original study, measurements of extracellular K+, hemoglobin, and lactic dehydrogenase from repeatedly mixed and centrifuged units that were stored in an inverted position were comparable to those of uncentrifuged conventionally-stored AS-3 RBCs (Strauss et al. 1995).

AS-1 Red Blood Cells

The method described previously for AS-3 RBCs has been modified to prepare AS-1 RBCs with hematocrits of approximately 85% to 87% after storage at 4°C for as long as 42 days (Strauss et al. 1996). For this method, RBCs are processed and stored by conventional method using AS-1 (Adsol, Baxter Healthcare Corp., Deerfield, IL) at hematocrits of approximately 60%. When a transfusion is requested, the AS-1 unit is centrifuged in an inverted position at 4000 × g for 4 minutes. Aliquots are removed, as before. During storage, the primary AS-1RBC units are remixed and stored until another transfusion was ordered and aliquoted.

These or similar methods for preparing volume-reduced RBCs for neonatal transfusions have been adopted by many hospitals. However, the subject of volume-reduced RBC transfusions remains controversial. Not all physicians agree that additional centrifugation(s) of conventional RBC components (hematocrit 60%) to prepare a syringe of volume-reduced RBCs (hematocrit 85% to 87%) is a clinically meaningful manipulation. The authors concur with the caution that "... this technique may not be desired by all centers, and if it is adopted, quality control studies should be performed to ensure the quality and sterility of the RBC aliquots" (Strauss et al. 1996).

Alternatives to Saline-Washed or Volume-Reduced Red Blood Cells

Depending on the clinical situation, there may be alternatives to saline-washed RBCs for reducing the risks of excessive K^+ or plasma proteins in RBC components. For rapid transfusion in a small infant, reconstituted whole blood (that is, RBCs resuspended in fresh frozen plasma shortly before transfusion) should dilute supernatant K^+ to a safe concentration. Accordingly, the Pediatric Hemotherapy Committee of the American Association of Blood Banks (AABB) recommended that "reconstituted whole blood ... or whole blood (≤5 days old) may be given [to infants and children <18 years] without the need for further justification in the setting of massive transfusions" (Blanchette et al. 1991). Also, selecting a RBC component stored in AS-1, AS-3, or AS-5, rather than as packed CPDA-1-anticoagulated RBCs in plasma, should reduce the concentration of plasma proteins to a level tolerable for most patients with nonspecific recurrent allergic reactions. Lastly, storing RBC components in the refrigerator "upside down" (inverted gravity sedimentation) will concentrate AS-1 units to hematocrits in the 68% range by 72-hours (Figure 10.1) (Sherwood et al. 2000). This simple manipulation provides a method for volume-reducing AS-1 RBC units in transfusion services that may not have access to an appropriate refrigerated centrifuge.

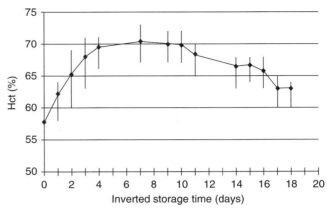

FIGURE 10.1 Hematocrits of 20-mL aliquots of AS-1 RBCs concentrated by gravity sedimentation. Reproduced with permission (Sherwood et al. 2000).

WASHED AND VOLUME-REDUCED PLATELET CONCENTRATES

Indications

Neonatal Alloimmune Thrombocytopenia

Neonatal alloimmune thrombocytopenia (NAIT) is a potentially fatal disease of the fetus or newborn that may occur when there is incompatibility for human platelet antigens (HPAs) between mother and fetus resulting in immune destruction of platelets. Approximately 90% of cases of NAIT in Caucasians of European ancestry present in an HPA-1a–positive newborn or fetus of an HPA-1a–negative mother who has developed anti-HPA-1a (Uhrynowska et al. 1997). Approximately 6% to 19% of cases of NAIT in Caucasians are due to anti-HPA-5b and 0% to 2% are due to anti-HPA-1b (McFarland 1998; Kroll et al. 1998). Mothers of Asian ancestry whose newborns have NAIT most often have antibodies specific for HPA-4 antigens (Stroncek 2002). More than 16 different HPAs have been implicated in NAIT (Skupski and Bussel 1999).

One approach to managing an at-risk pregnancy consists of using periumbilical blood sampling (PUBS) to monitor fetal platelet counts after 20 weeks of gestation and infusing the mother with intravenous immune globulin (IVIG) (1 gm/kg) weekly to ameliorate fetal thrombocytopenia (Bussel et al. 1996). In this protocol, adding 1.5 mg dexamethasone to weekly infusions of IVIG did not improve the response rate. An alternative approach first measures the fetus' platelet count at 20 weeks of gestation and recommends weekly infusions of IVIG (1 mg/kg) for the mother if the fetus is thrombocytopenic (Johnson et al. 1997). A second PUBS is performed at 26 weeks, and prednisone (60 mg daily) is added to weekly IVIG infusions if the fetus remains thrombocytopenic. PUBS is repeated before delivery, and the fetus is transfused with platelets if thrombocytopenic (Johnson et al. 1997).

Since maternal platelets are serologically compatible with the offending antibody in NAIT, they are a potential source of platelets for transfusion in the newborn (Adner et al. 1969). However, since maternal platelets are suspended in plasma that contains the antibody to the fetus' platelet antigen, it is necessary to wash the maternal platelets and resuspend them in compatible plasma or saline. Platelet components may be washed by manual or automated methods using 0.9% sodium chloride, with or without adding ACD-A (Pineda et al. 1989). Loss of up to 33% of original platelets may occur (Pineda et al. 1989). Also, since washing presently requires an open system, washed platelets must be transfused within 4 hours.

It is logistically preferable to supply HPA-matched platelet transfusions using panels of HPA-typed and qualified donors (Ranasinghe et al. 2001). Increasingly, large community blood centers have established panels of known HPA-1a–negative platelet donors. These platelet donors are scheduled for routine platelet apheresis donations at frequent intervals, and therefore the products qualify under emergency-release protocols to be released for transfusion before infectious disease testing is completed. Serologically compatible platelets collected from such call-in donors may be collected and emergency-released more quickly than if one collects, tests, and processes platelets collected from the mother.

Community donors of rarer HPA specificities are more difficult to recruit since maternal platelets or platelets from pretyped allogeneic donors are selected to be serologically compatible with the implicated antibody. Conventional doses of platelets (10 mL/kg) should be adequate for most clinical situations. Nevertheless, careful monitoring of posttransfusion platelet counts is essential given the serious consequences of inadequate dosing in thrombocytopenic newborns. In many cases of NAIT presenting at delivery following the first pregnancy, a clear-cut diagnosis cannot be made as serological test results will not be available. For this reason, it is advisable to have a policy in place for the inevitable emergency request to collect and/or wash maternal platelets for transfusion to an affected fetus or newborn.

Out-of-Group Platelets

Platelets from Group O donors may be transfused to Group A, B, or AB recipients if ABO-group specific platelets are not available. Usually, the small volume of donor plasma, representing a "minor-type" ABO incompatibility, does not cause overt hemolysis. Often recipients of out-of-group platelet transfusions develop

a weakly positive direct antiglobulin test result, but that is the only evidence of the incompatibility. Occasionally, acute intravascular hemolysis may occur, typically when ABO-incompatible apheresis platelets from a donor who has a high-titer anti-A titer or anti-B titers have been transfused to an A_1 recipient (Larsson et al. 2000; Pierce et al. 1985). Anti-A and anti-B titers are not routinely performed on apheresis platelets. As clinical practice shifts from transfusing pools of multiple random-donor platelet concentrates where there is dilution of the titer to transfusing more single-donor (apheresis) platelets (Larsson et al. 2000), more cases may be identified. In most situations, management consists of obtaining and transfusing platelets from donors who are ABO compatible. In situations where the demand for platelets is high and the inventory is low, volume reduction of platelets to decrease the amount of plasma transfused is a reasonable measure to reduce the risk of further hemolysis.

Febrile Nonhemolytic Transfusion Reactions

In adult transfusion recipients, the incidence of febrile nonhemolytic transfusion reactions (FNHTRs) is lowered when plasma-reduced platelet concentrates are used (Heddle et al. 1999; Heddle et al. 1994). This effect is attributed to decreasing leukocyte-derived proinflammatory cytokines that accumulate in the plasma of platelet components during storage (Heddle et al. 1994; Muylle et al. 1996). In children, a randomized, prospective, crossover study compared the frequency of acute reactions to poststorage plasma-removed platelets and standard platelets (Couban et al. 2002). Study platelets were prepared by removing the plasma from the stored component and replacing it with an equal volume of ABO-compatible fresh frozen and thawed plasma "to maintain a blind study design." While there was a trend toward a lower frequency of FNHTRs with poststorage plasma removal, the results were not statistically significant (Couban et al. 2002). An as yet unexplained finding of the study was that FNHTRs occurred less frequently in children than in adults. Based on the results of the study, we do not recommend removal of plasma with or without replacement with FFP as a method to reduce the incidence of FNHTRs in children.

Prevention of Circulatory Overload

The recentrifugation of platelet components to reduce their volume and prevent circulatory overload, particularly in low birth weight newborns, is a controversial topic. A typical argument to support the availability of such "superconcentrated platelets" is that a standard 50-mL unit of random-donor platelet concentrates represents more than half of the blood volume of a 1-kg infant and may precipitate circulatory overload (Moroff et al. 1984). In vitro studies of recentrifuged, volume-reduced platelet concentrates have demonstrated satisfactory viability and in vivo studies have demonstrated satisfactory survival of ^{51}Cr-labeled concentrated platelets and posttransfusion platelet count increments (Moroff et al. 1994; Simon and Sierra 1984). Arguments against routinely reducing the volume of platelet concentrates by recentrifugation focus on the reliability of transfusing 5 to 10 mL/kg of standard platelet concentrates or apheresis platelets to increase the platelet count to $>100 \times 10^9$/L (Strauss 2000b).

A secondary concern is that while satisfactory concentration of platelets may be achieved by experienced research technologists in the controlled environment of a limited study, routine transfusion service operations may not be able to achieve that level of quality control. A transfusion service undertaking recentrifugation of platelets must take special precautions to control for platelet loss, clumping, and dysfunction caused by additional handling (Strauss 2000b).

The Pediatric Hemotherapy Committee of the AABB published the results of a national survey of neonatal transfusion practices, noting "because of the potential for harm, institutions' transfusion volume-reduced platelets should monitor both the quality of the final product (i.e., the number of platelets, degree of clumping, and function) and in vivo effects such as post-transfusion increment in platelet count and adverse reactions, including altered vital signs and pulmonary distress" (Strauss et al. 1993). The Committee made the observation that 61% of respondents reported a final desired volume of 10 to 15 mL, and an additional 30% desired 18 to 25 mL, both volumes being within the range likely to achieve the targeted platelet count increase using an unmodified platelet concentrate.

Method to Volume-Reduce Platelet Concentrates

The previous discussion should leave the reader with the impression that routine recentrifugation of platelet concentrates is neither necessary nor prudent. As stated by the AABB Committee on Pediatric Hemotherapy, "... volume reduction of platelet concentrates ... should be reserved for special infants for whom marked reduction of all intravenous fluids is truly needed" (Strauss et al. 1993). For that special infant, we recommend the method of Moroff (Moroff et al. 1984). This method uses standard platelet concentrates that may be stored for as many as 5 days before recentrifugation, a standard blood bank refrigerated centrifuge, and manual resuspension of the centrifuged platelets.

Methods to Wash Platelets

Rarely, a patient may have a requirement for platelet transfusions and a contraindication for conventional plasma, such as those with IgA anaphylactic reactions or recurrent severe allergic reactions. While many blood centers maintain call-in lists of IgA-deficient donors for persons with IgA anaphylactic reactions, such special platelet components may not be readily available. Also, platelets collected from mothers of fetuses or newborns with NAIT require washing before transfusion to remove platelet alloantibodies.

Methods are available for manual (Mustard et al. 1972; Silvergleid et al. 1977) or automated washing using blood cell processors (Kalman and Brown 1982). Washing platelets requires skill and experience to avoid platelet loss, activation, and clumping. Automated cell processors are more likely to deliver a consistent result and are the recommended method. Kalman and Brown describe their protocol using the IBM 2991 Blood Cell Processor, which removed a mean of 99.6% of plasma protein following a 1500 mL wash using 0.9% sodium chloride. That cell processor is currently marketed as the COBE 2991 Blood Cell Processor (COBE Laboratories, Inc., Lakewood, CO).

References

Adner MM, Fisch GR, Starobin SG, and Aster RH. 1969. Use of "compatible" platelet transfusions in treatment of congenital isoimmune thrombocytopenic purpura. *N Engl J Med* 280:244–246.

Barnett EC, Dunlop JBW, and Pullar TH. 1951. Chronic haemolytic anemia with paroxysmal haemoglobinmuria (Marchiafava syndrome). Report of a case improved by splenectomy. *NZ Med J* 50:39.

Blanchette VS, Hume HA, Levy GJ, et al. 1991. Guidelines for auditing pediatric blood transfusion practices. *Am J Disease Child* 145:787–796.

Brecher ME. 2002. *Technical manual*, 14th ed. Bethesda, MD: American Association of Blood Banks.

Bull BS and Bull MH. 1990. The salvaged blood syndrome. *Blood Cells* 16:5–23.

Bussel JB, Berkowitz RL, Lynch L, et al. 1996. Antenatal management of alloimmune thrombocytopenia with intravenous gamma-globulin: a randomized trial of the addition of low-dose steroid to intravenous gamma-globulin. *Am J Obstet Gynecol* 174:1414–1423.

Couban S, Carruthers J, Andreou P, et al. 2002. Platelet transfusions in children: results of a randomized, prospective, crossover trial of plasma removal and a prospective audit of WBC reduction. *Transfusion* 42:753–758.

Dacie JV. 1948. Transfusion of saline–washed red cells in nocturnal haemoglobinuria (Marchiafava–Micheli disease). *Clin Sci* 7:65.

Eder AF and Manno CS. 2001. Does red–cell T activation matter? (Annotation). *Brit J Haematol* 114:25–30.

Estrin JA, Belani KG, Karnavas AG, et al. 1986. A new approach to massive blood transfusion during pediatric liver resection. *Surgery* 99:664–670.

Gilstad CW, Kessler C, and Sandler SG. 2002. Transfusion patients with anti-immunoglobulin A subclass antibodies. *Vox Sang* 83:363.

Hall TL, Barnes A, Miller JR, et al. 1993. Neonatal mortality following transfusion of red cells with high plasma potassium levels. *Transfusion* 33:606–609.

Heddle NM, Klama L, Meyer R, et al. 1999. A randomized controlled trial comparing plasma removal with white cell reduction to prevent reactions to platelets. *Transfusion* 39: 231–238.

Heddle NM, Klama L, Singer J, et al. 1994. The role of the plasma from platelet concentrates in transfusion reactions. *N Engl J Med* 331:625–628.

Hillmen P, Lewis SM, Bessler M, et al. 1995. *N Engl J Med* 333:1253–1258.

Hirsch J, Ungar B, and Robinson JS. 1964. Paroxysmal nocturnal haemoglobinuria: an acquired dyshaemopoiesis. *Aust Ann Med* 13:24.

Inaba S, Nibu K, Takano H, et al. 2000. Potassium-adsorption filter for RBC transfusion: a Phase III clinical trial. *Transfusion* 40: 1469–1474.

Jimenez DF and Barone CM. 1995. Intraoperative autologous blood transfusion in the surgical correction of craniosynostosis. *Neurosurgery* 37:1075–1079.

Johnson JA, Ryan G, al-Musa A, et al. 1997. Prenatal diagnosis and management of neonatal alloimmune thrombocytopenia. *Semin Perinatol* 1:45–52.

Kalman ND and Brown DJ. 1982. Platelet washing with a blood cell processor. *Transfusion* 22:125–127.

Kletzel M, Olszewski M, Danner-Koptik K, et al. 1999. Red cell salvage and reinfusion in pediatric bone marrow donors. *Bone Marrow Transplant* 24:385–388.

Kroll H, Keifel V, and Santoso S. 1998. Clinical aspects and typing of platelet alloantigens. *Vox Sang* 74(Suppl 2):345–354.

Kruger LM and Colbert JM. 1985. Intraoperative autologous transfusion in children undergoing spinal surgery. *J Pediatr Orthop* 5: 330–332.

Larsson LG, Welsh VJ, and Ladd DJ. 2000. Acute intravascular hemolysis secondary to out-of-group platelet transfusion. *Transfusion* 40:902–906.

Luban NLC. 1995. Massive transfusion in the neonate. *Transfusion MedicineReviews* 9:200–214.

Luban NLC, Strauss RF, and Hime HA. 1991. Commentary on the safety of red cells preserved in extended storage media for neonatal transfusions. *Transfusion* 31:229–235.

Manchester RC. 1945. Chronic hemolytic anemia with paroxysmal hemoglobinuria. *Ann Intern Med* 23:935.

McFarland JG. 1998. Platelet and neutrophil alloantigen genotyping in clinical practice. *Transfus Clin Biol* 5:13–21.

Moroff G, Friedman A, Robkin-Kline L, et al. 1984. Reduction of the volume of stored platelet concentrates for use in neonatal patients. *Transfusion* 24:144–146.

Mustard JF, Perry DW, Ardlie NG, and Packham MA. 1972. Preparation of washed platelets from humans. *Brit J Haematol* 22:193–204.

Muylle L, Wouters E, and Peetermans ME. 1996. Febrile reactions to platelet transfusion: the effect of increased interleukin 6 levels in concentrates prepared by the platelet–rich plasma method. *Transfusion* 36:886–890.

Pierce RN, Reich LM, and Mayer K. 1985. Hemolysis following platelet transfusions from ABO incompatible donors. *Transfusion* 25:60–62.

Pineda AA, Zylstra VW, Clare DE, et al. 1989. Viability and functional integrity of washed platelets. *Transfusion* 29:524–527.

Ramasethu J and Luban NLC. 2001. T activation. *Br J Hematol* 112:259–263.

Ramirez G, Romero A, Garcia-Vallejo JJ, and Munoz M. 2002. Detection and removal of fat particles from postoperative salvaged blood in orthopedic surgery. *Transfusion* 42:66–75.

Ranasinghe E, Walton JD, Hurd CM, et al. 2001. Provision of platelet support for fetuses and neonates affected by severe fetomaternal alloimmune thrombocytopenia. *Brit J Haematol* 117:482–483.

Rizk S, Ibrahim IY, Mansour IM, and Kandil D. 2002. Screening for paroxysmal nocturnal hemoglobinuria (PNH) clones in Egyptian children with aplastic anemia. *J Trop Pediatr* 48:132–137.

Rosse WF. 1990. *Clinical immunohematology: basic concepts and clinical applications.* Boston: Blackwell Scientific Publications.

Sandler SG, Eckrich R, Malamut D, and Mallory D. 1994. Hemagglutination assays for diagnosis and prevention of IgA anaphylactic transfusion reactions. *Blood* 84:2031–2035.

Sandler SG, Malloy D, Malamut D, and Eckrich R. 1995. IgA anaphylactic transfusion reactions. *Transfusion Medicine Reviews* 9:1–8.

Sandler SG, Yu H, and Rassai N. 2003. Risks of blood transfusion and their prevention. *Clinical Advances in Hematol and Oncol* 1:120–124.

Scanlon JW and Krakaur R. 1980. Hyperkalemia following exchange transfusion. *J Pediatr* 96:108–110.

Sherman SP and Taswell HF. 1977. The need for transfusion of saline-washed red blood cells to patients with paroxysmal hemoglobinuria: a myth (abstract). *Transfusion* 17:683.

Sherwood WC, Clapper C, and Wilson S. 2000. The concentration of AS-1 RBCs after gravity sedimentation for neonatal transfusion. *Transfusion* 40:618–619.

Silvergleid AJ, Hafleigh EB, Harbin MA, et al. 1977. Clinical value of washed-platelet concentrates in patients with nonhemolytic transfusion reactions. *Transfusion* 17:33–37.

Simon TL and Sierra ER. 1984. Concentration of platelet units into small volumes. *Transfusion* 24:173–175.

Skupski DW and Bussel JW. 1999. Alloimmune thrombocytopenia. *Clin Obstet Gynecol* 42:335–348.

Strauss RG. 2000a. Data-driven blood banking practices for neonatal RBC transfusions. *Transfusion* 40:1528–1520.

Strauss RG. 2000b. Neonatal transfusion. In *Scientific basis of transfusion medicine: implications for clinical practice*, 2nd ed. KC Anderson and PM Ness, eds. Philadelphia: W.B. Saunders Company.

Strauss RG, Burmeister LF, Johnson K, et al. 1996. AS-1 red cells for neonatal transfusions: a randomized trial assessing donor exposure and safety. *Transfusion* 36:873–878.

Strauss RG, Levy GJ, Sotelo-Avila C, et al. 1993. National survey of neonatal transfusion practices: II. Blood component therapy. *Pediatrics* 91:530–536.

Strauss RG, Villhauer PJ, and Cordle DG. 1995. A method to collect, store and issue multiple aliquots of packed red blood cells for neonatal transfusions. *Vox Sang* 68:77–81.

Stroncek D. 2002. Neonatal alloimmune neutropenia and alloimmune thrombocytopenia. In *Pediatric transfusion therapy*. JH Herman and CS Manno, eds. Bethesda, MD: AABB Press.

Uhrynowska M, Maslanka K, and Zupanska B. 1997. Neonatal thrombocytopenia: incidence, serological and clinical observations. *Am J Pernatol* 14:415–418.

van Loghem JJ, van der Hart M, and Land ME. 1955. Polyagglutinability of red cells as a cause of severe haemolytic transfusion reaction. *Vox Sang* 5:125.

Wainwright L, Brodsky RA, Erasmus LK, et al. 2003. Paroxysmal nocturnal hemoglobinuria arising from Fanconi anemia. *J Pediatr Hematol Oncol* 25:67–168.

Williams RA, Brown EF, Hurst D, and Franklin LC. 1989. Transfusion of infants with activation of erythrocyte T antigen. *J Pediat* 115:949–953.

Wynn RF, Stevens RF, Bolton–Maggs PH, et al. 1998. *Clin Lab Haematol* 20:373–375.

11

Technical Considerations/Mechanical Devices

PATRICIA T. PISCIOTTO, MD, AND EDWARD C.C. WONG, MD

INTRODUCTION

A plethora of mechanical devices are utilized for the collection as well as the infusion of blood and blood components. These include needles and catheters used to ensure vascular access and semiautomated cell processors for autologous blood collection and apheresis procedures. Because of the necessity for precise volumetric control of fluids and components to avoid rapid increases in intravascular volume, infants and small children create unique challenges. The need to minimize hemolysis is also of concern since this may lead to hyperkalemia and hyperbilirubinemia in infants who may already be metabolically compromised. This chapter will focus on the use and evaluation of mechanical devices and accessories used to deliver blood and blood components including manual exchange transfusion. Autotransfusion techniques and devices utilized in the pediatric setting will be discussed in Chapters 4 and 13.

INTRAVENOUS DELIVERY SYSTEMS FOR BLOOD AND COMPONENTS

When evaluating the impact of the system utilized to deliver blood and blood components, the entire transfusion apparatus setup needs to be considered, including needle gauge, length and diameter of the tubing, flow rate, type of filter, and the mechanism of the infusion device. When considering red cell transfusions, variables related to the component itself also may affect the life span of the red cells; they include the type of product, age and temperature of the unit, and whether the product has been manipulated before infusion by washing, centrifuging, or irradiation.

The vast majority of the literature has focused on the potential of these systems to induce red cell hemolysis. However, the literature is somewhat controversial since most studies have not controlled for all the variables that may be involved in delivery of the component that could affect the integrity of the red cell. There also are no established, definitive tests to evaluate hemolysis, and no consensus on the percent of hemolysis that is clinically acceptable. The U.S. Food and Drug Administration (FDA) has recommended a maximum of 1% hemolysis for deglycerolized red blood cells (RBCs) but has not established official guidelines for an acceptable level of hemolysis for transfusion of other red cell products (Sowemimo-Coker 2002). Various parameters have been used to characterize the degree of hemolysis, including plasma potassium, plasma free hemoglobin, plasma lactate dehydrogenase (LDH), and osmotic fragility. Wilcox et al. (1981) have suggested a hemolysis rate of 0.2% as being clinically acceptable, which some investigators have used as a gauge in analyzing their data, even though the recommendation is not based on objective evidence. Much less data evaluate the effects of delivery systems on the integrity of platelets and/or granulocytes.

Needle Size

The impact of needle size on hemolysis depends on the system examined and is based primarily on ex vivo studies. In 1959, McDonald and Berg reported the development of hemoglobinemia and hemoglobinuria in a 1-month-old infant when 55 mL of 10-day-old whole blood was injected manually in an intermittent fashion over an hour through a small caliber needle (23 gauge) into a scalp vein. This subsequently led to ex vivo laboratory studies of small gauge needles that demonstrated a noticeable increase in hemolysis at flow rates as low as 8 mL/min (23 gauge) up to 84 mL/min (21 gauge). Moss and Staunton (1970) showed that very small needles, from 20 to 25 gauge, could be used to draw blood without producing hemolysis. However, hemolysis (~15%) was observed when heparinized blood was forcibly (150 psi) ejected through a large bore needle (20 gauge) in contrast to the hemolysis observed (~0.17%) using a smaller bore needle (25 gauge). In this study a power injector and steel syringe were used to generate the constant delivery pressure. It was postulated that the turbulence generated most likely resulted in nonlaminar flow causing a higher degree of hemolysis. A similar trend was observed in a study by Eurenius and Smith (1973) when an external pressure device (up to 300 mmHg) was used to force whole blood (drawn into citrate, phosphate, dextrose [CPD] anticoagulant, Hct 41% to 43%) through various sized needles (18 to 26 gauge). In this study, however, the plasma hemoglobin concentration was minimally elevated even under the worst conditions (defined as 7-day-old blood infused through an 18-gauge needle at external pressure of 300 mmHg), and the elevation of plasma hemoglobin was not thought to be clinically significant. Herrera and Corless (1981) testing a syringe pump at various flow rates (20, 50, 100 mL/hr) and needle sizes (21 to 27 gauge) showed no evidence of hemolysis using fresh (<24 hr) whole blood or packed cells (hematocrit 45% vs. 75%). Wilcox and associates (1981), also using a similar pump and altering the flow rate while using a 25-gauge needle, showed hemolysis with older whole blood (9 days old, hematocrits ranging from 55% to 75%) especially at lower flow rates. In contrast, fresher sedimented RBCs (2 days old) did not demonstrate hemolysis at flow rates ranging from 10.6 to 70 mL/hr. The total degree of hemolysis observed, however, was approximately 0.2%. These studies together suggest that under constant flow rates hemolysis appears to be minimal even with administration through small gauge needles at least to a gauge of 23. However, if only fresh blood is used, transfusion using a needle with a gauge as low as 27 may also be associated with minimal hemolysis.

The one in vivo study that was performed (Humphrey et al. 1982) utilized a Harvard pump with blood administered through a 25-gauge needle over 2 hours. This study involved neonates receiving 10 mL/kg of either undiluted or diluted packed red cells as well as fresh blood (less than 5 days old) or older blood (greater than 14 days old). Plasma hemoglobin levels were significantly increased from baseline 2 hours after transfusion with a return to pretransfusion levels by 6 hours. There was significantly more hemolysis in the group receiving diluted blood with no difference observed based on the age of the blood. Neither the age of the blood nor the dilution significantly altered the 6 hour or 1 week posttransfusion hematocrits. So while hemolysis was observed, it cleared rapidly and had no detrimental effect on ultimate hematocrit level achieved. An in vitro study (Calkins et al. 1982) performed under laminar flow conditions without a needle could not demonstrate a hemolytic effect related to external pressure, flow rate, or length of administration tubing using older packed RBCs (19 to 24 days), suggesting that passage through the needle itself may account for more of the damage observed in other studies. In contrast to the Humphrey study, however, undiluted red cells hemolyzed more than diluted red cells, but the degree of hemolysis was not significant.

Little data are available on the effects of needle size on the infusion of platelets. One study that was designed to evaluate the effects of electromechanical infusion pumps on platelet function did utilize two different gauge needles (25 versus 18). There did not appear to be any affect on in vitro platelet protein discharge or release based at pump speeds ranging between 100 to 999 mL/hr using an 18-gauge needle as compared to a 25-gauge needle at a pump speed of 200 mL/hr. Furthermore, in vivo studies demonstrated similar platelet survival between platelets infused without a pump and those infused at 400 mL/hr through a pump system (Snyder et al. 1984).

Catheters

Several studies have examined the effect of catheter type and size on the administration of whole blood and packed RBC (PRBC) products. Oloya et al. (1991) studied the effect of a small gauge percutaneous catheter (24 gauge, 30.5 cm length) and a short catheter (Quike Cath, 24 gauge, 1.6 cm length) on PRBC transfusion. Three-day-old PRBCs (hematocrits ranging from 88% to 92%) were drawn through a PDF-20 Pediatric Transfusion Filter Set (Fenwal Products, Deerfield, IL) into an IVAC model 700 syringe (IVAC Co., San Diego, CA) and subsequently studied at a flow rate of 10 mL/hr (a flow simulating flow rates in preterm infants). Significant increases of similar magnitude in plasma hemoglobin were noted when PRBCs were infused through either catheter (median 179 and 133 mg/dL for short and percutaneous catheter, respectively). Plasma potassium also increased through either catheter (median 1.2 and 2.0 mmol/L for short and percutaneous catheter, respectively), but the increase was only significant for the short catheter. The findings in this study suggested that hematocrit and gauge and not length were major factors influencing hemolysis. Frelich and Ellis (2001) examined the effects of external pressure (150 versus 300 mmHg using an external pneumatic device), catheter gauge (16 to 22 gauge; Mediflon, Delhi, India) and storage time (10.8 versus 28.9 days old) on hemolysis associated with PRBC transfusion. Multiple regression analysis revealed that the age of the unit and external pressure applied were significant independent determinants of the degree of hemolysis on all parameters studied: hemoglobin, hematocrit, RBC count, plasma free hemoglobin concentration, LDH, and potassium. Catheter gauge was found to be a significant determinant of the hematocrit, RBC count, and free hemoglobin concentration only. However, there was no significant difference in the effect of the pressure applied or catheter diameter on the degree of hemolysis between the older and fresher units. Comparing the most stringent versus the most lenient conditions, the relative decrease in hemoglobin, RBC count, and hematocrit was relatively small (5%), with only plasma free hemoglobin demonstrating large differences (15.5% to 20%). Changes in potassium were not significant, and no trend toward increased potassium levels with the most stringent conditions was noted. Thus, hemolysis under these conditions is not felt to be clinically significant.

Several studies on central venous catheters for infants and neonates have been reported. Beebe et al. (1995) described the transfusion through Hickman (central venous, silastic, internal diameter 1.6 or 2.0 mm), Arrow (peripheral, teflon sheath, internal diameter 2.0 to 2.85 mm) and Jelco (peripheral, teflon sheath, internal diameter 1.65 and 2.11) catheters. Blood was warmed using a Level 1 H-500 fluid warmer (Level 1 Technologies Inc., Rockland, MA) and subject to external pressure of 300 mmHg. Although comparable rapid flow was possible through these devices (ranging from 168 to 755 mL/min) no evaluation of hemolysis was undertaken. A study by Soong et al. (1993), described the transfusion of fresh whole blood through either a small polyurethane catheter (Vialon, Becton, Dickinson, UT), median size silastic catheter (Dow Corning Corp., Midland, MI) secured to a Jelco 24-gauge cannula, or a large size silastic catheter (Dow Corning Corp., Midland, MI) secured to a Jelco 22 gauge cannula. Infusion rates ranged from 10 to 40 mL/hr. Transfusion through these catheters demonstrated no significant difference in RBC count, hematocrit, or potassium level suggesting that transfusion through a variety of silastic central venous catheters is not associated with significant hemolysis. Wong et al. (2000) studied PRBC transfusion through 1.9 Fr NeoPICC (Klein-Baker Medical, San Antonio, TX) central venous catheter at 2 and 20 mL/hr over a 2- to 4-hour period. This transfusion study simulated transfusion of PRBCs using a syringe pump and was unique in that a pressure transducer was incorporated into the experimental setup. Little if any hemolysis was noted despite tremendous changes in intraluminal pressure during the transfusion suggesting that transfusion using this apparatus is feasible. Further study, however, is necessary to determine if this central venous catheter can also be used with other parenteral fluids in the neonatal intensive care setting.

Filters

All blood and blood components must be filtered according to *AABB Standards* (Gorlin 2002). Some transfusion services dispense aliquots of blood for neonatal transfusion. There are a variety of systems available in which the primary red cell unit can be entered via a sterile connecting device and aliquots of blood drawn off either into an aliquot bag or a syringe (Roseff 2002). The blood is usually passed through an in-line filter ranging from 150 to 195 microns (standard filter) as it is withdrawn from the primary unit. This serves as the final filtration step in some blood banks. Many of the reported studies looking at transfusion practices on red cell hemolysis either did not use filters or did not specify whether a filter was used (Luban 1985). In the study by Herrera (1981), a standard filter (170 micron) was included in the administration apparatus with no evidence of hemolysis.

Under selected clinical or experimental conditions, filters other than the standard blood filters can result in undesirable transfusion complications. For example, there are no controlled clinical studies evaluating the use of microaggregate filters in the neonate; therefore, there is no evidence-based data for routine use. Often these filters were selected because of small priming volumes. Transfusion of blood through a screen microaggregate filter has been associated with hemoglobinemia and hemoglobinuria in two infants (Schmidt et al. 1982). In this report, follow-up in vitro studies demonstrated a twofold increase in hemolysis when the infusion rate was decreased from 200 mL/hr to 20 mL/hr. In another in vitro study of a microaggregate filter by Longhurst et al. (1983), negative pressure filtration and older blood units also demonstrated increased hemolysis. Therefore, if it is necessary to use a microaggregate filter, fresher units should be selected, and the faster rates of infusion should be considered with avoidance of negative pressure filtration. RBCs may also be damaged if forced quickly through leukocyte-reduction filters (Sowemimo-Coker 2002; Gambino 1992; Ma 1995). An increased filtration time and greater degree of hemolysis has been reported with older versus fresher additive solution red cells during leukocyte-reduction filtration (Gammon et al. 2000).

Standard blood filters appear to have little effect on either the quantitative or qualitative aspects of platelet transfusions. In addition, prestorage leukoreduction filters do not appear to affect random donor or apheresis platelets' viability or function and are associated with decreased incidence of febrile, nonhemolytic transfusion reactions secondary to decreased cytokine levels (Sweeney et al. 1995; Ogawa et al. 1998; Heddle 1994). However, leukocyte-reduction filtration of a single unit of random donor platelet concentrate can result in losses (secondary to filter and tubing) as high as 20% with a possible need for adjustment of the ordered dose.

Recently, several reports have documented clinically significant hypotensive reactions associated with platelet transfusions, with greater than 85% of the components involved having undergone leukocyte reduction by filtration (Hume et al. 1996). Negatively charged surfaces can activate the contact system of coagulation, which includes factor XII, prekallikrein, and high-molecular weight kininogen with the generation of bradykinin, a substance known to increase vasopermeability and induce vasodilatation. It has been shown that filtration through negatively charged filters has resulted in increased levels of bradykinin, which was not observed with the use of positively charged filters (Shiba et al. 1997). It has been suggested that increased production of bradykinin combined with decreased metabolism, such as may occur in patients receiving ACE inhibitors, may play a role in some of these reactions. Such reactions have been observed in children (Yenicesu et al. 1998).

Because of the nature of granulocyte concentrates, leukoreduction filters should not be used as they will clog the filter. In addition, depth-type microaggregate filters have been shown to retain 20% to 60% of the granulocytes' concentrate depending on the filter used. This is in contrast to screen microaggregate filters or standard blood filters, which appear to retain only 1% to 3% of the transfused product and do not affect granulocyte function (Snyder et al. 1983). Therefore, granulocyte transfusion should be transfused using either standard blood filter or screen microaggregate filters.

Electromechanical Infusion Pumps

Electromechanical pump systems have been used for several decades for the delivery of intravenous fluids and medications. Many of these systems have also been used to deliver cellular blood components. They are used when the accuracy in the volume to be delivered or the flow rate required cannot be provided by manually adjusted gravity systems. Infusion pumps are available that can deliver metered volumes at flow rates ranging from as low as 0.1 mL/hr up to 999 mL/hr. Infusion devices that are classified for neonates require high accuracy and consistency of flow and have a low-pressure occlusion setting (up to 300 mmHg) with low bolus on release of occlusion (Quinn 2000). Higher pressure infusion devices are usually more suitable for older children. The main difference between these devices is that the neonatal devices are more accurate in the short term (up to 60 minutes) when infusing at rates less than 5 mL/hour and the occlusion pressure is set lower (up to 300 mmHg versus 500 mmHg).

There are many types of pumps available that employ different mechanisms of delivery including: peristaltic mechanism, which squeezes the administration set tubing to create flow; piston-actuated syringe-type mechanism that uses a constant high speed refill cycle of the syringe followed by a slower ejection phase controlled by a check valve; and a piston-actuated diaphragm cassette mechanism in which the blood is gravity fed into the cassette and the flow is controlled by two pistons acting against two check valves (Denison et al. 1991). One of the main concerns with the use of these devices has been potential red cell damage caused by the shear force of the pumping apparatus.

Gibson and associates (1984) evaluated three types of pumps (peristaltic, diaphragm, and piston) maintaining a constant age (2 to 3 days) and temperature of the blood. They found that the peristaltic-type pump produced the greatest hemolysis of PRBCs being higher at lower flow rates (5 mL/hr versus 50 mL/hr). This was in contrast to whole blood, which showed greater hemolysis at the higher rate possibly reflecting the difference in the viscosity of the blood. They found little effect on hemolysis when altering needle size (25 versus 18 gauge) or tubing diameter or length as long as the flow rate was less than 50 mL/hr. They recommended that PRBCs should not be infused through either the peristaltic or diaphragm devices and that the piston device was suitable for both red cells and whole blood. Veerman et al. (1985) confirmed minimal hemolysis using two piston-type pumps (IMED 960 and IMED 965, IMED Corp, San Diego, CA) infusing 2- to 3-day-old blood at 5, 10, 25, 50, and 100 mL/hr through a 22-gauge needle.

There have been subsequent studies evaluating the effects of different linear peristaltic infusion devices in which minimal hemolysis was observed (Gurdack et al. 1989; Burch et al. 1991). In the latter study, which involved the Gemini PC-2 (IMED Corp, San Diego, CA), both fresh and stored packed red cells preserved in adenine-saline 3 (AS-3) and citrate-phosphate-dextrose-adenine-1 (CPDA-1) as well as CPDA-1 whole blood were evaluated and included flow rates of

5, 50, 100, and 999 mL/hr to simulate both neonatal and adult flow rates. A needle-free system was used to eliminate any potential effect of needle gauge on red cell integrity and no in-line filter was used. It appears that there is no clear relationship between the type of linear peristaltic device and the degree of red cell damage. Each new infusion system needs to be evaluated individually in order to establish its suitability for infusion of blood. While no in vivo studies have been performed simulating pediatric transfusions, 24-hour survival studies have been performed using various infusion pumps that utilize the piston-diaphragm mechanism showing no evidence of decreased viability (Linden et al. 1988; Simon et al. 1994).

Infusion pumps have also been used to administer platelet components. Snyder and associates (1984, 1990) passed platelet concentrates through different infusion pumps using a piston-diaphragm mechanism under standard transfusion conditions using a 170-micron filter and 18-gauge needle. In vitro and in vivo (adult studies) measures of platelet function showed no change. Studies simulating clinical conditions in the neonatal setting using 24- and 25-gauge needles and infusing platelet concentrates slowly (60 mL/hr) through both a diaphragm-type and syringe-type pump (mounted) showed no difference in in vitro platelet function (Luban 1985). Other studies by Norville and associates (1994, 1997) using either the piston syringe pump (IMED 980, IMED Corp., San Diego, CA) and linear peristaltic pump (Gemini, IMED Corp.) demonstrated no difference in platelet count or morphology count compared to gravity flow at flow rates of 150 mL/hr. Further studies using the Gemini pump also found no differences in platelet count or morphology at flow rates up to 999 mL/hr. No needles or catheters were used in the in vitro studies. Follow-up in vivo studies in pediatric oncology patients (age 2 to 21 years) using the Gemini infusion pump through the central venous access device in place demonstrated no difference in 1 hour corrected count increment at either 5 or 10 mL/kg/hr compared to gravity flow at the same rates using a randomized crossover study design. Therefore the use of these types of infusion devices is compatible for use with platelet concentrates.

Only one study has evaluated the effects of the passage of granulocyte concentrates through an electromechanical pump (Snyder et al. 1986). Granulocyte concentrates that had been stored at room temperature for 16 to 18 hours were infused using a diaphragm-type pump at a flow rate of 100 mL/hr through a standard 170-micron filter and either 19- or 23-gauge needle. There was no evidence of hemolysis; loss of granulocytes; or change in neutrophil function, complement activation, or interference with platelet function. No in vivo studies to determine posttransfusion survival were performed, but based on in vitro studies, the delivery system did not appear to be detrimental to granulocyte integrity.

Rapid Infusion Devices/Blood Warmers

If a child is to undergo a procedure that may lead to rapid blood loss and hypovolemia, the site and type of intravenous access becomes extremely important in order to allow for rapid infusion. A hemorrhage of 1.5 mL/kg/minute would require at least 5 mL/kg/minute of fluid to be infused in order to maintain adequate blood pressure (Barcelona and Cote 2001). Several factors may affect flow rate including venous pressure; resistance in lines, filters, valves, and canulae; and the intrinsic resistance or viscosity of the fluid being infused. According to Poiseuille's Law (see the following formula) the flow is directly proportional to the fourth power of the radius of the tube and the pressure gradient across the tube and inversely proportional to the length of the tube and viscosity of the fluid, the latter altered by its temperature. Maximum flow therefore should be achieved with high-pressure gradients through a large diameter and short length catheter.

Poiseuille's Law: $\text{Flow} = [3.14 \times (P_1 - P_2)R^4]/8nL$

Where P_1 and P_2 = Fluid pressures at either end of tube, R = Radius of the tube, n = Viscosity of fluid, L = Length of tube.

Studies looking at flow rates of both saline and blood through 18- to 24-gauge catheters confirmed that short, large diameter catheters had higher flow rates and that flow under pressure was 17 times greater than that in a longer smaller diameter (Hodge III and Fleisher 1985). While central veins can accept larger catheters, the studies showed that catheters designed for peripheral use had a higher flow rate than the same gauge catheter designed for central compartment insertions. Flow rates obtained by gravity through a 4-, 5-, and 6-French Teflon catheter have been shown to be much faster (>300%) than with 22-, 20-, or 18-gauge intravenous catheters (Idris and Melker 1992). In neonates when using small 22- or 24-gauge catheters, manual syringe injection has been found more efficient than manual pressurization of a pump chamber (Barcelona and Cote 2001).

Massive blood loss requires a system that is able to infuse blood and blood components at a sufficient flow rate while maintaining the patient's core body temperature at or near 37°C. Hypothermia is a concern when rapid infusions are needed for pediatric patients because of their large surface area to weight ratio. Heat loss may be accelerated in patients undergoing surgical procedures, particularly if associated with an open thoracic or abdominal cavity, thereby exacerbating hypothermia. There are several consequences to hypothermia including decreased oxygen delivery to tissues as a result of increased oxygen affinity of hemoglobin (Cote 1991), decreased tissue perfusion secondary to vasoconstriction, and impaired platelet function (Valeri et al. 1987). Hypothermia can also lead to decreased drug metabolism as well as increase the potential for hypocalcemia since a cold liver does not metabolize citrate as well. Cold blood that may be rapidly infused through a central line with the tip near the sinoatrial node can cause fatal arrhythmias (Boyan and Howland 1961).

In-line blood/solution warmers are used during a variety of clinical procedures to warm fluids in order to help maintain the patient's core body temperature and minimize adverse thermal reactions. These devices warm a fluid as it passes from its source to the patient. The devices presently available are usually categorized by the method of heat exchange: water bath, countercurrent flow, dry heat, forced air, and microwave. There is more experience with the first three technologies since the latter two are relatively new. The conventional water bath units warm blood as it passes through either a disposable bag or coil of tubing immersed in a bath, with some of the units incorporating agitation of the water to improve heat exchange. The countercurrent flow technology is an improvement on the conventional water bath in that blood will flow through a lumen in one direction with heated water pumped in surrounding tubing flowing counter to the direction of the fluid. The dry heat warmers usually have blood either in a disposable bag or cassette between heating plates or in a bag wrapped around a heating element.

Many of the devices in use are not appropriate for children because of the large priming volume. Also, at the lower flow rates, fluid in the set may cool substantially before reaching the patient. One study showed that a pediatric blood warmer utilizing a countercurrent heat exchange technology (System 250™ LEVEL1^R Technologies Inc., Marshfield, MA) provided greater warming and a lower resistance to blood flow at flow rates greater than 40 mL/min compared to a more conventional blood warmer utilizing dry heat (Presson et al. 1990). Since resistance to flow depends on both the blood warmer and the IV catheter, flow through the warmers was evaluated using a 16-gauge catheter and a 6-French catheter. While the resistance to flow was lower using the countercurrent warmer for both catheters, the use of the 6-French catheter permitted flow rates that exceeded the capacity of either blood-warming device to warm the blood adequately. Flow rates of cold blood through the countercurrent system needed to be restricted to less than 250 mL/min, since

warming was inadequate above that rate. Comparison studies evaluating flow rates and thermodynamics of rapid infusion systems with countercurrent heating coils have been performed with catheters as small as 20 gauge for room temperature crystalloid solutions but not blood (Barcelona et al. 2000). The rapid infusion system was found to provide better flow characteristics and heating capabilities for catheters greater than 18 gauge.

There have been several issues raised associated with blood/solution warmers and administration. Of particular concern is overheating of red cells from improper heater control. Excessive heat alters the red cell membrane, which in turn affects elasticity, deformability, and osmotic fragility resulting in red cell lysis. The damage that occurs will depend on a combination of the temperature as well as the exposure time (Anonymous 1996). According to AABB *Standards* (Gorlin 2002) warming devices shall be equipped with visible thermometers and with a warning system to detect malfunctions and prevent hemolysis. It is important to realize that exposure to heat during red cell administration can occur even without using a blood warmer. Hemolysis in the neonatal setting can occur from tubing exposed to radiant warmers and phototherapy units (Strauss et al. 1986; Opitz et al., 1988). It is recommended that tubing be covered with aluminum foil or routed in such a way as to avoid exposure to these potential heat sources.

EXCHANGE TRANSFUSION IN THE NEWBORN

Exchange transfusion in the neonate constitutes a massive transfusion since it involves the replacement of one to two whole blood volumes. The primary indication for exchange in the newborn period is to reduce unconjugated bilirubin levels to prevent kernicterus in those infants who have failed other therapy. The frequency of exchange transfusion in the postnatal period has decreased significantly as a result of the use of Rh immune globulin to prevent Rh isoimmunization, approaches to prenatal management of hemolytic disease, and more aggressive management of jaundice in the newborn. Before performing an exchange transfusion, it is important to balance the risk of the procedure to the benefit of potentially preventing kernicterus.

Multiple factors are considered in the decision to perform an exchange transfusion including: gestational age, evidence of hemolysis, degree of anemia, rate of rise of bilirubin, and concurrent clinical conditions, such as asphyxia, acidosis, and hypoalbuminemia that may exacerbate bilirubin entry and toxicity to the central nervous system. A double volume exchange (85 mL/kg × 2 for term infants up to 100 mL/kg × 2 for very low birth weight infants) should remove approximately 90% of the fetal circulating red cells and a single volume exchange approximately 70% to 75%. The bilirubin level, however, is only decreased by approximately 50% due to tissue re-equilibration with a rebound in bilirubin level to approximately 60% of the pre-exchange level (Valaes 1963).

Mechanics of Manual Exchange Transfusion

Since exchange transfusion is considered a massive transfusion, this is a situation in which a thermostatically controlled blood warmer should be utilized (see page 126). There are two general techniques for manual exchange transfusion. The isovolumetric method first described by Wallerstein (1946) requires two sites of vascular access since blood is simultaneously withdrawn and replaced in order to avoid sudden changes in arterial pressure. In this setting blood is usually withdrawn from the umbilical artery and replaced through the umbilical vein. While this method carries the additional hazard of a second catheter, it eliminates the swings in blood volume and pressure that contribute to hemodynamic changes that may cause other complications. Techniques have been described in which syringe pumps are used to mechanically infuse and withdraw blood (Goldman and Chung Tu 1983). If the umbilical vessel cannot be cannulated, which is a concern in older infants with nonpatent umbilical stumps or in infants already established to have necrotizing enterocolitis, peripheral blood vessels have been used to perform exchanges. Peripheral arterial catheters for blood withdrawal and peripheral venous catheters for replacement are used for this technique (Campbell and Stewart 1979). An in vitro study of catheter-related hemolysis comparing the use of a 22-gauge Medicut versus a 5-French umbilical catheter at flow rates of 5, 10, and 15 mL/min showed that blood replacement through the peripheral catheter showed a smaller rise in plasma hemoglobin than through the conventional umbilical catheter (Campbell and Stewart 1979). The overall increase in plasma hemoglobin was relatively small. In a composite of 60 exchange transfusions utilizing peripheral vessels in 47 infants from various studies there were no reported complications attributed to the cannulations (Luban 1985). There were instances, however, of losing arterial access requiring recannulation.

The more commonly used method for exchange transfusion is the discontinuous technique described by Diamond et al. (1951). Small aliquots of blood are withdrawn and replaced through a single catheter with a special four-way stopcock. Usually no more than 5 mL/kg body weight or 5% of the infant's blood volume are removed and replaced during a 3 to 5-minute cycle. It is important that the exchange is not performed rapidly since sudden hemodynamic changes may affect cerebral blood flow and intracranial pressure contributing to intraventricular hemorrhage (Bada et al. 1979). Negative pressure is induced by withdrawal of blood through the umbilical vein, which may be transmitted to the mesenteric veins that could contribute to ischemic bowel complication associated with exchange transfusions (Touloukin et al. 1973). The total time duration for a double-volume exchange is 90 to 120 minutes or 45 to 60 minutes for a single-volume exchange.

Practical Considerations

It is important to ensure that the infant is stable before initiating an exchange transfusion. Besides the potential complications related primarily to administration of blood components, one must be familiar with procedure-related complications (Table 11.1). The infant should be under a warmer for ready accessibility and needs to be monitored closely throughout the procedure. This should include temperature, cardiopulmonary, and pulse-oximetry monitoring to evaluate pressure fluctuations as well as laboratory monitoring (complete blood counts, electrolytes, glucose, calcium) to avoid potential metabolic complications or coagulopathies. If at all possible, the infant should not be fed for 4 hours before the procedure, gastric contents should be removed, and an orogastric tube placed to prevent aspiration.

How the procedure is initiated will be determined by the infant's clinical status: if hypovolemic (low central venous pressure [CVP]) then one would start with transfusing an aliquot; if hypervolemic (high CVP) then the start would be by withdrawing a precalculated aliquot. Several combinations of blood components have been used for exchange transfusions ranging from ABO compatible whole blood to RBCs resuspended in compatible plasma. In either scenario, relatively fresh blood (that is, not >5 to 7 days old) is used to avoid high potassium levels and ensure maximum red cell survival. Units of blood are usually cytomegalovirus (CMV) negative, irradiated, and tested for hemoglobin S. Irradiation can potentiate potassium leakage from red cells; therefore this process should either be performed shortly before dispensing the component or, if that is not feasible, it has been suggested that the supernatant should be removed or the red cells washed to avoid hyperkalemic cardiac complications (Luban 1995). When using reconstituted whole blood the final hemat-

TABLE 11.1   Complications of Exchange Transfusion

Infections	Bacterial
	Viral
	Fungal
Metabolic Complications	Hyperkalemia
	Hypocalcemia (during or after exchange)
	Hypoglycemia (during or after exchange)
	Hyperglycemia
	Hypernatremia
	Late onset alkalosis (citrate metabolism)
Hematologic Complications	Hemolysis (intrinsic red cell problem or mechanical)
	Anemia/polycythemia (inadequate mixing, inappropriate unit Hct, patient volume deficit/surplus)
	Thrombocytopenia
	Neutropenia
	Coagulopathy (alteration of factors, DIC, thromboembolism)
	Graft-versus-host reaction (delayed effect)
Cardiovascular	Arrhythmia or arrest (metabolic basis, hypothermia, catheterization)
	Volume overload
Catheter Complications	Umbilical vein/artery perforation
	Air embolism
	Thromboembolism
	Portal vein thrombosis
	Necrotizing enterocolitis
	Bowel perforation
	Cardiac arrhythmia
Other	Change in intracranial pressure
	Hypothermia/hyperthermia
	Emesis with aspiration

DIC = Disseminated intravascular coagulation, Hct = Hematocrit.

ocrit is usually adjusted between 45% and 60% depending on the infant's clinical condition. In order to maintain the appropriate hematocrit throughout the exchange cycles the blood should be kept well mixed. It is also important when using a stopcock to understand the working positions of the stopcock and to make sure that all junctions are tight in order to ensure a closed, sterile system and to avoid the introduction of air. While the goal of the exchange transfusion is to remove bilirubin, medication levels may be altered by the procedure and therefore should also be monitored and adjusted appropriately (Luban 1995).

The National Institute of Child Health and Human Development (NICHD) collaborative phototherapy study demonstrated a mortality rate associated with exchange transfusion of 0.5% and a rate of adverse clinical problems of 6.7% (Keenan et al. 1983). Jackson (1997) in a review of 106 infants undergoing exchange transfusion during a more recent time period (15 year span from 1980 to 1995) reported an incidence of

procedure-related complications leading to death of 2% and the rate of severe complications of 4%. Since this is a procedure that is not performed anymore with great frequency, it is important to have a good understanding of the dynamics and potential complications when evaluating the risks versus the benefits.

References

Anonymous. 1996. In-line blood/solution warmers. *Health Devices* 25(10):352–390.

Bada HS, Chua C, Salmon JH, and Hajjar W. 1979. Changes in intracranial pressure during exchange transfusion. *J Pediatr* 94: 129–132.

Barcelona SL and Cote CJ. 2001. Pediatric resuscitation in the operating room. *Anesthesiol Clin North America* 19:339–365.

Barcelona SL, Vilich F, and Cote CJ. 2000. Comparison of flow rates and warming capabilities of the Level-1 and Rapid Infusion System with various size intravenous catheters. *Anesthesiol* 93A:1277.

Beebe S, Beck D, and Belani KG. 1995. Comparison of the flow rates of central venous catheters designed for rapid transfusion in infants and small children. *Paediatr Anaesth* 5:35–39.

Boyan CP and Howland WS. 1961. Blood temperature: a critical factor in massive transfusions. *Anesthesiology* 22:559–563.

Burch KJ, Phelps SJ, and Constance TD. 1991. Effect of an infusion devise on the integrity of whole blood and packed red blood cells. *Am J of Hospital Pharmacy* 48:92–97.

Calkins JM, Vaughn RW, Cork RC, Barberii J, and Eskelson C. 1982. Effects of dilution, pressure and apparatus on hemolysis and flow rate of packed erythrocytes. *Anesth Analg* 61:776–780.

Campbell N and Stewart I. 1979. Exchange transfusion in ill newborn infants using peripheral arteries and veins. *J Pediatr* 94:820–822.

Cote CJ. 1991. Blood, colloid and crystalloid therapy. *Anesthesiol Clin North Am* 9:865–884.

Denison M, Paul U, Bell R, Schuldreich R, and Chaudhri MA. 1991. Effect of different pump mechanisms on transfusion. *Australas Phys Eng Sci Med* 14:39–41.

Diamond LK, Allen FH, and Thomas WO. 1951. Erythroblastosis fetalis. VII. Treatment with exchange transfusion. *N Engl J Med* 244:39.

Eurenius S and Smith RM. 1973. Hemolysis in blood infused under pressure. *Anesthesiology* 39:650–651.

Frelich R and Ellis MH. 2001. The effect of external pressure, catheter gauge, and storage time on hemolysis in RBC transfusion. *Transfusion* 41:799–802.

Gambino C, Craig D, Stiles M, and Dariotis J. 1992. The effects of Pall RC-50 filtration under pressure on red cell hemolysis. *Transfusion* 32(suppl):S98.

Gammon RR, Stayer SA, Avery NL, and Mintz PD. 2000. Hemolysis during leukocyte-reduction filtration of stored red blood cells. *Ann Clin Lab Sci* 30:195–199.

Gibson JS, Leff RD, and Roberts RJ. 1984. Effects of intravenous delivery systems on infused red blood cells. *Am J Hosp Pharm* 41:468–472.

Goldman SL and Chung Tu H. 1983. Automated method for exchange transfusion: a new modification. *J Pediatr* 102:119–121.

Gorlin JB, ed. 2002. Standards for blood banks and transfusion services, 21st ed. Bethesda, MD: American Association of Blood Banks.

Gurdak RG, Anderson G, and Mintz PD. 1989. Evaluation of IV AC Variable Pressure Volumetric Pump Model 560 for the delivery of red blood cells, adenine-saline added. *Am J Clin Pathol* 91:199–202.

Heddle NM, Klama L, Singer J, Richards C, Fedak P, Walker I, and Kelton JG. 1994. The role of the plasma from platelet concentrates in transfusion reactions. *N Engl J Med* 8:625–628.

Herrera AJ and Corless J. 1981. Blood transfusions: effect of speed of infusion and of needle gauge on hemolysis. *J Pediatr* 99:757–758.

Hodge D III and Fleisher G. 1985. Pediatric catheter flow rates. *Am J Emer Med* 3:403–407.

Hume HA, Popvsky MA, Benson K, Glassman AB, Hines D, Oberman HA, Pisciotto PT, and Anderson KC. 1996. Hypotensive reactions: a previously uncharacterized complication of platelet transfusion? *Transfusion* 36:904–909.

Humphrey MJ, Harrell-Bean HA, Eskelson C, and Corrigan J. 1982. Blood transfusion in the neonate: effects of dilution and age of blood on hemolysis. *J Pediatr* 101:605–607.

Idris AH and Melker RJ. 1992. High-flow sheaths for pediatric fluid resuscitation: a comparison of flow rates with standard pediatric catheters. *Pediatr Emerg Care* 8:119–122.

Jackson JC. 1997. Adverse events associated with exchange transfusion in healthy and ill neonates. *Pediatrics* 99:E7.

Keenan WJ, Novak KK, Sutherland JM, Bryla DA, and Fetterly KL. 1983. Morbidity and mortality associated with exchange transfusion. *Pediatrics* 75:417–421.

Linden JV, Snyder EL, Kalish RI, and Napychank PA. 1988. In vitro and in vivo evaluation of an electromechanical blood infusion pump. *Laboratory Medicine* 19:574–576.

Longhurst DM, Gooch WM, and Castillo RA. 1983. In vitro evaluation of a pediatric microaggregate blood filter. *Transfusion* 23:170–172.

Luban NLC. 1995. Massive transfusions in the neonate. *Transfus Med Rev* IX:200–214.

Luban NLC. 1985. Mechanical devices in pediatric transfusion. In *Hemotherapy in childhood and adolescence.* Luban NLC and Kolins J, eds. Arlington, VA: American Association of Blood Banks.

MacDonald WB and Berg RB. 1959. Hemolysis of transfused cells during use of the injection (push) technique for blood transfusion. *Pediatrics* 234:8–11.

Moss G and Staunton C. 1970. Blood flow, needle size and hemolysis—examining an old wife's tale. *N Engl J Med* 282:967.

Norville R, Hinds P, Wilimas J, Fischl S, Kunkel K, and Fairclough D. 1997. The effects of infusion rate on platelet outcomes and patients responses in children with cancer: an in vitro and in vivo study. [Clinical Trial. Journal Article. Randomized Controlled Trial] *Oncology Nursing Forum* 24(10):1789–1793.

Norville R, Hinds P, Wilimas J, Fairclough D, Fischl S, and Kunkel K. 1994. The effects of infusion methods on platelet count, morphology, and corrected count increment in children with cancer: in vitro and in vivo studies. [Clinical Trial. Journal Article. Randomized Controlled Trial] *Oncology Nursing Forum* 21(10): 1669–1673.

Ogawa Y, Wakana M, Tanaka K, Oka K, Aso H, Hayashi M, Seno T, Ishida T, Nomura S, and Fukuhara S. 1998. Clinical evaluation of transfusion of prestorage-leukoreduced apheresis platelets. *Vox Sang* 75:103–109.

Oloya RO, Feick HJ, and Bozynski MA. 1991. Impact of venous catheters on packed red blood cells. *Am J Perinatol* 8:280–283.

Opitz JC, Baldauf MC, Kessler DL, and Meyer JA. 1988. Hemolysis of blood in intravenous tubing caused by heat. *J Pediatr* 112:111–113.

Presson RG, Haselby KA, Bezruczko AP, and Barnett E. 1990. Evaluation of a new high-efficiency blood warmer for children. *Anesthesiology* 73:173–176.

Quinn C. 2000. Infusion devices: risks, functions and management. *Nursing Standard* 14:35–41.

Roseff SD. 2002. Pediatric blood collection and transfusion technology. In *Pediatric transfusion therapy.* Herman JH and Manno CS, eds. Bethesda, MD: AABB Press.

Schmidt WF, Tomassini N, Kim HC, and Schwartz E. 1982. RBC destruction caused by a micropore blood filter. *JAMA* 248:1629–1632.

Shiba M, Tadokoro K, Sawanobori M, Nakajima K, Suzuki K, and Juji T. 1997. Activation of the contact system by filtration of platelet concentrates with a negatively charged white cell-removal filter and measurement of venous blood bradykinin in patients who received filtered platelets. *Transfusion* 37:457–462.

Simon TL, McDonough W, and Warthen MK. 1994. Red cell viability with infusion systems. *Transfusion* 34:278–279.

Snyder EL, Ferri PM, Smith EO, and Ezekowitz MD. 1984. Use of an electromechanical infusion pump for transfusion of platelet concentrates. *Transfusion* 24:524–527.

Snyder EL, Malech HL, Ferri JP, and Kalish R. 1986. In vitro function of granulocyte concentrates following passage through an electromechanical infusion pump. *Transfusion* 26:141–144.

Snyder EL, Rinder HM, and Napychank PA. 1990. In vitro and in vivo evaluation of platelet transfusions administered through an electromechanical infusion pump. *Am J Clin Pathol* 94:77–80.

Snyder EL, Root RK, Hezzey A, Metcalf J, and Palermo G. 1983. Effect of microaggregate blood filtration on granulocyte concentrates in vitro. *Transfusion* 23:25–29.

Soong WJ, Sun TK, and Hwang B. 1993. Hemolytic impact of central venous catheters on fresh whole blood. *Chin Med J* (Taipei) 51:14–18.

Sowemimo-Coker SO. 2002. Red blood cell hemolysis during processing. *Transfus Med Rev* 16:46–60.

Strauss RG, Bell EF, Snyder EL, Elbert C, Crawford G, Floss A, Wilmoth P, Rios G, and Koonts F. 1986. Effects of environmental warming on blood components dispensed in syringes for neonatal transfusions. *J Pediatr* 109:109–113.

Sweeney JD, Holme S, Stromberg RR, and Heaton WAL. 1995. In vitro and in vivo effects of prestorage filtration of apheresis platelets. *Transfusion* 35:125–130.

Touloukian RI, Kadar A, and Spencer RP. 1973. The gastrointestinal complications of neonatal umbilical venous transfusion: a clinical and experimental study. *Pediatrics* 51:36–43.

Valaes T. 1963. Bilirubin distribution and dynamics of bilirubin removal by exchange transfusion. *Acta Paediatr Scand* 52S: 149.

Valeri CR, Cassidy G, Khuri S, Feingold H, Ragno G, and Altschule MD. 1987. Hypothermia induced reversible platelet dysfunction. *Ann Surg* 205:175–181.

Veerman MW, Leff RD, and Roberts RJ. 1985. Influence of two piston-type infusion pumps on hemolysis of the infused red blood cells. *Am J Hosp Pharm* 42:626–628.

Wallerstein H. 1946. Treatment of severe erythroblastosis by simultaneous removal and replacement of the blood of the newborn infant. *Science* 103:583.

Wilcox GJ, Barnes A, and Modanlou H. 1981. Does transfusion using a syringe infusion pump and small-gauge needle cause hemolysis? *Transfusion* 21:750–751.

Wong ECC, Schreiber S, Criss VR, LaFleur B, Rais-Bahrami K, Short B, and Luban NLC. 2000. Feasibility of red blood cell (RBC) transfusions through small bore central venous catheters in neonates. *Pediatr Res* 49:322A.

Yenicesu I, Tezcan I, and Tuncer AM. 1998. Hypotensive reactions during platelet transfusions. *Transfusion* 38:410.

12

Red Blood Cell Transfusions in the Neonate, Infant, Child, and Adolescent

RONALD G. STRAUSS, MD

INTRODUCTION

Transfusions of red blood cells (RBCs) are key to the successful management of many premature infants, children with cancer or hematologic diseases, recipients of hematopoietic progenitor cell transplants or organ allografts, and for children undergoing many surgical procedures. Although RBC transfusions can be lifesaving, they are not without risks. Accordingly, they should be given only when true benefits are likely (for example, to correct diminished oxygen-carrying capacity severe enough to cause a clinically significant problem). Because of the likely extended life span of infants and children following transfusions, it is critical to avoid posttransfusion complications that may lead to lifelong morbidity and mortality and to considerable expense over the years.

The principles of RBC transfusion therapy for older children and adolescents are similar to those for adults, but infants (particularly, neonates during the initial weeks of extrauterine life) have many special needs. Accordingly, each of these two age groups (that is, neonates and infants versus children and adolescents) will be discussed separately. General guidelines and recommendations will be given for RBC transfusions. However, it is important that they be adapted to fit local standards of practice. In particular, terms used to describe clinical conditions such as "severe" and "symptomatic" must be defined by local physicians in light of the patient problems being managed.

RBC TRANSFUSIONS FOR CHILDREN AND ADOLESCENTS

In children and adolescents the most frequently transfused blood component is RBCs, given to increase the oxygen-carrying capacity of the circulating blood with the goal to maintain satisfactory tissue oxygenation. Guidelines for RBC transfusions to this group (Table 12.1) are similar to those for adults (Roseff et al. 2002). However, transfusions may be given more conservatively (that is, at lower pretransfusion blood hemoglobin or hematocrit values) to children than adults because normal hemoglobin levels are lower in healthy children than in adults, and most children do not have the underlying cardiorespiratory or vascular diseases that develop with aging in adults and require more aggressive RBC transfusions (Wu et al. 2001). Thus, children as a group have greater abilities to compensate for anemia and safely tolerate lower hemoglobin or hematocrit levels than adults, particularly the elderly.

Surgery and Critical Care

In the perioperative period or following resuscitation from trauma, it is unnecessary to transfuse most children with hemoglobin levels of $\geq$80 g/L ($\geq$8 g/dL or hematocrit $\geq$24%), a level frequently desired for adults. There should be a compelling reason to administer any RBC transfusion during the critical care period, regardless of the blood hemoglobin value, because most children (without continued bleeding) can quickly restore their RBC mass if given iron and adequate nutritional therapy (Bratton and Annich 2003). As is true for adults, the most important measures in the treatment of acute

131

TABLE 12.1   RBC Transfusion Guidelines for Children
and Adolescents*

Pretransfusion Blood Level	Clinical Condition
• <80 g/L hemoglobin (<24% hematocrit)	Perioperative, critical care
• No specified hemoglobin/ hematocrit value	≥25% acute blood loss
• <10 g/L hemoglobin (<30% hematocrit)	Acute or chronic anemia and severe cardiorespiratory disease
• <80 g/L hemoglobin (<24% hematocrit)	Symptomatic chronic anemia
• <80 g/L hemoglobin (<24% hematocrit)	Bone marrow failure

*A RBC transfusion volume of 5 to 10 mL/kg of patient body weight will increase the blood hemoglobin level by 20 to 40 g/L (2 to 4 g/dL) or the hematocrit by 6 to 12 percentage points in stable patients without bleeding.

hemorrhage, occurring with surgery or injury in children, are first to control the hemorrhage and to restore blood volume and tissue perfusion with crystalloid and/or colloid solutions. Then, if the estimated blood loss is ≥25% of the estimated circulating blood volume of 65 mL/kg body weight (that is, approximately 16 mL/kg body weight) and the patient's condition remains unstable, RBC transfusions may be indicated.

In acutely ill children with severe cardiac or pulmonary disease, particularly those requiring assisted ventilation, it is common practice to maintain the hemoglobin close to the normal range at a level of 100 to 110 g/L (10 to 11 g/dL or hematocrit = 30% to 33%). Although this practice seems logical, its efficacy has not been documented by controlled scientific studies of children, and it has been challenged (Bratton and Annich 2003). Definitive studies are needed because liberal RBC transfusion practices in critically ill adults have been reported to have detrimental effects (Herbert et al. 1999). Overall transfusion supportive care in surgical and critical care settings is discussed in more detail in Chapter 13.

Chronic Anemia

With anemias that develop slowly, the decision to transfuse RBCs should not be based solely on blood hemoglobin or hematocrit levels because children with chronic anemias may be asymptomatic despite very low values. Children with dietary iron deficiency anemia, for example, often are treated successfully with oral iron alone, even at hemoglobin levels below 50 g/L (5 g/dL or hematocrit <15%). Factors other than hemoglobin concentration that must be considered in the decision to transfuse RBCs in the chronic anemia setting include:

(1) the patient's symptoms, signs, and functional capacities; (2) the presence or absence of cardiorespiratory and central nervous system disease; (3) the cause and anticipated course of the underlying anemia; and (4) alternative therapies such as iron and/or recombinant human erythropoietin (EPO) therapy. The latter of which has been demonstrated to reduce the need for RBC transfusions and to improve the overall condition of children with chronic renal insufficiency. When significant symptoms are present in the chronic anemia setting and the patient has significant underlying cardiorespiratory disease, it is common practice to transfuse RBCs (see Table 12.1) when the blood hemoglobin falls below 100 g/L (10 g/dL or hematocrit <30%). In the absence of cardiorespiratory disease, RBC transfusions often are not given for symptomatic chronic anemia until the hemoglobin is <80 g/L (8 g/dL or hematocrit <24%). When chronic anemia is asymptomatic, RBC transfusions are not recommended. Although in anemias that are likely to be permanent (for example, thalassemia and hemoglobinopathies), one must also balance the effects of anemia on impaired growth and development—adverse effects that might be ameliorated by RBC transfusions—versus the potential toxicities of repeated transfusions. Transfusion management of children with hemoglobinopathies is discussed in detail in Chapter 19.

Bone Marrow Failure

In the setting of marrow failure (for example, chemotherapy, hematopoietic progenitor cell transplant, acquired aplastic anemia, and congenital disorders of erythroid hypoplasia), it is customary to maintain the blood hemoglobin level above 80 g/L (8 g/dL or hematocrit ≥24%) for several reasons. These include: (1) to avoid symptomatic anemia; (2) to facilitate more normal growth and activities; (3) because rapid increases in erythropoiesis (if they occur) are unlikely to successfully compensate for anemia; and (4) because some investigators have shown that radiation therapy is more effective at relatively normal RBC levels and (albeit, somewhat theoretical and controversial) that the adverse effects of accompanying neutropenia and/or thrombocytopenia are less pronounced if the blood hemoglobin or hematocrit levels are closer to the usual normal range.

Selecting a RBC Product for Child and Adolescent Transfusions

For older pediatric patients, there is no concern over additives present in extended storage anticoagulant/ preservative solutions. Thus, any licensed RBC product is satisfactory. In children and adolescents, a transfusion of 5 to 10 mL/kg of RBCs rounded off to the nearest RBC unit (approximately, 300 mL each) to avoid wastage will increase the blood hemoglobin value by approximately 20 to 40 g/L (2 to 4 g/dL or hematocrit 6% to 12%).

RBC TRANSFUSIONS FOR NEONATES AND INFANTS

Pathophysiology of the Anemia of Neonates and Infants

All neonates experience a decline in circulating RBC volume that begins during the first weeks of life. This decline results both from physiological factors and, in sick preterm infants, from phlebotomy blood losses for laboratory monitoring. In healthy term infants, the nadir hemoglobin value rarely falls below 90 g/L (9 g/dL) at an age of 10 to 12 weeks. This decline is more rapid (that is, nadir at 4 to 6 weeks of age) and the blood hemoglobin falls to lower levels in infants born prematurely to approximately 80 g/L (8 g/dL) in infants with birth weights of 1.0 to 1.5 kg and to approximately 70 g/L (7 g/dL) in infants with birth weights <1.0 kg. Obviously, this drop can occur earlier or be made more pronounced by concomitant blood loss. Because this postnatal drop in blood hemoglobin level in term infants is well tolerated and requires no therapy, it is commonly referred to as the "physiological anemia of infancy." However, the pronounced decline in hemoglobin concentration that occurs in many extremely preterm infants (the "anemia of prematurity") may not be well tolerated as it is associated with abnormal clinical signs and, frequently, a need for RBC transfusions (Holland et al. 1987).

Many interacting physiological factors are responsible for the anemia of prematurity. One key reason that the hemoglobin nadir is lower in preterm than in term infants is the former group's diminished plasma EPO level in response to anemia. Although anemia provokes EPO production in premature infants, the plasma levels achieved in anemic infants, at any given hematocrit level, are lower than those observed in comparably anemic older persons (Stockman et al. 1984). One mechanism for diminished EPO output is that the primary site of EPO production in preterm infants is in the liver, rather than the kidney. This dependency on hepatic EPO is important because the liver is less sensitive to anemia and tissue hypoxia; hence, the relatively diminished EPO response to the falling hematocrit. Also, accelerated EPO catabolism may contribute to the low EPO plasma levels with the low plasma EPO in infants

likely being the combined effect of decreased synthesis and increased metabolism.

As a second factor, phlebotomy blood losses play a key role in the anemia of prematurity. The modern practice of neonatology requires critically ill neonates to be monitored closely with serial laboratory studies. The smallest preterm infants generally are the most critically ill, require the most frequent blood sampling, and suffer the greatest proportional loss of RBCs. Because the low EPO levels preclude a vigorous response to phlebotomy blood losses, the replacement of blood drawn for laboratory testing is a critical factor responsible for multiple RBC transfusions in critically ill neonates particularly for RBC transfusions given during the first 3 weeks of life.

Guidelines for transfusing RBCs to neonates and infants are controversial, and practices vary widely among institutions. Generally, RBC transfusions are given to maintain a level of hemoglobin or hematocrit believed to be most desirable for the existing clinical condition (Strauss 1995). Broad guidelines for RBC transfusions from birth to 1 year of age are listed in Table 12.2 and detailed in the text below. These guidelines are very general, and it is important that terms used to describe clinical conditions such as "severe" and "symptomatic" be defined to fit local practices and the needs of each infant being considered for transfusion.

Respiratory Distress and Cardiac Disease

In neonates with severe respiratory distress, such as those requiring high volumes of oxygen with ventilator support, it is customary to maintain the hemoglobin >130 g/L (13 g/dL or hematocrit >40%) particularly when blood is being drawn frequently from the infant for testing. This practice is based on the belief that transfused donor RBCs containing adult hemoglobin will

TABLE 12.2   RBC Transfusion Guidelines for Neonates and Infants

Pretransfusion Blood Level	Clinical Condition
<130 g/L hemoglobin (<40% hematocrit)	Severe respiratory distress
<100 g/L hemoglobin (<30% hematocrit)	Mild-moderate respiratory distress
<130 g/L hemoglobin (<40% hematocrit)	Severe cardiac disease
<100 g/L hemoglobin (<30% hematocrit)	Perioperative, critical care
<80 g/L hemoglobin (<24% hematocrit)	Symptomatic anemia*

*It is debated whether asymptomatic anemia is benefited by RBC transfusions. Nonetheless, RBC transfusions are recommended by some physicians at blood hemoglobin <60 g/L (<20% hematocrit).

provide optimal oxygen delivery throughout the period of diminished pulmonary function that is severe enough to require mechanical ventilation. For less severe respiratory distress (that is, lower volumes of oxygen or no need for mechanical ventilation), lower pretransfusion RBC values are permitted (see Table 12.2). Consistent with this rationale for ensuring optimal oxygen delivery in neonates with pulmonary failure, it seems logical, although unproven by controlled studies, to maintain the hematocrit above 40% in infants with cardiac disease that is severe enough to cause either cyanosis or congestive heart failure.

Perioperative and Critical Care

Definitive studies are not available to establish the optimal hemoglobin level for infants facing major surgery or during the management of severe illnesses. However, it seems reasonable to maintain the hemoglobin >100 g/L (10 g/dL or hematocrit >30%) because of the limited ability of infant heart, lungs, vasculature, and marrow to compensate for acute anemia. An additional factor is the inferior off-loading of oxygen by infant RBCs due to the diminished interaction between fetal hemoglobin and 2,3 diphosphoglycerate. This transfusion guideline is simply a recommendation for perioperative management—not a firm indication—and it should be applied with flexibility to individual infants facing surgical procedures of varying complexity. A more detailed discussion of RBC transfusion in critical care settings is presented in Chapters 13 and 16 dealing, respectively, with surgery and intensive care settings and with extracorporeal membrane oxygenation and cardiac bypass surgery.

Symptomatic Anemia

The clinical indications for RBC transfusions in infants who are not critically ill but nonetheless develop moderate anemia (hematocrit <24% or blood hemoglobin level <80 g/L) are extremely variable. Infants who are clinically stable despite significant anemia do not require RBC transfusions unless they exhibit clinical symptoms/problems ascribed either to the presence of anemia or predicted to be corrected by RBC transfusions. To illustrate, proponents of RBC transfusions to treat disturbances of cardiopulmonary rhythms believe that a low hematocrit contributes to tachypnea, dyspnea, or apnea and either tachycardia or bradycardia because of decreased oxygen delivery to the cardiorespiratory centers of the brain. If true, transfusions of RBCs should decrease the number of apneic spells by improving oxygen delivery to the central nervous system. However, results of clinical studies have been contradictory (Strauss 1995; Ramasethu and Luban 1999).

In practice, the decision whether or not to transfuse RBCs is based on the desire to maintain the blood hemoglobin or hematocrit at a level judged to best treat the infant's symptom/problem (for example, poor growth or the previously mentioned cardiorespiratory rhythm disturbances). Investigators who believe this "clinical" approach is too imprecise have suggested the use of "physiological" criteria for transfusions such as red cell mass, available oxygen, mixed venous oxygen saturation, and measurements of oxygen delivery and utilization to develop guidelines for transfusion decisions. However, these promising but technically demanding methods are, at present, difficult to apply in the day-to-day practice of neonatology—hence, the continued reliance on blood hemoglobin and/or hematocrit values plus clinical assessment (see Table 12.2). In the complete absence of symptoms, the efficacy of RBC transfusions is unclear regardless of the nadir blood hemoglobin or hematocrit value. Nonetheless, some physicians transfuse asymptomatic infants at blood hemoglobin levels <60 g/L (6 g/dL or hematocrit <20%).

Selecting a RBC Product for Neonatal and Infant Transfusions

The RBC products usually chosen for small-volume transfusions given to infants are RBCs suspended either in citrate-phosphate-dextrose-adenine (CPDA) solution at a hematocrit approximately 70% or in extended storage media (AS-1, AS-3, AS-5) at a hematocrit approximately 60%. Some centers prefer to centrifuge RBC aliquots before transfusion to prepare packed RBCs (PRBCs) at a hematocrit of 80% to 90% (Strauss et al. 1996). Most RBC transfusions are infused slowly over 2 to 4 hours at a dose of about 15 mL/kg body weight. Because of the small quantity of RBC preservative fluid infused and the slow rate of transfusion, the type of anticoagulant/preservative medium selected has been shown by many studies not to pose risks for the majority of premature infants given small volume transfusions (Strauss 2000). Accordingly, the historic use of relatively fresh RBCs (<7 days of storage) has been supplanted, in favor of diminishing donor exposure of multiply-transfused infants, by the repeated use of a dedicated unit of stored RBCs (that is, up to 42 days after collection) for each infant.

Neonatologists who object to stored RBCs and continue to insist on transfusing infants with fresh RBCs generally raise three objections: (1) the rise in plasma potassium (K^+) and (2) the drop in RBC 2,3 diphosphoglycerate that occurs during extended storage; and (3) the possible dangers of additives present in extended storage media.

1. After 42 days of storage in an extended storage medium, plasma K^+ levels in RBC units approximate 50 mEq/L (0.05 mEq/mL), a concentration that at first glance seems alarmingly high. By simple calculations, however, the dose of bioavailable K^+ transfused (that is, ionic K^+ in the extracellular fluid) is very small. An infant weighing 1.0 kg, given a 15 mL/kg transfusion of RBCs removed as an aliquot from the storage bag and directly infused at a hematocrit of 60%, will receive a K^+ dose of only 0.3 mEq. This dose is quite small compared to the usual daily K^+ requirement of 2 to 3 mEq/kg and, depending on the infusion rate, K^+ often is given at a lower rate (that is, mEq/min/kg infant body weight) during a RBC transfusion than by standard intravenous administration. However, it must always be remembered that this rationale does not apply to large-volume transfusions (>25 mL/kg) in which larger doses of K^+ may be harmful, especially if infused rapidly.

2. As for the second objection, 2,3 diphosphoglycerate is totally depleted from RBCs by 21 days of storage, and this is reflected by a P_{50} value that falls from about 27 mmHg in fresh blood to 18 mmHg at the time of storage outdate. The last value of older transfused RBCs corresponds to the "physiological" P_{50} measured using RBCs from the blood of many normal preterm infants at birth, reflecting the relatively high affinity for oxygen normally exhibited by infant RBCs. Thus, the P_{50} of older transfused RBCs is no worse than that of RBCs produced endogenously by the infant's own bone marrow. Moreover, these transfused older adult RBCs provide a benefit to the infant because the 2,3 diphosphoglycerate and the P_{50} of transfused RBCs (but not endogenous infant RBCs) increase rapidly after transfusion.

3. Regarding the third objection, the quantity of additives present in RBCs stored in extended storage media (for example, AS-1, AS-3, or AS-5) is believed not to be dangerous to neonates given small-volume (≤15 mL/kg) transfusions. The quantity of additives is quite small in the clinical setting in which infants are given small-volume transfusions of RBCs transfused over 2 to 4 hours (Table 12.3) and is far below doses believed to be toxic (Luban et al. 1991). Importantly, the efficacy and safety of these theoretical calculations have been confirmed by reports of many investigators successfully transfusing stored, rather than fresh, RBCs in the small-volume transfusion setting (Strauss 2000).

TABLE 12.3 Constituents (mg/kg) in a 15 mL/kg RBC Transfusion (60% Hematocrit)

Additive	CPDA	AS-1	AS-3	Toxic Dose*
NaCl	0	28	5	137 mg/kg/day
Dextrose	13	86	15	240 mg/kg/hr
Adenine	0.2	0.4	0.4	15 mg/kg/dose
Citrate	12	6.5	8.4	180 mg/kg/hr
Phosphate	9	1.3	3.7	>60 mg/kg/day
Mannitol	0	22	0	360 mg/kg/day

*Actual toxic dose is difficult to predict accurately because infusion rates are slow, permitting metabolism and distribution from the bloodstream into extravascular sites. Also, dextrose, adenine, and phosphate enter RBCs and are somewhat sequestered from other body tissues (Luban et al. 1991).

For physicians selecting RBCs stored in CPDA, it is important to remember that the volume of extracellular fluid is smaller in CPDA RBC units than in AS-1, AS-3, or AS-5 RBC units—hence, the concentration of extracellular K^+ and hemoglobin is higher in CPDA RBC units.

References

Bratton SL and Annich GM. 2003. Packed red blood cell transfusions for critically ill pediatric patients: when and for what conditions? *J Pediatr* 142:95–97.

Herbert PC, Wells G, Blajchman MA, Marshall J, Martin C, Pagliarello G, Tweeddale M, Schweitzer I, Yetisir E, and the Transfusion Requirements in Critical Care Investigators for the Canadian Critical Care Trials Group. 1999. A multicenter, randomized, controlled clinical trial of transfusion requirements in critical care. *N Engl J Med* 340:409–417.

Holland BM, Jones JG, and Wardrop CA. 1987. Lessons from the anemia of prematurity. *Hematol Oncol Clin North Am* 1:355–366.

Luban NLC, Strauss RG, and Hume HA. 1991. Commentary on the safety of red blood cells preserved in extended storage media for neonatal transfusions. *Transfusion* 31:229–235.

Ramasethu J and Luban NL. 1999. Red blood cell transfusions in the newborn. *Semin Neonatol* 4:5–16.

Roseff SD, Luban NLC, and Manno CS. 2002. Guidelines for assessing appropriateness of pediatric transfusion. *Transfusion* 42: 1398–1413.

Stockman JA, Graeber JE, Clark DA, McClellan K, Garcia JF, and Kavey RE. 1984. Anemia of prematurity: determinants of the erythropoietin response. *J Pediatr* 105:786–792.

Strauss RG. 2000. Data-driven blood banking practices for neonatal RBC transfusions. *Transfusion* 40:1528–1540.

Strauss RG. 1995. Red blood cell transfusion practices in the neonate. *Clin Perinatol* 22:641–655.

Strauss RG, Burmeister LF, Johnson K, James T, Miller J, Cordle DG, Bell EF, and Ludwig GA. 1996. AS-1 red blood cells for neonatal transfusions: a randomized trial assessing donor exposure and safety. *Transfusion* 36:873–878.

Wu RC, Rathmore SS, Wand Y, Radford M, and Krumholz HM 2001. Blood transfusion in elderly patients with acute myocardial infarction. *N Engl J Med* 345:1230–1236.

13

Transfusion of the Pediatric Surgery, Trauma, and Intensive Care Unit Patient

THOMAS J. RAIFE, MD, AND JORGE A. DI PAOLA, MD

ABSTRACT

Transfusion in surgery, trauma, and intensive care requires vigilant attention to the unique physiological status of pediatric patients, particularly in the neonatal period. Immature organ systems and limited capacities for regulation of thermal and metabolic challenges may combine to produce delicate and potentially dangerous vulnerabilities. Conditions sometimes require critical measures not generally encountered in adult medicine. Much of the challenge in acute care transfusion therapy is to maintain an appropriate balance of hemostatic processes. Both excessive bleeding and thrombosis are encountered in acute care and are of equal urgency in the need for expeditious and skilled management. The availability and appropriate use of pharmaceutical products that may supplement or replace hemostatic blood products must be part of the practitioner's repertoire. Facility with the combined use of new technology, new pharmaceutical products, and appropriate use of blood products is key to optimal outcomes.

Transfusion is a major component of pediatric patient care in surgery, trauma, and the intensive care unit (ICU). Although the medical needs of patients may be acute, the careful and judicious use of blood products is important in these circumstances, as in all others. Potential adverse effects of transfusion must be weighed against potential benefits. Overuse of transfusion should be avoided.

Red blood cells, platelets, granulocytes, plasma, and cryoprecipitate are all major elements of medical support of pediatric surgery, trauma, and ICU patients. There are detailed discussions of the preparation, dispensing, and clinical applications of cellular blood components in individual chapters. The specific requirements of transfusion support during extracorporeal membrane oxygenation (ECMO) and cardiac bypass surgery are given individual consideration in Chapter 17. This chapter will address matters of general concern regarding the use of cellular blood products in acute care settings and will focus more specifically on the goal of transfusion support of hemostasis.

AGE, SIZE, AND RATE OF TRANSFUSION

Age

In serological testing, pediatric patients are divided between those younger and older than 4 months of age. From the standpoint of physiological considerations in blood transfusion, the same age categories may be used. In general, transfusion of normally developing infants older than 4 months of age is similar to children and adults. Infants younger than 4 months of age, preterm, and low birth weight infants require special consideration of their immature hemostatic systems and their limited ability to tolerate thermal and metabolic alterations from transfusion.

Rate of Transfusion

Transfusion of red blood cells (RBCs) in volumes of 10 to 20 mL/kg over 2 to 4 hours is generally well tolerated in neonates and infants. Transfusion of over 25 mL/kg of RBCs, or greater than one blood volume in 24 hours, requires careful attention to possible metabolic and thermal alterations. As noted below, the effects of citrate, potassium, and 2,3 diphosphoglycerate (2,3 DPG) are of particular concern and must be anticipated and managed during massive transfusion. Chapter 11 contains detailed discussions of these considerations.

Coagulation

Because of immaturity of the liver, the production of hemostatic factors is reduced in infants compared to children and adults. The level of hemostatic factor activity decreases with degree of prematurity. In infants of 30 to 38 weeks gestation, coagulation factors range from about 12% to 50% of adult levels. In full-term newborns, ranges are about 30% to 95% of adult levels. Ranges of hemostasis regulatory factors are similar. Both procoagulant and anticoagulant hemostatic factors progress rapidly to 80% to 90% of adult ranges within 6 months to 1 year. Adult levels are reached in the middle to late teen years (Richardson et al. 2002).

Thermal Regulation

Newborns have a limited capacity to regulate body temperature and have a high body surface area to volume ratio. Premature and low birth weight infants are particularly vulnerable to thermal alterations from infusion of hypothermic fluids. The temperature of blood products arriving from a blood bank or stored in an operating room may range from approximately room temperature to 4°C. Rapid infusion (>1 mL/minute) of cold blood to young infants can result in significant hypothermia. Core body temperatures less than 35°C can slow hemostatic responses and impair oxygen delivery in tissues. Severe hypothermia (<33°C) can result in hypotension, apnea, and cardiac arrythmias. Therefore, careful monitoring of body temperature is important. When possible, use of a blood warmer is recommended for rapid or massive transfusion (>25 mL/kg). A device manufactured and approved for blood warming is the standard of care. The practice of warming blood products with lights, ordinary microwave ovens, or other nonapproved methods is strongly discouraged due to risk of thermal injury to blood products (Festa et al. 2002).

Immune Status

The normal immaturity of neonatal immune systems requires special consideration. Neonatal leukocyte activities are generally decreased compared to children and adults. Neonatal immune systems may be unable to resist untoward effects of transfusion, such as cytomegalovirus (CMV) transmission or transfusion-associated graft versus host disease (TA-GVHD). Sick or injured infants, or those with congenital or treatment-related immune compromise, may be especially vulnerable. Use of CMV low-risk (leukocyte reduced or CMV seronegative) and gamma irradiated cellular blood products for prevention of CMV transmission and TA-GVHD and are recommended in infants younger than 4 months of age and patients of any age with significant immune compromise. See Chapters 8 and 9 for detailed recommendations.

Metabolism

Liver and renal function immaturity renders infants vulnerable to metabolic imbalance from transfusion. Particular vigilance regarding calcium, glucose, and potassium alterations can be critical (Strauss 2002; McDonald and Berkowitz 1994).

Citrate and Calcium

Sodium citrate anticoagulant in blood products binds ionized calcium and in high volume transfusions, especially of fresh frozen plasma, may result in transient hypocalcemia. Metabolism of citrate produces bicarbonate that may result in alkalosis and subsequent hypokalemia. During surgery, hypocalcemia is characterized by hypotension despite adequate volume replacement. Severe decreases in ionized calcium result in narrowed pulse pressure and prolonged QT intervals. Hypothermia and immature liver function impair citrate metabolism. Frequent monitoring of ionized calcium, pH, and potassium may be necessary during high volume transfusion (>25 mL/kg). Corrective measures include calcium gluconate (100 mg/kg) for routine replacement and calcium chloride (10 to 20 mg/kg) for rapid or emergent replacement; caution is advised, as excessively rapid infusion of calcium chloride has been associated with death (Kevy and Gorlin 1998, Festa et al. 2002).

Potassium

Stored RBC units accumulate extracellular potassium. In our laboratory, Adsol 3 (AS-3) preserved RBCs contained mEq/L of extracellular potassium equal to approximately the number of days of storage plus 10 (for example, 15 mEq/L K+ on day 5). At routine rates of transfusion (10 to 20 mL/kg over 2 to 4 hours) K+ from transfusion should be well tolerated. Rapid transfusion of RBCs containing high concentrations of extracellular K+ can result in fatal cardiac disturbances in small infants, especially when accompanied by acidosis and administration of other K+ containing drugs or intravenous fluids. Use of washed, extracellular volume-reduced or fresh (<5 to 7 days old) RBC products may be necessary. Such products may require an additional 20 to 30 minutes preparation in the blood bank. Resuscitation with crystalloid or colloid solutions may be advisable in lieu of rapid infusion of high K+ containing RBCs. See Chapters 10 and 12 for detailed discussions.

2,3 Diphosphoglycerate

During several weeks of storage 2,3 diphosphoglycerate (2,3 DPG) is completely depleted from RBC units. The corresponding reduction in the P50 of hemoglobin results in a level of oxygen offloading similar to RBCs rich in fetal hemoglobin, as occurs in premature infants. Transfused RBC 2,3 DPG levels normalize within hours of transfusion in adults and probably also in children. Generally the effect of temporarily low levels of 2,3 DPG in vivo is insignificant; however, in massive exchange transfusion impaired oxygen offloading can occur. Although studies have not shown a clear advantage, fresh RBCs are sometimes used (Strauss 2002).

RED BLOOD CELL TRANSFUSION

Widely observed guidelines for transfusion triggers in pediatric populations have not been established. Both the extreme clinical variability of patients and the increasing knowledge of potential adverse physiological effects of RBC transfusion contribute to the lack of consensus.

Small Volume Booster Transfusions

One of the most controversial issues in pediatric transfusion is the appropriate use of small volume transfusions of RBCs to replace blood lost from laboratory testing or for other insidious causes of anemia. In general, the trend over several decades has been toward less use of RBC transfusions in neonatal nurseries. In critically ill adults a liberal transfusion threshold has been clearly shown to offer no survival advantage and in some patient subsets is disadvantageous compared to more restrictive transfusion practices (Hébert et al. 1999). Even in preterm and low birth weight infants in whom physical signs suggestive of anemia are common, physiology does not correlate well with hematocrit.

Transfusion may be necessary to maintain adequate oxygen-carrying capacity in critically ill neonates, especially those with cardiopulmonary compromise. In these patients a target hematocrit of 30% in children and 40%

in neonates is common. Measurements of arterial oxygenation, mixed venous oxygen tension, and cardiac output may be helpful in guiding transfusion decisions (Kevy and Gorlin 1998).

Overall, it is recommended that RBC transfusions for routine boosting of hematocrit levels be used judiciously and that the transfusion decision be based on demonstrable physiological benefit and not strictly on laboratory tests.

Erythropoietin

Comparatively low erythropoietin levels in critically ill low birth weight neonates and infants with anemia of prematurity have prompted exploration of the use of pharmaceutical erythropoietin. Although the ability of erythropoietin to enhance RBC production in neonates is well established, most studies of erythropoietin use in neonatal nurseries have not demonstrated a significant reduction in requirements for RBC transfusions.

Severely ill, low birth weight infants often do not respond to erythropoietin sufficiently to avoid transfusions altogether. Therefore, if a program of limited donor blood is used, the benefit of erythropoietin in reducing transfusion exposure risk is minimal. Larger, more stable infants are more likely to respond to erythropoietin, but they are also less likely to require transfusions. Therefore, clinical studies offer little support for the generalized use of erythropoietin in neonates and infants. Nevertheless, in individual cases of severely ill neonates with ongoing needs for RBC transfusions, pharmaceutical erythropoietin may be a consideration (Goodstein 2002). The availability of limited donor blood and input from parents as well as physicians should be considered in making a decision to use erythropoietin. In clinical trials, doses of 200 to 400 U/kg per week were required to affect RBC utilization. In our institution, erythropoietin is sometimes used to support pediatric patients who are unable to receive transfusions for religious reasons.

Acute Blood Loss

Volume Replacement

A fundamental tenet of both pediatric and adult transfusion medicine is that, in the event of acute hemorrhage, replacement of intravascular volume and measures to stop bleeding are the highest priority. Above hemoglobin levels of 3 to 4 g/dL, maintaining intravascular volume in acute blood loss is more important than measures to increase oxygen-carrying capacity with RBC transfusions.

Box 13.1   Signs of Moderate-to-Severe
            Acute Blood Loss

Irritability or stupor
Pallor or mottling
Tachycardia
Tachypnea
Decreased pulse intensity
Delayed capillary refill
Cool extremities
Hypotension
Metabolic acidosis
Oliguria or anuria

Both crystalloid solutions, such as normal saline or lactated Ringer's solution, and colloid solutions, usually 5% or 25% albumin, are used to expand intravascular volume. Crystalloid solutions redistribute up to 80% into the extravascular space and require quantities three to four times the actual blood loss to maintain intravascular volume.

Transfusion Trigger

Oxygen transport during massive hemorrhage becomes inadequate despite maintenance of intravascular volume at hematocrit levels below about 10%. Decisions to transfuse acutely bleeding patients must be made individually in consideration of ongoing and anticipated rates of blood loss, individual oxygen requirements, and status of tissue perfusion.

In children, signs of circulatory compromise and impending shock may not be apparent until 25% to 30% of blood volume is lost (Box 13.1). In some individuals, transfusion may be indicated after 10% to 15% of blood volume is lost. It is important to recognize that during massive blood loss, depending on the adequacy of volume replacement, measured hematocrit and hemoglobin levels may overestimate RBC mass.

Transfusion Target

Animal studies indicate that optimal oxygen transport occurs at hematocrit levels between 30% and 40%. Higher values offer more oxygen-carrying capacity but less favorable fluid dynamics. Although appropriate transfusion target hematocrit levels are not established, some practitioners consider a hematocrit of >30% as a reasonable target in resuscitation following massive blood loss. In this setting, as in others, overtransfusion should be avoided.

Pulmonary and Cardiac Failure

Neonates with acute respiratory distress syndrome being supported with mechanical ventilation often require RBC transfusion to replace blood lost from laboratory testing and to optimize tissue oxygenation. Transfusion replaces fetal hemoglobin RBCs with hemoglobin A-containing cells, providing improved oxygen offloading. In general, patients with pulmonary failure, cyanotic heart disease, or congestive heart failure are often transfused for hematocrit levels <40%. See Chapters 12 and 17 for detailed discussions.

PLATELET TRANSFUSIONS

A complete discussion of platelet transfusions in pediatric populations is presented in Chapter 22. The following are general considerations about platelet transfusion in acute care settings.

Massive Transfusion

The dilutional effect from massive replacement of intravascular volume can result in thrombocytopenia. Clinically significant thrombocytopenia usually does not occur until one or two blood volumes have been lost. In a setting of purely dilutional thrombocytopenia, platelet counts above 30,000/μL to 50,000/μL are usually sufficient for hemostasis. In premature infants and patients undergoing procedures using extracorporeal circuitry, platelet function may be impaired and transfusion of platelet concentrates may be needed for counts below 100,000/μL.

Dosage

In surgery and most other settings a standard dosage for platelet transfusions is 5 to 10 mL/kg. Consideration for selection and preparation of platelet dosages, leukocyte reduction, and irradiation are presented in Chapter 22.

Platelet Transfusion in Extracorporeal Membrane Oxygenation

Transfusion in ECMO is discussed in detail in Chapter 16. One of the most common complications in ECMO is intracranial hemorrhage. Platelet counts in children undergoing ECMO are often targeted at >80,000/μL. Transfusion for platelet counts below 50,000/μL is strongly recommended. The coagulation function of blood in the extracorporeal circuit during

ECMO is typically monitored by the activated clotting time (ACT). An ACT in the range of 220 to 250 seconds is a common target for adjustment of heparin, with correction using fresh frozen plasma.

Platelet Transfusion in Uremic Patients

Blood urea nitrogen (BUN) levels greater than 40 to 50 mg/dL are associated with platelet function defects. Mucocutaneous bleeding including oral mucosal bleeding and epistaxis may result. Excessive surgical bleeding in uremic patients may prompt consideration of platelet transfusions, however, since transfused platelets are rapidly impaired by uremia, their use should be considered secondary. Two primary corrective measures are recommended to improve hemostasis in uremic patients with excessive bleeding: transfusion of RBCs to a hematocrit value >30% and administration of 1-deamino-8-D-arginine (DDAVP). DDAVP can be administered either by nasal aerosol or intravenous infusion (Box 13.2).

The recommended dose interval of DDAVP is once every 24 hours. Responses of hemostatic factors to repeated doses decrease by about 30% after the first dose. In severely bleeding patients with uremia, platelet transfusions together with DDAVP are sometimes used. However, their added benefit is uncertain in the absence of significant thrombocytopenia (Mannucci 1997).

Immune Thrombocytopenic Purpura (ITP)

Clinically significant bleeding is a relatively rare event in patients with childhood ITP, and the appropriate treatment of childhood ITP is controversial. Intracranial hemorrhage is the most worrisome complication, but excessive surgical or traumatic bleeding is also possible. Most patients who develop intracranial hemorrhage have platelet counts less than 10,000/μL. Primary measures to correct very low platelet counts include use of corticosteroids and intravenous immunoglobulin (IVIG). Some practitioners utilize a target platelet count of 20,000/μL to prevent intracranial hemorrhage, however, supportive data are meager.

Treatment

In the case of life-threatening hemorrhage, high dose parenteral corticosteroids (30 mg/kg/day intravenous methylprednisolone) or high dose IVIG have been recommended. Early trials with IVIG used 0.4 g/kg/d for 5 days. A more recent trial showed success using a single dose of 0.8 g/kg. In some cases, platelet transfusions and emergency splenectomy may be warranted. In massive transfusion and acute blood loss, platelet transfusion may be the primary treatment option. It must be recognized that transfused platelets are susceptible to the same immune-mediated dysfunction affecting native platelets. A combination of hemostatic measures that may include platelet transfusions and other hemostatic agents is recommended (Di Paola and Buchanan 2002).

Platelet Dysfunction

Drugs

Platelet dysfunction may result from drugs such as aspirin, indomethacin, penicillins, and cephalosporins. In the setting of surgical and acute blood loss, hemostasis may or may not be significantly compromised. As with uremia and ITP, circulating drugs may affect transfused as well as native platelets, and platelet transfusion should be reserved for excessive bleeding and significant thrombocytopenia.

Congenital Defects

Rare congenital platelet function defects include the platelet receptor defects Glanzmann's thrombasthenia and Bernard-Soulier syndrome, and the storage granule defects gray syndrome, Chédiak-Higashi syndrome, and Hermansky-Pudlak syndrome.

Treatment

The degree of hemostatic compromise varies among syndromes and among individuals. In the acute care setting, platelet transfusions may be required for adequate hemostasis. In anticipation of surgical bleeding, a thorough clinical assessment should consider the

---

Box 13.2   Hemostatic Measures for Uremic Bleeding

Transfuse to hematocrit >30%
Nasal aerosol DDAVP (150 μg/squirt)

- <50 kg patient—1 squirt
- >50 kg patient—2 squirts

Intravenous DDAVP Infusion

- 0.3 μg/kg in 30 to 60 mL normal saline

Precautions

- Monitor vital signs for tachycardia
- Monitor laboratory values for hyponatremia
- Risk of seizure activity in children < 2 years of age
- Repeated use may result in tachyphylaxis

possible need for prophylactic platelet transfusions. Initial doses of approximately 5 to 10 mL/kg of platelet concentrates should raise transfused platelet counts to hemostatic thresholds. It should be recognized that in patients with platelet receptor defects, transfused platelets may stimulate production of platelet glyco-protein-specific alloantibodies and may cause future refractoriness to standard platelet transfusions. In some settings DDAVP or antifibrinolytic agents may augment platelet-dependent hemostasis (see Box 13.2). New in the armamentarium of hemostatic agents that shows promise in platelet function defects is recombinant activated factor VII (rFVIIa).

CRYOPRECIPITATE AND FRESH FROZEN PLASMA

Cryoprecipitate and fresh frozen plasma (FFP) remain central components in the treatment and resuscitation of acutely bleeding pediatric patients. However, the appropriate use of these products must also consider a potential role of pharmaceutical hemostatic agents. Effective and pathogen-safe hemostatic agents can greatly augment the efficacy and reduce the need for traditional blood products (DeLoughery 2002).

Cryoprecipitate

Although cryoprecipitate contains concentrated FVIII and von Willebrand factor (vWF), these hemostatic factors are widely available in virus-inactivated and potentially more effective pharmaceutical forms. With availability of these pharmaceutical products, the primary utility of cryoprecipitate is to supply a concentrated form of fibrinogen.

Premature and term newborns may have normal or modest reductions in fibrinogen levels compared to children and adults. However, as with other coagulation factors, reduced synthesis from immature liver function may render them more susceptible to consumptive depletion.

Indications

Fibrinogen levels below 80 to 100 mg/dL may be associated with excessive bleeding and warrant consideration of cryoprecipitate. The choice between cryoprecipitate or FFP should consider whether multiple hemostatic factors are in need of replacement. Plasma contains all hemostatic factors in proportion to their concentration in blood. Cryoprecipitate is concentrated to contain about 180 to 250 mg fibrinogen in a 15 mL volume. Therefore, cryoprecipitate is often used to rapidly restore fibrinogen levels, especially in the setting of intravascular volume constraints.

Dosage

A standard dose of cryoprecipitate is 5 to 10 mL/kg or 1 unit per 5 to 10 kg body weight. In infants, cryoprecipitate should be ABO blood group compatible to avoid introduction of incompatible RBC antibodies. Clinical assessments and fibrinogen levels are used to evaluate the efficacy of cryoprecipitate transfusion.

Factor XIII

Both FFP and cryoprecipitate contain coagulation factor XIII, and either can be used to treat severe factor XIII deficiency. A pharmaceutical form of factor XIII is available on a clinical trial basis but is not yet approved by the Food and Drug Administration (FDA). Because of its long half-life and minimal requirements for adequate hemostasis, only small doses of cryoprecipitate or plasma are necessary to treat severe factor XIII deficiency. A cryoprecipitate or plasma dose of 1 to 5 mL/kg is sufficient for 2 to 6 weeks of hemostatic FXIII levels.

Cryoprecipitate for Topical Administration

Historically, cryoprecipitate and bovine thrombin combinations have been used as "fibrin glue" for surgical hemostasis. However, bovine thrombin contains small amounts of bovine FV and has been associated with development of cross-reactive antiFV or antithrombin antibodies that have caused life-threatening bleeding. For safety reasons these formulations are not recommended because human thrombin-containing, virus-inactivated alternatives are available.

Fresh Frozen Plasma

FFP contains normal adult concentrations of both procoagulant and anticoagulant hemostatic factors. Its use is appropriate to restore hemostatic levels of coagulation proteins resulting from multiple factor deficiency or to provide hemostatic levels of specific coagulation factors, such as factor V, that are not available as pharmaceutical concentrates.

Historically, plasma is the most overused blood product. The longstanding message bears repeating that plasma transfusion is inappropriate for the sole purpose of volume replacement or as a source of immunoglobulins. For these purposes, crystalloid, colloid, and IVIG are preferable. Overuse of plasma may occur in attempts to correct minor elevations of prothrombin time (PT) or activated partial thromboplastin time (aPTT) before surgery. Minor prolongations of clotting times correlate poorly with surgical bleeding risk. Individual consideration of hemostatic risk in relation to patient age and other factors is preferable to reflexive attempts to correct minor laboratory test abnormalities with FFP.

Replacement

Hemostasis may be compromised by replacement of plasma volume with crystalloid or colloid solutions in massive transfusion. However, dilutional coagulopathy does not usually occur until replacement of one to two blood volumes within 24 hours. Newborns and premature neonates may be more sensitive to the effect of dilutional coagulopathy because of lower constitutive levels of coagulation factors. Thrombocytopenia may accompany dilutional coagulopathy. In some settings, attention to adequate platelet transfusion may be more beneficial to hemostasis than empirical FFP or cryoprecipitate transfusion.

Indications and Dosage

Standardized guidelines for use of FFP in massive transfusion have not been developed. Hemostatic compromise from multiple coagulation factor deficiency depends on the degree of factor depletion. FFP contains 1 U/mL of all hemostatic proteins normally found in plasma. Therefore, assuming a plasma volume of about 40 mL/kg in a neonate, infusion of 20 mL/kg of FFP provides about 50% of normal levels, which is adequate for hemostasis. Since factor levels are not completely depleted during massive transfusion, dosages of FFP of 10 to 15 mL/kg are usually adequate. To maintain short half-life factors, dosing may be necessary every 6 to 8 hours.

Use of clinical bleeding assessment together with coagulation studies is recommended to guide FFP transfusion. Empirical use of both FFP and cryoprecipitate in massive transfusion should be avoided. In cases of persistent excessive bleeding, despite reasonable measures to correct dilutional coagulopathy and thrombocytopenia, consideration of pharmacological hemostatic agents may be warranted. These agents are discussed in Chapter 20.

Liver Failure and Coumadin Treatment

Liver failure from congenital infectious hepatitis, sepsis, congenital hemochromatosis, or other factors is a major cause of coagulation laboratory test abnormalities and bleeding in neonatal patients. Nevertheless, most newborns with liver failure do not experience excessive bleeding, and empirical FFP utilization is not

---

> **Box 13.3   Hemostasis Assessment and Treatment in Liver Failure**
>
> - Careful analysis of clinical picture to determine causes of bleeding
> - Laboratory assessments
>   - PT, aPTT, fibrinogen
> - FFP for active bleeding or before invasive procedure
>   - 10 to 15 mL/kg every 4 to 6 hours
>   - Follow PT, aPTT for additional dosing

---

recommended. Box 13.3 details hemostasis assessment and treatment in liver failure.

Complete normalization of coagulation parameters with FFP is often difficult or impossible to achieve, especially if there are volume constraints. Moreover, modest elevations of coagulation parameters that persist after adequate dosing of FFP are unreliable indicators of bleeding risk. Therefore, unnecessary delays of invasive procedures or excessive transfusion of FFP to normalize coagulation parameters should be avoided.

Coumadin therapy in children may pose a risk of excessive bleeding in emergent acute bleeding. Vitamin K administration will reverse coumadin effects within 12 hours in a patient with normal liver function and is the appropriate treatment when time permits. A dosage of 5 to 10 mg by slow infusion (20 to 30 minutes) is recommended. Patients should be monitored carefully for the rare complication of anaphylaxis. In cases of trauma or emergency surgery, FFP can be used to correct the hemostatic defect of coumadin. As in liver failure, initial FFP doses of 10 to 15 mL/kg are recommended. Efficacy should be assessed by careful clinical assessment of bleeding and laboratory monitoring with the PT and aPTT.

Disseminated Intravascular Coagulation

Disseminated intravascular coagulation (DIC) is one of the major challenges of hemostatic transfusion therapy in acute care settings. Mortality in infants with severe DIC approaches 80%. Infants and children with DIC can experience very severe bleeding and often require multiple therapeutic approaches to sustain adequate hemostasis.

Causes of DIC

Causes of DIC in children are similar to those in adults (Box 13.4).

Among infectious causes of DIC are gram-negative sepsis, pneumococcal sepsis, Hemophilus influenza infections, systemic aspergillosis, and rickettsial infections. Identification of the cause of DIC is critical: The

---

Box 13.4   Causes of DIC in Pediatric Patients

- Sepsis (up to 45% of patients)
- Neoplasia
- Vascular disease
- Liver disease
- Obstetric complications
- Inherited coagulation disorders
- Transfusion reactions
- Surgery
- Envenomation
- Trauma
- Respiratory distress syndrome
- Drugs

Box 13.5   Clinical Manifestations of DIC

- **Hematological:** thrombocytopenia, microangiopathic hemolysis, leukocytosis
- **Skin:** purpura, oozing from venipuncture sites, focal necrosis, gangrene
- **Cardiovascular:** shock, acidosis, thromboembolism
- **Renal:** oliguria, azotemia, hemoglobinurea, acute tubular necrosis, cortical necrosis
- **Pulmonary:** hypoxemia, edema, hemorrhage, acute respiratory distress syndrome
- **Hepatic:** jaundice, parenchymal damage
- **Gastrointestinal:** hemorrhage, mucosal necrosis
- **Neurological:** seizure, stupor, coma, intracranial hemorrhage
- **Endocrine:** adrenal insufficiency

Box 13.6   Characteristic Coagulation Tests in DIC

- PT: ↑
- aPTT: ↑ or nl
- Thrombin time: ↑
- Fibrinogen: ↓ or nl
- FDP: >10 mg/dL
- D-dimer: present

(FDP) or fibrin D-dimer (Box 13.6). Accurate test results require properly collected samples. Heparin contamination from samples drawn through central venous catheters is a frequent source of errors in PT, aPTT and thrombin time tests.

Management of DIC

DIC represents a perturbation of the delicate balance between procoagulant and anticoagulant hemostatic mechanisms. The end result can be either excessive bleeding or microvascular thrombosis. For these reasons, treatment of the underlying causes of DIC is the definitive intervention. Treatment of bleeding or thrombosis must be considered a temporizing measure.

Thrombosis in DIC   Although bleeding is the most frequent complication of DIC, insidious microvascular thrombosis can be a devastating event. Significant microvascular thrombosis occurs most often in DIC associated with malignancy. A notable example is acute promyelocytic leukemia. End-organ dysfunction secondary to hypoxic and ischemic injury from widespread microvascular thrombosis is of paramount concern. If evidence of end-organ injury is present, especially encephalopathy and cardiopulmonary dysfunction, anticoagulant treatment may be warranted. A combination of FFP and heparin may be used with a goal of inhibiting thrombin activity while providing a balance of procoagulant and anticoagulant hemostatic factors. A target aPTT of about 1.5 times the upper limit of normal and FFP doses of 10 to 15 mL/kg are reasonable. Antithrombin III concentrates, when available, may be beneficial, but their use in DIC is considered experimental, and availability of ATIII is uncertain.

Bleeding in DIC   Bleeding in DIC results primarily from exhaustion and impaired function of procoagulant hemostatic proteins and thrombocytopenia. Accordingly, transfusion of FFP, cryoprecipitate, and platelets may be required (Box 13.7).

Although historically there has been concern that use of hemostatic blood products may drive DIC toward

---

first tenet of DIC management is treatment of the underlying disorder.

Clinical Manifestations of DIC

Clinical signs of DIC appear in many organ systems (Box 13.5).

The extent and severity of findings depend on the severity of DIC. Because severe DIC is associated with high mortality, especially when accompanied by septic shock, vigilant monitoring for early clinical signs in an infant or child with clinical risk factors can be critical. Identification and treatment of underlying causes and early intervention will optimize outcome.

Coagulation Test Features of DIC

Laboratory confirmation of DIC is essential to differentiate DIC from other disorders with similar features, such as autoimmune hemolysis, ITP, and thrombotic microangiopathy syndromes. Important laboratory tests include PT, aPTT, thrombin time, fibrinogen concentration, and fibrinogen degradation products

---

Box 13.7   Transfusion Management of Bleeding in DIC

FFP: 10 to 20 mL/kg as rapidly as tolerated

• Repeat every 6 to 8 hours as needed

Cryoprecipitate for low fibrinogen: 10 mL/kg

• Maintain fibrinogen >100 mg/dL

Platelets for thrombocytopenia: 5 to 10 mL/kg

• Maintain platelet count >50,000/μL with active bleeding

---

pathological thrombosis, this concern has not been validated by experience.

Purpura Fulminans

Purpura fulminans is a rare syndrome of widespread microvascular thrombosis primarily effecting skin and soft tissues. It is most often associated with acquired deficiency of the anticoagulant protein C in the setting of certain infections. *Neisseria meningitidis* and streptococcal pneumonia are the most common bacterial infections; viral infections including varicella are also sometimes associated with purpura fulminans. Rare congenital forms are associated with loss-of-function mutations of protein C or protein S. Widespread fibrin thrombosis of the microvasculature can have life-threatening and limb-threatening consequences. Mortality may be as high as 50% and a third of patients may require limb amputations.

The standard treatment for purpura fulminans is transfusion of FFP 10 to 20 mL/kg every 12 hours. FFP replaces deficient protein C. An investigational form of protein C concentrate and a pharmaceutical form of activated protein C recently approved for use in sepsis (drotrecogin alpha-activated) have been reported to be effective in treating purpura fulminans (Weisel et al. 2002; White et al. 2000). Consideration of activated protein C concentrates is recommended when purpura fulminans is encountered.

Thrombotic Microangiopathic Disorders

Childhood thrombotic thrombocytopenic purpura (TTP) and the hemolytic uremic syndrome (HUS) are rare disorders that may be encountered in acute care settings and may require FFP treatment.

TTP

Most cases of infant and childhood TTP are associated with inherited loss-of-function defects of the vWF cleaving metalloproteinase ADAMTS13 (Tsai 2003). It was recently discovered that the childhood hemostatic disorder Upshaw-Schulman syndrome has the same functional ADAMTS13 defect. Clinical features include severe microangiopathic hemolytic anemia, hyperbilirubinemia, thrombocytopenia, and bleeding. Onset may occur at birth or later in childhood, and the course is often relapsing. As in all cases of TTP, definitive treatment requires simple transfusion of FFP or plasma exchange to provide a source of functional ADAMTS13. Transfusions of whole blood, cryoprecipitate, cryoprecipitate supernatant, and FVIII/vWF concentrates have been reported to be effective in Upshaw-Schulman syndrome (Fujimura et al. 2002). All the latter products probably contain ADAMTS13 activity, which may account for their reported efficacy.

FFP transfusion and plasma exchange remain as the standard of care for TTP. In congenital deficiency of ADAMTS13, periodic transfusion of 10 to 20 mL/kg FFP may be sufficient to treat or prevent relapse episodes. If ADAMTS13 deficiency results from autoantibody inhibitor, as occurs more commonly in adult TTP patients, plasma exchange is warranted to remove antibody and replace ADAMTS13. The availability and reliability of assays for ADAMTS13 activity and inhibitor are limited, and systematic studies of the clinical utility of assays are lacking. Therefore when TTP is suspected, FFP treatment, preferably plasma exchange, should proceed without delay for ADAMTS13 activity assays.

HUS

Both diarrhea-associated and nondiarrhea-associated HUS occur in pediatric acute care settings. The clinical and laboratory features of these thrombotic microangiopathic syndromes can be difficult to distinguish from childhood TTP. A similar pattern of renal failure and microangiopathic hematological disturbances occurs in both TTP and HUS. A classic history of hemorrhagic enterocolitis followed by clinical and laboratory features of thrombotic microangioapathy is suggestive of shiga toxin-associated HUS. Epidemic exposure and detection of enteropathic bacteria or Shiga toxin in stool are supportive of the diagnosis.

In contrast to TTP, treatment of HUS with FFP is of uncertain value. The typically normal ADAMTS13 levels in HUS do not support the use of FFP as a source of that enzyme. Supportive care, including dialysis when necessary, is the current standard of care. In severe cases with encephalopathy, or otherwise life-threatening organ failure, many practitioners include plasma exchange treatment. However, convincing evidence of efficacy in Shiga's toxin-associated HUS is lacking. Heparin and fibrinolytic agents have not been shown to be effective in animal models or humans.

HEMOSTATIC AGENTS IN ACUTE CARE

Recombinant Activated Factor VII

Recombinant activated human factor FVII (rFVIIa) is approved by the Food and Drug Administration (FDA) for treatment of hemophilia A and B patients with inhibitor (see Chapter 20). Nonhemophilia patients with spontaneous FVIII inhibitor have also been successfully treated with rFVIIa. The detailed mechanisms for the hemostatic efficacy of rFVIIa remain under investigation, however, it has been shown to bind to platelets and enhance thrombin generation and to promote platelet adherence to endothelial cell matrices.

In addition to hemophilia inhibitor patients, rFVIIa has been explored in other challenging hemostasis settings (Hedner and Erhardtsen 2002). Case reports have shown efficacy of rFVIIa in severe thrombocytopenia and in patients with the platelet function defects Glanzmann's thrombasthenia and Bernard-Soulier syndrome. Several small case series have reported hemostatic efficacy—sometimes dramatic—in heavily bleeding surgery or trauma patients. Amid burgeoning off-label use of rFVIIa in trauma and surgery, it must be recognized that controlled trials to establish the efficacy and safety of rFVIIa in nonhemophilia inhibitor patients have yet to be conducted. The very high cost of rFVIIa is a relevant issue when its use is considered.

Considerations for Use of rFVIIa in Nonhemophilia Surgical and Trauma Patients

Potentially useful settings may include:

- Dilutional coagulopathy (massive transfusion) with uncontrolled traumatic or surgical bleeding
- Uremic bleeding
- Bleeding in the presence of antiplatelet or anticoagulant agents
- Platelet function defects (Glanzmann's thrombasthenia, Bernard-Soulier syndrome, and others)
- Idiopathic (immune) thrombocytopenia (reported efficacy with counts <5000/μL)
- Alloimmune platelet refractoriness (matched platelet support may be appropriate)
- Liver dysfunction with coagulopathy and uncontrolled bleeding, or preparation for liver biopsy
- Patients with high bleeding risk and volume restrictions undergoing invasive procedures

At this time most institutional guidelines for use of rFVIIa in nonhemophilia patients restrict use primarily to patients with active, usually severe bleeding. At our institution the following guidelines were devised. Before consideration of rFVIIa:

1. Transfusion support should be optimized. Attempts should be made to achieve blood levels of:
   - Fibrinogen >100 mg/dL
   - Platelet count >30,000/μL
   - INR <2.5
2. Other hemostatic agents should be considered:
   - DDAVP
   - Alpha aminocaproic acid (Amicar)
   - Tranexamic acid
   - Aprotinin
3. Consider rFVIIa for ongoing uncontrolled bleeding with potential for hemorrhagic shock and organ damage. We recommend administration of rFVIIa before substantial hemorrhage-related morbidity develops.
4. Dosage and administration:
   Firm guidelines have not been developed. Commonly used dosages are:
   - 90 μg/kg
   - Repeat dosage if needed every 2 hours up to three total doses—dosages beyond three may warrant hematology consultation
5. Overall complication rate is about 1/11,000 doses. Possible reported complications include:
   - Arterial or deep venous thrombosis
   - Pulmonary embolism
   - Catheter-associated thrombosis
   - Acute myocardial infarction

Complications are more common in elderly and cardiovascular disease patients.

rFVIIa in FVII-Deficient Patients

Patients with the rare condition of severe congenital FVII deficiency have been successfully treated with rFVIIa in lieu of FFP. In these patients a significantly lower dosage of rFVIIa has been used. Kinetic modeling of thrombin generation indicates that in the absence of endogenous FVII competition for substrate binding sites, markedly less FVIIa is required for equivalent thrombin generation.

Experience in a small number of patients indicates that dosages of 20 to 30 μg/kg are effective. FVII inhibitors have developed in a few patients, in one case after an exceedingly high dose.

Aprotinin

Aprotinin is a proteinase inhibitor with significant antifibrinolytic activity that has been shown to reduce surgical bleeding and RBC transfusion requirements in adults. Results of clinical studies in pediatric patients have been variable, however, overall evidence suggests that patients undergoing extensive procedures including cardiac reoperation, arterial switch, and heart and lung transplantation have less bleeding and blood loss with use of aprotinin (Pouard 1998).

Aprotinin is derived from a bovine source and has been associated with an allergic response following re-exposure in up to 6% of cases. The highest risk of allergic reactions occurs when re-exposure occurs within 6 to 8 months. Premedication with antihistamines or corticosteroids may be warranted in this setting.

Tranexamic Acid

The fibrinolytic inhibitor tranexamic acid has been demonstrated to reduce blood loss in adult surgery patients. Like aprotinin, results are more equivocal in pediatric patients. Doses larger than 50 mg/kg or continuous infusion may be required to optimize efficacy.

Alpha Aminocaproic Acid (Amicar)

Alpha aminocaproic acid (Amicar) is an antifibrinolytic agent widely used for promoting hemostasis in settings of oral mucosal or gastrointestinal mucosal bleeding. Dosages are the same for oral and intravenous administration (Box 13.8).

In the setting of massive hematuria Amicar has been associated with development of urethral clots. If Amicar is deemed necessary in a patient with significant hematuria, a urinary catheter should be placed and flushed frequently with normal saline.

Topical Hemostatic Agents

Several topical hemostatic agents are available as adjuncts to surgical hemostasis (Jackson 2002). Topical fibrin and thrombin combinations are most appropriate for minor and diffuse surgical bleeding not amenable to suturing. Application to larger vascular wounds may be disappointing, as the fibrin polymer may be washed away by hemorrhage (Jackson 2002).

Box 13.8   Dosage of Alpha Aminocaproic Acid (Amicar) for Oral or GI Mucosal Bleeding

- Oral or IV: 100 mg/kg initial dose then 50 mg/kg every 6 hours as needed
- 18 g total dose maximum per day
- Side effects of nausea and vomiting are common
- Potential thrombotic risk in DIC

The fibrin sealant Tisseel VH Kit (Baxter Healthcare) was recently approved by the FDA. The product is a ready-made application kit containing purified human fibrinogen and human thrombin, as well as aprotinin, to facilitate hemostasis. For topical hemostasis this product offers compelling benefits over homemade "fibrin glue," including viral inactivation, controlled concentrations of reagents, and use of human thrombin, which should avoid stimulation of potentially dangerous antiFV and antithrombin antibodies.

Other topical hemostatic products are available that contain bovine-derived thrombin. Antibovine FV and antibovine thrombin antibodies have been associated with some of these products.

References

DeLoughery T. 2002. Hemorrhagic and thrombotic disorders in the intensive care setting. In *Consultative hemostasis and thrombosis.* Kitchens C, Alving B, and Kessler C, eds. Philadelphia, PA: W.B. Saunders Company.

Di Paola J and Buchanan G. 2002. Immune thrombocytopenic purpura. *Pediatr Clin N Am* 49:911–928.

Festa C, Feng A, and Bigos D. 2002. Transfusion support in pediatric surgery, trauma, and the intensive care unit. In *Pediatric transfusion therapy.* Herman J and Manno C, eds. Bethesda, MD: AABB Press.

Fujimura Y, Matsumoto M, Yagi H, et al. 2002. Von Willebrand factor-cleaving protease and Upshaw-Schulman syndrome. *Int J Hematol* 75:25–34.

Goodstein M. 2002. Neonatal red cell transfusion. In *Pediatric transfusion therapy.* Herman J and Manno C, eds. Bethesda, MD: AABB Press.

Hébert P, Wells G, Blajchman M, Marshall J, et al. 1999. A multicenter, randomized, controlled clinical trial of transfusion requirements in critical care. *N Engl J Med* 340:409–417.

Hedner U and Erhardtsen E. 2002. Potential role for rFVIIa in transfusion medicine. *Transfusion* 42:114–124.

Jackson M. 2002. Topical hemostatic agents for localized bleeding. In *Consultative hemostasis and thrombosis.* Kitchens C, Alving B, and Kessler C, eds. Philadelphia, PA: W.B. Saunders Company.

Kevy S and Gorlin J. 1998. Red cell transfusion. In *Hematology of infancy and childhood.* Nathan D and Orkin S, eds. Philadelphia, PA: W.B. Saunders Company.

Mannucci PM. 1997. Desmopressin (DDAVP) in the treatment of bleeding disorders: the first 20 years. *Blood* 90:2515–2521.

McDonald T and Berkowitz R. 1994. Massive transfusion in children. In *Massive transfusion.* Jeffries L and Brecher M, eds. Bethesda, MD: American Association of Blood Banks.

Pouard P. 1998. Review of efficacy parameters. *Ann Thorac Surg* 65:S40–44.

Richardson M, Allen G, and Monahan P. 2002. Thrombosis in children: current perspective and distinct challenges. *Thromb Haemost* 88:900–911.

Strauss R. 2002. Additive solutions and product age in neonatal transfusion. In *Pediatric transfusion therapy.* Herman J and Manno C, eds. Bethesda, MD: AABB Press.

Tsai H. 2003. Deficiency of ADAMTS13 causes thrombotic thrombocytopenic purpura. *Arterioscler Thromb Vasc Biol* 23:388–396.

Weisel G, Joyce D, Gudmundsdottir A, et al. 2002. Human recombinant activated protein C in meningococcal sepsis. *Chest* 121: 292–295.

White B, Livingstone W, Murphy C, et al. 2000. An open-label study of the role of adjuvant hemostasis support with protein C replacement therapy in purpura fulminans-associated meningococcemia. *Blood* 96:3719–3724.

Pediatric Transfusion in Developing Countries

Kenneth A. Clark, MD, MPH

INTRODUCTION

This chapter discusses the practice of transfusion medicine in developing countries, so as to illustrate the considerable differences between the practice there and the practice in developed nations of the world. Perhaps the greatest differences are in the amount of material resources available to operate blood transfusion systems and in the distribution of diseases and their treatment. In some developing countries, the lack of reliable sources of electricity and refrigeration, coupled with the expenses of blood-bank equipment, testing kits, and reagents have hindered the establishment of good-quality blood banks. This is particularly the case in countries that have been compromised by poverty, political instability, and armed conflict. Restraints on material and financial resources have also hindered the development of a body of persons that are well trained in transfusion practice and blood-banking technology. As a result of these challenges, blood transfusion services among resource-restricted countries vary considerably in their levels of development, safety, and quality.

In many developing countries, the primary indication for transfusion is anemia in childhood or pregnancy. Such anemias would not likely occur in most developed countries, and if they did occur, they would not be treated with blood transfusion. Pediatric blood transfusions make up a large percentage of transfused patients in developing countries and represent over half of all transfusions in some regions, particularly in subSaharan Africa.

Effective and safe donor recruitment, reliable laboratory screening, avoidance of unnecessary transfusions, and proper storage and transport of blood are signifi-

cant problems in many developing countries. A source of safe and reliable voluntary blood donors is the cornerstone of high-quality blood services worldwide. However, most of the developing world, because of lack of effective mobilization, cultural attitudes, and lack of education, can rely on volunteer donors for less than half of its blood supply. Laboratory screening of donor blood is a serious problem in developing countries. In the year 2000, there were a total of about 90 million blood transfusions worldwide. According to the World Health Organization (WHO), 31% of all the blood units donated for these transfusions were not screened for one or more of the three most serious transfusion-transmitted viruses, human immunodeficiency virus (HIV), hepatitis B virus (HBV), and hepatitis C virus (HCV) (Rapiti et al. 2003). Almost all of these screening lapses occurred in the world's developing countries.

SPECIAL TRANSFUSION ISSUES IN RESOURCE-RESTRICTED COUNTRIES

Organization of Blood-Transfusion Systems

In most developing countries, blood-transfusion systems and services are much less fully established and organized than in developed countries. Most developing countries now have adopted a national approach to blood-transfusion systems in an attempt to centrally regulate and coordinate national safe-blood programs. These systems are still in development in most of these nations.

Many developing countries have at least some regional blood collection and processing centers.

However, these centers are often unable to supply the system's total blood need. Additionally, difficulties in transportation and distribution have led to the formation of numerous hospital-based collection centers. These hospital-based facilities vary considerably in their donor-recruitment methods and in the quality of their testing and processing. In some countries, small hospital-based collection centers represent a significant percentage of the total blood collected, tested, and processed.

Populations Requiring Transfusions

The patient groups at highest risk for needing blood transfusion in most resource-restricted countries include children with severe anemia, women with pregnancy-related hemorrhage or anemia, accident victims, and patients with hemoglobinopathies and thalassemias. Hemophiliacs are not, as a group, major users of blood products in most developing countries.

Anemia is a common problem in children in parts of Asia, Central and South America, and subSaharan Africa. In severe cases, it is commonly treated with blood transfusions. Studies have shown that over 50% of all hospitalized pediatric patients in subSaharan Africa receive blood transfusions for severe anemia during their admission (Greenberg et al. 1988). The majority of these transfused children are very young, as exemplified by data from Uganda, where 65% of all transfusions are given to children younger than 5 years of age (Coulter et al. 1993).

Conditions Requiring Transfusions

Childhood anemia in developing countries is frequently due to multiple underlying conditions. Very young children are particularly at risk due to a combination of high rate of red cell volume expansion during growth and a high frequency of underlying illnesses. The major causes of pediatric anemia in developing countries are listed in Box 14-1.

The prevalence and causes of pediatric anemia vary considerably among developing countries in different parts of the world. In regions where malaria, nutritional deficiencies, helminth infections, and thalassemia are common, children are at increased risk for developing severe anemia.

Malaria

Malaria causes anemia primarily through repeated episodes of hemolysis. In areas of high malaria endemicity, many adults have chronic mild anemia, having achieved some level of natural immunity as a result of

---

> **Box 14.1   Major Causes of Pediatric Anemia in Developing Countries**
>
> - Malaria infection
> - Nutritional anemia due to inadequate dietary iron, B_{12}, or folate
> - Hookworm infection
> - Thalassemia
> - Hemoglobinopathy
> - HIV infection
> - Chronic or recurrent infections
> - Accidents

---

frequent bouts of malaria. In young children who have not yet achieved a degree of natural immunity, anemia can become quite severe, particularly during the peak malaria season when children are exposed to repeated infectious mosquito bites. Malarial anemia can be especially severe when it is superimposed on an existing chronic anemia. Acutely ill patients typically have either *Plasmodium vivax* or *P. falciparum* infection, although a small percentage of infections are due to more than one species. *P. falciparum* infection, in particular, can be extremely severe, infecting as many as half of the host's red cells. *P. vivax* and *P. ovale*, which only infect reticulocytes, and *P. malaria* are not usually associated with such high levels of parasitemia.

In developing countries of subSaharan Africa and parts of Asia, malaria is a primary cause or a significant contributing factor for pediatric hospitalization due to severe anemia. In malaria-endemic regions, children form the largest group receiving blood transfusions (Greenberg et al. 1988). One study in a highly endemic area of western Kenya found that 25% of hospitalized children had Hgb levels less than 5 g/dL, 30% of whom died from their anemia (Lackritz et al. 1997).

Nutritional Anemia

Iron deficiency anemia is highly prevalent in many developing countries. The WHO estimates that as many as 50% of children in developing countries have iron-deficiency anemia (WHO 2001). Rural populations with more restricted resources and lower educational levels may be more severely affected than urban populations. Infants who have been breast-fed for more than 4 to 6 months are particularly susceptible if their diets are not supplemented with foods rich in iron or dietary supplements. In some areas, the use of iron supplements is limited by resources.

Inadequate dietary intake of vitamin B_{12} or folate also contributes to pediatric anemia. Dietary insufficiencies are often compounded by additional illnesses such as diarrheal diseases that further impair intestinal absorption.

Thalassemia

Thalassemia is an important public-health problem in resource-restricted countries in Southeast Asia, Africa, and the Middle East. Its treatment places major demands on the blood supply in these regions. Good data on the transfusion needs of children with thalassemia are not available for most resource-restricted countries. However, in India alone, it is estimated that 50,000 to 100,000 children with thalassemia receive blood transfusions each year (Konstenius 2003).

In many developing countries, more stringent indications for repeated transfusions should be implemented, as the risk of adverse events outweighs the benefits. Some authorities feel that there is little indication for repeated transfusions as long as the hemoglobin level remains stable (Wake et al. 1998).

Hemoglobinopathy

Sickle-cell disease is one of the most common hemoglobinopathies, with over 150,000 persons born with the disease each year. Most of the world's population with sickle-cell disease is born in developing countries. Over 80% is born in Africa, with most of the remainder born in the Middle East and India (WHO 2001).

Unfortunately, the regions of Africa where the hemoglobin gene is most common coincide with those where the HIV pandemic is greatest. Patients with sickle-cell disease are recognized as a group at risk of acquiring HIV infection, the risk being dependent on both the seroprevalence of HIV in the blood donor population and the patient-management plans (Fleming 1997).

In developing countries, transfusions in pediatric sickle-cell disease patients are reserved for those with severe anemia (Hgb <5 g/dL) in aplastic or sequestration crises and for severe anemia brought on by superimposed illnesses such as malaria and malnutrition. Management of sickle-cell disease with regular transfusions is discouraged in developing countries due to the shortage of available blood and the risk of HIV or hepatitis infection.

Other

A number of other illnesses may contribute to chronic anemia in children. Hookworm and other intestinal parasitic infections cause anemia through chronic blood loss. Chronic or recurrent bacterial or viral infections may decrease the rate of red cell production.

HIV infection, in particular, can result in anemia or pancytopenia.

Accidents are very common in many developing countries, due to inadequate resources for public-safety measures. Such accidents are a frequent cause of acute blood-loss anemia, often resulting in the need for blood transfusion.

Availability of Blood Products

Throughout the developing world there is a chronic shortage of blood products, which particularly affects the two groups most in need of transfusion services, children and pregnant women. This shortage has been compounded by the problem of inappropriate transfusion practice and the failure to use sound medical guidelines for transfusion. Inappropriate transfusion serves to put additional stress on an already short blood supply. Studies in Africa have estimated that from 13% to 47% of all pediatric transfusions are given unnecessarily (Jager et al. 1990; Lackritz et al. 1993). The blood inventory in many developing countries consists of, at most, only 1 or 2 days' supply. Often, there is no blood in inventory and the patient's family must locate a suitable donor to supply the blood on an as-needed basis (Greenberg 1988). Such donors are referred to as family/replacement donors.

Whole blood, usually collected in CPDA-1 anticoagulant, is the most commonly available blood product. Packed red blood cells (PRBCs) are frequently unavailable in many developing countries and are only rarely available in the least-developed countries. Fresh-frozen plasma and platelets are similarly scarce in many resource-restricted countries. In many regions, they are only available at private hospitals and at major teaching hospitals and then in short supply. Leukocyte-reduced blood products, washed red-cell products, and irradiated or CMV-screened blood products are usually beyond the resource restrictions of most developing countries.

BLOOD COLLECTION AND PROCESSING IN DEVELOPING COUNTRIES

Donor Recruitment, Selection, and Screening

In 1975, the World Health Assembly (WHA) adopted Resolution WHA 28.72 requiring all of its members to promote the development of national transfusion services based on the use of nonremunerated volunteer blood donors (WHO 2001). As in the developed world, such donors represent the safest blood donors from the lowest risk populations. A number of studies has shown that family/replacement donors and paid donor groups have a higher incidence and prevalence of transfusion-transmitted infectious diseases than volunteer donor groups.

Unfortunately, the WHA resolution has not been realized in many resource-restricted countries, despite its adoption more than a quarter century ago. In fact, as of 1999, less than 40% of the 30.2 million total blood donations in resource-restricted countries were collected from lower-risk volunteer donors. Sixty to seventy percent of the donations were given by family/replacement and paid donors, often in countries with relatively high prevalence rates of HIV, HBV, and HCV infection (WHO 2001).

The health status of the blood donor population is a major factor in most developing countries. The rate of donor deferral after laboratory screening may exceed 10%. This not only compounds the risk to the recipient but increases the overall cost of the transfusion product (Beal 1993).

In much of the developed world, the major transfusion-transmitted infectious diseases are localized in populations with certain risk behaviors such as intravenous drug use, sex with commercial sex workers, male-male sex, or travel to certain regions of the world. Persons with these risk behaviors represent a minority of all potential donors in the developed world and are usually readily identifiable by screening questionnaires; therefore, finding and targeting a group of lower-risk volunteer donors is not difficult. However, in areas with high HIV prevalence, such as many African countries, it is difficult to identify and target such a low-risk population for blood donation without eliminating most of the potential donors.

In spite of inherent difficulties in donor selection and deferral in high prevalence areas, recruiters in developing countries have attempted to identify effective risk factor deferral criteria using medical history, demographic information, and social history. One factor that has been used in subSaharan Africa as a surrogate marker for risk is age. The peak seroprevalence of HIV, while somewhat variable among different regions and populations, is typically 15 to 29 years of age in women and 25 to 39 years of age in men (Fleming 1997). Therefore, recruitment of donors has typically centered on students at secondary schools and on older adults.

Another factor that has been used as a surrogate marker for risk is demographic status. In general the HIV prevalence in rural populations is lower than that in urban centers (Barongo et al. 1992). However, recruitment of rural donors is sometimes difficult and is more expensive due to the poor conditions of roads in many developing countries. Also, the rural population

has a generally lower level of nutritional health, resulting in a higher rate of anemia and a proportionally lower rate of potentially acceptable donors (Jacobs 1994). To partially circumvent the problem of identifying a suitable low-risk donor population by risk-deferral screening, donor recruitment activity often includes targeted recruitment at secondary schools, selected businesses, churches, and repeat-donor clubs. As funding and resources for mobile donor recruitment are limited, there is often a forced reliance on family/replacement and directed donors to fill the supply deficit.

Laboratory Testing

Screening of Donated Blood for Transfusion-Transmitted Infections

The WHO strategy for laboratory screening of blood recommends HIV, HBV, and syphilis screening of all donated blood. Where appropriate, donated blood should also be screened for HCV, malaria, and Chagas' disease (WHO 2001). However, resource restrictions preclude implementation of 100% donor screening in many or most resource-restricted regions.

According to the WHO's Global Database on Blood Safety, less than 60% of developing countries fully screen their blood for HIV, and even fewer screen their blood for HBV or HCV infections (WHO 2001). Such lack of screening in countries with high HIV prevalence rates explains why transfusions are identified as the cause of 5% to 10% of HIV infections in developing countries (WHO 2002). In subSaharan Africa, blood transfusion is the third most important mode of HIV transmission, predominantly affecting children under age 5 (Jager et al. 1991).

Malaria screening is generally recommended in malaria-endemic areas. In some developing countries with very high prevalence rates of malaria, such as sub-Saharan Africa, screening is not practical. There, much of the donor population may have a low-level of chronic parasitemia due to repeated exposure to infectious mosquito bites and the subsequent development of some degree of natural immunity. In areas of lower malaria prevalence, such as Southeast Asia, screening for malaria with Giemsa-stained blood films is common but not universal.

Screening tests for HIV antibodies and HBsAg or HCV antibodies are usually performed with test kits using enzyme immunoassay (EIA) methodology. In rural areas, HIV screening is sometimes performed with rapid tests. Polymerase chain reaction (PCR) testing is not generally available or affordable in most developing countries, nor is HIV p24 antigen screening. Blood donor screening for Chagas' disease is usually performed in endemic areas using EIA methods to detect *T. cruzi* antibodies.

Pretransfusion Compatibility Testing

In many developing countries, there is limited ability to perform laboratory tests to ensure immunological compatibility. Many blood banks are able to perform only ABO grouping and Rh typing and to perform crossmatching. More extensive testing such as additional alloantibody or autoantibody identification is beyond the current capabilities of many developing-country blood banks due to lack of reagents and training. The laboratory methods are largely manual, as automated equipment is not readily available or affordable.

RISKS OF INFECTIOUS DISEASE TRANSMISSION BY TRANSFUSION

HIV

Transfusion-transmitted HIV infection is one of the most serious adverse consequences of blood transfusion in children in developing countries. Despite requirements of government legislation, routine testing for many infectious agents, including HIV, is not carried out in many developing countries. In the 1990s, most blood banks in one Southeast Asian country claimed to be universally screening blood donors for HIV antibodies. However, it was subsequently learned that testing was not being done on all donors, and when it was done, it was carried out with methods less sensitive than the third generation EIAs available elsewhere (Wake et al. 1998). The primary reason given by blood banks in many developing countries for not testing is lack of financial resources. EIA testing for HIV and hepatitis can double the basic processing costs for a unit of blood. The lack of knowledge about HIV and hepatitis in developing countries, combined with widespread poverty, prompt the practice taking chances with less expensive blood rather than paying the increased costs of testing (Wake et al. 1998).

Although many countries have national policies on blood-donor testing, the extent of testing actually carried out is somewhat variable. Exact figures are not available for many countries, and the surveillance data are not always reliable. In the Democratic Republic of the Congo (DRC), it has been estimated that as many as 25% of pediatric AIDS cases have resulted from transfusions (Heymann et al. 1992). With an estimated 10% to 15% of HIV transmissions in subSaharan Africa

being attributed to transfusions (WHO 2002), the proportion of infections attributable to transfusions in the pediatric populations is actually much higher, as over half of transfusions are administered to children. In India, nearly 7% of AIDS patients reported acquiring infection following transfusion of infected blood.

A retrospective study of blood transfusion was conducted in the pediatric population of Kinshasa, DRC, in 1988. Although this study took place over a decade ago, circumstances of transfusion practices may not be significantly different today in some least-developed countries. Sixty-nine percent of blood transfusions were administered to children with malaria. Thirteen percent of the hospitalized children were later found to be HIV positive, with the likelihood of seropositive status being closely related to receipt of transfusion. It was estimated that 561 HIV-infected transfusions had been given to children in a single year in a single large public hospital (Greenberg et al. 1988).

Implementation of HIV testing alone without providing proper education for technologists and without establishing quality-assurance programs is not effective in these countries. A study was conducted in the largest public hospital in Kinshasa, DRC, 1 year after universal HIV screening of blood donors was implemented. It was estimated that even with universal HIV screening, 25% of new pediatric HIV infections and 40% of new HIV infections occurring in children over age 1 were due to blood transfusion (Shaffer et al. 1990).

Little research has been performed in recent years to assess the exact risk of HIV infection by blood transfusion in developing countries. In 1994, the Kenya Ministry of Health performed a collaborative multicenter study to assess the risk of HIV transmission in screened blood and to identify ways to improve blood safety. During the study period, the overall prevalence of HIV among blood donors was 6.4%, ranging from 2% to 20% among the hospital blood banks in the study. Using donor-recipient pairs, investigators found that even after HIV antibody screening, there was a 2.0% residual risk of transfusion-transmitted HIV infection. The residual risk of transmission was thought to be the result of erroneous testing practices (Moore et al. 2001). Although Kenya has since undertaken a number of corrective interventions and has moved to create regional transfusion centers, this study serves to show that similar problems may exist in other resource-restricted countries, even with universal HIV antibody testing.

The risk of HIV infection from transfusion in most countries in Central and South America is generally lower than in resource-restricted countries from other continents. The lower rate is due to the lower prevalence rates of HIV infection in the donor population and the higher rates of HIV testing in most Central and South American countries. The probability of transmitting HIV infection by blood transfusion is generally much lower than the risk for transmitting hepatitis or trypanosomiasis (Schmunis et al. 1998).

Hepatitis

The issues of transmission of HBV and HCV are similar to HIV. The WHO recommends HBV screening on all donated blood, yet such screening is not universally performed. As of 2002, developing countries in subSaharan Africa report that only 55% of donated blood is screened for HBV, despite relatively high prevalence rates (Tapko 2003). The prevalence of chronic carriage of HBV in blood donors in subSaharan Africa ranges from 2% to 22%. Yet, the WHO estimates that no more than 50% of the blood donations in subSaharan Africa are screened for HBsAg. The low screening rate is due to both the lack of funds and to the low perceived utility (Allain et al. 2003).

HCV screening is performed on only 40% of donated blood units in Africa (Tapko 2003). The main barrier to implementation of hepatitis testing is the cost of test kits, which is currently prohibitive for many resource-restricted countries. The WHO estimates that unsafe blood transfusions contribute to at least 10% of the global burden of HCV (Rapiti et al. 2003).

The risk of HBV transmission by blood transfusion in developing countries is not currently well known, but in areas of high prevalence and lack of universal screening, it is most likely significant. In Central and South America the risk of acquiring HBV infection from blood transfusions is 1 to 17 per 10,000 transfused units and for HCV is 4 to 75 per 10,000 transfused units (Schmunis et al. 1998). In Southeast Asia, it has been estimated that there are 85 million carriers of HBV and 25 million carriers of HCV, making for an enormous potential for transfusion transmission (Kumari 2003).

Other

Malaria

Malaria screening of donated blood is recommended by WHO when considered appropriate. Such screening usually occurs in areas of low malaria endemicity. In highly endemic areas, such as most of subSaharan Africa, much of the donor population has a low level of chronic parasitemia, making donor screening for malaria impractical. In such highly endemic areas, the pediatric transfusion-recipient population is often being treated for acute malarial anemia, including treatment with antimalarial drugs. If the transfusion recipients in

these regions are being treated for conditions other than malarial anemia, malaria prophylaxis should then be considered for the recipients.

There are few data on the risk of transfusion-transmitted malaria in developing countries. Until recently, it has been difficult to attribute the source of a patient's malaria infection to transfusion, as the potential for acquiring malaria from environmental exposure is great. Newer genetic sequencing techniques will allow such studies in the future.

Chagas' Disease

Trypanosoma cruzi, the causative agent of Chagas' disease, is endemic in Central and South American countries. It is transmitted primarily by insect vectors; however, transfusion of infected blood is the second most important cause of transmission (WHO 2001).

Despite the implementation of screening efforts by most countries in these endemic regions, *T. cruzi* remains the infectious agent with the highest transfusion-transmission rate in Central and South America. The risk of transmission by blood transfusion ranges from 2/10,000 to 219/10,000 transfusions. The risk appears to be primarily due to the incomplete screening practices in some countries (Schmunis et al. 1998).

Crystal violet has been used as an additive to stored blood to inactivate *T. cruzi*. It is effective in amounts of 125 mg/unit of blood. However, additive crystal violet causes staining of skin and mucous membranes in transfusion recipients. Also, the additive process can lead to bacterial contamination if not done properly (WHO 2001).

Bacterial Contamination and Sepsis

Bacterial contamination and sepsis have not been widely studied in most resource-restricted countries. Even in developed countries, bacterial contamination is one of the more common transfusion-related adverse events. The lack of commonly available standard-operating-procedure manuals in many resource-restricted countries, the shortage of laboratory refrigeration and cold-transportation equipment, and the lack of rigorous quality-assurance systems raise the concern that bacterial contamination of blood products may be an even more significant problem.

Syphilis

Syphilis has the potential for transmission by blood transfusion. Studies have shown that treponemal spirochete survival is significantly decreased in blood that has been stored for at least 72 hours at 4°C (Chambers

1969). Therefore, the risk of transfusion-transmitted syphilis is greatest for blood transfused soon after collection or for platelets stored at room temperature. Most of the blood transfused in developing countries is given soon after collection, allowing for the possibility of syphilis transmission. The risk of transmission by transfusion in developing countries has not been well studied. Most laboratories in developing countries do perform syphilis screening serological tests on all blood donations. However, the quality of testing can be of concern, particularly when done in emergency situations.

CURRENT TRANSFUSION PRACTICES

Overview

As has been stated, transfusion practices in resource-restricted countries differ significantly from those in developed countries, in both the types of illnesses and their treatments. The majority of transfusions are given for basic, usually urgent or life-threatening conditions, rather than for support of tertiary care needs, such as the complex types of surgery or chemotherapy seen in developed countries.

Therefore, the greatest transfusion need is for RBC products, along with volume expanders. The need for platelet concentrates and for fresh frozen plasma or cryoprecipitate is much less than in developed countries. More specialized coagulation products are usually not available. The product most readily available in least-developed countries is whole blood, with PRBCs being only occasionally available. Even whole blood is often in short supply. Pediatric blood units are largely unavailable in least developed countries, due to cost restrictions. Leukocyte-reduced units or CMV-screened units are also scarce in most resource-restricted countries.

Transfusion Decision Issues

In many parts of the developing world, blood is in very short supply and may not be readily available for urgent transfusion needs. Family donors or other directed donors are often called to supply the needed blood. Under such circumstances, laboratory infectious disease testing may be incomplete before a transfusion is given, or may be performed under less than ideal circumstances. Therefore, clinicians are often faced with the difficult decision of ordering a transfusion to increase the chances of patient survival, or choosing a more conservative transfusion approach in order to prevent possible transfusion-transmitted infectious disease. In hospitals with ineffective or incomplete

screening of blood for HIV antibodies or hepatitis virus, the risk of transfusion of HIV or hepatitis may be considerable, determined largely by the prevalence of transfusion-transmissible infectious disease among blood donors.

Because of the combined problems of high risk of transfusion-transmitted infectious disease and acute blood-product shortage, prudent clinicians in developing countries are often more reluctant to transfuse than are their counterparts in developed countries.

Guidelines for pediatric transfusion are similar to those in developed countries but tend to be more conservative, due to the increased risk of adverse events. For example, in developing countries, a transfusion may not be recommended except for severe anemia (Hgb <5 g/dL), combined with signs of cardiac failure or respiratory distress. Typical guidelines for pediatric transfusion used in developing countries are shown in Box 14.2 below.

Transfusions to small children and neonates need to be administered slowly when whole blood is used. Otherwise, there is a risk of volume overload. Whole blood transfusions are often administered at a dose of 20 mL/kg over 2 to 4 hours. When PRBCs are available, they are typically given at a dose of 15 mL/kg. In cases of profound anemia and very high malaria parasitemia (>20% of red cells infected), a higher amount of red cell product may be needed. The rapid transfusion of whole blood has actually been shown to increase the death rate of small children and neonates with severe malaria with Hgb levels greater than 5 g/dL, perhaps due to volume overload (Lackritz et al. 1992).

Because of the high risk of transfusion-transmitted disease in developing countries, avoidance of unnecessary transfusions is critically important. As has already been mentioned, it has been found that as many as 47% of transfusions in developing countries may be performed unnecessarily (Lackritz et al. 1993). This high

---

Box 14.2 Typical Guidelines for Pediatric Transfusion in Developing Countries

If Hgb <4 g/dL, transfuse.*
If Hgb <5 g/dL, transfuse when signs of respiratory distress or cardiac failure are present.
If Hgb <5 g/dL and patient is clinically stable, monitor closely and treat the cause of the anemia.
If Hgb ≥5 g/dL, transfusion is usually not necessary. Consider transfusion in cases of shock or severe burns. Otherwise, treat the cause of the underlying anemia.
*20 mL/kg of whole blood or 15 mL/kg of PRBCs. In the presence of profound anemia or very high malaria parasitemia (>20% parasitemia), a larger amount may be needed.

rate is an indicator of the variability in the quality of transfusion practices in developing countries, which can be significant (Holzer et al. 1993).

Perhaps the best method to reduce inappropriate transfusions is to limit their use to only the most urgent conditions. Studies in Africa have documented that pediatric blood transfusions are associated with improved survival only when they are provided to children with severe anemia (Hgb <5 g/dL) and signs of cardiorespiratory failure such as forced respiration (grunting), intercostal retraction, or nasal flaring (Lackritz et al. 1992). In another study of children with profound anemia and malaria, prostration, along with respiratory distress, was found to be an additional strong indicator of transfusion need (English et al. 2002).

To be beneficial, the transfusions must be made available as soon as possible (English et al. 2002; Lackritz et al. 1992). The speed of response in providing blood for transfusion has been found to be critical in at least one study in a malaria-endemic region. In at least 40% of the cases where severe anemia contributed to a child's death, blood transfusion was either not possible or was incomplete before death occurred (English et al. 2002). In developing countries, obtaining blood for transfusion may take significantly longer than in the developed world, due to frequent lack of availability of blood and the need to collect and test blood from family member–directed donors (English et al. 2002). Since blood units are not usually available, a compatible donor must be found for every child requiring a transfusion. Due to the urgent nature of the conditions requiring treatment, the transfusion must be administered within hours of donation. HIV-antibody and HBsAg screening may not be routinely available under such conditions; therefore, the risk of disease transmission by transfusion is directly linked to the disease prevalence (Greenberg et al. 1988).

The attempts to minimize transfusion in developing countries perhaps place a greater emphasis on the use of volume expanders and intravenous (IV) replacement fluids than in developed countries. The very high rate of accidents in developing countries frequently leads to pediatric patients being treated for acute blood loss and hypovolemia. IV replacement fluids are the first line of treatment in such patients. The use of replacement fluids to stabilize a hypovolemic patient may decrease the need for a red cell transfusion. Guidelines for their use are similar to those in developed countries.

Administration of Transfusions

The use of pediatric blood units is recommended whenever they are available. However, since pediatric units are not commonly available, blood for transfusion is usually taken from adult blood units through a transfer pack. Removal of aliquots from the primary collection bag for small volume transfusion is sometimes performed in small volume bags, sterile syringe sets, or buret sets when available. Infusion pumps are not widely available, so infusion rates are determined by drip-rate methods. In this system, rates are calculated by counting drops per minute in the drip chamber and dividing this by the drops/mL rating of the infusion system.

Blood warming devices are not widely available. However, in tropical developing countries a short exposure time to the relatively high temperature of the ambient air quickly raises the temperature of the blood in the transfusion set. Neonatal exchange transfusions are uncommon in many resource-restricted countries; therefore, the lack of blood-warming devices is not usually of concern.

PREVENTION MEASURES TO REDUCE NEED FOR TRANSFUSION

Nutrition

The most effective way to eliminate the need for pediatric blood transfusion is through interventions to prevent anemia. Such interventions include the administration of oral iron supplements during pregnancy and the provision of maternal nutritional education. Small children should be given diets supplemented with iron. Health care workers should make efforts to detect childhood anemia at an early stage. Early identification and treatment of the cause of mild anemia will help reduce the number of cases of severe anemia and subsequently reduce the number of pediatric transfusions.

Malaria Prevention

In regions highly endemic for malaria, children may receive hundreds of infectious bites per year. In such areas, bed nets should be used to prevent exposure to mosquitoes. Children should have routine screening for anemia, followed by appropriate antimalarial therapy.

References

Allain JP, Candotti D, Soldan K, Sarkodie F, Phelps B, Giachetti C, et al. 2003. The risk of hepatitis B virus infection by transfusion in Kumasi, Ghana. *Blood* 101:2419–2425.

Barongo LR, Borgdorff MW, Mosha FF, Nicoll A, Grosskurth H, Senkoro KP, Newell JN, Changalucha J, Klokke AH, Killewo JZ, et al. 1992. The epidemiology of HIV-1 infection in urban areas, roadside settlements and rural villages in Mwanza Region, Tanzania. *AIDS* 6:1521–1528.

Beal R. 1993. Transfusion science and practice in developing countries: "...a high frequency of empty shelves...." *Transfusion* 33:276–278.

Blood Safety and Clinical Technology Progress 2000–2001. 2002. World Health Organization, Geneva, Switzerland.

Chambers RW, Foley HT, and Schmidt PJ. 1969. Transmission of syphilis by fresh blood components. *Transfusion* 9:32–34.

The clinical use of blood in medicine, obstetrics, paediatrics, surgery & anesthesia, trauma & burns. 2001. World Health Organization. Geneva, Switzerland.

Coulter JB. 1993. HIV infection in African children. *Ann Trop Paediatr* 13:205–215.

English M, Ahmed M, Ngando C, Berkleym J, and Rossm A. 2002. Blood transfusion for severe anaemia in children in a Kenyan hospital. *Lancet* 359:494–495.

Fleming AF. 1997. HIV and blood transfusion in sub-Saharan Africa. *Transfus Sci* 18:67–79.

Global Database on Blood Safety, Summary Report 1998–1999. 2001. World Health Organization, Geneva, Switzerland.

Greenberg AE, Nguyen-Dinh P, Mann JM, Kabote N, Colebunders RL, Francis H, et al. 1998. The association between malaria, blood transfusions, and HIV seropositivity in a pediatric population in Kinshasa, Zaire. *JAMA* 259:545–549.

Heymann SJ and Brewer TF. 1992. The problem of transfusion-associated acquired immunodeficiency syndrome in Africa: a quantitative approach. *Am J Infect Control* 20:256–262.

Holzer BR, Egger M, Teuscher T, Koch S, Mboya DM, and Smith GD. 1993. Childhood anemia in Africa: to transfuse or not transfuse? *Acta Trop* 55:47–51.

Jacobs B, Berege ZA, Schalula PJ, and Klokke AH. 1994. Secondary school students: a safer blood donor population in an urban with high HIV prevalence in east Africa. *East Afr Med J* 71:720–723.

Jager H, Jersild C, and Emmanuel JC. 1991. Safe blood transfusions in Africa. *AIDS* 5 (Suppl 1):S163–S168.

Jager H, N'Galy B, Perriens J, Nseka K, Davachi F, Kabeya CM, et al. 1990. Prevention of transfusion-associated HIV transmission in Kinshasa, Zaire: HIV screening is not enough. *AIDS* 4:571–574.

Konstenius T. 2003. Personal communication. American Red Cross International Services, Washington, DC.

Kumari S. 2003. Review of blood transfusion services in south-east asia region of World Health Organization. Meeting of the International Consortium for Blood Safety and Liaised Organizations and Institutions, February 15–17, 2003. Atlanta, GA.

Lackritz EM, Campbell CC, Ruebush TK, Hightower AW, Wakube W, Steketee RW, et al. 1992. Effect of blood transfusion on survival among children in a Kenyan hospital. *Lancet* 340:524–528.

Lackritz EM, Hightower AW, Zucker JR, Ruebush TK, Onudi CO, Steketee RW, et al. 1997. Longitudinal evaluation of severely anemic children in Kenya: the effect of transfusion on mortality and hematologic recovery. *AIDS* 11:1487–1494.

Lackritz EM, Ruebush TK, Zucker JR, Adungosi JE, Were JB, and Campbell CC. 1993. Blood transfusion practices and blood-banking services in a Kenyan hospital. *AIDS* 7:995–999.

Moore A, Herrera G, Nyamongo J, Lackritz E, Granade T, Nahlen B, et al. 2001. Estimated risk of HIV transmission by blood transfusion in Kenya. *Lancet* 358:657–660.

Rapiti E, Dhingra N, Hutin Y, and Lloyd S. 2003. 11th International Symposium on Viral Hepatitis and Liver Disease. Sydney, Australia.

Schmunis GA, Zicker F, Pinheiro F, and Brandling-Bennett D. 1998. Risk for transfusion-transmitted infectious diseases in Central and South America. *Emerg Infect Dis* 4:5–11.

Shaffer N, Hedberg K, Davachi F, Lyamba B, Breman JG, Masisa OS, Behets F, Hightower A, and Nguyen-Dinh P. 1990. Trends and risk factors for HIV-1 seropositivity among outpatient children, Kinshasa, Zaire. *AIDS* 4:1231–1236.

Strategy for safe blood transfusion. 1998. World Health Organization, Southeast Asia Region, New Delhi, India.

Tapko JB. 2003. Regional strategy: priority interventions for improving in the African region. Meeting of the International Consortium for Blood Safety and Liaised Organizations and Institutions, February 15–17, 2003. Atlanta, GA.

Wake DJ and Cutting WA. 1998. Blood transfusion in developing countries: problems, priorities and practicalities. *Trop Doct* 28: 4–8.

Exchange Transfusion in the Infant

NANCY ROBITAILLE, MD, ANNE-MONIQUE NUYT, MD, ALEXANDROS PANAGOPOULOS, MD, AND HEATHER A. HUME, MD

INTRODUCTION

Exchange transfusion in neonates is performed primarily to avoid kernicterus, a consequence of hyperbilirubinemia. In this chapter, the rationale and indications for exchange transfusion in the infant and the procedure itself will be reviewed. Recommendations for the choice of blood components will be discussed, with particular reference to blood types, preservative solutions, length of storage, gamma irradiation, and the cytomegalovirus (CMV) status of the blood products. Finally, potential complications associated with exchange transfusion will be briefly reviewed.

RATIONALE AND INDICATIONS

Exchange transfusion involves the replacement of the total blood volume with compatible donor red blood cells (RBCs) and plasma. The principal indication for exchange transfusion in newborns is severe unconjugated hyperbilirubinemia that is not controlled by phototherapy and places the infant at risk for developing kernicterus. The list of etiologies of neonatal unconjugated hyperbilirubinemia includes: prematurity, infections, disorders of conjugation (Gilbert syndrome and Crigler-Najjar syndrome types I and II), birth trauma, breast-feeding, and hemolysis due to either hemolytic disease of the newborn (HDN), or erythrocyte structural defect or enzymatic defects (Dennery et al. 2001). Kernicterus refers to the finding on autopsy of neuronal injury due to the accumulation of bilirubin at the levels of the basal ganglia, brainstem nuclei, and auditory nuclei (Volpe 1995). The clinical expression of ker-

nicterus is an acute phase characterized by hypertonia, opisthotonos, and a high pitched cry, evolving slowly in the majority of patients to the chronic form dominated by choreoathetosis, gaze abnormalities, and sensorineural hearing loss in children that usually conserve a normal intelligence thus "giving the appearance of a normal mind trapped in an uncontrolled body" (Bhutani and Johnson 2003). Based on different studies, it is estimated that about 1 in 650 healthy newborns can develop dangerous hyperbilirubinemia and be at significant risk of developing kernicterus (Bhutani and Johnson 2003). Bilirubin neurotoxicity depends mainly on unconjugated and free bilirubin levels. However, other factors also affect this neurotoxicity. These include the albumin level and its affinity to bind bilirubin, the presence of endogenous or exogenous competitors to the albumin binding sites for bilirubin, the state and permeability of the blood-brain barrier, and the metabolism of bilirubin in the central nervous system. It appears therefore that it is impossible to define a single bilirubin level that is safe for every infant (Hansen 2002). The kernicterus registry inaugurated by Brown et al. in 1990a and b identified the most frequent causes of excessive unconjugated hyperbilirubinemia leading to kernicterus in term infants. Glucose-6-phosphate dehydrogenase deficiency (G6PD) was found in 31.5% of cases, hemolysis (excluding sepsis and G6PD deficiency) in 14.7%, cephalhematoma and bruising in 9.9%, systemic infection in 6.6%, and Crigler-Najjar syndrome in 3.2%. In 31.5% of cases, the unconjugated hyperbilirubinemia was considered as idiopathic and only related to an excessive weight loss (>10% of total body weight) (Johnson et al. 2002). Risk factors for excessive unconjugated hyperbilirubinemia are prema-

TABLE 15.1 Indications for Exchange Transfusion from 1997 to 2002, Sainte-Justine Hospital, Montreal, Canada

Indications for Exchange Transfusion	Number Performed (1997–2003)
Rhesus alloimmunization	12
ABO alloimmunization	5
Immune hemolysis (other than Rh or ABO)	3
Prematurity	10
Hereditary hemolytic anemia	3
Inborn error of metabolism	1
Congenital leukemia	1
Hyperbilirubinemia of undetermined etiology	4
Other (unknown)	2

turity, exclusive breast-feeding, family history of a previous newborn with jaundice, cephalhematoma and bruising, Asian race, and advanced maternal age (Newman et al. 2000).

HDN is the most common indication for exchange transfusion; ABO incompatibility and RhD HDN being the entities most frequently encountered (Brecher 2002; Herman and Manno 2002). In our institution, a tertiary level neonatal intensive care unit (NICU) with approximately 1200 admissions per year, 41 exchange transfusions have been performed from 1997 to 2002. Rhesus alloimmunization was the most common indication. Indications for all 41 exchange transfusions are shown in Table 15.1.

In addition to the treatment of hyperbilirubinemia, exchange transfusion is also indicated to remove toxic agents such as boric acid, methyl salicylate, and naphthalene in infants showing signs of poisoning (Panagopoulos, Valaes, and Doxiadis 1969; Boggs and Westphal 1960).

Due to the morbidity and mortality associated with exchange transfusion and the recent developments in the management of neonatal hyperbilirubinemia, exchange transfusion is now used only when other treatment modalities have failed to control the rise in bilirubin. Phototherapy has become the standard of care. Intravenous gamma globulins (IVIGs), albumin, protoporphyrins, phenobarbital, and clofibrate protoporphyrins are potential alternatives to exchange transfusion (Hammerman and Kaplan 2000). IVIGs are used routinely in Europe for the treatment of neonatal jaundice due to Rh and ABO incompatibility. It has been postulated that IVIGs work by blocking Fc receptor, thereby inhibiting hemolysis and reducing the formation of bilirubin. It has also been proposed that IVIGs could accelerate the rate of immunoglobulin G catabolism (Hammerman and Kaplan 2000). Doses used vary between 0.5 and 1 g/kg (Rübo et al. 1992; Alpay et al. 1999; Sato et al. 1991). In two randomized

studies, IVIG therapy combined with phototherapy reduced the need for exchange transfusion and no side effects were observed (Rübo et al. 1992; Alpay et al. 1999).

Some earlier studies have shown that albumin infusion might increase the efficiency of exchange transfusion if given shortly before or during the procedure (Tsao and Yu 1972; Comley and Wood 1968). No study, however, has demonstrated the efficacy of albumin infusion for preventing exchange transfusion. The infusion of albumin during phototherapy has resulted in a more rapid decline in unconjugated, unbound bilirubin levels although it did not seem to result in a durable effect (Caldera et al. 1993; Hosono et al. 2001). Therefore the use of albumin in cases of dangerous unconjugated hyperbilirubinemia cannot be routinely recommended.

Metalloporphyrins act by competitively inhibiting the enzyme heme oxygenase, thereby reducing bilirubin production. They are administered by intramuscular injections. Prospective randomized clinical trials demonstrated that tin-mesoporphyrin reduced the requirement for phototherapy, and its only side effect was a transient erythema due to phototherapy (Kappas, Drummond, and Valaes 2001; Kappas et al. 1988; Martinez et al. 1999; Valaes, Drummond, and Kappas 1998). Although promising, metalloporphyrins remain experimental therapy. Phenobarbital is used to increase the conjugation and excretion of bilirubin by enhancing the action of the enzyme glucoronyl transferase, but it takes several days before being effective (Dennery et al. 2001; Hammerman and Kaplan 2000). Clofibrate is an experimental therapy. Its mechanism of action is similar to that of phenobarbital, but it is effective in a few hours (Hammerman and Kaplan, 2000).

Optimal timing for exchange transfusion varies according to gestational age, birth weight, the degree of anemia, the clinical status of the infant, and the etiology of the hyperbilirubinemia. Guidelines for the bilirubin threshold level at which exchange transfusion should be performed differ in the literature (AAP 1994; Canadian Paediatric Society [CPS] 1999). The American Academy of Pediatrics (AAP) recommends exchange transfusion in an otherwise healthy term newborn (≥37 weeks of gestation) with nonhemolytic hyperbilirubinemia when bilirubin levels are higher than 20 mg/dL before 48 hours of age and higher than 25 mg/dL thereafter and phototherapy has failed to lower these levels (AAP 1994). Phototherapy should produce a decline in serum bilirubin level of 1 to 2 mg/dL within 4 to 6 hours, and levels should continue to fall thereafter (AAP 1994; 1999). Guidelines for exchange transfusion suggested by the CPS are slightly different from the recommendations of the AAP. For term infants without risk factors, the CPS recommends that exchange transfusion be considered at bilirubin levels of 25 mg/dL; for term infants with risk factors, the recommended level is 20 mg/dL. Risk factors include gestational age younger than 37 weeks, birth weight less than 2500 g, hemolysis, jaundice at less than 24 hours of age, sepsis, and the need for resuscitation at birth (CPS 1999). Lower bilirubin levels are suggested for exchange transfusion in premature and low birth weight infants (Peterec 1995).

PROCEDURE

Two techniques for exchange transfusion have been described. The discontinuous method was described by Diamond et al. in 1951; it involves the removal and then replacement of small aliquots of blood through a venous umbilical catheter. In the continuous isovolumetric method described by Wallerstein in 1946, recipient blood is withdrawn through an arterial umbilical catheter while donor blood is infused simultaneously via the umbilical vein. The former method is the most commonly used. It appears to be the safer method because the quantities of blood removed and infused can be more reliably controlled, monitored, and recorded.

The infant should be fasting for 4 hours before beginning the exchange transfusion (otherwise, the gastric content must be aspirated before the exchange to prevent inhalation). The infant is placed in a supine position under a radiant warmer. Heart rate, blood pressure, respiratory rate, pulse oximetry, and temperature must be monitored throughout the procedure. Equipment for respiratory support and resuscitation must be immediately available. The venous umbilical catheter should be as large as possible (8 French for a term infant) and be inserted just far enough to permit a good blood return. If an arterial umbilical catheter is used, the tip should reside between T6 and T9 or at L3-L4. A 3,5 or 5 French catheter is the usual size for a term infant.

Twice the total blood volume is usually exchanged (2 × 85 mL/kg). A two-volume exchange transfusion is effective in controlling the hyperbilirubinemia by removing about 50% of the bilirubin, 75% to 90% of circulating RBCs and, in cases of hyperbilirubinemia due to HDN, 75% to 90% of the antibodies to erythrocytes (Brecher 2002). The exchange transfusion should be completed within 2 hours. Using the discontinuous method, a maximum of 5 mL/kg is replaced over 2 to 4 minutes during each cycle of the exchange. One should avoid performing the procedure too rapidly since an acute depletion of the infant's blood volume

could cause a detrimental decrease in cardiac output and blood pressure. A nurse should record exactly how much blood has been exchanged. If too much recipient blood is removed, anemia will ensue; conversely, if too much donor blood is infused, it will lead to congestive heart failure.

Donor blood is warmed to 37°C to prevent hypothermia. The blood may be warmed using in-line blood warmers or in a temperature-controlled waterbath. Some clinicians allow the blood to warm under the infant's radiant warmer. However, this method is not recommended as the temperature of the blood cannot be controlled, and there is a risk of overheating, which can result in the hemolysis of the RBCs to be infused. During the procedure, donor blood is gently agitated every 15 minutes to prevent red cell sedimentation in the bag.

Precautions must be taken to avoid metabolic and hematologic disturbances. A complete blood count (CBC), blood gas, and blood chemistry, including electrolytes, glucose, calcium, and magnesium, should be performed before and after the exchange transfusion. During the procedure, glucose and ionized calcium levels should be verified every 30 minutes or after every 100 mL of blood exchanged. Administration of calcium gluconate (1 mL of 10% calcium gluconate after every 100 mL of blood exchanged) to prevent a fall in ionized calcium due to the binding effect of citrate present in anticoagulants of blood components has been recommended (Maisels et al. 1974). However, there is not consensus concerning its routine use; for example, Maisels et al. (1974) demonstrated that calcium gluconate is not effective in preventing the fall in ionized calcium, which occurs during exchange transfusion with ACD-anticoagulated blood. Furthermore, episodes of bradycardia have been associated with calcium infusion (Keenan et al. 1985). If administered, calcium should be infused slowly via a peripheral vein; infusion through the catheter used for the exchange transfusion should be avoided as there is a risk of clot formation in the blood being infused.

Serum bilirubin levels are monitored at 2, 4, and 6 hours after the exchange transfusion and at every 6-hour interval thereafter. Since there is re-equilibration of the bilirubin between the intravascular and the extravascular spaces after the exchange transfusion, a rebound bilirubin level is to be expected (Valaes 1963). Phototherapy should be resumed immediately after the exchange transfusion.

Due to the high glucose concentration contained in some preservative/anticoagulant and additive solutions, a rebound hypoglycemia can occur after the procedure. Therefore glucose levels should also be monitored postexchange.

SELECTION OF DONOR BLOOD

Once the decision to perform an exchange transfusion is made, blood should be available as soon as possible. Whole blood (WB) or reconstituted WB (that is, a RBC unit mixed with a unit of fresh frozen plasma [FFP]) are the usual choices. Since exchange transfusion does constitute a massive transfusion (that is, transfusion of more than one blood volume in less than 24 hours) and some coagulation factors (for example, factor IX) are physiologically low in neonates, FFP is preferable to albumin as the reconstituting solution (Hume 1999). The reconstituted WB should have a hematocrit between 40% and 50%. The volumes of RBCs and FFP to be used can be calculated using the following formula (reproduced, with permission, from Herman and Manno, 2002).

Total volume (in mL)
  = Infant's weight in kg × 85* mL/kg × 2
Absolute volume of RBCs required (in mL)
  = Total volume × 0.45 (the desired hematocrit)
Actual volume of RBCs required (in mL)
  = Absolute volume/hematocrit of unit after
  any manipulation
Necessary volume of FFP = Total volume required
  − Actual volume of RBCs required

*85 to 100 mL/kg, depending on the estimated blood volume according to gestational age (that is, 85 mL/kg at term, 100 mL/kg for preterm infants)

Pretransfusional analyses include ABO and Rh typing, a direct antiglobulin test (DAT) and a screen for (and if positive an identification of) clinically significant unexpected red cell antibodies. For blood grouping it is preferable to use a specimen collected from the infant's peripheral blood; for antibody detection a peripheral blood or a cord blood specimen may be used. If an adequate blood specimen from the infant is not available, the antibody detection tests may be performed on maternal blood, and in the case of HDN, if at all possible blood grouping and antibody identification should be performed on maternal blood. If the DAT is positive, an elution should be performed and antibody detection/identification done on the eluate.

Special considerations need be taken with respect to blood group choices when hyperbilirubinemia is a consequence of HDN. In cases of ABO incompatibility, the recipient plasma must not contain antibodies (antiA/B) corresponding to antigens (A and/or B) found on donor RBCs, and the ABO group of the FFP should be compatible with the infant's RBCs. RBCs from group O donors and FFP from group AB donors are acceptable choices for every recipient blood group. For RhD incompatibility, RhD-negative blood RBC components must be used. When HDN is due to other clinically significant unexpected red cell antibodies, we recommend, if at all possible, using only RBC units negative for the corresponding antigen(s). However, the American Association of Blood Banks (AABB) standards do allow that such units be either negative for the corresponding antigen(s) or compatible by antiglobulin crossmatch (AABB 2002).

A screening test for hemoglobin S should be performed and found to be negative on all RBC or WB units in order to avoid the risk of intravascular hemolysis (Murohy, Malhorta, and Sweet 1980).

The safety of RBCs stored in additive solution has been evaluated for small-volume transfusions (≤15 mL/kg) in neonates (Luban, Strauss, and Hume 1991; Strauss et al. 1996; Strauss et al. 2000; Goldstein 1993). There are no such data for massive transfusion, and therefore questions as to the safety of additive solutions for large-volume transfusions in neonates remain unanswered. In that context, RBCs stored in CPDA1 solution remain a simple choice for exchange transfusion. However, they may not always be available. If RBCs stored in additive solution are used, it is recommended that the additive solution be removed either by washing the RBCs or by centrifuging the unit and removing the supernatant fluid (Luban, Strauss, and Hume 1991).

Due to the increased potassium content in stored WB or RBC units, fresh WB or RBCs (that is, units stored for less than 5 to 7 days) should be used. While the potassium content does not pose a problem in the setting of small-volume neonatal transfusions (≤15 mL/kg) administered slowly over 3 to 4 hours (Luban, Strauss, and Hume 1991; Strauss et al. 1996; Strauss et al. 2000), the potassium content of stored blood, when infused rapidly and in large volumes, may be lethal for an infant (Hall et al. 1993; Scanlon and Krakaur 1980; Brown et al. 1990a; Brown et al. 1990b). If RBCs stored for more than 5 to 7 days must be used, the unit should be centrifuged and the supernatant fluid removed.

Another potential disadvantage of RBCs stored for extended periods is the drop in 2,3 diphosphoglycerate (2,3-DPG) that occurs during storage. Intraerythrocyte 2,3-DPG plays a major role in the red cell capacity to release oxygen to the tissues (as reflected by the p50 level, the blood oxygen tension at which hemoglobin is 50% saturated with oxygen) (Benesch and Benesch 1967). 2,3-DPG is almost totally depleted from RBCs by 21 days of storage: at collection the p50 value of RBCs is 27 mmHg (approximately the normal value for adults), and this falls to 18 mmHg at outdate (Strauss 1999). In adults this decline in 2,3-DPG and p50 appears to have little significance in most clinical situations since the 2,3-DPG level increases to more than 50% of normal within several hours following transfusion (Heaton, Keegan, and Holme 1989). Even in the setting of massive transfusion, detrimental effects of the low level of 2,3-DPG in stored RBCs have not been demonstrated in adults (Falchry, Messick, and Sheldon 1996). Although these observations may not be generalizable to massive transfusions in the neonate, it should be remembered that the newborn has a physiologically low p50 value comparable to the p50 of stored RBCs (because of the effects of high fetal hemolgobin levels) and, assuming sufficient glucose and phosphate levels in the neonate's bloodstream, the p50 of stored transfused blood likely increases following transfusion.

Transfusion-associated graft-versus-host disease (TA-GVHD) has been reported following exchange transfusion in neonates (Przepiorka et al. 1996; Voak et al. 1996). TA-GVHD results from the engraftment of transfused immunocompetent donor T lymphocytes in a blood transfusion recipient whose immune system is unable to reject them. Clinical manifestations are characterised by fever, rash, pancytopenia, and, in some patients, diarrhea and/or liver dysfunction. Death occurs in more than 90% of reported cases and is usually due to the complications of bone marrow failure (Sanders and Graeber 1990). Gamma irradiation prevents TA-GVHD by prohibiting T-lymphocyte proliferation. Both the American Society of Clinical Pathology and the British Council for Standards in Haematology consider exchange transfusion an indication for the use of irradiated blood components (Ohto and Anderson 1996; Hume and Preiksaitis 1999). However, there is an increase in potassium concentration in stored irradiated RBC units as compared to unirradiated units (Hillyer, Tiegerman, and Berkman 1991). In order to avoid hyperkalemia, for neonatal transfusions it is recommended to perform irradiation of the blood components as close to the time as transfusion as possible. If irradiation of RBC units is performed more than 24 hours before an exchange transfusion, it would be prudent to centrifuge the unit and remove the supernatant fluid.

A final consideration in the choice of blood components is the necessity of providing components at reduced risk for transmitting CMV. CMV is transmitted by leukocytes in cellular blood components collected from (a not well-defined subset of) CMV seropositive donors. CMV antibody prevalence in blood donors in industrialized countries varies from 30% to 80% (Preiksaitis 1991). Two types of blood components are considered to be CMV "safe" or at reduced risk of CMV transmission, namely blood collected from CMV-seronegative donors or blood components that have been processed to have a residual leukocyte count below 5×10^6 (AABB 1997; Napier et al. 1998). Most guidelines do recommend the provision of CMV-reduced risk blood components for low birth weight infants, particularly if the mother is CMV seronegative or of unknown CMV serostatus (AABB 1997; Napier et al. 1998; CPS 2002). However the question of the necessity of providing CMV-reduced risk cellular blood components to term or near-term neonates undergoing massive transfusion is more controversial. A presumptive case of transfusion-transmitted CMV infection resulting in the death of a full-term infant undergoing massive transfusion has been reported (Preiksaitis 1991). Given the modest quantities of blood that are used for exchange transfusions and the relative ease of providing CMV-reduced risk components, it would seem reasonable in most cases to do so.

Other than the reduced risk for CMV transmission there is no evidence to suggest that the use of leukoreduced components reduces morbidity or mortality associated with exchange transfusion (Strauss 2000). A recent study did show a small decrease in neonatal morbidity in preterm infants who received prestorage leukoreduced cellular components for all transfusions as opposed to those who did not (Fergusson et al. 2003). One could therefore opt to use prestorage leukoreduced components to provide a CMV-reduced risk component and this may, in preterm infants at least, offer additional advantages.

COMPLICATIONS

Complications include those related to blood transfusion as well as those related to the procedure itself. Hypocalcemia, hyperkalemia, and bleeding from thrombocytopenia are potential complications related to massive transfusion. The former two can be life-threatening since they can lead to cardiac arrythmias and cardiac arrest. Prevention of these complications is discussed above. TA-GVHD has been reported following exchange transfusion but, as also discussed previously, can be prevented by gamma irradiation of cellular blood products. Anemia, hypothermia, apnea, bradycardia, hypoglycemia, and necrotizing enterocolitis have all been associated with exchange transfusion. Air embolus, portal vein thrombosis, and sepsis are inherent complications of an umbilical catheter. Vascular insufficiency of the lower limbs and thrombi in the abdominal aorta are potential complications when the exchange is done through an arterial umbilical catheter (Keenan et al. 1985).

Early studies defined mortality rate of exchange transfusion according to the definition suggested by

Boggs and colleagues, which refers to the number of infants who died during or within six hours following an exchange transfusion (1960). The mortality rate was between 0.79% and 3.2% per patient and between 0.6% and 1.9% per procedure (Panagopoulos, Valaes, and Doxiadis 1969; Boggs and Westphal 1960; Weldon and Odel 1968). These studies also demonstrated that the mortality rate appeared to be more closely related to the infant's clinical status at the beginning of the procedure than the procedure itself (Panagopoulos, Valaes, and Doxiadis 1969; Boggs and Westphal 1960; Weldon and Odel 1968). More recent studies show similar results. Keenan et al. (1985) reported a 0.53% mortality rate per patient and 0.3% per procedure using Bogg's definition (Keenan et al. 1985). Jackson (1997) demonstrated an overall mortality rate of 2%, with all the deaths occurring in ill infants (Jackson 1997). Considering the potential morbidity and mortality associated with exchange transfusion, this procedure should be used only after other modalities have failed and should be performed only by or under the supervision of experienced nurses and physicians.

References

AABB. *Standards for Blood Banks and Transfusion Services*, 21st ed. 2002. Bethesda, MD: American Association of Blood Banks.

AABB. 1997. Leukocyte reduction for the prevention of transfusion-transmitted cytomegalovirus. *American Association of Blood Banks, Association Bulletin 97-2.*

Alpay F, et al. 1999. High-dose intravenous immunoglobulin therapy in neonatal immune haemolytic jaundice. *Acta Paediatr* 88:261–219.

American Academy of Pediatrics, Committe for Quality Improvment and Subcommittee on Hyperbilirubinemia. 1994. Practice parameter: management of hyperbilirubinemia in the healthy term newborn. *Pediatrics* 94:558–565.

Benesch R, Benesch RE. 1967. The influence of organic phosphates on the oxygenation of hemoglobin. *Fed Proc* 26:673.

Bhutani VK and Johnson LH. 2003. Kernicterus: lessons for the future from a current tragedy. *Neoreviews* 4:30–32.

Boggs TR and Westphal MC. 1960. Mortality of exchange transfusion. *Pediatrics* 26:745–755.

Brecher ME, ed. 2002. Perinatal issues in transfusion practice; Neonatal and pediatric transfusion practice. In *AABB technical manual*. Bethesda, MD: AABB.

Brown KA, et al. 1990a. Hyperkalemia during massive blood transfusion in paediatric craniofacial surgery. *Can J Anaesth* 37:401–408.

Brown KA, et al. 1990b. Hyperkalemia during rapid blood transfusion and hypovolaemic cardiac arrest in children. *Can J Anaesth* 37:747–754.

Caldera R, Maynier M, Sender A, et al. 1993. The effect of human albumin in association with intensive phototherapy in the management of neonatal jaundice. *Arch Fr Pediatr* 50:399–402.

Comley A and Wood B. 1968. Albumin administration in exchange transfusion for hyperbilirubinaemia. *Arch Dis Child* 43:151–154.

Dennery PA, et al. 2001. Neonatal hyperbilirubinemia. *N Engl J Med* 344:581–590.

Diamond LK, Allen FH, Thomas WO. 1951. Erythroblastosis fetalis. VII. Treatment with exchange transfusion. *N Engl J Med* 244:39.

Falchry SM, Messick WJ, and Sheldon GF. 1996. Metabolic effects of massive transfusion. In *Principles of transfusion medicine*, Baltimore: Williams & Wilkins.

Fergusson et al. 2003. Clinical outcomes following institution of universal leukoreduction of blood transfusions for premature infants. *JAMA* 289:1950–1956.

Fetus and Newborn Committee, Canadian Paediatric Society and College of Family Physicians of Canada. 1999. Approach to the management of hyperbilirubinemia in term newborn infants. *Paedritr Child Health* 4:161–164.

Fetus and Newborn Committee, Canadian Paediatric Society. 2002. Red blood cell transfusions in newborn infants: revised guidelines. *Paediatrics and Child Health* 7:553–558.

Goldstein MH, et al. 1993. Comparison of two preservation solutions for erythrocyte transfusions in newborn infants. *J Pediatr* 123:783–788.

Hall TL, et al. 1993. Neonatal mortality following transfusion of red cells with high plasma potassium levels. *Transfusion* 33:606–609.

Hammerman C and Kaplan M. 2000. Recent developments in the management of neonatal hyperbilirubinemia. *NeoReviews* 1:e19–e24.

Hansen TW. 2002. Mechanisms of bilirubin toxicity: clinical implications. *Clin Perinatol* 29:765–778.

Heaton A, Keegan T, and Holme S. 1989. *In vivo* regeneration of red cell, 2,3-diphosphoglycerate following transfusion of DPG-depleted AS-1, AS-3, and CPDA-1 red cells. *Br J Haematol* 71:131–136.

Herman JH and Manno CS. 2002. Neonatal red cell transfusion. In *pediatric transfusion therapy*. Bethesda, MD: AABB Press.

Hillyer CD, Tiegerman KO, and Berkman EM. 1991. Evaluation of red cell storage lesion after irradiation in filtered packed red cell units. *Transfusion* 31:497–499.

Hosono S, Ohno T, Kimoto H, et al. 2001. Effects of albumin infusion therapy on total and unbound bilirubin values in term infants with intensive phototherapy. *Pediatr Int* 43:8–11.

Hume HA. 1999. Blood components: preparation, indications and administration. In *Pediatric Hematology*, London: Churchill Livingstone.

Hume HA and Preiksaitis JB. 1999. Transfusion associated graft-versus-host disease, cytomegalovirus infection and HLA alloimmunization in neonatal and pediatric patients. *Transfusion Science* 21:73–95.

Jackson JC. 1997. Adverse events associated with exchange transfusion in healthy and ill newborns. *Pediatrics* 99:e7–e13.

Johnson LH, Bhutani VK, et al. 2002. System-based approach to management of neonatal jaundice and prevention of kernicterus. *J Pediatr* 140:396–403.

Kappas A, Drummond GS, and Valaes T. 2001. A single-dose of Sn-mesoporphyrin prevents development of severe hyperbilirubinemia in glucose-6-phosphate dehydrogenase deficient newborns. *Pediatrics* 108:25–30.

Kappas A et al. 1988. Sn-protoporphyrin use in the management of hyperbilirubinemia in term newborns with direct Coombs positive ABO incompatibility. *Pediatrics* 81:485–497.

Keenan WJ et al. 1985. Morbidity and mortality associated with exchange transfusion. *Pediatrics* 75(suppl):422–426.

Luban NLC, Strauss RG, Hume HA. 1991. Commentary on the safety of red cells preserved in extended-storage media for neonatal transfusions. *Transfusion* 31:229–235.

Maisels JM, Li T, Piechocki JT, Wertman MW. 1974. The effect of exchange transfusion on serum ionized calcium. *Pediatrics* 53:683–686.

Martinez JC, et al. 1999. Control of severe hyperbilirubinemia in full-term newborns with the inhibitor of bilirubin production Sn-mesoporphyrin. *Pediatrics* 103:1–5.

Murohy RJC, Malhorta C, Sweet AY. 1980. Death following an exchange transfusion with hemoglobin SC blood. *J Pediatr* 96:110–112.

Napier A, et al. 1998. British Committee for Standards in Haematology, Blood Transfusion Task Force. Guidelines on the clinical use of leucocyte-depleted blood. *Transf Med* 8:59–71.

Newman T, Xiong BB, et al. 2000. Prediction and prevention of extreme neonatal hyperbilirubinemia in a mature health maintenance organization. *Arch Pediatr Adolesc Med* 154:1140–1147.

Ohto H and Anderson KG. 1996. Post-transfusion-graft-versus-host disease in Japanese newborns. *Transfusion* 36:117–123.

Panagopoulos G, Valaes T, and Doxiadis SA. 1969. Morbidity and mortality related to exchange transfusion. *J Pediatr* 74:247–254.

Peterec SM. 1995. Perinatal hematology: management of neonatal Rh disease. *Clin Perinatol* 22:561–592.

Preiksaitis JK. 1991. Indications for the use of cytomegalovirus-seronegative blood products. *Transfus Med Rev* 5:1–17.

Przepiorka D, et al. 1996. Use of irradiated blood components. Practice parameter. *Am J Clin Pathol* 106:6–11.

Rübo J, et al. 1992. High-dose intravenous immune globulin therapy for hyperbilirubinemia caused by Rh hemolytic disease. *J Pediatr* 121:93–97.

Sanders MR and Graeber JE. 1990. Post-transfusion-graft-versus-host disease in infancy. *J Pediatr* 117:159–163.

Sato K, et al. 1991. High-dose intravenous gammaglobulin therapy for neonatal immune haemolytic jaundice due to blood group incompatibility. *Acta Paediatr* 80:163–166.

Scanlon JW and Krakaur R. 1980. Hyperkalemia following exchange transfusion. *J Pediatr* 96:108–110.

Strauss RG. 2000. Data-driven blood banking practices for neonatal RBC transfusions. *Transfusion* 40:1528–1540.

Strauss RG. 1999. Blood banking issues pertaining to neonatal red blood cell transfusions. *Transfusion Science* 21:7–19.

Strauss RG, et al. 2000. Feasibility and safety of AS-3 red blood cells for neonatal transfusions. *J Pediatr* 136:215–219.

Strauss RG, et al. 1996. AS-1 red cells for neonatal transfusions: a randomized trial assessing donor exposure and safety. *Transfusion* 36:873–878.

Tsao YC and Yu VY. 1972. Albumin in management of neonatal hyperbilirubinaemia. *Arch Dis Child* 47:250–256.

Valaes T. 1963. Bilirubin distribution and dynamics of bilirubin removal by exchange transfusion. *Acta Paediatr Scand* 52S:149.

Valaes T, Drummond GS, and Kappas A. 1998. Control of hyperbilirubinemia in glucose-6-phosphate dehydrogenase deficient newborns using an inhibitor of bilirubin production, Sn-mesoporphyrin. *Pediatrics* 101:e1–e7.

Voak D, et al. prepared by the BCSH (Blood Transfusion Task Force). 1996. Guidelines on gamma irradiation of blood components for the prevention of transfusion-associated graft-versus-host disease. *Transf Med* 6:261–271.

Volpe JJ. 1995. *Neurology and the newborn*. 3rd ed. Philadelphia, PA: WB Saunders.

Wallerstein H. 1946. Treatment of severe erythroblastosis fetalis by simultaneous removal and replacement of the blood of the newborn infant. *Science* 103:583.

Weldon VV and Odel GB. 1968. Mortality risk of exchange transfusion. *Pediatrics* 41:797–801.

16

Granulocyte Transfusions in the Neonate and Child

MARIA LUISA SULIS, MD, LAUREN HARRISON, RN, BSN, AND MITCHELL S. CAIRO, MD

ABSTRACT

Unmobilized allogeneic granulocyte transfusions in neonates, children, and adults with severe neutropenia and sepsis have been associated with mixed success. A major limitation in the past with administering unmobilized allogeneic granulocyte transfusions has been the inability to collect a larger number of neutrophils during apheresis of allogeneic donors. The subgroups where the most success in reducing mortality has been demonstrated have been in neonates with severe neutropenia and sepsis because of the use of a higher dose of granulocytes per size of the recipient (neonate) and the use of repetitive transfusions over a minimum of 5 days. Recently, it has been demonstrated that the mobilization of allogeneic donors with dexamethasone and granulocyte-colony stimulating factor (G-CSF) before apheresis has significantly increased the yield of neutrophils by five- to tenfold. The use of dexamethasone and G-CSF to mobilize allogeneic donors induces a neutrophil collection in the range of 3 to 10×10^{10} neutrophils. The use of mobilized allogeneic granulocytes is associated with a significant increase in the patients' circulating absolute neutrophil count. It remains to be seen whether the use of higher doses of mobilized allogeneic donor granulocytes will significantly increase the survival rate of neutropenic septic neonates and children. Future prospective multicenter randomized trials will be required to accurately assess whether an increased granulocyte dose following mobilization of granulocyte donors will significantly improve survival compared to unmobilized granulocyte transfusions in severely neutropenic and septic children.

INTRODUCTION

The use of allogeneic granulocyte transfusions to treat patients with either severe neutropenia and/or neutrophil dysfunction with presumed or documented severe systemic infections has been limited in large part by the small quantity of granulocytes collected by leukopheresis from unstimulated donors and the minimal increment in the circulating absolute neutrophil count (ANC), especially in large recipients (Klein et al. 1996; Strauss 1998). Over 25 years ago, several investigators demonstrated some success in the use of unmobilized allogeneic granulocyte transfusions for adults with presumed or documented severe systemic infections (Alavi et al. 1977; Herzig et al. 1977; Vogler and Winton 1977). However, over the next 20 years there were few investigations demonstrating the benefit of unmobilized allogeneic granulocyte transfusions in adult recipients with presumed or documented severe systemic infection. However, Dale et al. more recently began to pursue methods of mobilization of allogeneic granulocytes and significantly renewed the interest in this potential therapeutic modality (Dale et al. 1997). Price et al. recently demonstrated the ability of mobilizing and collecting five- to tenfold more granulocytes by leukopheresis from allogeneic donors after mobilization with dexamethasone and G-CSF (Price et al. 2000). Neonates, who weigh approximately 1/25 of an average adult recipient, require significantly less granulocytes and therefore may benefit significantly more from allogeneic granulocyte transfusions from unmobilized allogeneic donors than larger adult recipients (Cairo et al. 1992). In this chapter we review the normal physiology of myelopoiesis; definitions of neutropenia in the neonate and child; indications for granulocyte transfusions in the child; methods of mobilization, collection, and functionality of allogeneic granulocytes for transfusion; and dosing administration and side effects of allogeneic granulocyte transfusions.

MYELOPOIESIS

The pluripotent stem cell in the bone marrow can self-replicate or ultimately differentiate into either a myeloid or lymphoid stem cell. The myeloid committed precursor cell proceeds to either self-replicate or differentiate into more committed precursor cells called colony-forming units (CFU). The myeloid stem cell can differentiate into either a CFU for the eosinophil development or into a CFU for the development of red cells, phagocytes, basophils, and megakaryocytes (CFU-GEMM, colony forming unit-granulocyte, erythrocyte, megakaryocyte, monocyte). Under the stimulus of several hematopoietic growth factors (HGF), the CFU-GEMM continues to differentiate into more mature and committed precursors. Colony forming unit-granulocyte (CFU-G) progenitor cells differentiate sequentially into myeloblasts, promyelocytes, myelocytes, metamyelocytes, and bands. Morphologically, the stages of maturation are characterized by a progressive decrease of the nuclear size, disappearance of nucleoli, and subsequent appearance of three different populations of granules containing various proteins and enzymes.

The large neutrophil pool in the bone marrow has been classified into a proliferating and a maturating compartment. The bone marrow neutrophil reserve is manyfold larger than the peripheral pool. The developmental time of myelopoiesis, from the more primitive myeloblast to the more mature neutrophil, is about 8 to 14 days, after which the mature granulocyte is released into the circulation. In the periphery, the granulocyte pool has been classified into two compartments: the circulating and the marginating pool. The neutrophil pool is under the influence of various specific chemotactic signals that induce neutrophils to migrate to sites of inflammation and infection. The life span of the granulocyte in the peripheral blood is approximately 6 to 10 hours and about 1 to 2 days in the tissues.

The proliferation and differentiation steps that lead to the formation of a mature neutrophil are regulated by HGFs. Among the various HGFs, the most important for these physiological processes include G-CSF and granulocyte and macrophage colony stimulating factor (GM-CSF). G-CSF is produced by monocytes, fibroblasts, and endothelial cells and appears to act on a more mature

and committed precursor cell, the CFU-G, regulating its growth and differentiation into the mature neutrophil. Initial in vitro studies showed that when human bone marrow cells were cultured in the presence of G-CSF, colonies of mature neutrophils and precursors would arise within 7 to 8 days of stimulation. Studies in primates confirmed the effect of G-CSF as an important stimulus for the production of granulocytes and opened the way for trials of rhG-CSF in humans. Initial studies in the late 1980s showed a dose-related increase in the number of circulating mature neutrophils following five to 6 days of administration of G-CSF to healthy subjects. Administration of G-CSF for 14 days following chemotherapy reduced the length of profound neutropenia, the number of infectious episodes, and the use of antibiotics. On the basis of these and other studies, the use of G-CSF following myelosuppressive chemotherapy that is associated with a high incidence of febrile neutropenia has become common medical practice. G-CSF also enhances granulocyte function by increasing the production of superoxide radicals, phagocytosis, and antibody-dependent cytotoxicity.

GM-CSF is produced by T lymphocytes, endothelial cells, fibroblasts, and monocytes. GM-CSF is not as lineage specific as G-CSF and affects both early and late myeloid progenitor cells. CFU-GEMM as well as the more committed CFU-GM and CFU-G require the activity of GM-CSF for growth and differentiation. Compared with G-CSF, bone marrow cells cultured in the presence of GM-CSF are able to induce mature neutrophil and monocyte development. GM-CSF also enhances neutrophil effective function in a similar way as G-CSF, but in addition it inhibits neutrophil migration.

NEONATAL NEUTROPENIA AND DYSFUNCTION

Bacterial sepsis is a significant cause of neonatal morbidity and mortality and is associated with a mortality rate that ranges between 25% to 75% (Siegel and McCracken 1981). The increased incidence and severity of bacterial sepsis in the neonate is in large part secondary to impaired neonatal host defense, specifically quantitative and qualitative abnormalities of phagocytic cellular immunity (Cairo 1989a). Preclinical studies in neonatal animals have demonstrated significantly decreased myeloid progenitor cells, an already high myeloid progenitor rate, a significant decrease in the bone marrow neutrophil storage pool of mature neutrophil effector cells, and a high propensity to develop peripheral blood neutropenia during experimental sepsis (Christensen et al. 1982a,b). In addition to reduced number of myeloid progenitor cells and mature neutrophil effector cells, neonates exhibit impaired neutrophil functional capacity at baseline and particularly during times of stress especially with respect to oxidative metabolism, chemotaxis, phagocytosis, bacterial killing, and impaired surface membrane expression of adhesion proteins (Cairo 1989b). Santos et al. demonstrated with the use of granulocyte transfusions versus placebo a significant reduction in the mortality rate (100% to 25%) in neonatal rats during experimental Group B streptococcal sepsis (1980). Subsequently, Christensen et al., utilizing a neonatal canine model infected with *Stapholococcus aureus*, also demonstrated a significant benefit of granulocyte transfusions with five of six neonatal pups surviving with granulocyte transfusions versus zero of six pups that did not receive granulocyte transfusions (1982a).

CHILDHOOD NEUTROPENIA

Neutropenia is, by definition, a decrease of neutrophils and bands in the peripheral blood below 1500 cells/μL in children older than 1 year of age and below 1000/μL between 2 months and 1 year of age. In the African-American population, childhood neutropenia is defined by a decrease in neutrophils and bands in the peripheral blood to values of about 200 to 600 cells/μL fewer when compared to Caucasians.

Neutropenia can be classified as mild (neutrophil and band count between 1500 and 1000 cells/μL), moderate (1000 to 500 neutrophils and bands/μL), or severe (less than 500 cells/μL). The degree of neutropenia is important in estimating the risk of developing severe bacterial and fungal infections, although factors other than the sole number of neutrophils are also important in assessing this risk (etiology of neutropenia, length of neutropenia, and so on). The most common infections encountered in neutropenic patients include bacterial cutaneous infections (cellulitis, furunculosis, abscess), pneumonia, otitis media, stomatitis, perirectal infection, and septicemia. Viral infections, however, are not increased in neutropenic patients. The most common infectious agents are *Staphylococcus aureus*, *Escherichia coli*, *Pseudomonas* species, and other gram-negative bacteria. In the following paragraph we classify neutropenia according to whether the defect is intrinsic or extrinsic to the myeloid cell (Table 16.1).

Neutropenia Secondary to Intrinsic Defects of the Myeloid Cell

The molecular mechanism responsible for this class of neutropenia is in most cases unknown. Bone marrow

TABLE 16.1   Classification of Neutropenia

Neutropenia Secondary to Intrinsic Defects of the Myeloid Precursors

Cyclic neutropenia
Familial benign neutropenia
Severe congenital neutropenia (Kostmann's syndrome)
Reticular dysgenesis
Dyskeratosis congenita
Shwachman syndrome
Aplastic anemia
Myelodysplastic syndrome
Fanconi's anemia

Neutropenia Secondary to Extrinsic Factors

Viral infections (Hepatitis A, B, C; influenza; RSV; EBV; CMV; HIV; measles; mumps)
Bacterial infections
Drug-induced causes
Radiation therapy
Immune neutropenia
Bone marrow malignant infiltration
Nutritional deficiencies

RSV = Respiratory syncytial virus, EBV = Epstein-Barr virus, CMV = cytomegalovirus, HIV = human immunodeficiency virus.

studies as well as peripheral blood findings can, in some cases, suggest the underlying defect. For example, in reticular dysgenesis, severe neutropenia together with lymphopenia and the absence of tonsil, lymph node, and splenic follicles suggest a defect in the stem cell before myeloid and lymphoid stem cell development. Bone marrow studies in the more benign cyclic neutropenia show a maturational arrest or hypoplasia at the myelocytic stage. A defect in the G-CSF receptor has been identified in some cases of severe congenital neutropenia (Kostmann syndrome). This subgroup of patients appears to be at greater risk of developing acute leukemia; it is still unclear whether the treatment with G-CSF has an additional role in the development of this malignancy. Generally, the symptomatology of severe neutropenia manifests in infancy or early childhood with a spectrum of severity according to the different entities, but having as a common feature recurrent infections. Severe, fatal bacterial infections, usually starting as cellulitis, cutaneous and perirectal abscesses, or stomatitis, frequently evolve into sepsis in patients with reticular dysgenesis or severe congenital neutropenia. However, in patients with cyclic neutropenia, dyskeratosis congenita, or Shwachman syndrome, the infectious episodes are frequent but rarely fatal. Laboratory studies usually reveal moderate to severe neutropenia with varying abnormalities in the red cell and platelet counts. Monocytosis and eosinophilia are common in severe congenital neutropenia. Anemia and thrombocytopenia are common in Shwachman syndrome, while fluctuations in reticulocyte and platelet counts accompany the change in the neutrophil count during the alternating phases of cyclic neutropenia. Supportive care is the standard treatment for these patients and G-CSF has been used successfully in the majority of these syndromes.

Neutropenia Secondary to Extrinsic Factors

A large variety of factors and conditions are responsible for secondary neutropenia secondary to extrinsic factors. Neutropenia that accompanies viral infection or autoimmune neutropenia is usually benign with mild and infrequent infections. On the contrary, neutropenia secondary to sepsis or following cytotoxic chemotherapy and/or radiation or secondary to infiltration of the bone marrow by malignant diseases is a very serious condition that places the patient at risk of severe, life-threatening infections. The risk of severe infections in this population is related to the degree of neutropenia, the length of severe neutropenia following therapy, and the impairment of granulocyte function due to "environmental" factors such as drugs or active cancer, inflammation, cell death, and so on.

Neutropenia Secondary to Infections

Viral infections are the most common cause of neutropenia in children. Typically, hepatitis A and B, respiratory syncytial virus (RSV), influenza A and B, Epstein-Barr virus (EBV), cytomegalovirus (CMV) infection, measles, mumps, and rubella cause neutropenia. This viral effect appears to be due to inhibition of proliferation of bone marrow myeloid precursors, redistribution of neutrophils from the circulating to the marginating pool, consumption in damaged tissues, and/or neutrophil immune destruction. The onset of neutropenia usually corresponds to the appearance of viremia but usually tends to last only a few days. Neutropenia can also occur following bacterial infection, usually typhoid, tuberculosis, brucellosis, and rickettsial infections, but is most frequently associated with sepsis. Neutropenia secondary to infection results from neutrophil destruction from endotoxins and neutrophil aggregation (mostly in the lungs) secondary to complement activation.

Neutropenia Secondary to Drugs

Numerous medications have previously been identified as causing neutropenia, however, the pathogenetic mechanism is frequently unknown. Possible mechanisms, however, include impaired drug metabolism with generation of toxic metabolites (for example, sulfasalazine) and immune destruction (for example, penicillin, phenytoin, quinidine). In this latter event, the drug may function as hapten or may promote the formation of an immune complex. In the majority of cases, especially when an immunologic mechanism is responsible, neutropenia tends to occur early on and lasts a few days or up to a week. Since drug-induced neutropenia can be a very serious disorder with frequent reports of fatal infections, discontinuation of the suspected drug is the most important therapeutic intervention.

Immune Neutropenia

This group of neutropenias comprises both autoimmune- and alloimmune-induced neutropenias. Neutrophils carry antigens common to other blood cells, such as the ABO blood antigen group, the I/i antigen, the Kx antigen of the McLeod group, and the antigens of the HLA-A and -B group, class I. Antigens specific to neutrophils and probably involved in the pathogenesis of immune neutropenias include the NA 1 and 2, the NB 1 and 2, the NC, the ND, and the NE 1. Antibodies against neutrophil antigens can be detected by immunofluorescence and agglutination tests, although their demonstration is not required for the diagnosis of immune-induced neutropenia.

Autoimmune neutropenia (AIN) can be idiopathic or secondary to infections, medications, or part of other generalized autoimmune disorders. The idiopathic form, also called chronic benign neutropenia or autoimmune neutropenia of childhood, tends to occur in the first 2 to 3 years of life. The neutrophil count is usually quite low (150 to 250 cells/µL) and is often associated with monocytosis and eosinophilia. Bacterial infections are common, typically manifested as skin infection, otitis media, and upper respiratory infection, but they are usually mild and easily treated. AIN is a benign disorder with spontaneous resolution in virtually all patients. Treatment with G-CSF (1–2 µg/kg), steroids, and/or immunoglobulins is indicated only in cases of severe and recurrent infections.

Alloimmune neutropenia occurs in newborn infants either following maternal sensitization with previous exposure to paternal disparate neutrophil antigen neutrophils or secondary to maternal AIN. Alloimmune neutropenia is usually severe and associated with serious, recurrent infections. Aggressive parenteral antibiotic therapy should be instituted as in any case of neonatal neutropenia and consideration should be given to the use of G-CSF (5 µg/kg/day until recovery) during episodes of severe infections.

In today's pediatric practice, the largest population of neutropenic patients at risk for developing severe infections is comprised of children receiving chemotherapy for treatment of malignancies or following myeloablative therapy before stem cell transplant. There has been marked improvement in supportive care offered to these patients, including routine use of red cell and platelet transfusions. However, despite the utilization of a variety of new and more potent antibiotic, antifungal, and antiviral medications and the advent of hematopoietic growth factors, bacterial and fungal infections remain a frequent cause of morbidity and mortality. The most important factor in determining a favorable outcome of severe infections in neutropenic patients is rapid myeloid reconstitution. A seminal study by Bodey et al. (1966) showed that the incidence of severe infection increased from 10% when the granulocyte count was above $1000/\text{mL}^3$ to 19% and 28% for granulocyte counts below 500 and $100/\text{mL}^3$, respectively. A similar correlation was found between duration of neutropenia and incidence and mortality from severe infection.

Following the Bodey et al. (1966) study and other studies and given the benefit observed from the routine use of platelet transfusions, several groups have investigated the use of granulocyte transfusions as prophylactic and/or therapeutic measures in severely neutropenic patients. Enthusiasm over the use of granulocyte transfusions has waxed and waned over the past three decades; however, major advances in the mobilization and collection strategies as well as results from several randomized trials have urged physicians to reconsider this treatment option as described below.

MOBILIZATION OF DONOR GRANULOCYTES

The most common reason for unsatisfactory results of granulocyte transfusions in severely neutropenic patients has been the very low dose of polymorphonuclear neutrophil leukocytes (PMNs) administered. Experimental studies done in the 1970s showed that the ability to clear *Pseudomonas* sepsis in neutropenic dogs was dependent on the number of granulocytes transfused (Applebaum et al. 1978). Considering that the normal human circulatory pool of PMNs is $3 \times 10^8/\text{kg}$ and the short life span of granulocytes, it is clear that the ability to collect large amounts of granulocytes is a major challenge. Granulocytes collected from unstimulated donors usually yield only 4 to 6×10^9 PMNs from each collection.

By the early 1970s, the introduction of the continuous flow centrifugation as a new collecting method and the administration of corticosteroids to donors as a mobilizing agent increased the collection to 10 to 20×10^9 granulocytes from a single donor. Corticosteroids increase the release of granulocytes from the bone marrow and increase the circulating neutrophil pool by decreasing neutrophil margination. Different types of corticosteroids have been used as mobilizing agents, although dexamethasone has been used more frequently in more recent studies. The most commonly used regimen of dexamethasone is 8 mg po given 12 hours before the collection of granulocytes (Dale et al. 1998; Price et al. 2000). A recent study showed that 8 mg of dexamethasone given 12 hours before the collection in concomitance with G-CSF was as effective as a 12-mg dose. Currently, 8 mg remains the recommended dose of dexamethasone, either alone or in combination with G-CSF, for neutrophil mobilization (Liles et al. 2000).

The addition of G-CSF has had a major impact in improving the yield of leukapheresis and the efficacy of granulocyte transfusions. A study by Bensinger et al. showed that administration of G-CSF at 5 µg/kg increased the yield of collected PMN from 6.8×10^9 in unstimulated donors to 41.6×10^9 (Bensinger et al. 1993). Lymphocyte counts increased slightly, monocytes remained unchanged, and a small number of immature granulocytes appeared in the peripheral blood. The platelet count decreased more markedly in the G-CSF-stimulated donors compared to controls (150 to $200,000/\text{mm}^3$ versus 200 to $250,000/\text{mm}^3$). The decrease in the hematocrit to 30% to 35% was similar in both treatment groups. The increase in the peripheral blood neutrophil count 24 hours after a neutrophil transfusion was significantly higher when G-CSF-mobilized granulocytes were used rather than unstimulated products (954 PMN/µL versus 50 PMN/µL). More patients receiving neutrophil transfusions had severe infections in the control group compared to the group receiving G-CSF-mobilized neutrophils.

The efficacy of several granulocyte mobilization methods has been investigated and is reported in Table 16.2. Regardless of the mobilization method, mild, transient anemia and thrombocytopenia were observed in the donors. Several different doses of G-CSF have been used in different trials, however, results from two studies comparing the mobilizing effect of 450 µg versus 600 µg and 600 µg versus 300 µg of G-CSF showed no significant difference in granulocyte yield (Liles et al. 2000). Similarly, the route of administration (intravenous or subcutaneous) of G-CSF did not appear to affect the neutrophil collection yield.

Dexamethasone and G-CSF mobilization as well as the apheresis procedure are well tolerated overall by healthy donors. Side effects experienced by donors secondary to G-CSF have included mild bone pain, myalgia, arthralgia, and headache. Despite the frequency with which these side effects were reported (up

TABLE 16.2 PMN Collection Yield Following Different Mobilization Strategies

Study	G-CSF Dose	G-CSF Schedule	Steroid Dose	Steroid Schedule	PMN Mobilized $\times 10^9$
Bensinger et al. 1993	5 μg/kg	Daily	—	—	41.6
Jendiroba et al. 1998	5 μg/kg	Daily	—	—	42
	5 μg/kg	Every other day	—	—	46
	—	—	Prednisone 60 mg	Daily	28.7
Dale et al. 1998	600 μg	Daily	Dexamethasone 8 mg	Daily	77.4
Stroncek et al. 2001	5 μg/kg	Daily	—	—	41.1
	5 μg/kg	Daily	Dexamethasone 8 mg	Daily	67.1
	—	—	Dexamethasone 8 mg	Daily	21

to 75% to 85% of patients), no donors in any of the studies had to discontinue G-CSF because of toxicity. Increased sodium, lactate dehydrogenase (LDH) uric acid, decreased potassium, phosphorus, and magnesium have also been attributed to the use of dexamethasone and G-CSF. Hyperglycemia has been attributed to corticosteroid administration, and hypocalcemia to the use of citrate as an anticoagulant. Weight gain and anemia, at least in part, seem to be due to hydroxyethyl starch (HES) that is used for red cell sedimentation. Although all the laboratory changes tended to return to normal values in the following weeks, it is clear that donors should be chosen cautiously and questioned about conditions that may represent contraindications to the use of steroids or volume expanders (that is, hypertension, diabetes, peptic ulcer).

Granulocyte Kinetics Following Donor Mobilization and Apheresis

One of the concerns in the use of granulocyte transfusions is whether the use of corticosteroids and G-CSF as mobilizing agents together with the process of apheresis would impair granulocyte function. Several studies have analyzed granulocytic function, including bactericidal activity, respiratory burst, chemotaxis, and so on, following donor mobilization and apheresis. Most studies have concluded that mobilized granulocytes tend to maintain their original functional activity. G-CSF/steroid-stimulated neutrophils exhibit increased expression of CD11b/CD18 and CD14, CD32, and CD64 surface adherence proteins while L-selectin expression is slightly diminished. The increased expression of these adhesion molecules is probably responsible for increased margination and decreased recovery of neutrophils following reinfusion into the allogeneic recipient (Dale et al. 1998). PMN respiratory burst of G-CSF/steroid-stimulated granulocytes as assessed with chemiluminescence is usually increased compared to unstimulated granulocytes but appeared to reach baseline levels following apheresis. These results suggest that

apheresis, probably from exposure to plastics or other substances, affects neutrophil activity (Dale et al. 1998). However, some procedures used to collect and separate granulocytes may impact on granulocytic function. The use of nylon columns to collect granulocytes is associated with decreased neutrophil recovery and half-life. Phagocytosis of *E. coli* and *S. aureus* was not significantly changed in apheresed, stimulated neutrophils compared to normal neutrophils, although increased activity was shown in mobilized granulocytes before apheresis. Normal neutrophil activity is maintained even after several doses of G-CSF are administered to the allogeneic donors.

A consistent finding in several studies is the prolonged survival, up to 20 hours, of mobilized granulocytes postinfusion compared to the half-life of normal granulocytes. These findings could be related to multiple factors: mobilization of relative immature cells, increased expression of adhesion molecules, and/or anti-apoptotic effects of both G-CSF and corticosteroids. It is also possible that G-CSF-stimulated neutrophils tend to accumulate in different tissues and redistribute at a later time point.

Methods of Granulocyte Collection

Based on different densities, granulocytes can be separated from other blood cells by centrifugation. Separation of granulocytes from red cells has been poor in the past because granulocytes and red cells have similar densities. Several agents can be used to sediment red cells in vitro, but the most commonly used agent in the United States is HES. HES promotes rouleaux formation and increased red cell density and is most effective in the separation of granulocytes. Granulocyte recovery doubled when HES was added to the leukapheresis system. HES, however, can cause blood volume overload that requires clinical management during the procedure.

Currently, granulocytes are collected by continuous flow centrifugation. The leukapheresis procedure takes

2 to 3 hours to process 6 to 8 liters of donor blood and extract about 20% to 40% of granulocytes. The final granulocyte concentrate is commonly about 200 mL but contains different amounts of granulocytes depending on whether mobilization has been used and what agents have been administered to the donors (see the previous section).

Granulocyte Concentrate

The granulocyte concentrate is a suspension of granulocytes in plasma. The number of granulocytes is variable in each concentrate, however, each concentrate (75% of the units) commonly contains at least 1×10^{10} granulocytes. Because granulocytes cannot be completely separated from red cells, a certain number of red cells (up to a hematocrit of 10%) are usually present in each granulocyte concentrate. Therefore, crossmatching is required before granulocyte concentration infusion. A small amount of platelets may also be present in the granulocyte concentrate, especially if continuous flow centrifugation is used as the separation method.

Given the short half-life of granulocytes, storage is a critical issue. Granulocytes can maintain bactericidal activity for 1 to 3 days if refrigerated, but chemotaxis decreases after 24 hours. Storage at room temperature for up to 8 hours seems to be safe, as recovery, survival, migration, and activity are maintained. Even so, there is some impairment of in vitro and in vivo PMN function. It is therefore recommended that granulocytes be transfused as soon as possible after collection. The American Association of Blood Banks (AABB) recommends storage of granulocytes for up to 24 hours at 20° to 24°C.

THERAPEUTIC GRANULOCYTE TRANSFUSIONS IN CHILDREN AND ADULTS

Following the study by Bodey et al. (1966) demonstrating a relationship between the degree of neutropenia and risk of infection, it appeared hypothetical that the transfusion of normal granulocytes would be beneficial for specific subsets of neutropenic patients. Initial studies both in animals and humans seem to support the use of granulocyte transfusions in specific settings such as bacterial sepsis and severe neutropenia. Most of the studies on the efficacy of granulocyte transfusions were conducted in the 1980s and early 1990s (Menitove and Abrams 1987; Strauss 1993; Vamvakas and Pineda 1996), and despite much criticism, their results still constitute the basis for the design of new trials today. It must be kept in mind though that many variables have

changed in the past 20 years. Granulocytes were previously collected from donors without any mobilization, therefore with limited yield. A frequently used collection method was filtration leukapheresis, which was later shown to impair much of the granulocytic function as well as to be responsible for several side effects. Additionally, supportive care available in the past for neutropenic patients was significantly inferior to what is currently available today. The utilization of HGF limits the degree, duration, and incidence of neutropenia. The broader choices of antimicrobials and antifungals certainly have contributed to the improved overall outcome of infected neutropenic patients, making the use of granulocyte transfusion less critical. However, fungal infections and some severe bacterial infections still remain a major risk in neutropenic patients, particularly following myeloablative therapy. The ability to mobilize a large number of granulocytes with steroids and G-CSF and the improved methods of neutrophil collection have generated new enthusiasm for the use of neutrophil transfusions in septic neutropenic patients.

From 1972 to 1982, seven controlled studies have been published that are worth considering in more detail (Graw et al. 1972; Fortuny et al. 1975; Higby et al. 1975; Alavi et al. 1977; Herzig et al. 1977; Vogler and Winton 1977; Winston et al. 1980a,b) (Table 16.3). Some of these trials included pediatric patients. In these studies, the outcomes of infected neutropenic patients who received antibiotic treatment and granulocyte transfusions were compared to matched patients who were treated with antibiotics only. Three studies showed a definite benefit from the use of neutrophil transfusions, two studies did not confirm these results, and two studies showed a benefit only in a subgroup of neutropenic patients. In the study by Graw et al. (1972), the advantageous effect of granulocyte transfusions was demonstrated in patients who had received at least three to four neutrophil transfusions. In the study by Alavi et al. (1977), the benefit was shown in patients who had persistent severe neutropenia.

Several conclusions can be obtained from these studies. The dose of granulocytes transfused is a fundamental and perhaps the most important factor in determining the success of granulocyte transfusions. In all of the studies that showed a benefit, a larger number of granulocytes were transfused. The method of collection, with preference for continuous flow filtration leukapheresis, is important for the preservation of neutrophil functional activity. Finally, if the antibiotic therapy is successful and the duration of neutropenia is short, there is no advantage in using granulocyte transfusions. The importance of leukocyte compatibility is still controversial, but at least in the seven mentioned controlled

TABLE 16.3   Summary of Controlled Studies of Therapeutic Granulocyte Transfusion in Neutropenic Patients
(Excluding Neonates)

Study	Randomized	Collection Technique	PMN $\times 10^{10}$	% Survival Study Group	% Survival Control Group	Infection
Graw et al. 1972	No	CFC/FL	0.6/2	46	40	Gram-negative sepsis
Fortuny et al. 1975	No	CFC	0.4	78	80	Clinical sepsis
Higby et al. 1975	Yes	FL	2.2	76	26	Clinical sepsis
Herzig et al. 1977	Yes	CFC/FL	0.4/1.7	75	36	Gram-negative sepsis
Vogler et al. 1977	Yes	CFC	2.7	59	15	Culture positive
Alavi et al. 1977	Yes	FL	5.9	82	62	Clinical sepsis
Winston et al. 1980	Yes	IFC	0.5	63	73	Culture positive

CFC = Continuous flow centrifugation, FL = Filtration leukapheresis, IFC = Intermittent flow centrifugation

studies, the success of neutrophil transfusion has been shown in the cases of leukocyte compatibility.

A recent study by Price et al. investigated the feasibility of transfusing neutropenic patients who had undergone myeloablative therapy and stem cell transplantation and had developed severe bacterial or fungal infection, with granulocytes obtained from community donors mobilized with dexamethasone and G-CSF (2000). This study demonstrated that there was no clear correlation between the dose of granulocytes transfused and the increment of neutrophil count in the recipients. However, in patients who had received $>2.0 \times 10^9$/kg, the increment in the peripheral neutrophil count was $>2 \times 10^3$/μL the morning following the transfusion. Although clinical outcome was not the objective of the study, it did report that all the patients with invasive bacterial infections and four out of seven patients with candidiasis cleared their infections versus none of the patients with invasive aspergillosis and mold infections.

The feasibility and benefit of transfusing granulocytes obtained from stimulated related or unrelated donors to septic neutropenic patients following myeloablative therapy and stem cell transplantation has also been investigated (Hubel et al. 2002). Despite a higher increment in the circulating neutrophil count in patients transfused with unrelated granulocytes, the number of fatal fungal infections and the survival was not different between the neutrophil transfused and untransfused patients.

Some useful advice can be offered to physicians considering the use of granulocyte transfusions. Granulocyte transfusions should probably be considered in:

- Severely neutropenic patients (ANC $<500 \times 10^9$) with bacterial sepsis or infection with yeast and fungi, in whom antimicrobials or antifungals have

failed and in whom bone marrow recovery is expected to be delayed at least 3 weeks. It is important therefore that physicians are aware of the outcome of bacterial, yeast, and fungal infections in this population in their medical center. Clearly, if the response to antibiotics and antifungals is very high, there is no need to transfuse granulocytes, given the potential risks associated with it.

- Allogeneic granulocyte donors should be mobilized with G-CSF (5 μg/kg/day) and dexamethasone (8 mg po), 12 hours before the scheduled collection.
- Granulocytes should be collected by continuous flow centrifugation techniques.
- At least 2 to 3×10^{10} granulocytes (not less than 1×10^{10}) should be transfused daily for a minimum of 5 days.
- Granulocytes should be transfused as soon as possible after the collection.
- Granulocytes should be obtained from compatible donors if possible (by HLA match and/or leukocyte crossmatching).

Fungal infections have become increasingly common in neutropenic patients and a major challenge for physicians. Prolonged periods of myelosuppression following intense chemotherapy regimens delivered today in the settings of bone marrow transplantation or salvage chemotherapy represent a major risk factor in the development of these infections. Not enough data supports the use of granulocyte transfusions in the setting of fungal or yeast infection in neutropenic patients. However, there is definite evidence of their benefit in treating these infections in patients with chronic granulomatous disease. The few recent studies that have been conducted to analyze the benefit of transfusing granulocytes to patients with fungal infec-

TABLE 16.4 Summary of Controlled Studies of Prophylactic Granulocyte Transfusion in Neutropenic Patients (Excluding Neonates)

Study	Randomized	Diagnosis	Collection Technique	PMN $\times 10^{10}$	% Infections Study Group	% Infections Control Group	% Survival Study Group	% Survival Control Group
Clift et al. 1978	Yes	BMT	CL/FL	1.6–2.2	6.9	43	100	98
Schiffer et al. 1979	Yes	AML	CL	1.2	22	67	100	78
Mannoni et al. 1979	Yes	AML	CL	1	4.5	39	100	91
Winston et al. 1980	Yes	BMT	FL	1.2	37	47	95	89
Strauss et al. 1981	Yes	AML	CL	0.34	52	65	78	88
Winston et al. 1981	Yes	AML	CL	0.56	76	62	84	90
Ford et al. 1982	Yes	AML	CL	1.5	30	44	70	78
Gomez-Villagran et al. 1984	Yes	AML	CL	1.24	58	81.3	100	75

CL = Centrifugation leukapheresis, FL = filtration leukapheresis, BMT = bone marrow transplantation, AML = acute myelogenous leukemia.

tions not responsive to standard antifungal treatment have not been concordant. Few studies that involved patients who developed fungal infections post bone marrow transplant did not show any benefit of the use of granulocytes, although the doses transfused were quite low. On the contrary, the study by Dignani et al. reported favorable responses in patients who had both shorter duration of neutropenia and known active infection (1997). Larger controlled randomized studies are needed to assess the potential benefit of granulocyte transfusions in these settings.

Prophylactic Granulocyte Transfusions

Learning from the success of prophylactic platelet transfusions in the prevention of hemorrhage, several controlled, randomized studies evaluated the use of granulocyte transfusions as a means to prevent infections in neutropenic patients. These studies were performed mostly in the setting of myeloablative therapy and bone marrow transplantation or remission induction chemotherapy for acute myelogenous leukemia. Granulocytes were transfused when the patients' ANC was below 500 or 200/µL. Granulocytes were obtained through either filtration or centrifugation leukapheresis from G-CSF stimulated donors and given daily or on an every-other-day schedule. In most of the studies, $>1 \times 10^{10}$ granulocytes were transfused. Several studies reported a decreased incidence of infection or septicemia in transfused patients compared to controls; however, none of these studies showed any effect on the overall survival (Table 16.4). Moreover, it appeared that recipients of prophylactic PMN transfusions had a higher incidence of cytomegalovirus infections as well as pulmonary infiltrates (Winston et al. 1980a). A meta-analysis of studies of prophylactic granulocyte

transfusion indicated dose of granulocytes transfused, assessment of leukocyte compatibility, and shorter duration of neutropenia as major determinants for prevention of bacterial infection in neutropenic recipients (Vamvakas and Pineda 1996). In conclusion, based on the available data and considering the improved supportive care delivered to neutropenic patients, granulocyte transfusions cannot be generally recommended as a prophylactic intervention.

Alloimmunization Following Granulocyte Transfusion

The antigens present on granulocytes can be broadly divided into two categories: antigens common to other blood cells and antigens restricted to the granulocytes. The first group includes HLA antigens, Ii, Jk, and Kx that are all present on other blood cells. Antigens such as NA1 and 2, NB 1 and 2, NC, ND, and NE 1 are more unique to granulocytes. Several methods are used to detect these antigens: granulocyte agglutination test (GAT), the lymphocytotoxicity assay (LCA), and the lymphocyte and granulocyte immunofluorescence test (GIFT, LIFT). None of these tests is 100% accurate, so usually a combination of them is used (Menitove and Abrams 1987).

The incidence of alloimmunization following granulocyte transfusion in neutropenic patients varies from 12% to 88% in different studies; part of the reason for the wide range in incidence relies in the difference in assays used and their limited accuracy, as stated previously. Moreover, the effect of alloimmunization on recovery and functionality of transfused granulocytes is still unclear. The majority of studies tend to support a decreased effect following transfusion of noncompatible granulocytes. Initial studies conducted in the 1970s demonstrated that transfused granulocytes obtained from patients with chronic myeloid leukemia had decreased white blood cell recoveries in the setting of HLA incompatibility or positive leukoagglutination and/or lymphocytotoxic assays. Studies using ^{111}In-labeled granulocytes showed failure of these cells to localize to sites of inflammation when transfused in patients with granulocyte-agglutinating antibodies (McCullough et al. 1981; Dutcher et al. 1983; McCullough et al. 1986). These results were recently confirmed by Adkins et al. (2000). In the Adkins study, patients received G-CSF-mobilized granulocytes on days two, four, six, and eight following myeloablative therapy and autologous peripheral blood stem cell transplantation. The ANC increment and the clinical outcome were compared between patients with positive or negative screening LCA. Patients with positive s-LCA had a lower ANC increment on days six and eight compared with patients with negative s-LCA. Despite comparable amounts of granulocytes transfused, delayed neutrophil engraftment and higher number of febrile days were reported in patients with a positive s-LCA. These results are in disagreement with a recent report by Price et al., who showed that the existence or development of leukocyte antibodies had no significant effect in the recovery of granulocytes transfused to infected neutropenic patients (2000).

Considering that granulocytes should be transfused as soon as possible after collection, it is frequently impractical for the blood bank to perform a panel of tests to accurately assess leukocyte compatibility. A practical approach has been to periodically screen the patients' serum for alloimmunization, especially if adverse reactions or refractoriness to platelet transfusion occur. It is clear that granulocyte concentrates must be ABO compatible with the recipient, given the significant presence of red blood cells in the granulocyte concentrate.

Adverse Reactions to Granulocyte Transfusions

Frequent adverse reactions are one of the reasons for the decreased use of granulocyte transfusions. The majority of episodes of adverse reactions are actually mild, consisting of fever occasionally associated with chills. The incidence of fever and chills varies in different studies and was probably higher in earlier studies. It has been reported that up to 72% of patients receiving prophylactic granulocytes transfusions have at least one adverse reaction, mostly fever and chills. In more recent studies (Adkins et al. 2000; Price et al. 2000), the incidence of adverse reactions ranged from 5.7% to 13%. Mild febrile reactions can be prevented by premedica-

tion with antipyretics, antihistamines, or corticosteroids. It is possible that the higher incidence of adverse effects in earlier studies was due to the collection method. Filtration leukapheresis was associated with the most frequent adverse reactions, likely due to activation and damage of neutrophils. No clear association exists between development of leukocyte antibodies and adverse reactions.

Pulmonary Complications Following Granulocyte Transfusions

One of the most severe reactions that may occur after granulocyte transfusion is respiratory distress with development of pulmonary infiltrates. The incidence of pulmonary infiltrates has been reported to be as high as 57% in neutropenic patients receiving prophylactic granulocyte transfusions, compared to 27% in the non-transfused group (Strauss et al. 1981). It is generally difficult to assess the incidence of pulmonary reactions solely related to granulocyte transfusions because other factors such as pulmonary infections or interaction between transfused neutrophils and amphotericin B can mimic these reactions.

Several mechanisms are potentially responsible for the development of pulmonary reactions: fluid overload due to high oncotic volume, especially in patients with congestive heart failure; sequestration of granulocytes in infected areas; leukoagglutination; and formation of intravascular aggregates and/or microembolization of granulocyte aggregates formed during collection.

It is well known that the use of amphotericin B and granulocyte transfusions in neutropenic patients is potentially associated with severe pulmonary reactions. Wright et al. in 1981 presented a review of 22 neutropenic, infected patients who had received amphotericin B and transfused granulocytes in a short time interval. 65% of the patients had respiratory compromise compared to only 6% in patients who had received only granulocytes transfused (Wright et al. 1981). It appeared that the worst reactions occurred when amphotericin B was administered within 4 hours from the granulocyte infusion and the transfusions were initiated before amphotericin B. In vivo animal studies have demonstrated that amphotericin B can induce granulocyte aggregation and enhance pulmonary leukostasis, and in vitro studies showed that amphotericin B can cause marked aggregation of granulocytes at concentrations achievable in vivo. Toxicity of amphotericin B in general is due to its ability to bind to membrane sterols, causing potassium leakage and cell lysis; in the case of neutrophils, this would lead to the release of proteases that would damage the pulmonary tissue (Wright et al. 1981).

The results by Wright et al. (1981) have not been consistently confirmed; several investigators have suggested that the respiratory compromise following concomitant administration of amphotericin B and granulocytes is probably due to underlying presence of pulmonary infections or to bacteremia or fungemia. In the absence of clear data, it is recommended that amphotericin B be administered at least 8 hours apart from the infusion of granulocytes. Newer formulations of amphotericin B such as the liposomal amphotericin B, which are associated with significantly lower incidence of side effects, are now available. We recently demonstrated that liposomal amphotericin B (AmBisome), compared with amphotericin B, induced less in vitro aggregation of neutrophils obtained from G-CSF and dexamethasone-mobilized donors. These findings were also observed following the addition of FMLP. Randomized clinical trials investigating the incidence of pulmonary reactions following administration of granulocytes and amphotericin B or liposomal amphotericin B are needed to confirm our findings in vivo (Sulis et al. 2002).

CMV Infection Associated with Granulocyte Transfusion

Transmission of CMV infection through blood transfusion can result in severe and fatal outcome. This is particularly true for severely immunocompromised patients such as those undergoing myeloablative therapy and stem cell transplantation or receiving high dose chemotherapy regimens. Several studies conducted in the early 1980s in recipients of bone marrow transplantation as well as in patients receiving treatment for acute myelogenous leukemia showed a significantly higher incidence of CMV infection in CMV-seronegative patients who had received granulocytes from seropositive donors compared to CMV-seropositive recipients or nontransfused controls. The mortality related to CMV interstitial pneumonitis in the transplanted population was also higher in the seronegative group compared to the untransfused controls (Winston et al. 1980a; Hersman et al. 1982).

Transfusion-related CMV infection is usually prevented by transfusing products obtained from CMV-seronegative donors or by leukodepletion. This latter option is obviously not possible when considering granulocyte transfusion, therefore, the use of CMV-seronegative donors becomes necessary in this setting. In the case where granulocyte transfusion is considered indicative but no CMV-seronegative donor is available, it would be prudent to closely monitor high-risk recipients for CMV infection in order to intervene early in the treatment.

GRANULOCYTE TRANSFUSIONS IN THE NEONATE

Granulocyte Transfusions in Neonates

We previously demonstrated a significant increase in survival in septic neonates with unmobilized leukopheresed granulocyte transfusions versus supportive therapy (Cairo et al. 1987). Neonates with presumed or proven sepsis were randomized to receive supportive therapy alone or supportive therapy plus granulocyte transfusions. Granulocytes were obtained by continuous-flow centrifugation leukopheresis, with a median dose of 0.5×10^9 PMNs per transfusion, and neonates received a total of five transfusions, two on the first day, two on the second day, and one on the third day. Survival was significantly improved in the unmobilized granulocyte transfusion group (95% versus 64%, $P < 0.05$) in the 21 treated neonates compared to the 14 neonates treated with only supportive care (Table 16.5) (Cairo et al. 1987). In a subsequent study we further demonstrated a significant increased survival in septic neonates treated with a similar regimen of unmobilized leukopheresed granulocyte transfusions versus supportive therapy plus intravenous gamma globulin (Cairo et al. 1992). Presumed or proven septic neonates received supportive therapy plus five unmobilized continuous-flow centrifugated leukopheresed granulocytes as mentioned previously with five granulocyte transfusions over a period of 3 days compared to supportive care plus intravenous gamma globulin (1.0 g/kg/day) for 3 consecutive days (Cairo et al. 1992). Septic neonates treated with unmobilized granulocyte transfusions had a 100% survival compared to only 64% survival in the intravenous gamma globulin treated group ($P < 0.03$) (see Table 16.5) (Cairo et al. 1992).

In a previous study, Laurenti et al. retrospectively analyzed 20 neonates with sepsis who received anywhere between two and 15 continuous-flow centrifugated leukopheresed granulocyte transfusions compared to 18 untransfused septic neonates and demonstrated a significant improvement in survival in the neonatal group receiving unmobilized granulocyte transfusions (90% versus 28%) (see Table 16.5) (1981). Additionally, Christensen et al. (1982a,b,c) reported on the results of a randomized controlled study in neonates with overwhelming sepsis, neutropenia, and neutrophil storage pool depletion who either received continuous-flow centrifugated leukopheresed granulocytes versus supportive care and demonstrated a significant improvement in survival in the subgroup of neonates receiving unmobilized granulocytes versus the subgroup receiving supportive care only (100% versus 11%) (see Table 16.5).

TABLE 16.5   Granulocyte Transfusions in Neonates

Author	Randomized	Source	# of GTX	Dose PMN per GTX	Survival GTX versus Other
Cairo et al. 1987	N = 21—GTX N = 14—Supportive care	CFCL	5	0.75×10^9	95% versus 64%
Cairo et al. 1992	N = 21—GTX N = 14—IVIG	CFCL	5	$0.75–1.0 \times 10^9$	100% versus 64%
Laurenti et al. 1981	N = 20—GTX N = 18—Supportive Care	CFCL	2–15	$0.5–1.0 \times 10^9$	90% versus 28%
Christensen et al. 1982a,b,c	N = 20—GTX N = 18—Supportive Care	CFCL	1	0.7×10^9	100% versus 11%
Baley et al. 1987	N = 12—BC N = 13—Supportive Care	BC	1–3	0.35×10^9	88% versus 69%
Wheeler et al. 1987	N = 4—BC N = 5—Supportive Care	BC	1	0.4×10^9	50% versus 40%

GTX = Granulocyte transfusions, CFCL = continuous flow centrifigation leukopheresis, BC = whole blood buffy coat, IVIG = intravenous gammaglobulin.

Whole Blood Buffy Coat Transfusions in Neonates

Baley et al. alternatively reported the results of a randomized study administering cryopreserved whole blood buffy coat transfusions in neonates with neutropenia and documented neutrophil storage pool depletion (1987). In this study Baley et al. (1987) administered an average of 0.35×10^9 PMN from cryopreserved stored buffy coat in the range of one to three buffy coat transfusions to 12 septic neonates and demonstrated a 58% survival rate compared to 13 similar septic neonates who had a survival rate of 69% with supportive care only (see Table 16.5). Similarly, Wheeler et al. (1987) also administered cryopreserved buffy coats (one transfusion only) in four septic neonates compared to five similar septic neonates who received supportive care only. The average PMN transfusion dose in this study was 0.4×10^9 PMNs. In this small number of patients, Wheeler et al. was unable to demonstrate a significant improvement in overall survival between the buffy coat-treated and the supportive care only-treated subgroups (50% versus 40% survival) (see Table 16.5) (1987).

SUMMARY AND RECOMMENDATIONS OF GRANULOCYTE TRANSFUSIONS

The previously mentioned studies utilized unmobilized allogeneic granulocytes either obtained following continuous-flow centrifugation leukopheresis (four studies) or cryopreserved whole blood buffy coat (two studies). In none of these studies in neonates with presumed or proven sepsis with or without neutropenia were mobilized allogeneic granulocytes administered. Furthermore, these studies suggested that PMN cell dose may be a critically important factor in the efficacy of allogeneic granulocyte transfusions in the neonate. In the studies in which allogeneic granulocytes were obtained by continuous-flow centrifugation leukopheresis, the cell dose ranged between 0.5 and 1.0×10^9 PMN per transfusion. In those studies in which this dose of PMN was administered, there appears to be a significant improvement in overall survival in the small number of randomized studies published to date. However, in those small studies in which the cell dose was less than 0.5×10^9 PMN per transfusion and the cells were obtained by cryopreserved whole blood buffy coat, there was no significant improvement in survival. These findings suggest that with the current ability to mobilize allogeneic granulocytes with G-CSF and dexamethasone stimulation of the allogeneic granulocyte donor, a significant number of granulocytes would likely be obtained.

Future studies are required to determine what is the critical PMN cell dose required for treatment in septic neonates and children, the number of transfusions required to significantly improve survival without increasing toxicity and/or morbidity from transfused allogeneic granulocytes, and the subgroups of neonates and children that would most likely benefit from this transfusion cell therapy. The current published literature suggests that neonates and children with severe neutropenia (ANC $\leq 200/mm^3$), especially those with bone marrow neutrophil storage depletion with presumed or proven sepsis, are the subgroups that would most likely benefit from allogeneic granulocyte transfusions. Furthermore, a minimum PMN transfusion dose of between 0.75 and 1.0×10^{10} PMNs is likely to be the required dose for improvement in survival. Granulocyte transfusions, if administered to neonates and children with presumed or proven sepsis, should likely be continued until recovery of neutropenia and/or presumed neutrophil functional defect to be most effective. Lastly, allogeneic granulocyte transfusions should contain no more than 10 to 15 mL/kg per transfusion and include premedication therapy with appropriate doses of diphenyhydramine and hydrocortisone.

Acknowledgments

The authors would like to thank Linda Rahl for her expert editorial assistance in the development of this chapter.

References

Adkins DR, Goodnough LT, et al. 2000. Effect of leukocyte compatibility on neutrophil increment after transfusion of granulocyte colony-stimulating factor-mobilized prophylactic granulocyte transfusions and on clinical outcomes after stem cell transplantation. *Blood* 95(11):3605–3612.

Alavi JB, Root RK, et al. 1977. A randomized clinical trial of granulocyte transfusions for infection in acute leukemia. *N Engl J Med* 296(13):706–711.

Applebaum FR, Bowles CA, et al. 1978. Granulocyte transfusion therapy of experimental Pseudomonas septicemia: study of cell dose and collection technique. *Blood* 52:323–331.

Baley JE, Stork EK, et al. 1987. Buffy coat transfusions in neutropenic neonates with presumed sepsis: a prospective randomized trial. *Pediatrics* 80:712–720.

Bensinger WI, Price TH, et al. 1993. The effects of daily recombinant human granulocyte colony stimulating factor administration on normal granulocyte donors undergoing leukapheresis. *Blood* 81:1883–1888.

Bodey G, Buckley M, et al. 1966. Quantitative relationship between circulating leukocytes and infection in patients with acute leukemia. *Ann Int Med* 64:328–340.

Cairo M, 1989a. Neonatal neutrophil host defense. Prospects for immunologic enhancement during neonatal sepsis. *Am J Dis Child* 143:40–46.

Cairo M, 1989b. Neutrophil transfusions in treatment of neonatal sepsis. *Am J Pediatr Hematol Oncol* 11(2):227–234.

Cairo MS, Worcester C, et al. 1987. Role of circulating complement and polymorphonuclear leukocyte transfusion in treatment and outcome in critically ill neonates with sepsis. *J Pediatr* 110(6):935–941.

Cairo MS, Worcester CC, et al. 1992. Randomized trial of granulocyte transfusions versus intravenous immune globulin therapy for neonatal neutropenia and sepsis. *J Pediatr* 120(2 Pt 1):281–285.

Christensen RD, Bradley PP, et al. 1982a. Granulocyte transfusion in septic canine neonates. *Pediatr Res* 16:571–575.

Christensen RD, MacFarlane JL, et al. 1982b. Blood and marrow neutrophils during experimental group B streptococcal infection: quantification of the stem cell, proliferative, storage, and circulating pools. *Pediatr Res* 16:549–553.

Christensen RD, Rothstein G, et al. 1982c. Granulocyte transfusions in neonates with bacterial infection, neutropenia, and depletion of mature neutrophils. *Pediatr* 70:1–6.

Clift RA, Sanders JE, et al. (1978). Granulocyte transfusions for the prevention of infection in patients receiving bone-marrow transplants. *N Engl J Med* 298(19): 1052–1057.

Dale DC, Liles WC, et al. 1998. Neutrophil transfusions: kinetics and functions of neutrophils mobilized with granulocyte-colony-stimulating factor and dexamethasone. *Transfusion* 38(8):713–721.

Dale DC, Liles WC, et al. 1997. Renewed interest in granulocyte transfusion therapy. *Br J Haematol* 98(3):497–501.

Dignani MC, Anaissie EJ, et al. 1997. Treatment of neutropenia-related fungal infections with granulocyte colony-stimulating factor-elicited white blood cell transfusions: a pilot study. *Leukemia* 11(10):1621–1630.

Dutcher JP, Schiffer CA, et al. 1983. Alloimmunization prevents the migration of transfused indium-111-labeled granulocytes to sites of infection. *Blood* 62(2):354–360.

Ford, J. M., Cullen, M. H., Roberts, M. M., et al. (1982). Prophylactic granulocyte transfusions: results of a randomized controlled trial in patients with acute myelogenous leukemia. *Transfusion* 22(4):311–316.

Fortuny IE, Bloomfield CD, et al. 1975. Granylocyte transfusion: a controlled study in patients with acute nonlymphocytic leukemia. *Transfusion* 15(6):548–558.

Gomez-Villagran, J. L., Torres-Gomez, A., et al. (1984). A controlled trial of prophylactic granulocyte transfusions during induction chemotherapy for acute nonlymphoblastic leukemia. *Cancer* 54(4):734–738.

Graw RG, Jr., Herzig G, et al. 1972. Normal granulocyte transfusion therapy: treatment of septicemia due to gram-negative bacteria. *N Engl J Med* 287(8):367–371.

Hersman J, Meyers JD, et al. 1982. The effect of granulocyte transfusions on the incidence of cytomegalovirus infection after allogeneic marrow transplantation. *Ann Intern Med* 96(2):149–152.

Herzig RH, Herzig GP, et al. 1977. Successful granulocyte transfusion therapy for gram-negative septicemia. A prospectively randomized controlled study. *N Engl J Med* 296(13):701–705.

Higby DJ, Yates JW, et al. 1975. Filtration leukapheresis for granulocyte transfusion therapy. Clinical and laboratory studies. *N Engl J Med* 292(15):761–766.

Hubel K, Carter RA, et al. 2002. Granulocyte transfusion therapy for infections in candidates and recipients of HPC transplantation: a comparative analysis of feasibility and outcome for community donors versus related donors. *Transfusion* 42(11):1414–1421.

Jendiroba DB, Lichtiger B, et al. 1998. Evaluation and comparison of three mobilization methods for the collection of granulocytes. *Transfusion* 38(8):722–728.

Klein HG, Strauss RG, et al. 1996. Granulocyte transfusion therapy. *Semin Hematol* 33(4):359–368.

Laurenti F, Ferro R, et al. 1981. Polymorphonuclear leukocyte transfusion for the treatment of sepsis in the newborn infant. *J Pediatr* 98:118-l22.

Liles WC, Rodger E, et al. 2000. Combined administration of G-CSF and dexamethasone for the mobilization of granulocytes in normal donors: optimization of dosing. *Transfusion* 40(6):642–644.

Mannoni P, Rodet M, et al. (1979). Efficiency of prophylactic granulocyte transfusions in preventing infections in acute leukaemia. *Rev Fr Transfus Immunohematol* 22(5):503–518.

McCullough J, Clay M, et al. 1986. Effect of leukocyte antibodies and HLA matching on the intravascular recovery, survival, and tissue localization of 111-indium granulocytes. *Blood* 67(2):522–528.

McCullough J, Weiblen BJ, et al. 1981. Effect of leukocyte antibodies on the fate in vivo of indium-111-labeled granulocytes. *Blood* 58(1):164–170.

Menitove JE and Abrams RA. 1987. Granulocyte transfusions in neutropenic patients. *Crit Rev Oncol Hematol* 7(1):89–113.

Price TH, Bowden RA, et al. 2000. Phase I/II trial of neutrophil transfusions from donors stimulated with G-CSF and dexamethasone for treatment of patients with infections in hematopoietic stem cell transplantation. *Blood* 95(11):3302–3309.

Santos JJ, Shigeoka AO, et al. 1980. Functional leukocyte administration in protection against experimental neonatal infection. *Pediatr Res* 114:1408–1413.

Schiffer CA, Aisner J et al. (1979). Alloimmunization following prophylactic granulocyte transfusion." *Blood* 54(4):766–774.

Siegel J and McCracken G. 1981. Sepsis neonatorum. *N Engl J Med* 304:642–647.

Strauss R. 1993. Therapeutic granulocyte transfusion in 1993. *Blood* 81:1675–1678.

Strauss RG. 1998. Neutrophil (granulocyte) transfusions in the new millennium. *Transfusion* 38(8):710–712.

Strauss RG, Connett JE, et al. 1981. A controlled trial of prophylactic granulocyte transfusions during initial induction chemotherapy for acute myelogenous leukemia. *N Engl J Med* 305(11):597–603.

Stroncek DF, Yau YY, et al. 2001. Administration of G-CSF plus dexamethasone produces greater granulocyte concentrate yields while causing no more donor toxicity than G-CSF alone. *Transfusion* 41(8):1037–1044.

Sulis ML, Van de Ven C, et al. 2002. Liposomal amphotericin B (AmBisome) compared with amphotericin B +/– FMLP induces significantly less in vitro neutrophil aggregation with granulocyte-colony-stimulating factor/dexamethasone-mobilized allogeneic donor neutrophils. *Blood* 99(1):384–386.

Vamvakas EC and Pineda AA. 1996. Meta-analysis of clinical studies of the efficacy of granulocyte transfusions in the treatment of bacterial sepsis. *J Clin Apheresis* 11(1):1–9.

Vogler W and Winton E. 1977. A controlled study of the efficacy of granulocyte transfusions in patients with neutropenia. *Am J Med* 63:548–555.

Wheeler JG, Chauvenet AR, et al. 1987. Buffy coat transfusions in neonates with sepsis and neutrophil storage pool depletion. *Pediatr* 79:422–425.

Winston DJ, Ho WG, et al. 1980a. Cytomegalovirus infections associated with leukocyte transfusions. *Ann Intern Med* 93(5):671–675.

Winston DJ, Ho WG, et al. 1980b. Prophylactic granulocyte transfusions during human bone marrow transplantation. *Am J Med* 68(6):893–897.

Winston DJ, Ho WG, et al. (1981). Prophylactic granulocyte transfusions during chemotherapy of acute nonlymphocytic leukemia. *Ann Intern Med* 94(5):616–622.

Wright DG, Robichaud KJ, et al. 1981. Lethal pulmonary reactions associated with the combined use of amphotericin B and leukocyte transfusions. *N Engl J Med* 304(20):1185–1189.

17

Extracorporeal Membrane Oxygenation and Cardiopulmonary Bypass

DAVID F. FRIEDMAN, MD, AND LISA M. MONTENEGRO, MD

INTRODUCTION

The technology to provide artificial oxygenation of the blood to supplement or entirely take over the gas exchange functions of the lungs was conceived and developed in animals in the 1930s and 1940s, first used in humans in the 1950s, and has been available for clinical use since the 1960s. When coupled with a blood pump to support cardiac output, this technology has two major medical applications: cardiopulmonary bypass (CPB) and extracorporeal membrane oxygenation (ECMO). CPB is fundamentally a technique in anesthesiology to take over blood circulation and oxygenation so that surgery can be performed on a nonbeating, bloodless heart and great vessels. In pediatrics, CPB is required to perform surgical repairs of a wide range of congenital heart disease, malformations of the great vessels, transplantation of thoracic organs, and other thoracic surgery. On the other hand, ECMO is an aggressive supportive care intervention used in the intensive care unit to support critically ill patients when conventional ventilatory support, circulatory support, or both have failed. In pediatrics, ECMO has applications in the neonatal intensive care unit for newborns with severe lung disease, in the cardiac intensive care unit for patients with severe heart failure, including complications after cardiothoracic surgery, and in the general pediatric intensive care unit for other causes of very severe circulatory or pulmonary failure.

Since both techniques involve large extracorporeal circulatory volumes in critically ill patients, CPB and ECMO programs depend heavily on support from the blood bank to provide blood products both at the time the extracorporeal circuits are first established, through the duration of the "run," and during the recovery period. This chapter will first review the basic layout of CPB and ECMO circuits. Then, because CPB and ECMO are used in different clinical situations, it will review the uses and outcomes of the two techniques separately. Their impact on the blood bank will be discussed at the end of each section.

CIRCUIT FOR ECMO AND CPB

The fundamental elements of the extracorporeal circuit for CPB or ECMO are cannulae, the pump, a blood reservoir, the oxygenator, and a heat exchanger. There are many additional components that provide for safety and monitoring. The cannulae are the connections of the patient's circulation to the extracorporeal circulation, drawing deoxygenated blood from the patient's venous circulation and returning oxygenated blood to either the arterial or venous side of the patient's circulation. The pump drives the blood through the components of the extracorporeal circuit, and if the blood return is to the arterial side of the circulation, the pump also delivers systemic arterial blood flow to supplement or replace the cardiac output. The oxygenator provides a large surface area over which gas exchange between the blood and an exogenous mixture of oxygen, air, carbon dioxide, and possibly anesthetic agents can occur. The heat exchanger permits manipulation of the temperature of the blood returned to the patient.

Cannulae

Access to the patient's circulation for CPB or ECMO must be large enough to support blood flow equivalent to the patient's entire cardiac output, as much as 5 L/min in an adult. A cannula is placed in a jugular or femoral vein by direct surgical cut down or within the superior vena cava or right atrium when placed for CPB during cardiothoracic surgery. To return oxygenated blood, the common carotid is used most commonly for arterial access for veno-arterial (VA) ECMO and an aortic cannula is placed at surgery most commonly for CPB. ECMO support can be delivered entirely through venous access (veno-venous [VV] ECMO is described further), in which case the return cannula may be in the femoral vein, the contralateral jugular vein, or in the second lumen of a double lumen venous cannula (Foley 2000). In many cases, the jugular vein and carotid arteries used for ECMO access are sacrificed (ligated) when ECMO support is discontinued.

Pump

A roller pump is most commonly used to move blood through the extracorporeal circuit and to provide arterial pressure for the patient. An alternate technology is a centrifugal pump. The roller pump consists of a disk that rotates within a semicircular raceway and moves rollers, which compress the flexible tubing against the inside wall of the raceway, propelling the blood forward. The adjustable rate of rotation of the pump determines the flow rate of blood through the pump and ultimately the amount of arterial flow contributed by the circuit. This flow is not pulsatile. The pump must have a mechanism to detect elevated pressure related to occlusion anywhere in the return side of the circuit downstream of the pump and to reduce or stop flow before the elevated pressure leads to a catastrophic disconnection in the circuit.

Blood Reservoir

Blood is drawn through the venous cannula into the extracorporeal circuit by gravity, entering a collection reservoir located well below the level of the access cannula. The reservoir provides a volume of blood that acts as a buffer to allow for discrepancies between the rate of blood coming from the patient and the pump rate. The reservoir must be equipped with a sensor to detect emptying. This sensor is part of a servo control mechanism that will slow or stop the pump to ensure that the pump flow rate does not exceed the rate at which blood comes from the patient. This helps prevent introduction of air in the circuit by cavitation and air embolism in the patient.

In an ECMO circuit, this reservoir is small, typically 30 mL, while in a CPB circuit the reservoir is larger with a capacity of a liter or more of blood. The reservoir in the CPB circuit serves other functions, including collection of the large volume of blood that is in the heart and vessels after CPB flow is established and salvaging blood removed from the surgical field by suction. The large blood reservoir provides a margin of safety (10 to 15 seconds of blood flow) in case the return side of the CPB circuit should leak or become dislodged. However, returning the red cells in the large volume of blood from the reservoir to the patient at the end of the CPB run presents a problem, as will be discussed.

Oxygenator

The oxygenator is the central component of the CPB or ECMO circuit since it performs the function of the lungs, which is to saturate the hemoglobin of the venous blood with oxygen. The oxygenator provides a large surface area for an interface between blood and the gas phase, either with hollow fibers or a folded silicone membrane. An oxygen rich gas mixture is passed through the oxygenator in the direction opposite to blood flow, establishing a countercurrent that promotes diffusion of oxygen into the blood. The total surface area, and therefore the capacity of the oxygenator, is selected based on the size of the patient and the blood flow that must be fully oxygenated. Exposure of blood to this large artificial surface is also thought to be responsible for many of the coagulation and inflammatory abnormalities associated with ECMO and CPB.

Heat Exchanger

Since flow rates equivalent to the entire cardiac output are involved, the infusion of blood at a temperature below normal body temperature would rapidly result in hypothermia. Thus, a heat exchanger is a necessary component of an ECMO circuit to ensure that the temperature of the blood returning to the patient is at 37°C. In some cardiothoracic surgery, controlled hypothermia is used to reduce the metabolic demands of the brain and other organs, especially if a period of circulatory arrest is planned. The heat exchanger can be used to cool the body temperature and, at the end of the procedure, to restore it to normal.

Extracorporeal Volume and Priming

The components of the ECMO or CPB circuit and the filters and tubing that connect them combine to form a large extracorporeal volume. The extracorporeal volume of the smallest circuit is 400 to 500 mL, about twice the blood volume of a neonate for whom it could be used. For larger patients, the extracorporeal volume of the circuit may be larger, principally because a larger oxygenator is required. However, the ratio of circuit volume to patient blood volume is greatest for neonates and generally decreases with larger patients.

The large extracorporeal volume of ECMO or CPB circuits means that, for many pediatric patients, these circuits must be primed with red cells before they are connected to the patient. A "bloodless" prime may be used if the patient is large enough and the hemoglobin high enough to tolerate a dilution effect. The red cells are typically combined with a buffer solution and may also be combined with 5% albumin solution or fresh frozen plasma (FFP), and the pH and electrolyte concentrations of the circuit prime are checked before initiation of support. In some situations, whole blood may be used as the priming fluid. Several considerations go into the decision for how the extracorporeal circuit will be primed, as listed in Table 17.1. These considerations are among the major determinants of how blood bank support for the ECMO or CPB program will be organized.

Anticoagulation

Because of the large area of artificial surface in contact with the blood in the oxygenator, patients on ECMO and CPB circuits experience activation of coagulation factors and platelets, and therefore must receive aggressive anticoagulation. This is usually accomplished using infusions of unfractionated heparin with dose adjustments based on whole blood activated clotting time (ACT) measurements. The typical target value for the ACT is 180 to 200 seconds for a patient on ECMO support, and >500 seconds for a patient on CPB. The typical protocol calls for hourly monitoring using a "point of care" instrument, at the bedside or in the operating room. Since the ACT is a less sensitive assay than the activated partial thromboplastin time (aPTT), the doses of heparin and level of anticoagulation for ECMO and CPB are typically higher than for medical

anticoagulation therapy. For this reason, patients on ECMO support are at risk for hemorrhagic complications (ELSO Registry 1995). At the end of a surgical procedure using CPB, the heparinization is reversed using protamine, in order to promote postoperative hemostasis.

ECMO

Indications

ECMO is a complex, risky, and expensive life-support measure that is usually reserved for patients whose underlying disease process is associated with a mortality of >80%, which has not responded to conventional ventilatory support and medical therapies, but that is still potentially reversible (Bartlett et al. 2000). Among term neonates, respiratory failure due to severe meconium aspiration syndrome or persistent pulmonary hypertension and among premature infants, severe respiratory distress syndrome are the most common indications for initiation of ECMO support (Table 17.2). In the ELSO database, these three indications accounted for 60% of reported ECMO for neonates (ELSO Registry 1995; Shanley et al. 1994). ECMO support limits the barotrauma associated with aggressive conventional or high frequency ventilation and provides a period of "lung rest" during which inflammatory processes in the lung may subside. Respiratory failure from lung hypoplasia with congenital diaphragmatic hernia (CDH) accounts for another 20% of ECMO utilization in neonates, but it is less clear that "lung rest" provided by ECMO has a net beneficial effect on the outcome of CDH.

TABLE 17.1 Physiological and Medical Considerations in Priming of ECMO and CPB Circuits

Level of urgency
Ratio of patient blood volume to circuit volume
"Cardiac stun"—related to pH, potassium, calcium, electrolytes
Desired hematocrit—degree of hemodilution
Pre-existing coagulopathy
Renal function
Number of blood donor exposures

TABLE 17.2 Indications for ECMO support

Neonatal	Pediatric
Meconium aspiration	Bacterial pneumonia
Respiratory distress syndrome	Viral pneumonia
Persistent pulmonary hyptertension	Acute respiraton distress syndrome (ARDS)
Congenital diaphragmatic hernia	Burns
Sepsis	Inhalation injuries
	Near drowning
	Sepsis
	Myocarditis—bridge to transplantation
	Failure to wean from CPB after surgery
	Arrest or heart failure after cardiac surgery

The decision to place a patient on ECMO support is a complex medical and ethical judgment. The ELSO guidelines for neonatal ECMO listed in Table 17.3 demonstrate that the patient must be both ill enough to warrant the risks of ECMO, including heparinization, and also free of irreversible complications and lethal disease processes so that the potential benefit of ECMO may be realized. There are calculated indices of severity of lung disease (Kim and Stolar 2000), which can be used to predict mortality and eligibility for ECMO support. The criterion that places a limit on the duration of conventional ventilation is intended to exclude patients who have already been exposed to prolonged barotrauma and have already developed bronchopulmonary dysplasia.

Beyond the neonatal period, in the pediatric intensive care unit, the indications for ECMO are more varied and include respiratory failure associated with sepsis, pneumonia, acute respiratory distress syndrome (ARDS) or systemic inflammatory response syndrome (SIRS), burns, inhalation injury, drowning, complications of bone marrow transplantation, and complications of cardiothoracic surgery, as listed in Table 17.2 (Bartlett et al. 2000; Montgomery et al. 2000; Green et al. 1996). Overall survival to hospital discharge is dependent on the initial diagnosis (ELSO Registry 1995) and is better if respiratory failure is not complicated by multiorgan system failure or infection or a protracted period of conventional ventilation (Masiakos et al. 1999). In cardiothoracic surgery, survival from postoperative ECMO is also better if the course is not prolonged and not complicated by infection (Montgomery et al. 2000; Aharon et al. 2001). ECMO support has been successfully used in patients with sickle cell disease with severe acute chest syndrome (Trant et al. 1996; Pelidis et al. 1997).

Veno-arterial Versus Venovenous ECMO

When the primary indication for ECMO support is respiratory failure and the patient's cardiovascular function is preserved, it may be preferable to use VV

TABLE 17.3 ELSO Guidelines for Neonatal ECMO (1997)

Gestational age ≥34 weeks and birth weight ≥2000 g
No significant coagulopathy or bleeding complications
No major intracranial hemorrhage
Mechanical ventilation less than 10–14 days and reversible lung
   disease
No uncorrectable cardiac lesions
No lethal congenital anomalies
No evidence of irreversible brain damage

ECMO support rather than VA ECMO. The VV ECMO circuit returns oxygenated blood to the venous circulation rather than the arterial side and relies on the heart to provide all systemic arterial blood flow. This strategy has the advantages of not sacrificing a carotid artery; avoiding the risk of arterial embolization, of air or particulates; providing better perfusion to the lungs and coronary arteries; and providing pulsatile blood flow from the heart, which improves renal function. Since VV ECMO relies on the native cardiac function, there is no ready mechanism to compensate for loss of cardiac function by "turning up" the level of support from the ECMO circuit pump. This feature of VV ECMO is especially critical for the startup phase if there is an acute reduction of cardiac output related to pH or electrolyte abnormalities in the priming fluid, a phenomenon referred to as "cardiac stun."

VA ECMO is required when both pulmonary and cardiac function must be supported by the ECMO circuit. This applies when the underlying indication for ECMO support is primary heart disease or when cardiovascular failure is a prominent feature of the disease process, as in septic shock.

Course and Outcomes

Once initiated, a course of ECMO support can be continued for days to weeks although the chance of survival from prolonged courses is diminished (Masiakos et al. 1999). ECMO support is weaned by incremental reduction of the pump flow rate while monitoring oxygenation, acidosis, and lung compliance (Kim and Stolar 2000; Hirschl 2002). Survival rates for patients on ECMO vary from 40% for adults to 80% for neonates (ELSO Registry 1995). The survival rates for neonatal ECMO for meconium aspiration syndrome (MAS), respiratory distress syndrome (RDS), persistent pulmonary hypertension (PPH) are in the range of 83% to 94%. For CDH, there was no difference in survival rates before and after ECMO support became available in one institution (Keshen et al. 1997) and no difference in survival in a study that compared two institutions caring for CDH patients, one with ECMO and one without. The overall survival rates of 53% with no ECMO support and 44% survival with ECMO (Wilson et al. 1997; Azarow et al. 1997) are similar to the 58% survival rate noted in the ELSO registry for infants with CDH. In older children, one study demonstrated a mortality of 28.6% in pediatric intensive care unit patients with a predicted mortality of 50% to 75% when supported with ECMO as compared to 71.4% in similarly high-risk patients who were not treated with ECMO (Green et al. 1996).

Blood Bank Support for ECMO Programs

The blood bank plays two roles in supporting the needs of ECMO patients: providing blood products for the initiation of ECMO support and providing blood products for the ongoing needs of a patient on ECMO. As discussed previously, the initiation of ECMO support for a pediatric patient usually requires red blood cells (RBCs) for priming the circuit and often plasma. Patients on ECMO support also have ongoing consumption of coagulation factors and plasma and may have ongoing bleeding and thus require frequent transfusions of FFP, cryoprecipitate, platelets, and booster transfusions of packed red blood cells (PRBCs).

Blood Products for Priming

Under ideal circumstances, the typical ECMO priming protocol would use one to two units of ABO and Rh group-specific and crossmatch-compatible PRBCs, depending on the extracorporeal volume (ECV) of the ECMO circuit, and possibly one unit of group-specific FFP. PRBCs should also be negative for sickle hemoglobin, since the red cells may be exposed to hypoxia and severely abnormal metabolic conditions. Since the ECV of an ECMO circuit is often a significant fraction of the total blood volume of the pediatric patient, the red cells used for the prime would ideally conform to guidelines for massive transfusion, including avoidance of PRBC units in additive solutions and use of relatively fresh red cells (Luban et al. 1991). Additional considerations include provision of cytomegalovirus (CMV)—safe blood products for low birth weight neonates and for thoracic organ transplant candidates who will be immunosuppressed.

In practical application, the single most important determinant in providing blood product support for ECMO is the level of urgency or the amount of time available for preparation. Extreme urgency may require compromise of the ideal blood product preparation as previously described. Examples of clinical scenarios for which ECMO circuits might be started are shown in Table 17.4. Provisions for the most extreme degrees of urgency require storage of group O, Rh(D)-negative PRBCs in a monitored refrigerator in the intensive care unit. These units may be somewhat older than ideal and will be transfused uncrossmatched. For a neonate transferred to a tertiary care center for consideration of ECMO support, there may be several hours of forewarning before the arrival of the patient, permitting time to find fresher PRBCs. However, the ABO/Rh blood group of the patient may not be established, and the PRBCs may not be crossmatched before the circuit is initiated. For the least urgent scenario in which a patient is electively transferred to ECMO support, preparation of the ideal blood product may be possible.

Blood Products for Coagulopathy and Ongoing Needs

The coagulation complications of ECMO are significant: since the patient is heparinized, a prolongation of the aPTT and a tendency to bleeding are expected complications of ECMO support. The Extracorporeal Life Support Organization (ELSO) registry gives an incidence of 15% intracranial hemorrhage and 7% other bleeding for neonates on ECMO (ELSO Registry 1995). Patients on ECMO circuits who also have chest tubes or recent surgery may have significant ongoing blood loss from those sites. Despite this aggressive

TABLE 17.4   Examples of Blood Product Protocols for ECMO Startup

Clinical Scenario	Urgency	Products	Blood Group	Storage*	Crossmatch
Cardiac arrest in cardiac ICU	5–10 minutes	2 units PRBCs	O-neg PRBCs	Additive solution, stored in ICU, <14 days	Retrospective
Disruption of ECMO circuit	5–10 minutes	2 units PRBCs	O-neg PRBCs	Additive solution, stored in ICU, <14 days	Retrospective
Progressive septic shock, not neonatal	30 minutes	2 units PRBCs	O-neg PRBCs	<10 days old, any preservative	Immediate spin
Neonate transferred for ECMO	1–2 hours	2 units PRBCs 1 unit FFP	O-neg PRBCs AB plasma and platelets	<10 days, CPD or CPDA	Retrospective
Cardiac ICU	30–60 minutes	2 units PRBCs	Type specific	<7 days old, additive solution	Immediate spin or full
Gradual respiratory or cardiac failure on conventional support	Hours to days	2 units PRBCs	Type specific	<10 days, CPD or CPDA for neonate, otherwise any	Full

*Based on the protocols of The Children's Hospital of Philadelphia.

anticoagulation, there is activation and consumption of platelets and procoagulant factors related to exposure of blood to artificial surfaces. Clotting within the ECMO circuit occurs frequently, with an incidence of 25% of ECMO circuits for neonates. Fibrin deposition within the oxygenator and other parts of the circuit may reduce its efficiency for gas exchange and may obstruct flow. During prolonged ECMO "runs," fibrin may accumulate within an ECMO circuit to a level that requires transferring the patient to a new, freshly primed circuit. The rate at which fibrin accumulates may be influenced by the materials used in the circuit (Grossi 2000) as well as the patient's underlying disease process.

Surveys of blood product utilization for pediatric patients on ECMO support generally report more frequent transfusion of platelet concentrates than plasma or PRBCs and also a wide range of transfusion requirements for individual patients. The number of transfusions will vary significantly with the transfusion trigger employed in the ECMO support protocol (Minifee et al. 1990). For example, the transfusion trigger for platelets for patients on ECMO is reported from 50,000 to 110,000/μL (Minifee et al. 1990; Chevuru et al. 2002). One survey of 91 neonates on ECMO support for a mean of 4.6 days found that these patients received 0.6 to 1.0 red cell transfusions, 0.1 to 0.3 FFP transfusions, and 1.8 platelet transfusions per day on ECMO (McCoy-Pardington et al. 1990). However, the range of numbers of transfusions per ECMO patient was wide: 0 to 17 transfusions for red cells, 0 to 8 for FFP, and 1 to 32 for platelets. Another survey reported a mean of 1.3 platelet transfusions per day on ECMO, with variation related to institution and ECMO technique (VV versus VA) (Chevuru et al. 2002). Sepsis while on ECMO clearly increases the platelet requirements (Chevuru et al. 2002; Zavadil et al. 1998). For comparison, mean daily blood product requirements for adults on ECMO were 4.6 units of red cells, 0.5 units of FFP, 15 units of random donor platelet concentrates, and 1.0 units of cryoprecipitate (Butch et al. 1996). The numbers of blood products required for an ECMO program for adults are predictably greater than for pediatric patients.

Although not described in these surveys, cryoprecipitate is commonly used to supplement the fibrinogen level in pediatric patients on ECMO circuits, with a transfusion trigger to maintain the fibrinogen above 100 mg/dL.

CPB

CPB is used to facilitate heart surgery by supporting the systemic blood flow and oxygenation artificially,

permitting the heart to be asystolic and drained of blood. In pediatric cardiothoracic surgery, CPB is necessary for surgical correction of congenital heart disease that requires opening the ventricles or great vessels. This includes lesions of low complexity such as atrial septal defect and ventricular septal defect, as well as more complex lesions such as tetralogy of Fallot, transposition of the great vessels, single ventricles, hypoplastic left heart syndrome, and cardiac or pulmonany transplantation. The patient population varies widely in age, physiology, and complexity and includes cyanotic newborns undergoing arterial switch procedures and stage 1 Norwood procedures to school-age children undergoing elective repairs. In some centers, the patient population may also include young adults who have developed late complications of cardiac operations in childhood and adults with symptoms related to long-standing unrepaired congenital heart disease. In some cases, cardiac malformations are corrected in stages, with a palliative procedure first and a definitive correction later. In addition, there are patients who require surgical revision of previous repairs because of incompetent valves or vascular stenosis. Patients in the latter two categories will undergo second or third cardiac operations with CPB.

Overview of CPB Run

The CPB circuit is typically primed with red cells shortly before the start of surgery and the pH, hematocrit, potassium, and calcium levels of the circuit prime checked. The circuit is also heparinized. The cannulae for CPB are placed in the vena cava and aorta by the surgeon and handed from the surgical field to the bypass perfusionist to be connected to the CPB circuit. CPB support is initiated after the surgical field is exposed, just as the intracardiac portion of the surgical procedure is begun. A cardioplegia solution containing a high concentration of potassium is administered. Time on CPB is kept to a minimum because risk of bleeding and neurological complications are correlated with the duration of the CPB run (Menache et al. 2002). The actual time may vary from 20 minutes to over 60 minutes depending on the complexity of the repair and the patient's anatomy. Additional anesthesia techniques such as hypothermia, cooling of the head, and periods of circulatory arrest may also be used. Transfusions, fluids, and medications may be administered via the CPB circuit, depending on the amount of blood loss, and red cell salvage techniques can also be incorporated into CPB. CPB is discontinued when the repair is complete, cardiac function has been re-established, and surgical hemostasis is obtained. Before the patient leaves the operating room, the CPB circuit may be used for additional manipulations as discussed later. After the use of the CPB circuit is complete and the cannulae have been removed, the patient receives a dose of protamine calculated to reverse the heparin received during CPB.

Postoperative Hemorrhage

Transient mediastinal bleeding is an expected result of cardiac surgery, and chest tubes are routinely left in place to decompress the mediastinum and facilitate measurement of blood loss. Chest tube drainage varies widely from 16 to 110 mL/kg/24 hours (Manno et al. 1991) but is a cause for re-exploration of the surgical site in about 4% to 9% of cases (Williams, Bratton, and Ramamoorthy 1999; Chambers et al. 1996). CPB has multiple effects on the coagulation proteins, platelets, and inflammatory cytokines, causing both activation and consumption, which may contribute to postoperative bleeding (Williams, Bratton, and Ramamoorthy 1999; Williams et al. 1999; Williams et al. 1998; Despotis and Hogue 1999; Dietrich 1996). The factors associated with bleeding following cardiothoracic surgery with CPB are many and include the patient's age, preoperative condition, the type of procedure, whether the procedure is a first or repeat sternotomy, the length and number of suture lines, the duration of CPB, type of blood product support, anesthesia techniques used, adequacy of reversal of heparin, adequacy of surgical hemostasis, platelet counts, platelet function, coagulation tests for procoagulants and fibrinolysis, and thromboelastogram parameters (Williams, Bratton, and Ramamoorthy 1999). Because of this complexity, controlled studies of bleeding after CPB are difficult to perform in order to isolate, for example, the relative effect of blood product support on postoperative bleeding.

Blood Bank Support for CPB Programs

Fresh Whole Blood

A well-controlled, although not strictly randomized study of the effect of using fresh whole blood in the immediate post-bypass care of children undergoing cardiothoracic surgery with CPB was conducted by Manno et al. (1991). This study compared three blood products: very fresh whole blood, stored at room temperature for less than 6 hours after donation; fresh whole blood, stored refrigerated for 24 to 48 hours; and reconstituted whole blood, consisting of one unit each of PRBCs stored less than 5 days, plasma, and whole blood-derived platelets. These blood products were administered following the CPB run. The study found a significant reduction in postoperative blood loss for very fresh

and fresh whole blood versus reconstituted whole blood for the entire group of 161 patients. The mean postoperative blood loss was approximately doubled for patients younger than 2 years and for patients undergoing complex procedures who received reconstituted blood versus fresh whole blood. No hemostatic advantage was noted for patients older than 2 years or for simple procedures. There was no advantage of the 6-hour fresh whole blood versus the 24- to 48-hour fresh whole blood. The most significant laboratory finding correlated with the use of fresh whole blood compared to reconstituted whole blood was an improvement in platelet function, and the overall superiority of fresh whole blood was attributed primarily to better preservation of platelet function in the fresher product and therefore better restitution of platelet function in vivo after CPB.

Fresh whole blood has advantages beyond improved hemostasis in the care of children after surgery with CPB. Using whole blood to replace acute blood loss as measured by chest tube drainage is logistically simpler than component therapy and may therefore be safer in a busy intensive care unit. The leak of potassium from red cells into plasma during storage is primarily a function of storage age, and the potassium concentration in the supernatant of whole blood is lower than in packed cells of the same storage age (Michael et al. 1975). These features of fresh whole blood may also be safety advantages for patients who have recently undergone cardiac surgery and whose transfusions may be administered rapidly through the CPB cannulae or other central lines.

Providing fresh whole blood less than 48-hours old is a logistic challenge for blood centers, transfusion services, and the cardiac anesthesia and surgery services, and it is not universally available. Since it contains both red cells and plasma, whole blood must be type specific—there is no "universal donor" type. A reliable supply of fresh whole blood requires a commitment from the blood supplier to retain a portion of the daily collections of at least the common ABO and Rh blood groups types as whole blood, to perform the infectious disease testing of those units on an expedited schedule, and to facilitate delivery of those units to the transfusion service such that they can be crossmatched and made available within 48 hours of collection. The blood center must also be willing to recruit blood donations for less common blood types. Family members of the patient may be valuable committed donors for this program (Manno et al. 1991). The transfusion service must be in frequent contact with both the cardiothoracic surgery service and the blood center with updates on changes in the surgery schedule, adequacy of specimens for crossmatch, unanticipated serologic findings such as maternal isohemaglutinins in newborns, and difficulties with blood availability. The costs associated with the use of fresh whole blood include the loss of plasma and platelet components that would otherwise have been made from units retained as whole blood, the labor costs of expedited processing, and the cost of repeated blood ordering for the same patient because of inevitable last-minute changes in the operating room schedule for this critically ill patient population. The transfusion service must also develop mechanisms to manage the untransfused whole blood in its inventory. Despite these difficulties, providing fresh whole blood for pediatric cardiothoracic surgery is feasible (Kwiatkowski and Manno 1999), and the costs associated with it may be viewed as a part of the total cost of caring for this group of patients.

Ultrafiltration and Aprotinin

Two recent developments in cardiac anesthesia offer the possibility of reducing complications after cardiac surgery with CPB. Conventional ultrafiltration (CUF) involves the incorporation of a device with a permeable membrane with a defined pore size into the CPB circuit, allowing removal of water and solutes, and small molecules from the CPB circuit with retention of the cellular elements of the blood. This technique serves as a hemoconcentration step, removing fluid introduced during priming and raising the hematocrit of the bypass circuit, but may have additional benefits in reducing postoperative edema, improving immediate post-operative cardiac function, improving cerebral metabolic recovery after CPB, and possibly removing soluble inflammatory mediators generated during CPB (Ramamoorthy and Lynn 1998; Montenegro and Greeley 1998; Gaynor 2001). Modified ultrafiltration (MUF) (Friesen et al. 1997; Quattro et al. 2002) is a variation of this technique in which the ultrafiltration is performed after discontinuation of CPB rather than during, with blood flow through the CPB circuit reversed, that is, drawing from the arterial side and returning to the venous side. MUF prolongs the duration of the surgical procedure by 5 to 20 minutes (Ungerleider 1998) but may be more efficient in removing fluid. MUF also permits concentration and return of blood in the CPB reservoir to the patient. One study comparing CUF and MUF in pediatric cardiothoracic surgery patients found no difference in a variety of outcomes including post-operative hematocrit, hemodynamics, and blood product use (Thompson et al. 2001), but the study design may not have employed MUF to its fullest advantage. The value of MUF in CPB for pediatric cardiac surgery is still under debate (Ramamoorthy and Lynn 1998; Gaynor 2001).

Aprotinin is a nonspecific serine protease inhibitor that has effects both in blocking fibrinolysis and attenuating contact inhibition as measured by inhibition of kallikrein (Mossinger and Dietrich 1998). These actions of aprotinin might serve to counteract the activation of coagulation proteins and inflammatory mediators generated during CPB (Dietrich 1996). The use of aprotinin during and/or immediately following CPB has beneficial effect in adults undergoing coronary bypass surgery, and some studies have suggested that aprotinin may improve surgical closure time, blood product utilization, and postoperative hemostasis in pediatric patients (Costello et al. 2003; Mossinger et al. 2003), while others have not (Davies et al. 1997).

Blood Product Utilization

The overall blood product requirements for pediatric cardiothoracic surgery patients are difficult to estimate because of the many factors that come into play, including age of patients and complexity of procedures attempted (Manno et al. 1991; Chambers et al. 1996), priming volume of the equipment in use, CPB technique (Friesen et al. 1997; Davies et al. 1997), availability of fresh whole blood, target hematocrit for the end of CPB, and the multiple factors that affect postoperative bleeding and the algorithms that are used to manage it (Despotis and Hogue 1999). One survey from an institution that used whole blood less than 48-hours old and intraoperative cell salvage but not CUF, MUF, or aprotinin and that had defined transfusion guidelines, which were not monitored for the study, reported that pediatric cardiothoracic surgery patients received a mean of 3.1 whole blood or red cell units, 1.4 platelet units, and 1.1 plasma units, for a total of 5.6 +/− 5.1 donor exposures (Chambers et al. 1996). Total blood product utilization may be reduced by practices such as dividing whole blood units when the CPB circuit can be primed with half of a unit.

SUMMARY

Blood support for CPB and ECMO programs places special demands on a pediatric transfusion service because of the variety and critical nature of the disease processes involved, the time-sensitive demands for specialized blood products, and the absolute requirement for blood products to initiate these therapies in small children.

References

Aharon A et al. 2001. Extracorporeal membrane oxygenation in children after repair of congenital cardiac lesions. *Ann Thorac Surg* 72:2095–2102.

Azarow K et al. 1997. Congenital diaphragmatic hernia—a tale of two cities: the Toronto experience. *J Ped Surgery* 32(3):395–400.

Bartlett R et al. 2000. Extracorporeal life support: the University of Michigan experience. *JAMA* 283(7):904–908.

Butch SH et al. 1996. Blood utilization in adult patients undergoing extracorporeal membrane oxygenated therapy. *Transfusion* 36(1):61–63.

Chambers L, Cohen D, and Davis J. 1996. Transfusion patterns in pediatric open heart surgery. *Transfusion* 36:150–154.

Chevuru SC, et al. 2002. Multicenter analysis of platelet transfusion usage among neonates on extracorporeal membrane oxygenation. *Pediatrics* 109(6).

Costello JM et al. 2003. Aprotinin reduces operative closure time and blood product use after pediatric bypass. *Ann Thorac Surg* 75(4):1261–1266.

Davies M et al. 1997. Prospective, randomized, double-blind study of high-dose aprotinin in pediatric cardiac operations. *Ann Thorac Surg* 63:497–503.

Despotis G and Hogue C. 1999. Pathophysiology, prevention, and treatment of bleeding after cardiac surgery: a primer for cardiologists and an update for the cardiothoracic team. *Am J Cardiol* 83:15B–30B.

Dietrich W. 1996. Reducing thrombin formation during cardiopulmonary bypass: is there a benefit of the additional anticoagulant action of aprotinin? *J Cardiovasc Physiol* 27 (Suppl 1):S50–S57.

ELSO Registry. 1995. ECMO Registry of the Extracorporeal Life Support Organization. http://www.med.umich.edu/ecmo/registry.htm.

Foley D et al. 2000. Percutaneous cannulation for pediatric venovenous extracorporeal life support. *J Pediatr Surg* 35(6):943–947.

Friesen RH et al. 1997. Modified ultrafiltration attenuates dilutional coagulopathy in pediatric open heart operations. *Ann Thorac Surg* 64(6):1787–1789.

Gaynor J. 2001. Use of ultrafiltration during and after cardiopulmonary bypass in children. *J Thorac Cardiovasc Surg* 122(2):209–211.

Green T et al. 1996. The impact of extracorporeal membrane oxygenation on survival in pediatric patients with acute respiratory failure. *Crit Care Med* 24:323–329.

Grossi E et al. 2000. Impact of heparin bonding on pediatric cardiopulmonary bypass: a prospective randomized study. *Ann Thorac Surg* 70:191–196.

Hirschl R. 2002. Support of respiratory failure in the pediatric surgical patient. *Curr Opin Pediatr* 14:459–469.

Keshen TH et al. 1997. Does extracorporeal membrane oxygenation benefit neonates with congenital diaphragmatic hernia? Application of a predictive equation. *J Ped Surg Perinatol* 32(6):818–822.

Kim E and Stolar C. 2000. ECMO in the newborn. *Am J Perinatol* 17(7):345–356.

Kwiatkowski JL and Manno CS. 1999. Blood transfusion support in pediatric cardiovascular surgery. *Transfus Sci* 21(1):63–72.

Luban N, Strauss R, and Hume H. 1991. Commentary on the safety of red cells preserved in extended-storage media for neonatal transfusions. *Transfusion* 31:229–235.

Manno C, Hedberg K, and Kim H. 1991. Comparison of the hemostatic effects of fresh whole blood, stored whole blood, and components after open heart surgery in children. *Blood* 77:930–936.

Masiakos P et al. 1999. Extracorporeal membrane oxygenation for nonneonatal acute respiratory failure. *Arch Surg* 134(4):375–380.

McCoy-Pardington D et al. 1990. Blood use during extracorporeal membrane oxygenation. *Transfusion* 30:307–309.

Menache C et al. 2002. Current incidence of acute neurological complications after open-heart operations in children. *Ann Thorac Surg* 73(6):1752–1758.

Michael J et al. 1975. Potassium load in CPD-preserved whole blood and two types of packed red blood cells. *Transfusion* 15:144–149.

Minifee P et al. 1990. Decreasing blood donor exposure in neonates on extracorporeal membrane oxygenation. *J Pediatr Surg* 25:38–42.

Montenegro L and Greeley W. 1998. Pro: the use of modified ultrafiltration during pediatric cardiac surgery is a benefit. *J Cardiothorac Vasc Anesthesia* 12(4):480–482.

Montgomery V, Strotman J, and Ross M. 2000. Impact of multiple organ system dysfunction and nosocomial infections on survival of children treated with extracorporeal membrane oxygenation after heart surgery. *Crit Care Med* 28(2):526–531.

Mossinger H and Dietrich W. 1998. Activation of hemostasis during cardiopulmonary bypass and pediatric aprotinin dosage. *Ann Thorac Surg* 65(6 Suppl):S45–S51.

Mossinger H et al. 2003. High-dose aprotinin reduces activation of hemostasis, allogeneic blood requirement, and duration of postoperative ventilation in pediatric cardiac surgery. *Ann Thorac Surg* 75(2):430–437.

Pelidis MA et al. 1997. Successful treatment of life-threatening acute chest syndrome of sickle cell disease with venovenous extracorporeal membrane oxygenation. *J Pediatr Hematol/Oncol* 19(5):459–461.

Quattro L, Bowser M, and Schwendt A. 2002. Performing modified ultrafiltration on pediatric patients. *AORN Journal* 76(2):300–302.

Ramamoorthy C and Lynn A. 1998. Con: the use of modified ultrafiltration during pediatric cardiac surgery is not a benefit. *J Cardiothorac Vasc Anes* 12(4):483–485.

Shanley C et al. 1994. Extracorporeal life support for neonatal respiratory failure. A 20-year experience. *Ann Surg* 220(3):269–280.

Shapira O et al. 1998. Reduction of allogeneic blood transfusions after open heart operations by lowering cardiopulmonary bypass prime volume. *Ann Thorac Surg* 65:724–730.

Thompson LD et al. 2001. A prospective randomized study comparing volume-standardized modified and conventional ultrafiltration in pediatric cardiac surgery. *J Thorac Cardiovasc Surg* 122(2):220–228.

Trant CA, Jr, et al. 1996. Successful use of extracorporeal membrane oxygenation in the treatment of acute chest syndrome in a child with severe sickle cell anemia. *ASAIO Journal* 42(3):236–239.

Ungerleider RM. 1998. Effects of cardiopulmonary bypass and use of modified ultrafiltration. *Ann Thorac Surg* 65(6 Suppl):S35–S39.

Williams GD, Bratton SL, and Ramamoorthy C. 1999. Factors associated with blood loss and blood product transfusions: a multivariate analysis in children after open-heart surgery. *Anesthesia and Analgesia* 89:57–64.

Williams GD et al. 1999. Coagulation tests during cardiopulmonary bypass correlate with blood loss in children undergoing cardiac surgery. *J Cardiothorac Vasc Anes* 13(4):398–404.

Williams GD et al. 1998. Fibrinolysis in pediatric patients undergoing cardiopulmonary bypass. *J Cardiothorac Vasc Anes* 12(6):633–638.

Wilson JM et al. 1997. Congenital diaphragmatic hernia—a tale of two cities: the Boston experience. *J Pediatr Surg* 32(3):401–405.

Zavadil DP et al. 1998. Hematological abnormalities in neonatal patients treated with extracorporeal membrane oxygenation (ECMO). *J Extra Corporeal Tech* 30(2):83–90.

CHAPTER

18

Hemolytic Disease of the Newborn

JAYASHREE RAMASETHU, MD, FAAP

INTRODUCTION

The average life span of red cells in full term neonates is 60 to 90 days, decreasing to 35 days with decreasing maturity. Hemolytic disease of the newborn (HDN), characterized by decreased life span of red blood cells (RBCs) due to accelerated destruction, may be due to many causes (Table 18.1). All these conditions share common features that point to a hemolytic process. These include anemia in the absence of hemorrhage or other evidence of blood loss, jaundice with predominantly unconjugated hyperbilirubinemia, and increased excretion of endogenous carbon monoxide, a byproduct of heme breakdown (Herschel et al. 2002a).

A remarkable interaction of the fetomaternal unit gives rise to alloimmune HDN, with no ongoing hemolysis after early infancy. However, other conditions, such as hereditary disorders of the red cell membrane or disorders of hemoglobin synthesis, with onset in fetal life, may have lifelong implications.

Apart from the anemia, which may be severe and life-threatening in many of the conditions associated with HDN, the immaturity of the neonatal liver in handling the bilirubin load from breakdown of RBCs, and the unique susceptibility of the neonatal brain to toxic effects of unconjugated bilirubin make hemolytic disease a therapeutic emergency in newborn infants.

Deposition of free, lipid-soluble, unconjugated bilirubin in the basal ganglia and cerebellum of the neonatal brain causes neuronal necrosis and gives rise to a clinical syndrome known as bilirubin encephalopathy or kernicterus. The acute phase is characterized by lethargy, progressing to hypertonia, opisthotonus, irregular respiration, and death. The hypertonia becomes less pronounced in surviving infants, who then develop any or all of the classic sequelae of choreoathetoid cerebral palsy, upward gaze palsy, and sensorineural hearing loss. The clinical features in the acute stage are often less distinct in preterm infants, who nevertheless are at increased risk from this complication, due to immaturity of both the liver and the blood-brain barrier.

The pathogenesis, clinical features, and laboratory diagnosis of the different entities causing HDN are discussed in the first part of this chapter; therapeutic interventions are discussed later.

IMMUNE-MEDIATED HEMOLYTIC DISEASE

Alloimmune Hemolytic Disease of the Newborn

Alloimmune HDN is due to the action of transplacentally transmitted maternal IgG antibodies on paternally inherited antigens present on fetal red cells, but absent on the maternal red cells. The antigen-negative mother may have naturally occuring antibodies or may have developed antibodies to fetal red cell antigens by exposure to the antigens by blood transfusion or, more often, by silent fetomaternal hemorrhage during pregnancy (iso- or alloimmunization).

There are three main classes of alloimmune HDN based on the antigen responsible: Rh, ABO, and hemolytic disease due to other red cell antigens (Ramasethu and Luban 2001). The incidence of severe Rh hemolytic disease has decreased significantly in countries with immunoprophylaxis programs, but the

TABLE 18.1   Causes of Hemolytic Disease in the Newborn

Immune Mediated

Alloimmune—Rh, ABO, other red cell antigens
T activation
Maternal autoimmune hemolytic disease

Red Cell Enzyme Deficiencies

Glucose-6-phosphate dehydrogenase deficiency
Pyruvate kinase deficiency
Other red cell enzyme deficiencies

Red Cell Membrane Defects

Hereditary spherocytosis
Hereditary elliptocytosis
Hereditary pyropoikilocytosis

Disorders of Hemoglobin Synthesis

Homozygous alpha thalassemia
Hemoglobin H disease

Other Causes of Hemolytic Disease

Infections—malaria, bacterial, congenital viral infections
Microangiopathic—DIC, hemangiomas

condition is an excellent prototype to understand the pathogenesis of alloimmune HDN. It will be discussed in greater detail, and the distinguishing features of the other classes of HDN will be emphasized.

Rh HEMOLYTIC DISEASE

Genetics

The inheritance of Rh antigens is determined by two genes, one encoding the protein carrying the D antigen and the other encoding the protein carrying the C or c and E or e antigens (Moise 2002). There is no "d" antigen; the letter "d" is commonly used to designate the absence of D. The presence or absence of D determines the Rh-positive or Rh-negative status of the individual. In Caucasian D-negative individuals, the RHD gene is deleted, whereas the D-negative phenotype in other populations is associated with an inactive or partial RHD gene at the locus. There is significant racial variation in the prevalence of Rh-negativity. About 15% of Caucasians are Rh-negative, compared to 7% to 8% of American blacks, 5% of Asian Indians, and 0.3% of the Chinese. Rh-positive individuals may be homozygous for D (DD), having inherited the D antigen from both parents, or heterozygous for D (Dd), having inher-

ited a D-containing set from one parent and a non–D-containing set of Rh antigens from the other parent. All the offspring of a homozygous Rh-positive (DD) man and an Rh-negative (dd) woman will be Rh or D positive (Dd), whereas a fetus produced by a heterozygous Rh-positive (Dd) father with an Rh-negative mother (dd) could be either Rh-positive (Dd) or Rh-negative (dd).

Pathogenesis

Alloimmunization of Rh-negative women occurs most often by exposure to Rh-positive fetal red cells, secondary to fetomaternal hemorrhage. Alloimmunization by transfusion of Rh-positive cells is now rare. However, the repetitive exposure to minute amounts of Rh-positive cells, by the sharing of contaminated needles between Rh-negative intravenous drug-abusing women with Rh-positive partners has been reported to lead to severe Rh sensitization.

Fetomaternal Hemorrhage

Asymptomatic transplacental passage of fetal red cells occurs in 75% of pregnant women at some time during pregnancy or during labor and delivery. The incidence of fetomaternal transfusion increases with advancing gestation: from 3% in the first trimester, 12% in the second trimester, 45% in the third trimester, to 64% at the time of delivery (Bowman et al. 1986). The average volume of fetal blood in the maternal circulation following delivery is less than 1 mL in 96% of women. Intrapartum fetomaternal hemorrhage of more than 30 mL may occur in up to 1% of pregnancies. However, even minute amounts of antigen-positive blood may cause sensitization in the antigen-negative mother. Fetomaternal transfusion can also result from invasive obstetrical procedures such as chorionic villus sampling, amniocentesis, funipuncture, therapeutic abortion, cesarean section, manual removal of the placenta, and from pathological conditions such as abdominal trauma, spontaneous abortion, or ectopic pregnancy (Sebring and Polesky 1990).

Maternal Alloimmunization

The presence of D-positive red cells in the D-negative mother initially provokes a primary immune response that initially consists of IgM antibodies, which do not cross the placenta, followed by the production of anti-D IgG antibodies capable of crossing the placenta. Repeated exposure to Rh-positive fetal RBCs, as in a second Rh-positive pregnancy in a sensitized Rh-negative woman produces a secondary immune response, which is marked by the rapid production of large amounts of anti-D IgG antibody. Primary sensitization has been reported in 80% of individuals injected with 0.5 mL of Rh-positive cells; secondary immune responses may occur with as little as 0.03 mL of Rh-positive cells. In the absence of Rh immunoglobulin prophylaxis, sensitization occurs in 7% to 16% of women at risk within 6 months after delivery of the first Rh-positive ABO-compatible fetus and in 2% after delivery of an ABO-incompatible fetus. Fetomaternal ABO incompatibility offers some protection against primary Rh immunization because incompatible fetal red cells are destroyed rapidly by maternal anti-A and anti-B antibodies, reducing the maternal exposure to Rh D antigenic sites. However, ABO incompatibility confers no protection against the secondary immune response once sensitization has occurred. It is not clear why many women at risk for sensitization do not appear to become alloimmunized.

Hemolysis

Although Rh antigens are found on fetal cells as early as the seventh week of gestation, the active transport of IgG across the placenta is slow until 24 weeks of gestation. The binding of transplacentally transferred maternal anti-D IgG antibodies to D-antigen sites on the fetal red cell membrane is followed by adherence of the coated red cells to the Fc receptors of macrophages with rosette formation, leading to extravascular non–complement-mediated phagocytosis and lysis, predominantly in the spleen. The degree of hemolysis may be influenced by the functional immaturity of the fetal reticuloendothelial system before 20 weeks of gestation, maternal IgG levels, the IgG subclass, and the rate of transplacental transfer. Antibodies of the IgG1 and IgG3 subclasses, often produced in Rh alloimmunization, have a high affinity for Fcγ receptors and are associated with severe disease.

Fetal anemia secondary to hemolysis results in increased fetal erythropoietin levels and a striking increase in erythropoiesis, associated with marked extramedullary hematopoiesis in the liver, spleen, adrenal glands, and placenta. Down-modulation of platelet and neutrophil production may occur. Extensive extramedullary hemopoiesis in the liver and spleen leads to portal and umbilical venous hypertension, and this may result in ascites and pleural effusions, with compression of the developing lung and ensuing pulmonary hypoplasia. Trophoblastic hypertrophy and placental edema cause impaired placental function. Hypoproteinemia due to liver dysfunction results in generalized edema. "Hydrops fetalis," a state of anasarca, is the end result of a combination of anemia,

cardiac failure, hypoproteinemia, increased capillary permeability, and impaired lymphatic clearance (Phibbs 1998). Before the institution of intrauterine transfusions, most of these fetuses died in utero or soon after birth.

Although fetal bilirubin levels are elevated secondary to hemolysis, the placenta effectively transports most of the lipid soluble unconjugated fetal bilirubin, so the infant is not clinically jaundiced at birth. At birth, the newborn infant's immature liver is incapable of handling the large bilirubin load that results from the ongoing destruction of antibody-coated neonatal red cells and unconjugated bilirubin levels rise.

Some infants with hemolytic disease develop anemia beyond the immediate neonatal period lasting up to 8 to 12 weeks of age. Delayed anemia is partly due to continuing hemolysis because of persistence of maternal antibodies and partly due to the natural decline in red cell production after birth.

Clinical Features

Anemia, jaundice, and hepatosplenomegaly are the hallmarks of alloimmune HDN, but there may be wide variation in the clinical spectrum. The severity of Rh HDN either remains the same or worsens in subsequent affected pregnancies.

Half of the infants with Rh HDN have very mild disease, with cord blood hemoglobin levels only slightly lower than normal; about 25% of affected infants are born at term with moderate anemia and develop severe jaundice. In the days before intrauterine intervention, hydrops developed in utero in the remaining one-quarter with half becoming hydropic before 34 weeks gestation. Hydrops recurs in 90% of affected pregnancies, often at an earlier gestation. Hemoglobin values may continue to fall after birth in all affected infants, with hemolysis continuing until all incompatible red cells and/or circulating maternal alloantibody are eliminated from the circulation.

Most infants with hemolytic disease are not jaundiced at birth. The umbilical cord and vernix caseosa may be stained with bilirubin from the amniotic fluid in severely affected infants. Clinical icterus usually develops during the first day of life, often in the first few hours of life in such infants, progressing in a cephalopedal direction with rising bilirubin levels. In patients with mild disease, the serum indirect bilirubin parallels physiological jaundice in the newborn, peaking by the fourth or fifth day and then declining slowly. Premature infants may have greater levels of serum bilirubin due to lower activity of hepatic glucuronyl transferase activity.

Hepatosplenomegaly is usually present, with marked enlargement being seen in newborn infants with hydrops. Infants with hydrops fetalis also have peripheral edema, ascites, pleural and pericardial effusions often complicating respiratory distress due to surfactant deficiency. Purpura associated with thrombocytopenia is sometimes seen in severely affected infants and may be a bad prognostic sign. The placenta is thickened, enlarged, and pale.

Infants who have received intrauterine transfusions may still have significant hepatosplenomegaly and anemia and may develop hyperbilirubinemia. In a retrospective review of the outcome of 75 newborn infants born alive after fetal intravascular transfusions for severe blood group antagonism, Janssens et al. (1997) found that all 75 needed exchange transfusions after birth, 72 of the 75 neonates needed erythrocyte transfusions (range 1 to 9), and 73 required phototherapy (range 1 to 8 days) for hyperbilirubinemia. A positive correlation was noted between the number of intrauterine transfusions and the number of erythrocyte transfusions required after delivery, probably secondary to the sustained suppression of fetal erythropoiesis by the fetal intrauterine transfusions, but intrauterine transfusions led to a decrease in the number of exchange transfusions.

Laboratory Diagnosis

See Laboratory Diagnosis in Alloimmune Hemolytic Disease.

ABO HEMOLYTIC DISEASE

ABO HDN is limited to mothers who are blood group type O and whose babies are group A or B. ABO HDN is much more common than Rh HDN but is usually milder and rarely responsible for fetal deaths (Table 18.2). A higher incidence and greater severity is reported in Southeast Asians, Latin Americans, Arabs, South African and American Blacks.

Pathogenesis

Unlike Rh disease, ABO HDN may affect the first-born ABO-incompatible infant since anti-A and anti-B antibodies are present normally in Group O adults. These naturally occuring antibodies are probably secondary to sensitization against A or B antigens in food or bacteria. Although ABO incompatibilty is estimated to be present in about 15% of O group pregnancies, ABO hemolytic disease occurs only in about 3% of all births. The low incidence of ABO HDN may be due to the fact that most anti-A and anti-B antibodies are of the IgM type and do not cross the placenta, with only a a small proportion of Group O individuals producing anti-A and anti-B antibodies of the IgG type capable of crossing the placenta. The severity of the disease in the infant may relate in part to the level of anti-A or anti-B IgG in the mother and the IgG subclass. IgG2, a significant component of anti-A and anti-B antibody, is

TABLE 18.2 Comparison of Rh and ABO Incompatibility

	Rh	**ABO**
Blood Groups		
Mother	Negative	O
Infant	Positive	A or B
Type of antibody	IgG1 and /or IgG3	IgG2
Clinical aspects		
Occurence in first-born child	<5%	40% to 50%
Predictable severity in subsequent pregnancies	Usually	No
Stillbirth and/or hydrops	Frequent	Rare
Severe anemia	Frequent	Rare
Degree of jaundice	+++	+
Hepatosplenomegaly	+++	+
Laboratory Findings		
Maternal antibodies	Always present	Not clear cut
Direct antiglobulin test infant	++	+
Peripheral blood picture	Nucleated red blood cells	Microspherocytes
Treatment		
Antenatal measures	Yes	Not indicated
Exchange transfusion frequency	Approximately two-thirds	Rare
Donor blood type	Rh-negative, group specific when possible	Group O only
Incidence of late anemia	Common	Rare

Modified from Ramasethu 2002, with permission.

transported less readily across the placenta compared to IgG1 or IgG3, and is a less-efficient mediator of macrophage-induced red cell clerance. In addition, there are only a small number of fully developed A or B antigen sites on fetal and neonatal RBCs. The effect of anti-A and anti-B antibodies on red cells is also diluted by other tissues bearing these surface antigens. The incidence and severity of hyperbilirubinemia in ABO incompatible neonates in certain populations may be increased by the presence of a variant UDP glucuronyltransferase gene promoter (Kaplan et al. 2000).

Clinical Features

Early neonatal jaundice, becoming clinically visible before 36 hours of age, is usually seen, but milder cases may be indistinguishable from physiological jaundice. Hepatosplenomegaly is not marked but may be detected on careful examination in neonates with moderate and severe disease. Severe fetal anemia and hydrops is very rare (McDonell et al. 1998). ABO hemolytic disease has been reported to recur in subsequent pregnancies, but the recurrence is not as predictable in its severity as is Rh HDN. A recurrence rate of 88% has been reported in subsequent pregnancies (with the same blood type as the index baby), with 62% of the affected infants requiring therapy (Katz et al. 1982).

Laboratory Diagnosis

See Laboratory Diagnosis in Alloimmune Hemolytic Disease.

HEMOLYTIC DISEASE DUE TO OTHER RED CELL ANTIBODIES

Antenatal screening programs detect clinically significant antibodies in 0.24% to 1% of pregnant women. When D and ABO are excluded, non-D Rh antibodies and those belonging to the Kell, Duffy, Kidd, and MNS systems are most frequently involved. Anti-c, anti-Kell, and anti-E may cause HDN as severe as that seen in anti-D HDN.

Kell hemolytic disease accounts for 10% of the cases of antibody-mediated severe fetal anemia. The Kell blood group system is composed of at least 24 discrete antigens. The Kell antigen (also called KEL1 or K1) is expressed in only about 10% of individuals. Alloimmunization in Kell-negative women is more often the result of blood transfusion rather than sensitization by fetomaternal hemorrhage from a Kell-positive fetus (Bowman et al. 1992). Kell hemolytic disease is rare in alloimmunized pregnancies because fetal anemia due to

transplacentally transmitted antibodies can occur only in a Kell-positive fetus. The partners of Kell-negative women are likely to be Kell-positive only in 10% of pregnancies, and only half of these pregnancies are likely to be incompatible because of paternal heterozygosity. Therefore, only 2.5% to 10% of Kell immunized pregnancies end in the delivery of affected infants, with about half the infants requiring intervention. However, anemia in infants with Kell HDN may be more severe than expected and disproportionate to the degree of hemolysis, because of suppression of erythropoiesis at the progenitor cell level by anti-Kell antibodies (Vaughan et al. 1998). Frank hydrops has been described with Kell HDN.

Laboratory Diagnosis in Alloimmune Hemolytic Disease

Maternal Testing

Antenatal Serology

The aim of antenatal serological testing is to identify maternal alloimmunization and to determine the risk to the fetus from alloimmune hemolytic disease. Antenatal testing is not usually indicated for ABO hemolytic disease.

Every obstetrical patient should have ABO and Rh(D) typing and be tested for irregular serum antibodies, irrespective of Rh type, at the initial prenatal visit, preferably by 12 to 16 weeks gestation. Women who initially test as Rh-negative should be tested for the weak-D phenotype, also termed as D^u or D^{+w}. Screening tests should include a 37°C incubation phase and the use of anti-IgG (Coombs' reagent, indirect antiglobulin test). Tests with enzyme-treated red cells or polyspecific antiglobulin sera are not recommended since they may detect clinically insignificant antibodies. Antibody screening is repeated at 28 to 30 weeks gestation in Rh-negative women before the administration of Rh immunoglobulin.

Identification of positive red cell antibody screens in pregant women should be followed by determination of the specific antibody and the clinical significance of the identified antibody. The presence of an irregular antibody does not always connote that HDN will occur. The fetus may not have the antigen corresponding to the maternal antibody and not all antibodies cross the placenta.

Once identified, antibody quantification is usually performed by titration using the indirect antiglobulin test, with different laboratories establishing "critical titers." These vary from eight to 32 for Rh-D antibodies. A weakly reactive anti-D (titer of four or less) may be demonstrated in women who have received antenatal Rh Immunoglobulin and should not be mistaken for sensitization. The trend in sequential antibody levels, together with the previous obstetrical history, is considered more important than any isolated level in predicting disease severity. Serological tests in alloimmunized women may be measured every 2 to 4 weeks from 18 weeks gestation, with rapidly rising levels or a critical titer or level dictating further investigation. Specimens are frozen and successive titrations performed using the same methods (AABB 2002). The significance of titer levels for antibodies other than D have not been defined. Maternal anti-Kell titers in particular correlate poorly with fetal outcome.

Quantification of the concentration of anti-D is carried out in some countries using an AutoAnalyzer; anti-D concentrations below 4 IU/mL are considered to be benign whereas concentrations exceeding 15 IU/mL are associated with moderate to severe hemolytic disease. AutoAnalyzer quantification of anti-c antibodies is also available. Enzyme-linked immunosorbent assays and flow cytometric methods with the additional advantage of measuring IgG subclasses may be performed in serological reference laboratories but are not yet widely used in clinical practice (Engelfriet and Reesink 1995).

Functional cellular assays measure the ability of maternal antibodies to cause red cell destruction and may be useful in predicting the severity of HDN or in special circumstances, for example, if in a previous pregnancy the severity of HDN was much worse than expected from the anti-D titre. In these assays, RBCs sensitized with maternal antibodies are incubated with effector cells carrying Fcγ receptors, such as lymphocytes or monocytes. Cellular interaction such as binding, phagocytosis, or cytotoxic lysis may be measured by monocyte monolayer assay, chemiluminescence tests or by antibody-dependent cellular cytotoxicity assay.

Testing for Fetomaternal Hemorrhage

Testing for fetomaternal hemorrhage is performed in all unsensitized Rh-negative women approximately one hour after delivery of an Rh-positive baby to determine the dose of Rh immunoglobulin necessary to prevent isoimmunization. The standard 300 μg dose of RhIG affords protection against 30 mL of Rh-positive blood. However, fetomaternal hemorrhage in excess of 30 mL may occur in women without predisposing risk factors. Testing for fetomaternal hemorrhage is also indicated during the antenatal period, after 20 weeks gestation, if clinical circumstances suggest the possibility of excessive transplacental hemorrhage, for example, abdominal trauma or abruptio placentae. A rosette test or an enzyme-linked antiglobulin test may be used as a screening test for excessive fetomaternal hemorrhage at delivery (AABB 2002). If the rosette test is positive, the number of fetal red cells should be determined. The Kleihauer-Betke test, the standard test in use in most laboratories, permits quantification of fetal hemoglobin-containing red cells in a maternal blood sample. The test is based on the resistance of fetal hemoglobin to acid elution, unlike adult hemoglobin. A smear of a sample of maternal blood is exposed to acid buffer and then counterstained to expose the dark red fetal RBCs against the light pink to white nonstained maternal "ghost" cells. The volume of fetomaternal hemorrhage is calculated by counting the ratio of fetal to adult cells. The manual performance and interpretation of the test make the reliability of the Kleihauer-Betke test subject to the experience of the personnel performing the test. False-positive results may be obtained in conditions that are associated with increased fetal hemoglobin such as hereditary persistance of fetal hemoglobin, sickle cell disease and trait, and other hemoglobinopathies. Flow cytometric methods offer increased accuracy and reliability (Bromilow and Duguid 1997).

Fetal Testing

Determination of Fetal Blood Type

The child of an Rh-negative mother and a heterozygous Rh-positive father has a 50% chance of being Rh-negative, and thus being unaffected by prior maternal Rh alloimmunization. When the father is heterozygous, or when paternal zygosity is unknown, the determination of fetal blood type early in pregnancy allows the early institution of monitoring and therapy in RhD-positive fetuses who are at risk, and the avoidance of invasive procedures if the fetus is Rh-negative. Fetal blood sampling for serological blood typing is asssociated with a 40% risk of fetomaternal hemorrhage and worsening maternal sensitization and up to 2% risk of fetal loss. RhD-positive fetal cells may be detected rapidly in chorionic villus samples by flow cytometry, but chorionic villus sampling may also increase sensitization. The fetal RhD genotype may be determined from fetal cells obtained by chorionic villus sampling or amniocentesis. Recently noninvasive methods of prenatal diagnosis have been refined to the extent that the fetal Rh-D status may be determined from fetal cells isolated from the maternal blood and even by using fetal DNA extracted from maternal plasma early in the second trimester of pregnancy.

Testing for fetal genotype may rarely not be concordant with fetal phenotype (expression of the RhD antigen on the fetal red cells as determined by serology). These inconsistencies may be due to rearrangement in the paternal RhD gene locus or the presence of an RhD pseudogene, which would test positive by genotype, but negative by serology. Testing of paternal and maternal samples will help rule out potential errors (Moise 2002).

Prenatal determination of the Kell genotype, necessary for analyzing the possible risk to the fetus in Kell alloimmunization, may be performed either by flow cytometry or by DNA amplification of fetal tissue obtained by chorionic villus sampling or from amniocytes. Antenatal testing is not indicated for ABO HDN.

Amniotic Fluid Spectrophotometry

Elevations of optical density at 450 nm (ΔOD_{450}) on spectrophotometric analysis of amniotic fluid reflect the concentration of amniotic fluid bilirubin derived from fetal tracheal and pulmonary secretions. The change in optical density is quantified by measuring the elevation of the optical density at 450 nm above a line connecting the optical density values obtained at 375 and 550 nm, and then plotted against gestational age. Contamination of amniotic fluid samples with blood or meconium make ΔOD_{450} readings impossible. Liley, in 1965, defined three zones, with readings in zone one, the lowest zone indicating mild or no hemolytic disease with a 10% risk of needing a postnatal exchange transfusion, zone two indicating moderate disease and zone three, the upper zone, indicating severe fetal disease with hydrops or impending fetal death. Serial determinations of ΔOD_{450} can achieve a sensitivity of 95% in detecting the severity of fetal anemia in the third trimester of pregnancy but are unreliable during the second trimester. Modifications of the Liley zones before 27 weeks gestation may help to determine if fetal blood sampling is indicated for definitive diagnosis and treatment (Queenan et al. 1993).

Ultrasonography

Ultrasonography is noninvasive, can be performed serially, and may be combined with other diagnostic studies to assess the fetal condition, estimate the need for further aggressive management, and obtain a biophysical profile of the fetus to determine fetal well-being. Hepatosplenomegaly, ascites, edema, or frank hydrops can be detected. In the absence of hydrops, ultrasonographical parameters such as intra- and extra-hepatic vein diameters, abdominal and head circumference, head/abdominal circumference ratio, and intraperitoneal volume have been unreliable in distinguishing mild from severe fetal anemia. Doppler monitoring of flow velocity indices in the fetal middle cerebral artery is more accurate in predicting fetal

anemia up to 35 weeks gestation and is replacing ΔOD_{450} determinations for fetal surveillance (Mari et al. 2002).

Percutaneous Umbilical Blood Sampling

Percutaneous umbilical blood sampling (PUBS) allows direct measurement of blood indices to evaluate the severity of fetal hemolytic disease as early as 17 to 18 weeks gestation. Indications for PUBS in alloimmunized pregnancies include fetal Rh typing, ΔOD_{450} measurement in Liley zone three or rising through zone two, when an anterior placenta precludes amniocentesis in a fetus where maternal history or antibody titers indicate high fetal risk or ultasonographical evidence of severe anemia or early or frank hydrops (Moise 2002). High-resolution, real-time ultrasound scanning with a linear or curvilinear transducer is used to locate the umbilical cord insertion site on the placenta, and the umbilical vein (umbilical artery occasionally) is punctured 1 to 2 cm from the placental insertion using a 20- to 25-gauge 10- to 16-cm spinal needle. Temporary fetal muscle relaxation may be induced to prevent displacement of the needle by intravenous or intramuscular pancuronium or vecuronium. Specimens of fetal blood are obtained for direct measurement of complete blood count, reticulocyte count, red cell antigen phenotyping, direct antiglobulin test, bilirubin, blood gases, and lactate to assess acid-base status. To exclude maternal blood contamination, fetal blood should be examined using a number of fetal-specific markers, such as red cell size, hemoglobin F, and/or expression of the "i" red cell antigen (Forestier et al. 1988).

For a woman with a previous alloimmunized pregnancy, umbilical blood sampling with transfusion should be timed 10 weeks before the time of the earliest previous fetal or neonatal death, fetal transfusion, or birth of a severely affected baby, but not before 18 weeks gestation unless hydrops is evident. Defined management protocols can reduce the need for multiple invasive procedures while providing specific information about fetal status. Fetal blood samples with reticulocyte counts >97.5 percentile for gestation, a strongly positive direct antiglobulin test, or anemia predict fetuses at high risk of having significant antenatal anemia, thus requiring frequent ultasonographical monitoring and repeated cordocentesis at 1- to 2-week intervals to determine if intrauterine transfusion is warranted. Complications of fetal blood sampling include fetal loss with procedure-related rates ranging from 0% to 4.9%, umbilical cord bleeding, fetal bradycardia, chorioamnionitis, and a significant risk of fetomaternal hemorrhage with anamnestic maternal sensitization (Ghidini et al. 1993).

Neonatal Testing

A sample of cord blood should be collected at the time of delivery from all newborns. However, specific testing of cord blood samples is performed only if the mother is Rh-negative, or when the maternal serum contains red cell alloantibodies of potential clinical significance, or if the neonate develops signs of hemolytic disease. Tests should include ABO and Rh typing and a direct antiglobulin test (DAT). Occasionally, high titers of maternal antibody may block Rh antigenic sites on the neonatal red cells, leading to false-negative Rh typing.

Although the DAT is usually positive in all forms of alloimmune HDN, it cannot reliably predict the degree of clinical severity. This is especially true for cases due to ABO sensitization. The cord blood DAT may be only weakly positive in ABO disease, because the A and B antigenic sites are weak on neonatal RBC membranes and very little anti-A or anti-B antibody is attached to the cells. Gel tests are more sensitive than formerly used procedures. In ABO incompatible infants who are DAT-negative, other causes of hemolysis may be present and should be pursued (Herschel et al. 2002a and b). The DAT may be negative in neonates who have received multiple intrauterine transfusions, since most of their circulating red cells are transfused antigen-negative cells, but the indirect antiglobulin test will be strongly positive. A weakly positive DAT may be present at birth in infants of mothers who have received antepartum RhIG. Contamination of the cord blood sample with Wharton's jelly during collection can also result in a false-positive DAT result. If the DAT is positive and the maternal antibody screen is negative, HDN may be caused by ABO incompatibility, or by antibody directed against a low incidence antigen, not present on reagent cells (AABB, 2002).

Antibody elution tests to determine the specificity of maternal antibody attached to the infants red cells may be useful when several antibodies are present in the maternal serum, or when the maternal antibody screen is negative.

Measurements of cord blood hemoglobin and indirect bilirubin reflect disease severity in Rh HDN. Cord hemoglobin levels of less than 11 g/dL and/or a cord indirect bilirubin levels greater than 4.5 to 5 mg/dL indicate severe HDN, which often necessitates an early exchange transfusion. Early exchange transfusion may also be indicated if the rate of rise of serum bilirubin, measured every 4 to 6 hours, exceeds 0.5 mg/dL per hour.

The reticulocyte count is usually more than 6% and may approach 30% to 40% in severe Rh disease. Low reticulocyte counts, disproportionate to the low hemat-ocrit, may be noted in Kell HDN. Neonates who have received multiple intrauterine transfusions also have very low reticulocyte counts at birth.

The peripheral blood smear in Rh HDN is characterized by increased nucleated RBC counts, polychromasia, and anisocytosis. Severely affected infants may develop thrombocytopenia with platelet counts below 30,000/mL. The peripheral blood smear in ABO HDN is marked by the presence of microspherocytes. Increased osmotic fragility and autohemolysis, similar to hereditary spherocytosis, may be demonstrated in ABO HDN, but unlike hereditary spherocytosis, the autohemolysis in ABO HDN is not corrected by the addition of glucose. These tests are usually not necessary to diagnose ABO HDN.

Hypoglycemia, secondary to hyperinsulinemia, is also seen in severely affected infants. Arterial blood gas analysis may reveal metabolic acidosis and/or respiratory decompensation. Hypoalbuminemia is often present. Cardiomegaly and pleural and pericardial effusions may be evident on radiological investigation. Cardiac hypertrophy with disproportionate septal hypertrophy has been noted by echocardiography in severely affected infants.

Infants who have received intrauterine transfusions may have mild or moderate anemia at birth with relative reticulocytopenia.

T ACTIVATION

T activation is a form of polyagglutination characterized by alteration of the red cell membrane in certain pathological states, resulting in agglutination and hemolysis of such altered cells in the presence of most ABO-compatible adult plasma (Ramasethu and Luban 2001). T activation is reported in 11% to 27% of infants with necrotizing enterocolitis (NEC), a serious gastrointestinal condition afflicting preterm neonates in the intensive care unit.

Pathogenesis

The enzymatic action of bacterial neuraminidases alters red cell membrane structure by removal of N-acetyl neuraminic acid residues from the normally disialylated tetrasaccharides of the MN, Ss, and other RBC membrane sialoglycoproteins. This exposes the normally masked beta-linked galactosyl residue, the T cryptantigen. The unmasked antigen binds to anti-T IgM antibodies normally present in adult sera but absent in neonatal blood, leading to intravascular hemolysis that is sometimes severe and occasionally fatal. It is postulated that in preterm infants with NEC,

T activation is caused by exposure of neonatal red cells to bacterial neuraminidases released locally from devitalized bowel. Although there are reports of anerobic infection, particularly clostridial infection, associated with NEC and T activation, this is not always clearly established. In some patients, bacterial hemolysins and disseminated intravascular coagulation (DIC) may also contribute to hemolysis.

Clinical Features

T activation has been reported mainly in newborn infants with NEC but may also be seen in clinically septic infants with other surgical problems. The clinical signs of NEC are abdominal distension, feeding intolerance, occult or gross blood in the stools, together with systemic signs ranging from temperature instability and lethargy to shock and collapse. Many authors have noted a strong association between T activation and the severity of NEC, particularly with intestinal perforation and gangrene.

T activation should be suspected in neonates at risk who have evidence of intravascular hemolysis with hemoglobinuria and hemoglobinemia following transfusion of blood products, or unexplained failure to achieve the expected posttransfusion hematocrit increment.

Laboratory Diagnosis

There is substantial variation in the management of T activation in centers around the world, with some centers screening all infants with NEC for this condition, others testing only in cases with hemolysis, and others not testing for T activation at all. Polyagglutination will not be detected by routine crossmatching techniques when monoclonal ABO antiserum is used. Discrepancies in forward and reverse blood grouping and evidence of hemolysis on smear may indicate T antigen activation in infants at risk. The diagnosis is confirmed by specific agglutination tests using peanut lectin *Arachis hypogea* and *Glycine soja*.

AUTOIMMUNE HEMOLYTIC DISEASE

IgG antibodies may cross the placenta in women with autoimmune hemolytic disease and occasionally cause hemolytic jaundice in the neonate, similar to alloimmune HDN. Maternal systemic lupus erythematosus is often implicated. Autoantibodies developing de novo in newborn infants and causing hemolysis are extremely rare.

RED CELL ENZYME DEFICIENCIES

Glucose-6-Phosphate Dehydrogenase Deficiency

Glucose-6-phosphate dehydrogenase (G6PD) deficiency is the most common inherited enzyme defect and may be responsible for almost half the morbidity and mortality associated with neonatal jaundice globally (Kaplan and Hammerman 2002). Although G6PD deficiency is most prevalent in Africa, Southeast Asia, and the Mediterranean region, immigration making this condition universal. There are numerous variants of the enzyme, but only few (A-, Mediterranean) are associated with acute hemolysis and neonatal jaundice.

Genetics

The G6PD gene has been localized to Xq28. G6PD deficiency is an X-linked disorder predominantly affecting male children. Heterozygotic females may have a normal or a deficient enzyme in a cell, since only one X chromosome is active in any given cell (Lyon hypothesis), and hence, the overall G6PD activity may be normal or reduced to a greater or lesser extent, depending on the degree of lyonization.

Pathogenesis

The major function of G6PD is to protect the cell from oxidative damage. G6PD catalyses the first step of the hexose monophosphate shunt (HMP) that converts nicotinamide adenine dinucleotide phosphate (NADP) to its reduced form NADPH, which is necessary for the regeneration of reduced glutathione, which has a critical role in neutralizing oxidizing agents. G6PD is present in all cells in the body, but red cells are the most vulnerable to oxidative damage because the HMP shunt is their only source of NADPH. Oxidation of sulfhydryl groups of hemoglobin produces intracellular precipitates called Heinz bodies, which attach to and damage the red cell membrane, making it more susceptible to opsonization by the reticuloendothelial system. In addition, G6PD-deficient red cells undergo peroxidation of membrane phospholipids and oxidative crosslinking of spectrin, decreasing RBC deformability, resulting in splenic trapping and destruction.

Clinical Features

There is a higher incidence of neonatal hyperbilirubinemia in populations with G6PD deficiency than in populations with normal G6PD levels. There are two common clinical presentations of G6PD deficiency in neonates, a severe acute hemolytic anemia accompanied by severe jaundice, or a more gradual onset of significant jaundice without severe anemia. The acute hemolytic anemia syndrome has been associated with hydrops fetalis or severe neonatal anemia. An oxidant stress such as maternal ingestion of fava beans, ascorbic acid, or sulpha drugs may be elicited, but in many cases there appears to be no definite trigger. Serum bilirubin levels may rise rapidly despite phototherapy. In other G6PD-deficient infants, jaundice appears on the second or third day and rises more gradually to peak on the fourth or fifth day, sometimes at extremely high and dangerous levels. Jaundice in these cases may be out of proportion to the anemia, suggesting a hepatic component to the illness.

Laboratory Diagnosis

Erythrocyte morphology is relatively normal. Anisocytosis, reticulocytosis, poikilocytosis, and bite cells may be present. Heinz bodies may be detected by supravital staining but are often cleared rapidly from the circulation. The definitive diagnosis is made by quantitating G6PD activity in red cells using spectrophotometric measurement of the reduction of NADP to NADPH. Semiquantitative methods used in neonatal screening indirectly measure G6PD activity by changes in color or fluoresence resulting from the activity of NADPH. False-negative results may be obtained in heterozygous females or in infants with acute hemolysis in whom older, more G6PD-deficient cells have been destroyed, and younger reticulocytes with adequate enzyme levels are in circulation.

Pyruvate Kinase Deficiency

Pyruvate kinase (PK) deficiency is the most frequent enzyme abnormality of the Embden-Meyerhof pathway, causing hereditary nonspherocytic hemolytic anemia.

Genetics

The gene encoding for PK (PK-LR) has been localized to the long arm of chromosome 1. More than 130 different mutations have been described; 1529A and 1456T are the most common mutations in Caucasians. The condition has a worldwide geographical distribution, most commonly in Northern European populations. PK deficiency is inherited in an autosomal recessive manner; affected individuals are homozygous or compound heterozygotes with two mutant alleles (Zanella and Bianchi 2000).

Pathogenesis

PK catalyses the conversion of phospho*enol*pyruvate to pyruvate with the generation of ATP. The mechanism of hemolysis in PK deficiency has been postulated to be ATP depletion leading to potassium and water loss from erythrocytes, making them less deformable, with resulting early destruction in the reticuloendothelial system.

Clinical Features

The degree of hemolysis varies widely, ranging from barely detectable to life-threatening fetal and neonatal anemia and jaundice. Hydrops fetalis has been reported. Slight to moderate splenomegaly may be noted. Early onset of symptoms often predicts a severe course, with those requiring exchange transfusions in the neonatal period often needing chronic transfusion therapy and ultimately splenectomy.

Laboratory Diagnosis

Anemia, reticulocytosis, and unconjugated jaundice are evident. The red cell morphology is characterized by normochromic, normocytic erythrocytes, with some poikilocytosis, anisocytosis, and a variable proportion of small densely staining spiculated cells called echinocytes. Definitive diagnosis is by quantitative enzymatic assay or by identification of the gene mutation.

Other Red Cell Enzyme Deficiencies

Deficiencies of hexokinase, glucose phosphate isomerase, triose phosphate isomerase, phosphoglycerate kinase, and other enzymes in red cells have been associated with varying degrees of hemolytic aemia in the newborn. These deficiencies are uncommon with few case reports.

RED CELL MEMBRANE DEFECTS

Hereditary Spherocytosis

Hereditary spherocytosis (HS) is the most common cause of nonimmune hemolytic anemia in people of Northern European ancestry, with a prevalence of one in 2000.

Genetics

Inheritance is autosomal dominant in 75% of cases; 25% are sporadic, and about half of these are thought to be caused by a recessive form and the rest by spontaneous mutations.

Pathogenesis

Family-specific mutations lead to combined spectrin and ankyrin deficiency in almost half the patients; 30% have isolated spectrin deficiency, and 20% have band 3 deficiency, with a small number of patients exhibiting protein 4.2 deficiency. These protein defects lead to defective vertical interactions between the red cell protein complex membrane and lipid bilayer, leading to increased fragility of the red cell membrane, microvesiculation, loss of membrane surface area, spherocytosis, and trapping and destruction of red cells in the reticulendothelial system (Tse and Lux 1999). The clinical severity is determined not only by the underlying defect, but also by evolving splenic function and compensatory increase in erythropoiesis, which compensates for the ongoing hemolysis. It has been proposed that anemia is aggravated soon after birth because of increased splenic microvascular filtration combined with the expected decrease in erythropoiesis that occurs at birth.

Clinical Features

Most newborn infants with HS have normal hemoglobin levels at birth, which then decreases sharply over the next 20 days, leading to transient and sometimes severe anemia (Delhommeau et al. 2000). Pallor, dyspnea, or both may be seen. Significant hyperbilirubinemia develops in most infants. Splenomegaly may be noted within the first 6 weeks. Very severe cases, including hydrops fetalis, have been reported. A family history of anemia, jaundice, splenectomy, or gall stones may be present.

Laboratory Diagnosis

Spherocytes lacking central pallor are pathognomonic for HS, but the numbers of spherocytes may vary. Increased osmotic fragility may be detected in most, but not all patients. Incubation at 37°C for 24 hours intensifies osmotic fragility, and the incubated osmotic fragility test is the procedure of choice in neonates. The autohemolysis test and acidified glycerol lysis test are unreliable in the newborn. Structural and functional studies of red cell membrane proteins or DNA analysis may be performed, if necessary.

Hereditary Elliptocytosis/Ovalocytosis/ Pyropoikilocytosis

Hereditary elliptocytosis (HE) is common in African-Americans and in persons of Mediterranean ancestry. HE and hereditary pyropoikilocytosis (HPP) are caused by mutations, which lead to failure of spectrin heterodimers to self associate into heterotetramers, the basic building blocks of the RBC membrane skeleton network. Defects in protein 4.1 and glycophorin C have also been described (Tse and Lux 1999).

The clinical manifestations of HE vary from asymptomatic carrier status to severe hemolytic disease, with occasional reports of hydrops fetalis. A variable number of cigar-shaped elliptocytes may be seen in the peripheral circulation but are not often seen in the neonatal period. Southeast Asian ovalocytosis is a unique form of HE, with similar clinical presentation, and 20% to 50% of characteristic oval-shaped cells in circulation. HPP is rare but strongly associated with HE. Parents and siblings of children with HPP are found to have HE, and HPP appears to evolve into HE in many children later in life. Erythrocyte morphology in HPP is characterized by microspherocytosis, pyknocytes, elliptocytes, fragmented red cells.

DISORDERS OF HEMOGLOBIN SYNTHESIS

Alpha Thalassemia

The alpha thalassemia (αTH) syndromes are the most common inherited disorders of hemoglobin synthesis in Southeast Asia and are common in regions of the United States with large Southeast Asian populations.

Genetics

αTH results from deletions or point mutations (or both) in the one or more of the four α globin genes, normally present in pairs on chromosome 16. Individuals with three intact alleles are clinically silent carriers, those with two normal alleles have the αTH trait, whereas those with one normal allele develop HbH disease. Patients with deletion of all four alleles have homozygous αTH also known as Bart's hemoglobinopathy. When a couple are both heterozygous for the αTH gene deletion, each pregnancy carries a 25% risk of homozygous αTH.

Pathogenesis

Early in gestation, the production of embryonic hemoglobins Gower and Portland protects the fetus, but later, the absence of α-globin expression in homozygous αTH prevents the synthesis of hemoglobin F ($\alpha2,\gamma2$) or hemoglobin A ($\alpha2\beta2$). Unstable Bart's hemoglobin ($\gamma4$ tetramers) produced instead has a very high oxygen affinity and is ineffective in oxygen transport, resulting in tissue hypoxia, edema, congestive heart failure, and hydrops fetalis. Destruction of abnormal red cells in the marrow and peripheral circulation contributes to the anemia.

Deletion of three α genes leads to HbH disease, which is characterized by tetramers of β chains. HbH precipitates in red cells and is cleared from the circulation, leading to microcytic hemolytic anemia in the neonatal period. The level of Hb Barts in such patients may vary from 15% to 25%. When the switch from γ to β chains occurs at about 3 months of age, Hb Barts disappears and is replaced by HbH.

Clinical Features

Hydropic changes may occur as early as 12 weeks of pregnancy but are usually detected at approximately 20 weeks in fetuses with homozygous αTH. The affected fetus often dies in utero between 23 and 38 weeks or soon after birth, despite resuscitative efforts. Homozygous αTH was thought to be uniformly lethal in the past, but advances in antenatal diagnosis, intrauterine transfusions, and postnatal interventions in recent years have led to a few survivors (Singer et al. 2000). Congenital urogenital and limb defects have been associated with homozygous αTH.

Infants with hemoglobin H disease may present with significant hemolytic anemia at birth, but infants with αTH trait are usually asymptomatic in the neonatal period.

Laboratory Diagnosis

The hemoglobin level ranges between 3 to 10 g/dL. The peripheral blood smear shows marked hypochromia and microcytosis. DNA analysis can confirm deletion of α globin genes. Hemoglobin electrophoresis reveals very high levels of hemoglobin Barts (usually 80% to 90%) with varying levels of hemoglobin Portland in homozygous αTH. Neonates with HbH disease have 25% or more of hemoglobin Barts.

OTHER CAUSES OF HEMOLYTIC DISEASE IN NEONATES

Infections in the newborn may cause significant hemolysis and anemia by a number of mechanisms. Congenital malaria causes hemolysis by direct invasion of RBCs and presents as anemia, fever, and hepatosplenomegaly in the newborn. Severe intravascular hemolysis has been decribed with clostridial infections, secondary to the release of hemolysins.

Bacterial neuraminidases may cause T activation and immune-mediated hemolysis (see section on T activation). Hemolytic anemia may be seen in newborn infants with congenital syphilis, cytomegalovirus infection, rubella, and toxoplasmosis. Microangiopathic hemolytic anemia may be associated with DIC, a serious complication of sepsis in neonates. Microangiopathic hemolysis may also be seen in neonates with hemangiomatous malformations. Kasabach-Merritt syndrome is the association of giant hemangiomas, often located in the gastrointestinal system, with DIC and microangiopathic hemolytic anemia.

Vitamin E deficiency was a common cause of hemolytic anemia in premature infants, because of iron-induced lipid peroxidation of red cell membranes, but this complication is rarely seen now with adequate nutritional support. However, infants with cystic fibrosis continue to be vulnerable to vitamin E deficiency, and there are numerous case reports of hemolytic anemia associated with vitamin E deficiency in these infants.

THERAPY

Therapy in HDN is directed at correcting anemia in the fetus and/or the newborn and potentially neurotoxic hyperbilirubinemia in the newborn infant. Communication between the mother's obstetrician, the neonatologist, and the blood bank is crucial in the care of the infant with severe HDN since fetal interventions, early delivery, and intensive care for the newborn may be necessary. Results of antenatal monitoring and obstetrical interventions during pregnancy, together with the history of the outcome of previous pregnancies allow the neonatal team to anticipate the needs of the infant born with HDN. The management of hydrops fetalis is particularly challenging.

Intrauterine Fetal Transfusion

Intrauterine fetal transfusions initially used to treat severe fetal anemia associated with Rh HDN are now also used to correct severe fetal anemia from other causes, such as Kell HDN, or alpha thalassemia (Liley 1965; Schumacher and Moise 1996; Carr et al. 1995). Intraperitoneal fetal transfusion was first introduced by Liley in 1963. Absorption of red cells from the peritoneal cavity occurs through lymphatic channels, but this may be ineffective in hydropic fetuses. Direct intravascular fetal transfusion by funipuncture is the procedure of choice at present, but intraperitoneal transfusions may still be necessary when intravascular access is difficult, as in early pregnancy when the umbilical vessels are narrow, or later when increased fetal size prevents access to the umbilical cord. Other techniques of fetal transfusion reported include intrahepatic venous puncture, combinations of intravascular with intraperitoneal transfusions, and even intracardiac transfusion as a last resort.

The intravascular technique also offers precise diagnostic evaluation of the fetal status (see section on percutaneous umbilical blood sampling). The first umbilical blood sampling with transfusion ideally should be performed when the fetus is anemic, but before hydrops has developed. Transfusions are performed at hematocrits of 25% to 30% or less. Generally, the hematocrit drops by 1% to 2% per day in the transfused hydropic fetus; the fall in hematocrit is rapid in fetuses with severe hemolytic disease, necessitating a second transfusion within 7 to 14 days; the interval between subsequent transfusions is usually 21 to 28 days. Very low pretransfusion fetal hematocrits, rapid large increases in posttransfusion hematocrits, and increases in umbilical venous pressure during intravascular transfusion are associated with fetal death posttransfusion. Blood may be transfused either as a direct simple intravascular transfusion or as an intravascular exchange transfusion, but most centers find the shorter procedure time and technical ease of direct transfusions expedient.

The red cells selected for intrauterine transfusion should be Group O, D-negative or negative for the antigen corresponding to any identified maternal antibody. The blood should be cytomegalovirus seronegative or leukodepleted, irradiated, and crossmatched against the mother's blood. Some centers use washed maternal blood for repeated intrauterine transfusions, supporting maternal erythropoiesis with iron and folate supplemention. Blood that is washed free of the anticoagulant citrate and other additives is recommended. The blood should be as fresh as possible for maximal in vivo survival, warmed, and packed to a hematocrit of 70% to 85% in a volume calculated to increase the fetal hematocrit to between 40% to 45%, based on estimated fetal placental blood volume, fetal hematocrit, and hematocrit of donor blood. A dramatic increase in hematocrit is to be avoided in the extremely anemic fetus, since this may be associated with changes in viscosity and cardiovascular decompensation. In such cases, two smaller transfusions, 48 hours apart, may allow better fetal adaptation. The volume of red cells to be transfused to achieve a hematocrit increment of 10% may be roughly estimated by multiplying estimated fetal weight in grams (determined by ultrasound evaluation) by a factor of 0.02 (Moise 2002). Other formulas for determining the volume have been published (AABB 2002).

The decision about the ideal time to deliver the fetus is based on gestational age, fetal weight and lung maturity, fetal response to the transfusions, and the ease of performing the transfusion combined with the antenatal ultrasound and Doppler studies. Intrauterine transfusions are provided up to 33 to 34 weeks, with delivery as soon as lung maturity is achieved by antenatal steroid therapy. Less severely affected fetuses may be allowed to proceed to term before delivery.

Infants who have received multiple intrauterine transfusions are delivered closer to term and usually need fewer exchange transfusions in the neonatal period. However, some still have significant hemolytic anemia at birth requiring exchange transfusions and phototherapy, and many require additional simple transfusions for severe and prolonged hyporegenerative anemia secondary to suppression of fetal erythropiesis (Janssens et al. 1997). Perinatal survival rates of over 90% have been achieved with intrauterine transfusions in nonhydropic fetuses with severe Rh hemolytic disease. The survival rate for hydropic fetuses is lower at 74%, despite intrauterine transfusions.

Hydrops Fetalis

In infants with hydrops, severe anemia is frequently accompanied by perinatal asphyxia, surfactant deficiency, hypoglycemia, acidosis, and thrombocytopenia. The resuscitation and stabilization of hydropic infants is complicated and often difficult. Prompt endotracheal intubation, positive-pressure ventilation with oxygen, and surfactant therapy at birth is usually essential. Drainage of pleural effusions and ascites may be required to facilitate gas exchange. Inotropic support is frequently necessary. Metabolic acidosis and hypoglycemia require correction. A partial exchange transfusion using packed cells may be performed to improve hemoglobin levels and oxygenation; a double-volume exchange transfusion is considered only after the initial stabilization.

Exchange Transfusion

Indications

Exchange transfusions correct severe anemia and decrease serum levels of potentially neurotoxic unconjugated bilirubin. In addition to correcting anemia and jaundice, in alloimmune HDN, exchange transfusions remove free maternal antibody in the plasma and replace a large proportion of sensitized RBCs with antigen-negative RBCs that should have normal in vivo survival. In infants with severe hyperbilirubinemia secondary to T activation-induced hemolysis, exchange transfusion has been recommended to decrease bilirubin levels and prevent further hemolysis by replacing T-activated RBCs with donor cells not expressing the cryptantigen.

"Early" exchange transfusions are performed within 9 to 12 hours of birth in infants with severe HDN. Cord hemoglobin levels ≤10 g/L, cord bilirubin levels ≥5.5 mg/dL, and rapidly rising bilirubin levels ≥0.5 mg/dL/h despite phototherapy are commonly used criteria for early exchange transfusions. "Late" exchange transfusions are performed when serum bilirubin levels threaten to exceed 20 mg/dL in term infants with severe HDN, the level at which the risk of kernicterus is approximately 10%. Exchange transfusions are performed at lower bilirubin levels in premature infants, particularly those with hypoxemia, acidosis, and hypothermia.

A partial exchange transfusion is often performed in hydropic infants or those with severe anemia associated with cardiac failure or volume overload. The volume of blood required for the exchange transfusion may be calculated by the following formula:

$$\frac{\text{Infants blood volume (mL)} \times (\text{Desired HCT} - \text{Observed HCT})}{(\text{HCT of donor blood} - \text{Observed HCT})}$$

A double volume exchange transfusion

(calculated as 2 × 80 – 85 mL/kg in a term newborn, and 2 × 90 – 100 mL/kg in preterm newborn)

replaces approximately 85% to 90% of the infant's blood volume. A single-volume exchange transfusion replaces approximately 60% of the infant's blood volume. There is no advantage to performing an exchange transfusion exceeding two blood volumes, and similarly there is no major disadvantage in exchanging slightly less than two blood volumes (an issue that sometimes arises when arranging blood for exchange transfusions in large term neonates), since it may avoid the addition of another unit of blood and consequently an extra donor exposure.

Although a double volume exchange transfusion replaces 85% to 90% of the blood volume, when the exchange transfusion is performed for severe hyperbilirubinemia, the drop in serum bilirubin is often less than 50%, because of equilibration from a tissue-bound pool. The use of albumin before exchange transfusion in an effort to mobilize tissue bilirubin is controversial. Equilibration of extravascular and intravascular bilirubin and continued breakdown of sensitized and newly formed red cells by persisting maternal antibodies results in a rebound of bilirubin following initial exchange transfusion, often necessitating repeated exchange transfusions in severe hemolytic disease.

Selection of Blood for Exchange Transfusion

Group O Rh-negative blood is typically used for exchange transfusions. Blood chosen for the exchange should be ABO compatible and negative for the antigen responsible for the hemolytic disease in case of alloimmune HDN (AABB 2002). Maternal serum/plasma is preferred for crossmatching blood for exchange transfusions in alloimmune HDN because the larger volume available for testing permits accurate determination of the antibody, and it facilitates the crossmatching of blood for potential exchange transfusion *before* the birth of a severely affected fetus. Use of maternal serum/plasma for crossmatching may be perplexing if the mother has antibodies directed against antigens not present on the infants red cells or when there are IgM antibodies. If maternal blood is unavailable for crossmatching, the infant's serum/plasma or an eluate from the infant's red cells may be used instead. In ABO disease, the red cells for exchange transfusion must be Group O, but may be Rh-positive if the infant is Rh-positive. Group O red cells resuspended in AB plasma are often used but this results in exposure to two donors. Group O blood with low anti-A or anti-B titers may be used instead. In Rh HDN, the red cells must be Rh-negative but may be either Group O or group specific. Donor blood should be screened for hemoglobin S and G6PD deficiency in populations endemic for these conditions.

Irradiated blood is recommended for all exchange transfusions to prevent graft versus host disease, particularly if the neonate has received intrauterine transfusions. The blood should be as fresh as possible (<7 days) to maximize the in vivo survival of the transfused red cells. Additive solution anticoagulants are avoided. Citrate-phosphate-dextrose (CPD or CPDAI) blood is used as whole blood or reconstituted whole blood (red cells suspended in saline, albumin, or plasma). Frozen deglycerolized red cells may be reconstituted with FFP for rare blood types. Heparinized fresh whole blood supplies platelets but is not readily available (Petaja et al. 2000).

The hematocrit of the donor blood is adjusted within the range of 45% to 60% percent, depending on the desired end result, with the higher ranges being used for partial exchange transfusions for correction of severe anemia in hydropic infants.

Procedure of Exchange Transfusion

The newborn infant should be stabilized adequately before the exchange transfusion, with appropriate correction of respiratory or metabolic complications. Continuous cardiorespiratory monitoring during the procedure is essential.

Access for the exchange transfusion is usually through umbilical arterial and venous lines, although peripheral lines may be used if central access cannot be obtained. Blood is warmed through a temperature-controlled in-line blood warmer, and the exchange transfusion performed by either a traditional push-pull method with a single vascular access, usually the umbilical vein, or by an isovolumetric method with simultaneous infusion of donor blood through a venous line and removal of the infant's blood though a central or peripheral arterial line (Ramasethu 2002). The isovolumetric technique may be better tolerated in small sick preterm neonates because there is less fluctuation of blood pressure and cerebral hemodynamics. This technique is also preferable when only peripheral access is available.

In the push-pull method, aliquots of 5 to 20 mL with a maximum of 5 mL/kg are withdrawn and infused in turn. The stages of withdrawing and infusing blood from and into the infant should be done slowly, taking at least a minute each, to avoid rapid fluctuations in arterial pressure, which are accompanied by changes in intracranial pressure. Rapid withdrawal of blood from the umbilical vein induces a negative pressure, which may be transmitted to the mesenteric veins and contribute to ischemic bowel complications. When an isovolumetric exchange is being done, volumes to be removed/infused should not exceed 2 mL/kg/min.

The duration of the exchange is usually one to two hours. The blood bag should be agitated gently every 10 to 15 minutes during the procedure to prevent red cell sedimentation, which may lead to exchange with blood with a lower hematocrit towards the end of the transfusion. Calcium supplementation during the exchange transfusion is recommended only if there is documented hypocalcemia or there are changes in the Q-Tc interval. A dose of 1 mL/kg body weight of 10% calcium gluconate is recommended. Care should be taken to flush the umbilical venous line with normal saline if calcium gluconate is administered through the line, since it will reverse the effect of anticoagulant citrate in the donor blood and may cause clotting of the line.

Following the exchange transfusion, vital signs should be closely monitored for at least 4 to 6 hours. Serum ionized calcium and platelet levels should be monitored closely in preterm or sick neonates. Serum glucose levels should be monitored every 2 to 4 hours for 24 hours. Intravenous drug dosages may need to be adjusted to compensate for removal by the exchange.

Complications of Exchange Transfusion

Potential complications of exchange transfusion are enumerated in Table 18.3. Complications may be related

TABLE 18.3 Potential Complications of Exchange Transfusions in the Newborn

Cardiorespiratory

Apnea
Bradycardia
Cardiac arrhythmias
Hypotension
Hypertension

Metabolic

Hypoglycemia
Hyperglycemia
Hypocalcemia
Hyperkalemia

Hematologic

Dilutional coagulopathy
Thrombocytopenia
Neutropenia
Disseminated intravascular coagulation
Graft-versus-host disease

Vascular Catheter-Related Complications

Vasospasm
Thrombosis
Embolization
Malposition of catheter
Perforation of blood vessel

Gastrointestinal

Feeding intolerance
Necrotizing enterocolitis
Portal hypertension
Hepatic infarction/necrosis

Infection

Blood related infections—hepatitis B, C; HIV; CMV
Omphalitis
Septicemia

to the blood as well as to complications of vascular access. The most common adverse effects noted during or soon after exchange transfusions, and usually seen in infants who are preterm and/or sick are apnea, bradycardia, hypocalcemia, thrombocytopenia, and vascular spasm. Rebound hypoglycemia may occur after the exchange in infants with severe Rh HDN who often have hyperinsulinemia. The risk of death or permanant serious sequelae has been estimated less than 1% in healthy infants, but as high as 12% in sick infants, although there may be difficulty in ascribing adverse events to this particular procedure in infants who are already critically ill (Jackson 1997).

Other Transfusion Support

Platelet transfusions may be required after exchange transfusions in infants with severe HDN.

Packed red blood cell (PRBC) transfusions may be required to correct anemia in the first few weeks of life in infants with significant hemolysis. Most infants who have received intrauterine transfusions for alloimmune HDN require postnatal transfusions in early infancy, because of persistent hemolysis and suppression of erythropoiesis (Janssens et al. 1997). Care should be taken in choosing appropriate blood even for these transfusions in alloimmune HDN because of persisting maternal antibodies in the neonatal circulation.

In a series of 34 infants with hereditary spherocytosis, 26 (76%) required transfusions during the first year of life, with 24 of the 26 needing the first transfusion during the initial 2 months of life, and 14 needing multiple transfusions in early infancy (Delhommeau et al. 2000). Transfusion requirements decreased as compensatory erythropoiesis improved, but six of the infants remained transfusion dependent beyond 2 years of age and underwent subtotal splenectomy.

Special transfusion protocols have been recommended to prevent further hemolysis in infants with T activation (Ramasethu and Luban 2001). These include the use of washed or plasma-reduced RBCs and platelets and the use of low anti-T titer plasma if fresh frozen plasma is essential, although there is no clear definition of what constitutes a "low titer." Low titer has been defined as not greater than 1 in saline, agglutination titers less than 8, or an optical density value of 0.300 at 405 nm using a hemolysis test. Donor plasma units may be selected by using a minor crossmatch.

Phototherapy

Phototherapy decreases bilirubin levels and the risk of bilirubin toxicity by converting lipid-soluble unconjugated bilirubin to less toxic water-soluble structural and configurational isomers, which are easily excreted, without hepatic conjugation. Bilirubin adsorbs visible light in the blue wavelength range of 400 to 500 nm, with peak absorption between 450 to 460 nm. The effectiveness of phototherapy may be influenced by the wavelength and irradiance of light, the surface area of exposed skin, and the duration of exposure. A number of devices are available to deliver phototherapy (Bagchi 2002).

Intensive phototherapy has been found to effectively reduce bilirubin levels and decrease the need for exchange transfusions for hyperbilirubinemia in ABO and Rh HDN. The early institution of phototherapy in all infants with hemolytic disease may result in the unnecessary treatment of large numbers of infants with mild hemolytic disease whose bilirubin levels would not have risen to nonphysiological levels even without treatment. The American Academy of Pediatrics guidelines (1994) for management of hyperbilirubinemia in healthy term newborns *do not* apply to neonates with hemolytic disease who need close monitoring and initiation of phototherapy at lower bilirubin levels. Early and intensive phototherapy should be instituted in infants with moderate or severe hemolysis or in infants with rapidly rising bilirubin levels (>0.5 mg/dL/h). Phototherapy is indicated at lower levels for preterm or sick infants.

Erythropoietin

Recombinant human erythropoietin has been shown to decrease the need for postnatal transfusions in infants with late hyporegenerative anemia of Rh hemolytic disease, the hyporegenerative anemia noted in neonates with Kell HDN (Dhodapkar and Blei 2001), and even in infants with anemia due to hereditary spherocytosis (Tchernia et al. 2000).

Other Treatment

Intravenous immunoglobulin has been shown in a meta-analysis to reduce the need for exchange transfusions in term and preterm infants with Rh and ABO hemolytic disease (Alcock and Liley 2002). Selection of patients for intravenous immunoglobulin (IVIG) therapy, protocols for administration, and criteria for institution of exchange transfusions have varied among the different studies limiting generalizability of the results.

Sn-protoporphyrin, a potent heme oxygenase inhibitor, has been shown to blunt the postnatal rise and peak bilirubin levels in term newborns with ABO hemolytic disease and in G6PD deficiency. Further documentation of safety and effectiveness of this form of treatment is required, particularly since the relatively benign option of phototherapy for the treatment of neonatal hyperbilirubinemia is widely available.

In infants with alloimmune hemolytic disease, active hemolysis ceases with the disappearance of maternal antibody from the circulation, usually by 1 to 3 months of age, even in the most severely affected infants. Infants with hereditary spherocytosis may need occasionally prolonged transfusion support and ultimately require splenectomy to prevent further hemolysis. Homozygous alpha thalassemia was thought to be lethal and incompatible with extrauterine life in the past, but intrauterine transfusions have improved survival in some of of these infants. The survivors of such interventions continue to need transfusion support, chelation therapy, and eventual bone marrow transplantation (Singer et al. 2000).

Prevention

Transfusion of blood compatible with not only the D-antigen but also Kell and other Rh antigens has been advocated for premenopausal women to prevent alloimmunization.

The use of Rh immunoglobulin (RhIg) has dramatically decreased the incidence of Rh HDN. RhIg is derived from plasma pools and is produced by immunizing male plasma donors with repeated injections of RhD-positive red cells. RhIg is postulated to prevent sensitization to the D-antigen by attaching to D-antigen sites on Rh-positive RBCs in the circulation and interfering with the host's primary immune response to the foreign antigen. RhIg may also inhibit antigen induced B-cell responsiveness by stimulating an increase in suppressor T cells.

The postpartum administration of RhIg to all unsensitized Rh-negative women who deliver an Rh-positive infant decreases the incidence of Rh isoimmunization from 12% to 13% to approximately 2%. However about 1.8% of Rh-negative women are apparently sensitized during pregnancy from small asymptomatic transplacental hemorrhages. Further reduction in the incidence of Rh-isoimmunization to 0.1% has been achieved by additional antepartum RhIg prophylaxis at 28 to 30 weeks gestation. Despite appropriate Rh prophylaxis, about 0.1% of Rh-negative women may be sensitized before 28 weeks gestation. The recommended dose of RhIg should be administered within 72 hours of delivery of an Rh-positive fetus. If RhIg is accidentally omitted, some protection may still be obtained with administration up to 13 days and possibly up to 28 days after delivery. Once alloimmunization to RhD antigen has occurred, RhIG is ineffective.

The standard dose in the United States, 300 μg RhIg (1500 IU), affords protection against 15 mL of Rh-positive RBCs or 30 mL of Rh-positive whole blood (Hartwell 1998). A smaller dose of 50 μg of RhIg is effective until 12 weeks gestation because of the small volume of red cells in the fetoplacental circulation. Recommendations for the routine prophylactic dose vary around the world. Kleihauer-Betke test (to determine the presence of fetal cells in the maternal circulation) is recommended as routine in the postpartum period, and antenatally if clinical circumstances suggest the possibility of excessive fetomaternal hemorrhage, to determine if additional doses of RhIg are indicated. If testing indicates the need for a large dose of RhIG (more than five units), the entire calculated dose may be given intra-venously using a new FDA-approved form of RhIG (WinRhoSDF, Cangene Corp., Winnipeg, Manitoba, Canada).

Prophylaxis similar to RhIg does not yet exist for alloimmunization to antigens other than D. Predischarge bilirubin screening may help to predict which neonates may be at risk of developing hyperbilirubinemia, even when no blood group incompatibility is identified. In addition, screening programs help identify infants with G6PD deficiency, so that medications and substances that may lead to serious hemolysis may be avoided.

References

American Academy of Pediatrics. 1994. Provisional Committee for Quality Improvement and subcommittee on Hyperbilirubinemia. Practice Parameter. Management of hyperbilirubinemia in the healthy term newborn. *Pediatrics* 94:558–564.

American Association of Blood Banks. 2002. Technical manual, 14th ed. Brecher ME, ed. Bethesda, MD: AABB Press.

Alcock GS and Liley H. 2002. Immunoglobulin infusion for isoimmune haemolytic jaundice in neonates. *Cochrane Database Syst Rev* 3:CD 003313.

Bagchi A. 2002. Phototherapy. In *Atlas of procedures in neonatology*. MG MacDonald, J Ramasethu, eds. Philadelphia, PA: Lippincott, Williams & Wilkins.

Bowman JM, Pollack JM, Manning FA, Harman CR, and Menticoglou S. 1992. Maternal Kell blood group alloimmunization. *Obstet Gynecol* 79:239–244.

Bowman JM, Pollack JM, and Penston LE. 1986. Fetomaternal transplacental hemorrhage during pregnancy and after delivery. *Vox Sang* 51:117–121.

Bromilow IM and Duguid JKM. 1997. Measurement of fetomaternal haemorrhage: a comparative study of three Kleihauer techniques and two flow cytometry methods. *Clin Lab Haem* 19:137–142.

Carr S, Rubin L, Dixon D, Star J, and Dailey J. 1995. Intrauterine therapy for homozygous α-thalassemia. *Obstet Gynecol* 85:876–879.

Delhommeau F, Cynober T, Schischmanoff PO, Rohrlich P, Delaunay J, Mohandas N, and Tchernia G. 2000. Natural history of hereditary spherocytosis during the first year of life. *Blood* 95:393–397.

Dhodapkar KM and Blei F. 2001. Treatment of hemolytic disease of the newborn caused by anti-Kell antibody with recombinant erythropoietin. *J Pediatr Hematol Oncol* 23:69–70.

Eder AF and Manno CS. 2001. Does red cell T activation matter? *Br J Haematol* 114:25–30.

Engelfriet CP and Reesink HW. International Forum. 1995. Laboratory procedures for the prediction of the severity of hemolytic disease of the newborn. *Vox Sang* 69:61–69.

Forestier F, Cox WL, Daffos F, and Rainaut M. 1988. The assessment of fetal blood samples. *Am J Obstet Gynecol* 158:1184–1188.

Ghidini A, Sepulveda W, Lockwood CJ, and Romero R. 1993. Complications of fetal blood sampling. *Am J Obstet Gynecol* 168:1339–1344.

Hartwell EA. 1998. Use of Rh immune globulin. ASCP practice parameter. *Am J Clin Pathol* 110:281–292.

Herschel M, Karrison T, Wen M, Caldarelli L, and Baron B. 2002a. Evaluation of the direct antiglobulin Coombs' test for identifying newborns at risk for hemolysis as determined by end tidal carbon monoxide concentration ETCOc and comparison of the Coombs' test with ETCOc for detecting significant jaundice. *J Perinatol* 22:341–347.

Herschel M, Karrison T, Wen M, Caldarelli L, and Baron B. 2002b. Isoimmunization is unlikely to be the cause of hemolysis in ABO-incompatible but direct antiglobulin test negative neonates. *Pediatrics* 110:127–130.

Jackson JC. 1997. Adverse events associated with exchange transfusion in healthy and ill newborns. *Pediatrics* 99:E7.

Janssens HM, deHaan MJJ, van Kamp IL, Brand R, Kanhai HHH, and Veen S. 1997. Outcome for children treated with fetal intravascular transfusions because of severe blood group antagonism. *J Pediatr* 131:373–380.

Kaplan M and Hammerman C. 2002. Glucose-6-phosphate dehydrogenase deficiency: a potential source of severe neonatal hyperbilirubinemia and kernicterus. *Semin Neonatol* 7:121–128.

Kaplan M, Hammerman C, Renbaum P, et al. 2000. Gilbert's syndrome and hyperbilirubinemia in ABO incompatible neonates. *Lancet* 356:652–653.

Katz MA, Kanto WP, and Korotkin JH. 1982. Recurrence rate of ABO hemolytic disease of the newborn. *Obstet Gynecol* 59:611–614.

Liley AW. 1965. The use of amniocentesis and fetal transfusion in erythroblastosis fetalis. *Pediatrics* 35:836.

McDonnell M, Hannam S, and Devane SP. 1998. Hydrops fetalis due to ABO incompatibility. *Arch Dis Child Fetal Neonatal Ed* 78:F220–221.

Mari G, Detti L, Oz U, Zimmerman R, Deurig P, and Stefos T. 2002. Accurate prediction of fetal hemoglobin by Doppler ultrasonography. *Obstet Gynecol* 99:589–593.

Moise Jr KJ. 2002. Management of rhesus alloimmunization in pregnancy. *Obstet Gynecol* 100:600–611.

Petaja J, Johansson C, Anderson S, and Heikinheimo M. 2000. Neonatal exchange transfusion with heparinized whole blood or citrated composite blood: a prospective study. *Eur J Pediatr* 159:552–553.

Phibbs RH. 1998. Hydrops fetalis and other causes of neonatal edema and ascites. In *Fetal and neonatal physiology*, 2nd ed. Polin RA and Fox WW, eds. Philadelphia, PA:WB Saunders Company.

Queenan JT, Tomai TP, Ural SH, and King JC. 1993. Deviation in amniotic fluid optical density at a wavelength of 450 nm in Rh-immunized pregnancies from 14 to 40 weeks gestation: a proposal for clinical management. *Am J Obstet Gynecol* 168:1370–1376.

Ramasethu J and Luban NLC. 2001. Alloimmune hemolytic disease of the newborn. In *William's hematology*, 6th ed. E Beutler, MA Lichtman, BS Coller, TJ Kipps, U Seligsohn, eds. New York: McGraw-Hill.

Ramasethu J and Luban NLC. 2001. T activation. *Br J Haematol* 112:259–263.

Ramasethu J. 2002. Exchange transfusions. In *Atlas of procedures in neonatology*, 3rd ed. MG MacDonald, J Ramasethu, eds. Philadelphia: Lippincott, Williams & Wilkins.

Schumacher B and Moise Jr KJ. 1996. Fetal transfusion for red blood cell alloimmunization in pregnancy. *Obstet Gynecol* 88:137–150.

Sebring ES and Polesky HF. 1990. Fetomaternal hemorrhage: incidence, risk factors, time of occurance, and clinical effects. *Transfusion* 30:344–357.

Singer ST, Styles L, Bojanowski J, Quirolo K, Foote D, and Vichinsky EP. 2000. Changing outcome of homozygous alpha-thalassemia: cautious optimism. *J Pediatr Hematol Oncol* 22:539–542.

Tchernia G, Delhommeau F, Perrotta S, et al. 2000. Recombinant erythropoietin therapy as an alternative to blood transfusions in infants with hereditary spherocytosis. *Hematol J* 1:146–152.

Tse WT and Lux SE. 1999. Red cell membrane disorders. *Br J Haematol* 104:2–13.

Vaughan JI, Manning M, Warwick RM, Letsky EA, Murray NA, and Roberts IAG. 1998. Inhibition of erythroid progenitor cells by anti-Kell antibodies in fetal alloimmune anemia. *N Engl J Med* 338:798–803.

Zanella A and Bianchi P. 2000. Red cell pyruvate kinase deficiency: from genetics to clinical manifestations. *Baillieres Best Pract Res Clin Haematol* 13:57–81.

19

Hemoglobinopathies

KRISTA L. HILLYER, MD

INTRODUCTION

This chapter outlines the role of and methods for transfusion in children with sickle cell disease (SCD) and thalassemia. For both categories of hemoglobinopathies, the goals of transfusion are reviewed. A thorough listing of appropriate indications for transfusion in children with SCD is provided, and controversial and nonindications for transfusion in children with SCD are also mentioned. Methods of transfusion therapy, including acute simple transfusion, chronic simple transfusion, and both manual and automated red blood cell (RBC) exchange therapy, are explained, and formulae are provided for calculating the appropriate transfusion and exchange RBC volumes. For both SCD and thalassemia patients, adverse effects of transfusion and strategies for their prevention and management are also discussed.

SICKLE CELL DISEASE

Pathogenesis

As 8% of African-Americans are heterozygous carriers of hemoglobin S (HbS), one in every 300 African-Americans has SCD. The homozygous state (HbSS), also known as sickle cell anemia, is the most common type of SCD in the United States; one in 600 African-Americans has HbSS. Compound heterozygote states, such as sickle cell/hemoglobin C (HbSC) disease, together with the combination of HbS and β-thalassemia (sickle cell/β-thalassemia), account for most

of the remaining African-American cases of SCD (Lane 1996; Bunn 1997).

Clinical Pathology

Patients with SCD typically have HbS levels >50% of their total hemoglobin concentrations. This abnormal sickle hemoglobin, when deoxygenated, forms polymers within the erythrocyte that distort its shape and decrease its deformability, leading to vaso-occlusion. The vaso-occlusive events and other factors, resulting from the polymerization of HbS, cause hemolytic anemia in these patients and increased susceptibility to infection, organ damage, and recurrent episodes of pain (Lane 1996; Bunn 1997).

Role of Transfusion in Therapy

Many of the complications of SCD can be limited or prevented by transfusion of normal, donor RBCs. Because of the proven clinical benefits from transfusion, there are many goals, indications, and methods for transfusion in pediatric patients with SCD.

Goals of Transfusion Therapy

The goals of RBC transfusion in SCD patients are (1) to improve oxygen-carrying capacity by increasing the total hemoglobin concentration; (2) to decrease blood viscosity and improve blood flow by diluting RBCs that contain sickle hemoglobin; and (3) to suppress endogenous erythropoiesis by increasing tissue oxygenation.

Indications for Particular Types of Transfusion Therapy

RBCs may be transfused to patients with SCD by simple RBC infusion or by RBC (or partial RBC) exchange (termed "erythrocytapheresis," when performed in an automated manner). These types of transfusion may be performed either *episodically* (for the relief of acute symptoms of SCD), or *chronically* (for the prevention of long-term complications of SCD). See Table 19.1 for a summary of the major clinical indications for RBC transfusion in children with SCD.

Indications for Episodic Transfusion

Acute Symptomatic Anemia

Because children with SCD have chronic anemia, they are often asymptomatic from this anemia, despite having very low hemoglobin levels. However, SCD patients may become acutely symptomatic, making simple transfusion necessary, if they experience a rapid decrease in hemoglobin, hypoxia, or acute cardiac decompensation.

Acute symptomatic anemia can result from bleeding, suppression of erythropoiesis, RBC sequestration, or increased destruction of RBCs. Pulmonary or cardiac disease may cause acute decompensation, requiring acute episodic RBC transfusion to increase the patient's hemoglobin level above his or her stable baseline (Eckman 2001).

Acute Chest Syndrome

According to a report by the Cooperative Study of Sickle Cell Disease (Vichinsky 2001), acute chest syn-

TABLE 19.1   Clinical Indications for RBC Transfusion in Children with SCD

Type of Transfusion	Indication
Episodic	Acute symptomatic anemia
	Acute chest syndrome
	Cerebrovascular accident (stroke)
	Aplastic crisis
	Acute splenic sequestration
	Acute multi-organ system failure
	Surgery requiring general anesthesia
	Eye surgery
	Bacterial or malarial infections with severe anemia
Chronic	Prevention of recurrent strokes
	Prevention of first stroke
	Complicated pregnancy
	Chronic organ failure
	Frequent pain episodes

Modified from Hillyer and Hillyer 2001.

drome (ACS) occurs at least once in approximately 30% of patients with SCD, and half of these patients will experience one or more recurrent episodes in their lifetimes. ACS is responsible for 25% of all SCD patient deaths. RBC transfusion therapy is critically important in the treatment of ACS early in its course, and exchange transfusion has been shown to result in rapid improvement of ACS, if instituted within 48 hours after its diagnosis.

Simple transfusion often causes improvement of symptoms in children with ACS, especially when begun soon after diagnosis. Therefore, in patients with less serious pulmonary compromise, simple transfusion is preferred.

However, for those pediatric patients with either (1) a progressive decline in arterial oxygen pressure (PaO$_2$ < 70 mm Hg), or (2) rapid clinical deterioration, RBC exchange transfusion is recommended, by some authorities.

Cerebrovascular Accident (Stroke)

Approximately 10% of patients with SCD each year will experience one or more cerebrovascular accidents (strokes) (Adams 2000). The occurrence of infarctive stroke, rather than hemorrhagic stroke, is greatest in children with SCD. RBC exchange transfusion is often indicated upon diagnosis of stroke, as the neurological sequelae of strokes may be devastating. Dramatic recovery of neurological function has been documented after RBC exchange therapy in SCD pediatric patients following an acute stroke.

Aplastic Crisis

Aplastic crisis, typically defined as a decrease in hemoglobin of more than 3.0 g/dL from baseline with concomitant reticulocytopenia, is a relatively common occurrence in children with SCD. Aplastic crisis occurs in SCD pediatric patients after marked suppression of erythropoiesis for 7 to 10 days. This erythropoietic suppression is typically a result of infection of RBC precursors in the bone marrow by human parvovirus B19 (Wierenga et al. 2001). Because mean RBC survival time in most SCD patients is only 12 to 15 days, the acute and life-threatening anemia caused by human parvovirus B19 infection necessitates acute RBC transfusion, in order to sustain oxygen delivery to the tissues, until the infection subsides and the bone marrow recovers.

In the case of aplastic crisis, simple RBC transfusion is usually administered slowly (1 mL/kg/hr), because these children have expanded plasma volumes due to chronic anemia, and care must be taken to avoid volume overload and subsequent congestive heart failure. Partial exchange transfusion, which is performed by manually removing whole blood and returning RBCs to the patient (without replacing the plasma), is a method that may be preferred, if congestive heart failure from acute volume overload is of significant concern (Eckman 2001).

Acute Splenic Sequestration

When sickled RBCs are trapped within splenic sinusoids, the spleen enlarges and traps circulating RBCs, leading to increased anemia and subsequent circulatory failure. The fatality rate from acute splenic sequestration is ≤10%. Simple RBC transfusion immediately upon diagnosis has been shown to produce rapid resolution of acute splenic sequestration (Emond et al. 1985).

Acute Multi-Organ System Failure

Acute multi-organ failure syndrome (AMOFS) occurs can following episodes of severe pain crisis in pediatric patients with SCD. AMOFS may affect the lungs, liver, and/or kidneys. Immediate RBC transfusion has been recommended for the treatment of AMOFS. Clinical improvement and rapid reversal of organ dysfunction has been seen after prompt, aggressive RBC transfusion therapy (Hassell et al. 1994).

Alternatively, simple transfusion may be ordered for pediatric SCD patients with AMOFS who have severe anemia and rapidly falling hemoglobin levels. RBC exchange transfusion offers another option for those patients with higher hemoglobin levels or more severe manifestations of AMOFS.

Transfusion in the Peri-Operative Period

Hypoxia, volume depletion, hypotension, hypothermia, and acidosis are all complications surrounding the peri-operative period that can lead to intravascular sickling and vascular occlusion, resulting in high rates of morbidity and mortality among SCD patients who receive general anesthesia.

For pediatric patients with SCD who are to undergo general anesthesia, simple transfusion to increase the preoperative hemoglobin level to 10 g/dL, together with intra-operative or postoperative RBC transfusion is apprapriate and appears appropriate, based on the studies published to date (Vichinsky et al. 1995).

Although eye surgery is typically performed under local anesthesia, the microvascular nature of the surgery and the importance of avoiding damage to the eye justify the use of blood transfusion in children with SCD who undergo eye surgery. It is reasonable to follow transfusion guidelines similar to those used for SCD patients undergoing surgical procedures performed under general anesthesia.

Bacterial or Malarial Infections with Severe Anemia

Bacterial and malarial infections are common throughout the world and create life-threatening complications in SCD patients. Some experts do consider it acceptable medical practice to transfuse SCD patients with severe anemia who have serious infections. However, infection without concomitant anemia is not a widely accepted indication for RBC transfusion in SCD patients (Ohene-Frempong 2001).

Indications for Chronic Transfusion

Prevention of Recurrent Strokes

Approximately 10% of SCD patients will experience one or more cerebrovascular accidents (strokes). Without therapeutic intervention, strokes will recur in more than two-thirds of these patients within 2 to 3 years. Chronic RBC transfusions provide a reduction of up to 90% in the risk of recurrent stroke in these children with SCD (Adams 2000; Riddington and Wang 2002).

The chronic transfusion regimen (Cohen and Martin 2001; Ohene-Frempong et al. 1998) most often recommended for prevention of recurrent stroke is as follows:

- Simple RBC transfusion every 3 to 4 weeks
- Maintain HbS at <30%
- Maintain hematocrit at ≤30%

The optimal duration of chronic transfusion, for pediatric patients with SCD, to prevent recurrent stroke is not known. Wang and coworkers (1991) reported that five of 10 patients had recurrent strokes within 1 year after discontinuation of chronic transfusions, after a median period without recurrence (since first stroke) of 9.5 years. This rate is significantly greater than the 10% estimated risk of recurrent stroke among patients receiving chronic transfusion therapy (Riddington and Wang 2002). Therefore, most centers recommend continuation of chronic transfusion therapy indefinitely to prevent recurrent stroke.

Prevention of First Stroke

High blood flow velocity in the internal carotid and middle cerebral arteries, detected by transcranial Doppler ultrasonography, has been found to be predictive of subsequent stroke in children with SCD (Adams et al. 1992; Adams et al. 1997). Chronic transfusion greatly reduced the risk of first stroke in children with SCD who have abnormal transcranial Doppler ultrasonography results in the STOP trial. Because this diagnostic method is safe, and the prevention of stroke in these patients can greatly reduce the morbidity and mortality of this complication, many sickle cell centers have implemented regular transcranial Doppler screening and subsequent chronic transfusion in children with SCD who have repeated high Doppler flow rates, in order to prevent first strokes (Vichinsky et al. 2001; Ohene-Frempong 2001).

Complicated Pregnancy

Although normal pregnancy in SCD patients is not an indication for prophylactic transfusion (Koshy et al. 1989; Koshy 1995), certain complications of either the pregnancy itself or the underlying disease process are indications for simple or exchange transfusion (including pre-eclampsia/eclampsia, twin pregnancy, previous perinatal mortality, acute renal failure, sepsis, bacteremia, severe acute anemia, ACS, hypoxemia, anticipated surgery, and preparation for infusion of angiographic dye). For each of these indications, simple transfusion is most often utilized, if the patient's hemoglobin concentration is <5 g/dL and the reticulocyte count is <3%.

If the patient's hemoglobin is ≥8 to 10 g/dL, then RBC exchange transfusion is indicated, with the goal of a posttransfusion hemoglobin concentration of 10 g/dL and a posttransfusion HbS of ≤50%.

Chronic Organ Failure

SCD patients who have renal failure develop progressive anemia, due to the resultant loss of erythropoietin production by the kidney (Morgan et al. 1982). Many of these patients need regular RBC transfusions to avoid severe symptomatic anemia. Older children (and adults) with severe pulmonary disease may also benefit from chronic RBC transfusion to prevent symptoms (Weil et al. 1993).

Frequent Pain Episodes

Patients with frequent episodes of severe pain who are unable to engage in activities of normal daily living may benefit from chronic RBC transfusion. The transfusion programs used for stroke prevention will usually prevent recurrent pain episodes (Miller et al. 2001).

Controversial Indications for Acute or Chronic Transfusion

Recurrent Acute Chest Syndrome

Recurrent episodes of ACS in SCD patients have been associated with worsening pulmonary function and eventually restrictive lung disease. These pulmonary abnormalities can lead to severe pulmonary fibrosis, pulmonary hypertension, and cor pulmonale by the time the child reaches adulthood (Weil et al. 1993). Some evidence suggests that the use of chronic RBC transfusion protocols may be beneficial for certain pulmonary and cardiac complications (Nifong and Domen 2002; Miller et al. 2001), but the use of chronic transfusion to prevent recurrent episodes of ACS is still controversial.

Priapism

In some patients with acute priapism after RBC exchange or simple transfusion therapy, improvement of the priapism has been described (Walker et al. 1983; Seeler 1971). However, no controlled trial has been performed to establish the effectiveness of transfusion therapy for the treatment of acute priapism. Some centers will consider RBC exchange if simple transfusion or surgery fail. In fact, an *a*ssociation among *S*CD, *p*riapism, *e*xchange transfusion, and subsequent *n*eurological events has been described in several patients; this constellation of events and symptoms has been given the acronym "ASPEN" syndrome (Miller et al. 1995). As a result of the recognition of ASPEN syndrome, many centers conservatively manage priapism, using hydration and analgesia, and do not transfuse RBCs until a single episode has persisted for longer than 24 to 48 hours.

Acute Pain Crises

Episodes of severe pain, also known as pain crises, are the most common reason for hospital admission of SCD patients. Platt and colleagues (1991) showed that occurrences of pain crises have a direct correlation with hematocrit levels. There is evidence that chronic transfusion regimens reduce the frequency of acute pain episodes (Miller et al. 2001). However, simple transfusion is not routinely recommended for the treatment of acute pain crises.

Nonindications for Transfusion

Normal Pregnancy

Due to a randomized, controlled trial (Koshy et al. 1988), which concluded that prophylactic transfusion of the pregnant SCD patient could be omitted without harm to mother or fetus, normal pregnancy is not considered an indication for prophylactic RBC transfusion.

Leg Ulcers

Statistical analysis of various methods for the treatment of leg ulcers in SCD patients (including transfusion) detected no differences in rate of ulcer healing (Koshy et al. 1989). Therefore, the routine use of transfusion for the treatment of leg ulcers in SCD patients is not recommended.

Methods of Transfusion Therapy

Several methods of RBC transfusion therapy are used for SCD patients. Limitations exist for the administration of acute simple transfusions, including the risks of acute volume overload and increased blood viscosity.

Children with SCD have normal or increased total blood volume, because the plasma volume is increased in response to their chronic anemic states. Acute increases in total blood volume after transfusion of RBCs may increase cardiac work to the point of precipitating congestive heart failure (Schmalzer et al. 1987).

RBC transfusion significantly increases blood viscosity in studies where there are ≥60% HbS-containing cells present (Davies and Roberts-Harewood 1997). When the proportion of HbS-containing cells is ≤40%, the increase in blood viscosity is minimal. Clinically relevant reductions in tissue oxygenation occur in SCD pediatric patients who are transfused to a hematocrit >30%, forming the rationale for the use of acute RBC exchange transfusion for many SCD complications (Eckman 2001).

Acute Simple Transfusion

In simple transfusion, normal donor RBCs are infused into the child with SCD, without removal of his or her own sickled RBCs. In general, acute simple transfusion is indicated when the immediate need for oxygen-carrying capacity is increased, but no dramatic decrease in the percentage of HbS in the patient's blood is necessary. The volume of RBCs to be transfused can be calculated by the following formula:

Volume of RBCs to be transfused (mL)
    = [(desired % hct − starting % hct) × total blood volume] / % hct of RBCs units transfused

This is the formula for calculation of the volume of RBCs to be transfused to children with sickle cell disease, using the method of acute simple transfusion (Hillyer and Eckman 2002).

When transfusing children with SCD, it is desirable to maintain the hematocrit at ≤30% (see previous discussion). Once the patient's hematocrit rises above 30%, oxygen delivery to the tissues decreases, owing to increases in viscosity. Because the primary goal of RBC transfusion is to rapidly improve oxygen delivery to tissues, in order to prevent ongoing ischemia, the maximum beneficial hematocrit to be achieved by RBC transfusion is 30%.

For simple transfusion, donor RBCs may be infused through an 18- to 23-gauge needle or catheter, depending on the size of the child and the accessibility of his or her peripheral veins, using a standard blood infusion set (Brecher et al. 2002).

Chronic Simple Transfusion

Chronic simple transfusion is indicated when it is desirable to increase oxygen-carrying capacity and to simultaneously chronically depress the percentage of HbS in the patient's blood (Wayne et al. 1993). When a specific decrease in HbS percentage is desired, the following calculation estimates the dilutional effects of normal RBC transfusion on % HbS.

Final % HbS desired = [1 − (transfused RBC volume in mL × % hct of RBC units transfused)] / [(total blood volume × starting % hct) + (transfused RBC volume in mL × % hct of RBC units transfused)] × starting % HbS

This is the formula for calculation of the volume of RBCs to be transfused to children with SCD, using the method of chronic simple transfusion, when a specific decrease in HbS percentage is desired (Hillyer and Eckman 2002).

As is the case for acute simple transfusion, the typical goal of chronic RBC transfusion therapy is to maintain the % HbS at <30% of the total hemoglobin concentration. For children, the amount of RBC to be transfused may be calculated from the Figure 19.2 formula above. The total volume of RBCs to be transfused at each regular interval in a chronic simple transfusion protocol is determined by the pretransfusion level of hemoglobin A (HbA). If the HbA percentage is too low (in most cases, ≥70% HbA is the goal), either the volume of RBCs to be transfused should be increased, or the time interval in weeks between transfusions should be decreased.

In the case of chronic simple RBC transfusion therapy, peripheral catheters may be surgically implanted and used for multiple transfusion events. Catheter placement allows for improved venous access and less discomfort for the child. Peripheral veins may also be used for chronic transfusion, depending on the size of the child and the accessibility of his or her veins.

Blood Cell Exchange Transfusion

In RBC exchange therapy, donor RBCs are infused while the patient's own RBCs are simultaneously removed. The major benefits of RBC exchange are rapid adjustments that may be achieved in the patient's hematocrit and HbS levels. This ability to make rapid adjustments to hematocrit and % HbS is a critically important factor during acute sickling episodes, the major set of indications for RBC exchange therapy.

During certain acute ischemic crises, the replacement of sickled RBCs with nonsickle trait donor RBCs (improving oxygen delivery to ischemic tissues) must be accomplished quickly, to prevent further tissue damage. Importantly, unlike simple transfusion, RBC exchange allows for removal of the same volume of RBCs as is replaced, decreasing the risk of iron overload associated with chronic simple transfusion (Danielson 2002; Ohene-Frempong, 2001).

The usual measurable endpoints of RBC exchange therapy are:

- HbA level ≥70%
- HbS level <30%
- Overall hematocrit ≤30%

RBC exchange transfusion is typically performed with the use of an automated apheresis instrument (erythrocytapheresis). This instrument removes the patient's blood, separates the RBCs from the platelet-rich plasma by centrifugation, returns the plasma to the patient, and discards the patient's sickled RBCs. At the same time, normal donor RBCs are infused, maintaining a stable blood volume throughout the procedure (Pepkowitz 1997).

In the past, RBC exchange transfusion could only be performed manually, and a variety of formulae were used to estimate the appropriate RBC exchange volumes. Indications still exist for manual RBC exchange (including its use in infants whose blood volumes are too small to be placed on the automated apheresis machine), but erythrocytapheresis is currently the most common method of performing RBC exchange therapy for children with SCD.

The automated apheresis instrument has an internal programmable computer that calculates both the volume of the patient's RBCs to be removed and the volume of the donor's RBCs to be infused. The following formula provides a general, practical estimate of the number of RBC units to be used in automated RBC exchange transfusion:

$$\text{Volume of RBCs to be exchanged (mL)}$$
$$= \text{desired \% hct} \times \text{total blood volume}$$

This is the formula for the calculation of a practical estimate of the volume of RBCs to be transfused to children with SCD using the erythrocytapheresis method (Hillyer and Eckman 2002).

Having a preliminary estimate of the number of RBC units to be exchanged before initiation of the procedure is helpful, because the blood bank may need extra time to procure the appropriate RBC units.

In most cases of RBC exchange, central venous access must be established using a hemodialysis-grade, rigid-wall, large-lumen, double-bore catheter that can withstand high flow rates. In larger children with easily accessible peripheral veins, the antecubital veins may be acceptable for use, with 16- to 18-gauge needles being used for blood removal and 18- to 20-gauge catheters for blood return.

The apheresis instrument, after being programmed with the appropriate patient data and the desired endpoints for therapy, is used to remove the patient's RBCs and replace them with donor RBCs. Of note, it is desirable to replace the patient's sickled RBCs with HbS-negative (sickle trait–negative) donor RBCs, primarily to appropriately calculate and achieve the desired percentages of HbS and HbA.

In infants and very small children, the amount of blood in the extracorporeal circuit of the automated apheresis machine represents a significant percentage of the child's total blood volume, which may lead to hypotension, or a critically low hematocrit, or both, if the machine is primed with saline.

Instead, the automated apheresis instrument should be primed with RBCs if: (1) the extracorporeal circuit represents 12% or more of the child's blood volume, (2) the child weighs less than 20 kg, or (3) the child is anemic or unstable.

Occasionally, certain patients require manual RBC exchange transfusions. Indications for manual exchange include those infants who have very small total blood volumes and in emergency situations involving patients for whom the additional time required to establish appropriate venous access or mobilize the apheresis team would be deleterious to the patient's health.

A rapid manual partial RBC exchange may be performed by withdrawing blood from a peripheral (usually antecubital) vein and infusing RBCs via a stopcock into the same vein (or directly into a different peripheral vein), using the methods outlined in the following formula:

(1) Calculate RBC exchange volume, using 60–75 mL/kg (to calculate total blood volume) as a practical estimate.
(2) Divide the calculated RBC exchange volume into four equal aliquots of RBCs.
(3) Withdraw whole blood from the child, equal to one exchange aliquot.
(4) Infuse normal saline to the child, equal to one exchange aliquot.
(5) Withdraw whole blood from the child, equal to one exchange aliquot. In the cases of very small children/infants, or when emergent situations arise, understanding the procedure for manual RBC exchange is critically important for the appropriate treatment of these patients.
(6) Transfuse a volume of RBCs to the child, equal to *two* exchange aliquots.
     Repeat steps three through six.

This is the method for manual RBC exchange transfusion of children with sickle cell disease (Hillyer and Eckman 2002).

In the past, CDPA-1 RBC units were reported to have average hematocrits of 80% or more, and it was recommended that RBCs be "reconstituted" to the volume and hematocrit of whole blood (30% to 40%), by adding albumin or saline to the blood bag or syringe before manual exchange transfusion to the SCD patient (Piomelli et al. 1990). However, the current average hematocrit of an RBC unit is approximately 50% to 65% (Hillyer et al. 2001), and the alternating administration to an SCD patient of RBCs and saline in equal volumes (see previous formula) should theoretically deliver a product essentially identical to "reconstituted" whole blood, as described previously.

The requirements that must be honored to ensure a safe and effective manual exchange transfusion for the child with SCD include the following: (1) marked increases in blood viscosity should be avoided; (2) blood volume should be maintained throughout the exchange; and (3) the rate should be determined through consultation.

Chronic RBC Exchange Transfusion

Iron overload is one of the serious, long-term complications of chronic transfusion in SCD patients (Harmatz et al. 2000; Ballas 2001). As a result, many investigators have suggested that chronic erythrocytapheresis should be used in place of chronic simple transfusion for the prevention of complications of SCD, in order to decrease long-term iron accumulation in these patients (Hilliard et al. 1998; Singer et al. 1999; Lawson et al. 1999).

The procedure for chronic erythrocytapheresis is the same as that used for acute erythrocytapheresis. Similarly, target postpheresis HbS levels are <30%, with desirable postpheresis hematocrits of ≤30%. Currently, chronic erythrocytapheresis is not universally used in the United States for the routine management of SCD patients, although recent reports suggest that the potential benefits of this method may outweigh its risks and costs. A detailed discussion of the risks and benefits of chronic erythrocytapheresis in the treatment of children with SCD follows later in the chapter.

Adverse Effects of Transfusion Therapy and Strategies for Their Prevention and Management

Although transfusion is in many ways beneficial and is most often necessary for treatment of SCD in children, adverse effects resulting from transfusion of donor blood can lead to serious long- and short-term complications in SCD patients. Both immune and nonimmune complications of RBC transfusion may occur.

Immune-Related Adverse Effects

Febrile Nonhemolytic Transfusion Reactions

White blood cells (WBCs) synthesize and release cytokines during the storage of cellular blood products. These cytokines may cause fever and chills (febrile nonhemolytic transfusion reactions, or FNHTRs) in the transfusion recipient. Current technologies in leukoreduction filtration reduce the leukocyte counts in cellular blood products (for example, RBCs) to less than 5×10^6 WBCs/unit, effectively preventing the development of FNHTRs in recipients of leukoreduced blood products (Brecher et al. 2002).

Because SCD patients typically receive large numbers of transfusions, and FNHTRs occur in association with 0.5% to 1% of all transfusions, most experts recommend the use of leukoreduced blood products for all SCD patients.

In addition, an acute infection or a pain crisis in an SCD patient can manifest with the same symptoms (that is, fever, chills, malaise) as an FNHTR, confounding the clinical picture and possibly delaying appropriate treatment of underlying disorders in these patients, illustrating an additional important reason to transfuse leukoreduced blood components to children with SCD.

Alloimmunization to RBC Antigens

Alloimmunization to RBC antigens is a common problem in transfused SCD patients, leading to difficulty in obtaining compatible RBC units and the development of delayed hemolytic transfusion reactions (DHTRs) (Talano et al. 2003). DHTRs are especially problematic in SCD patients, in that the symptoms of a DHTR can mimic those of a pain crisis and can even lead to a pain crisis, complicating the clinical diagnosis and thus the appropriate treatment of these patients.

The most comprehensive study of the frequency of and risk factors associated with alloimmunization in SCD patients was published by Vichinsky and colleagues in 1990. These authors prospectively determined the transfusion history of, RBC antigen phenotype of, and alloantibody development in African-American patients with SCD. These results were compared with those from similar studies in two other groups: (1) nontransfused African-American SCD patients, and (2) Caucasian patients who had undergone multiple RBC transfusions for other forms of chronic anemia.

Vichinsky's results showed that the average alloimmunization rate for transfused SCD patients was 30%, compared with only 5% for the multiply transfused Caucasian patients who had other forms of anemia (p < 0.001). The alloimmunization rate observed in individual SCD patients increased exponentially with increasing numbers of RBC transfusions.

After conducting an RBC phenotyping study of local blood bank donors and comparing those phenotypes with the phenotypes of SCD patients and of Caucasian patients who had other forms of chronic anemia, the authors of this study suggested that the increased alloimmunization rate in SCD patients most likely resulted from RBC antigenic differences between the SCD patients (primarily African-Americans) and the volunteer blood donors (the majority of whom were Caucasian). Other authors confirmed these findings (Luban 1989; Rosse et al. 1990).

Because of this lack of phenotypical compatibility between the majority of available donor RBCs and SCD patients, many centers have suggested that SCD patients receive RBC units matched for those antigens most commonly associated with alloimmunization (Table 19.2). In a recent study of SCD patients who were transfused with nonantigen matched RBCs, the alloimmunization rate in children was 29% (Auygun et al. 2002), confirming Vichinsky's findings of more than a decade ago.

Currently, in many centers, all children with SCD undergo extensive RBC antigen phenotyping at the time of diagnosis, a policy supported by most experts (Rosse et al. 1990). When and if RBC transfusion is required, these SCD patients routinely receive RBC units phenotypically matched for (at least) C, E, and K antigens.

Once an SCD patient has made an RBC alloantibody, he or she may benefit from RBC units matched for C, E, K, Fy[a], and Jk[b] antigens, the five most common and most clinically significant antibodies made by SCD patients (see Table 19.2) (Vichinsky et al. 1990). The antigen of the newly formed antibody also must be respected.

These concepts are supported by many experts, and antigen-matched RBC transfusion results in an alloimmunization rate of only 1% to 5%, a significant decrease compared with the rates observed among SCD patients transfused with RBCs not matched for these common RBC antigens (29% to 30%) (Vichinsky et al. 2001).

Autoimmunization to RBC Antigens

Development of autoantibodies to RBC antigens in association with transfusions in SCD patients has been described in multiple case reports and in small series of patients. The rate of warm-reactive RBC autoantibody formation in transfused children with SCD is 8%, and 29% of patients with erythrocyte autoantibodies have clinically significant hemolysis thought to be caused by the RBC autoantibody (Castellino et al. 1999). There is also a strong association between autoantibody formation and the prior presence of RBC alloantibodies.

Whatever the cause, physicians should be aware that a syndrome of clinically significant posttransfusion hemolysis may occur in SCD patients in which both autologous and transfused RBCs are destroyed (also known as "bystander hemolysis") and that hemolysis may be exacerbated by further transfusions (Win et al. 2001). Serological findings may be negative or not helpful in identification of the autoantibody. In most cases, corticosteroids with or without intravenous immunoglobulin (IVIG) are beneficial in slowing hemolysis and allowing for successful continuation of necessary transfusions (Cullis et al. 1995).

Alloimmunization to Human Leukocyte Antigens or Platelet-Specific Antigens

As bone marrow/stem cell transplantation (BMT) and cord blood transplantation become viable options for selected patients with SCD, alloimmunization to platelets will present more serious problems for this group (Walters et al. 1996; Gore et al. 2000; Vermylen et al. 1998). Friedman and coworkers (1996) reported that 85% of SCD patients receiving 50 or more transfusions, 48% of SCD patients receiving one to 49 transfusions and demonstrated alloimmunization to human leukocyte antigen (HLA) or platelet-specific antigens.

Because platelet refractoriness is a serious complication following BMT, prevention of platelet alloimmunization appears prudent in this group of children with SCD. The authors support the use of leukoreduction of cellular blood products to prevent or reduce platelet alloimmunization and refractoriness in SCD patients, a

TABLE 19.2 Most Common RBC Alloantibodies Produced by Transfused Children with SCD

Alloantibody	Average frequency (%)
E	21
K	18
C	14
Le[a]	8
Fy[a]	7
Jk[b]	7
D	7
Le[b]	7
S	6
Fy[b]	5
M	4
e	2
c	2

From Luban 1989; Vichinsky et al. 1990; and Rosse et al. 1990.

practice employed for a variety of multiply transfused patient groups (Sniecinski et al. 1988).

Nonimmune-Related Adverse Effects

Iron Overload

Iron overload resulting in hemosiderosis is a serious long-term complication of chronic transfusion in children with SCD (Files et al. 2002). Patients who develop iron overload may be treated with long-term chelation therapy in the form of deferoxamine (DFO) (Silliman et al. 1993). However, DFO therapy is expensive (Wayne et al. 2000), and due to multiple side effects, the compliance rate is poor (Cohen and Martin 2001).

One potential transfusion methodology for the prevention of iron overload in children with SCD that is currently being investigated is chronic erythrocytapheresis. Chronic erythrocytapheresis procedures may be performed at 3- to 4-week intervals. In contrast to simple (additive) transfusions, the child's own sickled RBCs are removed, while an equal volume of normal donor RBCs is infused. The obvious potential benefit of chronic erythrocytapheresis compared with simple transfusion is the prevention of long-term iron accumulation and hemosiderosis.

Although it has not been universally implemented, chronic erythrocytapheresis may be clinically effective in reducing iron overload in chronically transfused SCD pediatric patients. Some reports suggest that erythrocytapheresis does limit iron accumulation in SCD patients. In addition, at-risk patients who were started on erythrocytapheresis without a long history of previous transfusions maintained very low serum ferritin levels and did not require chelation therapy (Hilliard et al. 1998; Singer et al. 1999).

Therefore, it appears that chronic erythrocytapheresis may be most beneficial when it is initiated early in the course of chronic transfusion therapy, before significant iron accumulation occurs. However, chronic erythrocytapheresis does appear to stabilize or decrease serum ferritin levels in patients who have already developed significant iron overload and also continue on chelation therapy.

The primary potential problems with the chronic erythrocytapheresis transfusion protocol (compared with chronic simple transfusion protocols) are: increased blood product exposure (with concomitant increased risks of alloimmunization to RBCs and platelets and of transfusion-transmitted infection), issues related to venous access and increased cost.

Published reports indicate that SCD patients' blood-product exposures do increase when chronic erythrocytapheresis protocols are performed, with reported increases in blood-utilization rates ranging from 52% to 100%. However, since the majority of SCD patients studied did receive RBC units matched for at least the C, E, and K RBC antigens, alloimmunization rates for these protocols have remained very low (≤1%) (Lawson et al. 1999).

The high cost of erythrocytapheresis is also an important issue. Hilliard and colleagues (1998) compared the total cost of erythrocytapheresis with the total cost for simple transfusion, and found an economically significant difference. However, these authors suggested that the added cost of chelation therapy when using simple transfusion (calculated to be approximately $400,000 per patient decade by Wayne and associates in 2000) makes erythrocytapheresis without chelation therapy a much less expensive alternative. This cost comparison provides further evidence that, if it is technically feasible, early initiation of chronic erythrocytapheresis in SCD patients, before significant iron accumulation occurs, may be preferable to long-term chronic simple transfusion and the subsequent complications of iron overload (and the necessity of chelation therapy).

Transfusion-Transmitted Infections

The risk of transmission (via transfusion) of infectious agents, particularly viruses, has become substantially reduced since the 1990s (Schrieber et al. 1996), largely owing to improved screening tests for donated blood products. Because the risk of contracting a transfusion-transmitted disease is quite low (Dodd et al. 2002), blood products should not be withheld (if an appropriate clinical indication for transfusion exists) for the sole purpose of preventing a transfusion-transmitted disease. However, the risks of all adverse effects of transfusion should be balanced against the clinical need for transfusion, on a case-by-case basis.

THALASSEMIAS

Pathogenesis

Thalassemias are among the most prevalent genetic disorders caused by a single gene (Weatherall and Clegg 1996). α-thalassemias are caused by mutations that reduce the synthesis of the α-globin chain of hemoglobin, and β-thalassemias from mutations that reduce β-globin synthesis (Olivieri 1999).

Clinical Pathology

One of the many serious clinical manifestations of severe β-thalassemia is transfusion-dependent anemia

(Olivieri 1999), resulting from severely ineffective erythropoiesis and RBC hemolysis.

Complications

Two major consequences of this ineffective erythropoiesis in children with thalassemia are (1) increased iron absorption and (2) progressive accumulation of iron in body tissues. Anemia, increased erythropoiesis, and hypersplenism also cause marked expansion of the plasma volume and blood volume in children with thalassemia.

Role of Transfusion in Therapy

Chronic RBC transfusion is required for those patients with homozygous thalassemia and in compound heterozygotes with β-thalassemia major. Individuals with β-thalassemia intermedia may have severe anemia requiring episodic transfusion.

Individuals with β-thalassemia major are transfusion-dependent from infancy. Chronic transfusion support has markedly improved the prognosis of these children. However, transfused thalassemic children routinely develop iron overload from increased iron absorption caused by repeated RBC transfusions. Cardiac, hepatic, and endocrine failure from iron overload often results in death in the patient's teens or 20s, in the absence of effective therapy to remove excess iron.

Although no ideal approach to treatment of transfusion-related iron overload exists in children with thalassemia, subcutaneous infusion of DFO over 8 to 12 hours, 5 to 7 days a week, appears to control iron accumulation and prevent cardiac and liver damage, thereby improving life expectancy. Bone marrow transplantation has cured many children with β-thalassemia major. Bone marrow transplant may be considered in thalassemic children who meet criteria and have a suitable bone marrow or stem cell donor (Lo and Singer 2002).

Other patients with variant forms of thalassemia may require transfusion therapy for specific reasons during their lifetimes. Hemoglobin H disease is caused by loss of globin synthesis from three of the normal four α-globin genes; these individuals are not transfusion-dependent in most cases, but they may require transfusion support for complications of the disease. Compound heterozygotes with β-thalassemia and hemoglobin E disease can have clinical manifestations, ranging from mild microcytic anemia to severe transfusion-dependent β-thalassemia major (Weatherall 1998).

Goals of Transfusion Therapy

The goals of transfusion therapy in children with β-thalassemia major are to (1) increase oxygen-carrying capacity by correcting the anemia, (2) prevent progressive hypersplenism, (3) suppress erythropoiesis, and (4) reduce increased gastrointestinal absorption of iron (Olivieri 1999). Transfusion therapy is begun in early childhood, to ameliorate the symptoms and signs of anemia, to reduce hypersplenism, to support normal growth and development, to prevent skeletal changes and complications of extramedullary hematopoiesis, and to reduce pathological fractures and other complications from osteopenia (Fosberg and Nathan 1990).

Indications for Transfusion

Symptoms and signs of anemia in children with thalassemia include growth retardation and failure to thrive, and RBC transfusion is initiated when these symptoms and signs present themselves, typically in early childhood. Transfusions are occasionally initiated in β-thalassemia intermedia and hemoglobin H disease patients to prevent facial and skull deformities from expansion of the medullary bone space. Progressive hypersplenism may require transfusion, in order to postpone splenectomy in β-thalassemia intermedia patients (Cazzola et al. 1997).

Methods of Transfusion

Simple Transfusion

Simple transfusion of leukoreduced RBCs, to maintain a hemoglobin level >9.5 g/dL, is the standard approach to transfusion in thalassemic children (Olivieri 1999). Splenectomy is recommended when hypersplenism increases the transfusion requirement >200 to 250 mL/kg/yr (Olivieri and Brittenham 1997).

Erythrocytapheresis

Chronic RBC exchange has also been applied to the transfusion therapy of children with thalassemia (Valbonesi and Bruni 2000). The use of automated chronic RBC exchange (erythrocytapheresis) resulted in a 30% reduction in RBC transfusion requirement and a 43% increase in transfusion intervals (Berdoukas and Moore 1986). A more detailed discussion of chronic erythrocytapheresis and its beneficial effects on iron overload are described in the previous section.

References

Adams R, McVie V, Nichols F, et al. 1992. The use of transcranial ultrasonography to predict stroke in sickle cell disease. *N Engl J Med* 326:605–610.

Adams RJ. 2000. Lessons from the stroke prevention trials in sickle cell anemia (STOP) study. *J Child Neurol* 15:344–349.

Adams RJ, McVie VC, Carl EM, et al. 1997. Long-term stroke risk in children with sickle cell disease screened with transcranial Doppler. *Ann Neurol* 42:699–704.

Auygun B, Padmanabhan S, Paley C, et al. 2002. Clinical significance of RBC alloantibodies and autoantibodies in sickle cell patients who received transfusions. *Transfusion* 42:3743.

Ballas SK. 2001. Iron overload is a determinant of morbidity and mortality in adult patients with sickle cell disease. *Semin Hematol* 38:30–36.

Berdoukas VA and Moorew RC. 1986. A study of the value of red cell exchange transfusions in transfusion dependent anemias. *Clin Lab Haematol* 8:209–220.

Brecher M, Combs MR, Drew MJ, et al. eds. 2002. *AABB Technical manual*, 14th ed. Bethesda, MD: AABB Press.

Bunn HF. 1997. Mechanisms of disease: pathogenesis and treatment of sickle cell disease. *N Engl J Med* 337:762–769.

Castellino SM, Combs MR, Zimmerman SA, et al. 1999. Erythrocyte autoantibodies in paediatric patients with sickle cell disease receiving transfusion therapy: frequency, characteristics, and significance. *Br J Haematol* 104:189–194.

Cazzola M, Borgna-Pignatti C, Locatelli F, et al. 1997. A moderate transfusion regimen may reduce iron loading in β-thalassemia major without producing excessive expansion of erythropoiesis. *Transfusion* 37:135–140.

Cohen AR and Martin MB. 2001. Iron chelation therapy in sickle cell disease. *Semin Hematol* 38:69–72.

Cullis JO, Win N, Dudley TM, et al. 1995. Post-transplant hyperhaemolysis in a patient with sickle cell disease: use of steroids and intravenous immunoglobulin to prevent further red cell destruction. *Vox Sang* 69:355–357.

Danielson CF. 2002. The role of red blood cell exchange transfusion in the treatment and prevention of complications of sickle cell disease. *Ther Apher* 6:24–31.

Davies SC and Roberts-Harewood M. 1997. Blood transfusion in sickle cell disease. *Blood Rev* 11:57–71.

Dodd RY, Notari EP, IV, and Stramer SL. 2002. Current prevalence and incidence of infectious disease markers and estimated window-period risk in the American Red Cross blood donor population. *Transfusion* 42:975–979.

Eckman JR. 2001. Techniques for blood administration in sickle cell patients. *Semin Hematol* 38:23–29.

Emond AM, Collis R, Darvill D, et al. (1985). Acute splenic sequestration in homozygous sickle cell disease: natural history and management. *J Pediatr* 107:201–206.

Files B, Brambilla D, Kutlar A, et al. 2002. Longitudinal changes in ferritin during chronic transfusion: a report from the Stroke Prevention Trial in Sickle Cell Anemia (STOP). *J Pediatr Hematol Oncol* 24:284–290.

Fosburg MT and Nathan DG. 1990. Treatment of Colley's anemia. *Blood* 76:435–444.

Friedman D, Lukas M, Jawad A, et al. 1996. Alloimmunization to platelets in heavily transfused patients with sickle cell disease. *Blood* 88:3216–3222.

Gore L, Lane PA, Quinones RR, et al. 2000. Successful cord blood transplantation for sickle cell anemia from a sibling who is human leukocyte antigen-identical: implications for comprehensive care. *J Pediatr Hematol Oncol* 22:437–440.

Harmatz P, Butensky E, Quirolo K, et al. 2000. Severity of iron overload in patients with sickle cell disease receiving chronic red blood cell transfusion therapy. *Blood* 96:76–79.

Hassell KL, Eckman JR, and Lane PA. 1994. Acute multiorgan system failure syndrome: a potentially catastrophic complication of severe sickle cell pain episodes. *Am J Med* 96:155–162.

Hilliard LM, Williams BF, Lounsbury AE, et al. 1998. Erythrocytapheresis limits iron accumulation in chronically transfused sickle cell patients. *Am J Hematol* 59(1):28–35.

Hillyer KL and Eckman JR. 2002. Transfusion in the hemoglobinopathies. In *Blood banking and transfusion medicine*. Hillyer CD, Silberstein, LE, Ness PM, et al., eds. Philadelphia, PA: Churchill Livingstone.

Hillyer KL and Hillyer CD. 2001. Packed red blood cells and related products. In *Handbook of transfusion medicine*. Hillyer CD, Hillyer KL, Strobl FJ, et al., eds. Philadelphia, PA: Academic Press.

Hillyer KL, Hare VW, Eckman JR, et al. 2001. Decreased alloimmunization rates in chronically-transfused sickle cell disease patients in a directed-donor, red blood cell antigen-matching program entitled "Partners for Life." *Blood* 98:2274.

Koshy M. 1995. Sickle cell disease and pregnancy. *Blood Rev* 9:157–164.

Koshy M, Burd L, Wallace D, et al. 1988. Prophylactic red cell transfusions in pregnant patients with sickle cell disease. *N Engl J Med* 319:1447–1452.

Koshy M, Entsuah R, Koranda A, et al. 1989. Leg ulcers in patients with sickle cell disease. *Blood* 74:1403–1408.

Lane PA. 1996. Sickle cell disease. *Pediatr Clin North Am* 43:639–664.

Lawson SE, Oakley S, Smith NA, et al. 1999. Red cell exchange in sickle cell disease. *Clin Lab Haem* 21:99–102.

Lo L and Singer ST. 2002. Thalassemia: current approach to an old disease. *Pediatr Clin North Am* 49:1165–1191.

Luban NL. 1989. Variability in rates of alloimmunization in different groups of children with sickle cell disease: effect of ethnic background. *Am J Pediatr Hematol Oncol* 11:314–319.

Miller ST, Rao SP, Dunn KE, et al. 1995. Priapism in children with sickle cell disease. *J Urology* 154:844–847.

Miller ST, Wright E, Abboud M, et al. 2001. Impact of chronic transfusion on incidence of pain and acute chest syndrome during the Stroke Prevention Trial (STOP) in sickle-cell anemia. *J Pediatr* 139:785–789.

Morgan AG, Gruber CA, and Sergeant GR. 1982. Erythropoeitin and renal function in sickle cell disease. *Br Med J* 285:1686–1688.

Nifong TP and Domen RE. 2002. Oxygen saturation and hemoglobin A content in patients with sickle cell disease undergoing erythrocytapheresis. *Ther Apher* 6:390–393.

Ohene-Frempong K. 2001. Indications for red cell transfusion in sickle cell disease. *Semin Hematol* 38:5–13.

Ohene-Frempong K, Weiner SJ, Sleeper LA, et al. 1998. Cerebrovascular accidents in sickle cell disease: rates and risk factors. *Blood* 91:288–294.

Olivieri NF. 1999. The β-thalassemias. *New Engl J Med* 341:99–109.

Olivieri NF and Brittenham GM. 1997. Iron-chelation therapy and treatment of thalassemia. *Blood* 89:739–761.

Pepkowitz S. 1997. Red cell exchange and other therapeutic alterations of red cell mass. In *Apheresis: principles and practic*. B. McLeod, et al., eds. Bethesda, MD: AABB Press.

Piomelli S, Seaman C, Ackerman K, et al. 1990. Planning an exchange transfusion in patients with sickle cell syndromes. *Am J Pediatr Hematol Oncol* 12:268–276.

Platt OS, Thorington BD, Brambilla DJ, et al. 1991. Pain in sickle cell disease. Rates and risk factors. *N Engl J Med* 2(4), 275–280.

Riddington C and Wang W. 2002. Blood transfusion for preventing stroke in people with sickle cell disease. *Cochrane Database Syst Rev* 1, CD003146.

Rosse WF, Gallagher D, Kinney T. 1990. Transfusion and alloimmunization in sickle cell disease. *Blood* 76:1431–1437.

Schmalzer EA, Lee JO, Brown AK, et al. 1987. Viscosity of mixtures of sickle and normal red cells at varying hematocrit levels. *Transfusion* 27:228–233.

Schrieber G, Busch M, Kleinman S, et al. 1996. The risk of transfusion-transmitted viral infections. *N Engl J Med* 334:1685–1690.

Seeler RA. 1971. Priapism in children with sickle cell anemia. *Clin Pediatr* 10:418–419.

Silliman CC, Peterson VM, Mellman DL, et al. (1993). Iron chelation by deferoxamine in sickle cell patients with severe transfusion-induced hemosiderosis: a randomized, double-blind study of the dose-response relationship. *J Lab Clin Med* 122:48–54.

Singer ST, Quirolo K, Nishi K, et al. 1999. Erythrocytapheresis for chronically transfused children with sickle cell disease: an effective method for maintaining a low hemoglobin S level and reducing iron overload. *J Clin Apheresis* 14:122–125.

Sniecinski I, O'Donnell MR, Nowicki B, et al. 1988. Prevention of refractoriness and HLA-allimmunization using filtered blood products. *Blood* 71:1402–1407.

Tahhan HR, Holbrook CT, Braddy LR, et al. 1994. Antigen-matched donor blood in the transfusion managment of patients with sickle cell disease. *Transfusion* 34:562–569.

Talano JA, Hillery CA, Gottschall JL, et al. 2003. Delayed hemolytic transfusion reaction/hyperhemolysis syndrome in children with sickle cell disease. *Pediatrics* 111:661–665.

Valbonesi M and Bruni R. 2000. Clinical application of therapeutic erythrocytapheresis (TEA). *Transfus Sci* 22:183–194.

Vermylen C, Cornu G, Ferster A, et al. 1998. Haematopoietic stem cell transplantation for sickle cell anaemia: the first 50 patients transplanted in Belgium. *BMT* 22:1–6.

Vichinsky EP. 2001. Current issues in blood transfusion in sickle cell disease. *Semin Hematol* 38:14–22.

Vichinsky EP, Earles A, Johnson RA, et al. 1990. Alloimmunization in sickle cell anemia and transfusion of racially unmatched blood. *N Engl J Med* 322:1617–1621.

Vichinsky EP, Haberkern CM, Neumayr L, et al. 1995. A comparison of conservative and aggressive transfusion regimens in the peri-operative management of sickle cell disease. *N Engl J Med* 333:206–213.

Vichinsky EP, Luban NL, Wright E, et al. 2001. Prospective red blood cell phenotype matching in a stroke-prevention trial in sickle cell anemia: a multicenter transfusion trial. *Transfusion* 41:1086–1092.

Walker EM, Mitchum EN, Rous SN, et al. 1983. Automated erythrocytopheresis for relief of priapism in sickle cell hemoglobinopathies. *J Urology* 130:912–916.

Walters MC, Patience M, Leisenring W, et al. 1996. Bone marrow transplantation for sickle cell disease. *N Engl J Med* 335:369–376.

Wang WC, Kovnar EH, Tonkin IL, et al. 1991. High risk of recurrent stroke after discontinuance of five to twelve years of transfusion therapy in patients with sickle cell disease. *J Pediatr* 118:377–382.

Wayne AS, Kevy SV, and Nathan DG. 1993. Transfusion management of sickle cell disease. *Blood* 81:1109–1123.

Wayne AS, Schoenike SE, and Pegelow CH. 2000. Financial analysis of chronic transfusion for stroke prevention. *Blood* 96:2369–2372.

Weatherall DJ. 1998. Hemoglobin E beta-thalassemia: an increasingly common disease with some diagnostic pitfalls. *J Pediatr* 132:765–767.

Weatherall DJ and Clegg JB. 1996. Thalassemia—a global public health problem. *Nat Med* 2:847–849.

Weil JV, Castro O, Malik AR, et al. 1993. Pathogenesis of lung disease in sickle hemoglobinopathies. *Am Rev Respir Dis* 148:249–256.

Wierenga KJ, Sergeant BE, and Sergeant GR. 2001. Cerebrovascular complications and parvovirus infection in homozygous sickle cell disease. *J Pediatr* 139:438–442.

Win N, Doughty H, Telfer P, et al. 2001. Hyperhemolytic transfusion reaction in sickle cell disease. *Transfusion* 41:323–328.

The Bleeding Child: Congenital and Acquired Disorders

BRIAN M. WICKLUND, MD, CM, MPH

ABSTRACT

This chapter provides an overview of inherited and acquired hemorrhagic disorders that do not involve abnormalities of platelets, their diagnosis, and treatment. Diagnostic approaches and laboratory investigation of these disorders are featured, with specific attention to the treatment of hemophilia, von Willebrand's disease, and other inherited disorders. Discussion includes the choice of clotting factor concentrates and adjunctive therapies. Acquired coagulation abnormalities reviewed include disseminated intravascular coagulation (DIC), hepatic failure, renal disorders, vitamin K deficiency, and acquired inhibitors of coagulation.

INTRODUCTION

The intent of this chapter is to review the causes of bleeding in infants and children that are related to defects in circulating coagulation factors, their diagnosis, and treatment. Platelet disorders will be covered in Chapter 22.

COAGULATION OVERVIEW

The coagulation or hemostatic system arrests bleeding from injured blood vessels via a system of proteins, platelets, and blood vessels. This serves to amplify the stimulus that activated the coagulation system and acts rapidly to stop the loss of blood from an injured vessel. The initial stimulus to the formation of clot comes from tissue injury that disrupts endothelial cells, exposing collagen and subendothelial tissues. Platelets adhere to the exposed endothelium mediated by platelet glycoprotein Ib and von Willebrand factor (vWF), release their granular contents and expose platelet surface glycoprotein IIb/IIIa. This glycoprotein in combination with fibrinogen and vWF recruits more platelets and binds them into a primary hemostatic plug (Colman et al. 2001).

Secondary hemostasis occurs in concert with the primary hemostatic plug, with the injury to the vessel wall producing tissue factor, which activates factor VII to factor VIIa. The tissue factor/factor VIIa complex then activates both factor X and factor IX. The direct activation of factor X by the extrinsic pathway serves to initiate the activation of the common pathway, with production of small amounts of factor Xa and then thrombin. The activation of factor IX by the factor VIIa/tissue factor complex then amplifies the amount of factor Xa produced to activate the common pathway, which forms the major proportion of the thrombin that is produced. Thrombin cleaves fibrinogen to form fibrin, which is crosslinked by factor XIII to form a stable fibrin clot. This process of secondary hemostasis serves to lay down an insoluble fibrin mesh that stabilizes the mass of platelets and prevents it from disaggregating (Esmon 2003).

The fibrinolytic system is activated by the release of tissue plasminogen activator from the injured vessel wall. Fibrinolysis deals with excess hemostatic material formed by the coagulation system and, once healing has taken place, begins the process of resolution of formed clot so that vascular patency can be restored. Defects that result in excessive fibrinolysis can cause bleeding. Inhibitors exist to control the amount and the location of activated clotting factors. These controlling systems include antithrombin, protein C, protein S, heparin cofactor II, and alpha-2-macroglobulin.

Developmental issues influence the functions of the coagulation system and some of the presentations of bleeding disorders. The levels of vitamin K-dependent clotting factors (factors II, VII, IX, and X) are low at birth and increase over the first 6 months of life. Factor VIII and vWF levels are elevated at birth and decrease over the first few months of life. Interpretation of factor levels must take into account the child's age, with the use of age-adjusted normal ranges. Reference ranges for preterm infants, gestational ages 28 to 31 weeks, and for term infants show the prothrombin (PT) normal range to be 14.6 to 16.9 seconds for the preterm infant and 10.1 to 15.9 seconds for the term infant. The activated partial thromboplastin time (aPTT) normal range for the preterm infant is 80 to 168 seconds and for the term infant is 31.3 to 54.5 seconds. PT normal range for the preterm infant is 19% to 54%, increases to 26% to 70% at birth, and is 34% to 102% at 1 month of age. Factor IX normal range for the preterm infant is 17% to 20%, for the term infant is 15% to 91%, and at 1 month of age is 21% to 81%. Factor VIII normal range for the preterm infant is 37% to 126%, increases to 50% to 178% at term, and drops to 50% to 157% at 1 month (Andrew et al. 1990).

Defects in the coagulation and fibrinolytic systems can be associated with bleeding disorders. Defects in the formation of the initial platelet plug give immediate bleeding. An example is the bleeding seen in von Willebrand's disease, with poor formation of the primary hemostatic plug causing bruising and mucous membrane bleeding. Defects in the secondary phase of coagulation cause delayed bleeding, such as is seen in hemophilia. There, the coagulation factor defect causes poor creation of the fibrin mesh leading to hemarthrosis and deep muscle bleeding. If there is excessive fibrinolysis, then bleeding results. If there is a defect in the strength of the clot that is formed by the coagulation system, such as when factor XIII fails to crosslink and stabilize fibrin clot, then delayed bleeding also occurs.

CLINICAL EVALUATION OF BLEEDING PROBLEMS

A careful history, including a full family history, is the best means of uncovering bleeding problems in a patient. These disorders may easily escape standard laboratory screening procedures, such as PT, aPTT, fibrinogen, and platelet count. However, patients with mild hemophilia without previous surgical procedures may have no history of bleeding problems. It is very important to consider the history as the most sensitive of the testing procedures and to thoroughly investigate any story of unusual bleeding, even if the screening tests are normal. However, pre-operative coagulation testing done in the absence of a suggestive history can result in false-positive results (Roher et al. 1988). There are several studies in patients undergoing tonsillectomy and adenoidectomy of the utility of the pre-operative PT and PTT, and based on those studies, routine screening with a PT and aPTT of all patients cannot be recommended (Zwack and Derkay 1997; Close et al. 1994).

In obtaining a history from a patient and parents, positive answers to any of the following questions should indicate the need for further evaluation (Rappaport 1983; Sramek et al. 1995).

Is there:

- History of easy bruising, bleeding problems, or an established bleeding disorder in either the patient or any family members?
- Excessive bleeding following any previous surgical or dental procedures in the patient, parents, or siblings; specifically tonsillectomy and/or adenoidectomy?
- Frequent nosebleeds and/or has nasal packing or cautery been needed due to bleeding?
- Bleeding without trauma into any joint or muscle?
- Bleeding or bruising after aspirin ingestion?
- Gingival bleeding following tooth brushing?
- Any medication that might affect platelets or the coagulation system?
- History of bleeding or prolonged oozing after circumcision or when the umbilical cord separated?
- History of excessive menstrual flow?
- Need for previous transfusions of blood or blood products; reason for the transfusion?

If there is a history of abnormal bleeding, the following points need to be established. The type of bleeding (petechiae, purpura, ecchymosis, and single or generalized bleeding sites) can give an indication of the underlying defect. Petechiae and purpura are most frequently associated with platelet abnormalities, either of function or numbers. Von Willebrand's disease is most frequently associated with mucosal bleeding, including epistaxis, whereas hemophilia is most often associated with bleeding into joints and or soft-tissue ecchymosis. Bleeding when the umbilical cord separates is most often associated with factor XIII deficiency (Lorand et al. 1980). A single bleeding site, such as repeated epistaxis from the same nostril is frequently indicative of a localized anatomical problem and not a system-wide coagulation defect.

The course or pattern of the bleeding (spontaneous or posttrauma), its frequency, duration of problems, and severity can provide clues to the etiology of the problem. A family history of bleeding is important to define, and the pattern of inheritance (X-linked, autosomal recessive, or dominant) can help in narrowing the differential diagnosis. Hemophilia A and B are X-linked recessive diseases, while von Willebrand's disease is an autosomal dominant trait. Any previous or current drug therapy must be fully documented, and a search made for any over-the-counter medications that the patient might be taking but does not consider "medicine" and therefore has not mentioned. Repeated doses of ibuprofen, cough medications containing guaifenesin, and antihistamines can uncover a pre-existing bleeding disorder such as von Willebrand's disease, when they would not cause sufficient platelet dysfunction by themselves to result in clinically apparent bleeding (George and Shattil 1991). The presence of other medical problems is important to establish, because problems such as renal failure with uremia, hepatic failure, malignancies, or collagen vascular diseases may have associated coagulation dysfunctions.

The physical examination is used to help narrow the differential diagnosis and guide the laboratory investigation of hemostatic disorders. Certain physical findings may be associated with a specific coagulation abnormality, while others may be indicative of an underlying systemic disease with an associated coagulopathy. Petechiae and purpuric bleeding occur with platelet and vascular abnormalities. If the petechiae are raised, a vasculitis is likely, while petechiae due to thrombocytopenia are not elevated and initially occur on extremities or mucosa. Acquired coagulation defects usually result in widespread ecchymotic bleeding with or without gastrointestinal (GI) or urinary tract bleeding. Bleeding into joints and bleeding that stops and restarts are characteristic of congenital coagulation factor deficiencies. Hemophilia patients often have bruises with a raised central nodule, called palpable purpura. Findings compatible with a collagen vascular disorder include the body habitus of Marfan syndrome, blue sclera, skeletal deformities, hyperextensible joints and skin, poor "cigarette paper" scar formation, and nodular, spiderlike or pinpoint telangiectasias. Hepatosplenomegaly and lymphadenopathy may suggest an underlying malignancy, and jaundice plus hepatomegaly may indicate hepatic dysfunction (Lusher 2003).

LABORATORY EVALUATION

At the present time, the frequently used tests for screening the hemostatic system are the platelet count, PT, aPTT, and fibrinogen. Additional tests can be done to assess the thrombin clotting time, screen for inhibitors of specific coagulation factors, and measure specific factor levels. Evaluation for evidence of disseminated intravascular coagulation (DIC) can also be done using multiple assays to test for the presence of various fibrinopeptides, depletion of anticoagulation factors, and the presence of products from the breakdown of fibrin or fibrinogen.

Platelet Count

The platelet count is obtained as part of a complete blood count and measures the adequacy of platelet numbers to provide initial hemostasis. Platelet counts are usually done using an automated hematology counter. The normal range is between 150,000 and 450,000/μL.

Bleeding Time and PFA-100

The bleeding time is the length of time a standardized incision takes to stop oozing blood that can be absorbed onto filter paper. A variety of procedures have been used, including the Duke method with a stab incision of the earlobe and the Ivy method with a standardized cut on the forearm, but both have been difficult to reproduce accurately. At present, standardized tests use a spring-loaded blade to make a controlled cut of a specific length and depth on the volar surface of the forearm after application of a blood pressure cuff. The normal values vary, depending on the procedure used and the individual laboratory, but are usually less than 9 minutes. Questions regarding the sensitivity, specificity, reliability, and predictability of the bleeding time have resulted in much less use of this test (Rodgers and Levin 1990).

The PFA-100 platelet function screen (Dade Behring) is a potential replacement for the bleeding time and is becoming available in many standard coagulation laboratories. It is an automated device that aspirates a small amount of citrated whole blood through an aperture in a membrane coated with collagen and epinephrine or adenosine diphosphate (ADP). This simulates a high shear condition and results in the vWF-dependent attachment, activation, and aggregation of platelets, resulting in a closure time that is dependent on occlusion of the aperture by a stable platelet plug. Studies show that the closure time is superior to the bleeding time in screening for most cases of von Willebrand's disease, aspirin effect, and for some causes of platelet dysfunction, giving a very high negative predictive value. Test results can be influenced by the sample's hematocrit (Kundu et al. 1995; Mammen et al. 1998).

Prothrombin Time

The PT screens the function of the extrinsic and common coagulation pathways. It is the time required to clot platelet-poor plasma after the addition of tissue factor, calcium, and phospholipid, the tissue factor being the material that is "extrinsic" to the plasma-based coagulation system. Isolated prolongations of the PT are seen in factor VII deficiency, fibrinogen defects, and in patients on warfarin anticoagulation. It is also considered the most sensitive screening test for liver dysfunction coagulopathies (Lusher 2003). The thromboplastin reagents used in the PT assay vary in their sensitivity to warfarin-induced reductions in clotting factor activity. Consequently, a patient could receive substantially different doses of warfarin depending on the thromboplastin used in the PT test. To control for this, the International Normalized Ratio (INR) was developed. Each thromboplastin reagent is assigned an International Sensitivity Index (ISI), which is used to adjust the PT ratio (patient PT divided by the mean normal PT for the lab) to correct for the sensitivity of the thromboplastin. Normal value for an INR is 1 and therapeutic values for warfarin anticoagulation are 2 to 3 (DeLoughery 2001).

Partial Thromboplastin Time

The aPTT screens the function of the intrinsic and common coagulation pathways. Platelet-poor plasma is incubated with kaolin, Celite, or ellagic acid to form activated factor XII. Then calcium and phospholipid are added, and the time to formation of clot is measured. Factor deficiencies of <30% to 40% of normal levels are needed to produce an abnormal test result, and the level at which the test becomes abnormal depends on the reagents and testing equipment used. The aPTT will detect deficiencies in factors XII, XI, IX, and VIII and the common pathway, but it is important to remember that mild factor deficiencies may be missed. The aPTT is also used to monitor anticoagulation with heparin (Thompson and Poller 1985). Several inherited disorders of coagulation will not be detected by the preceding tests. Factor XIII deficiency is detected by a urea clot solubility test (Lorand et al. 1980). Patients with von Willebrand's disease may have either normal or prolonged aPTTs, and patients with alpha 2-plasmin inhibitor deficiency have a normal aPTT. Both the PT and aPTT will be prolonged in patients with deficiencies of factors X, V, prothrombin, and fibrinogen, and in patients with DIC or severe liver disease (Lusher 2003; Goodnight and Hathaway 2001).

Thrombin Time (Thrombin Clotting Time)

The thrombin time, also called the thrombin clotting time (TCT), is a measure of fibrin formation, the final reaction of the clotting cascade. It is performed by adding thrombin to platelet-poor plasma and measuring the time to the formation of a fibrin gel. The thrombin time is normal in patients with defects in the intrinsic or extrinsic pathway but is prolonged with low levels of fibrinogen, with a dysfunctional fibrinogen, or when inhibitors of thrombin are present, such as heparin or fibrin-split products. This test is extremely sensitive to the presence of heparin (Greaves and Preston 2001).

Fibrinogen

The standard method for fibrinogen determination measures clottable fibrinogen using a kinetic assay. Normal levels of fibrinogen are 150 to 400 mg/dL. Since fibrinogen is the substrate for the final reaction in the formation of a clot, and all plasma-based screening tests depend on the formation of a clot as the endpoint of the reaction, fibrinogen levels below 100 mg/dL prolong the PT, aPTT, and thrombin time and make the results uninterpretable. Large amounts of fibrin degradation products will interfere with the formation of fibrin and cause an artificially low level of fibrinogen to be measured. Partially clotted samples will also cause a low level of fibrinogen to be assayed. An immunological-based assay for fibrinogen is available and is used to measure both clottable and nonclottable fibrinogen. This is most often used in identifying patients with a dysfibrinogenemia, where the functional level of fibrinogen is low and the immunological level is normal (Goodnight and Hathaway 2001).

Inhibitor Screening Tests

Repeating the PT or aPTT using a 1 : 1 mix of patient plasma with normal plasma is a useful procedure to do in the investigation of a prolonged PT or aPTT. Normal plasma has, by definition, 100% levels of all factors. When mixed with an equal volume of patient plasma, there will be a minimum of 50% of any given factor present, which should normalize the PT or aPTT. If the test normalizes, it suggests the presence of a factor deficiency, while lack of normalization suggests the presence of an inhibitor that interferes with either thrombin or fibrin formation.

Two types of acquired inhibitors prolong the aPTT. One blocks or inactivates one of the factors in the intrinsic pathway while the other is an antiphospholipid antibody, such as a lupus anticoagulant, that interferes with phospholipid-based clotting reactions. The first type of inhibitor occurs in 5% to 15% of hemophiliacs and can occur spontaneously but is extremely rare in nonhemophiliac children (DiMichele 2001). The strength of a specific factor inhibitor is measured with the Bethesda assay. This measures the amount of antibody that will inactivate 50% of the normal factor VIII or factor IX in 2 hours, and the dilution of normal plasma with test plasma that inhibits this amount of factor is determined. If the 1 : 10 dilution has this effect, then the inhibitor plasma has 10 Bethesda units of activity (Montgomery et al. 2003). The lupus anticoagulant or the "lupuslike" inhibitor is not associated with bleeding problems, but with an increased risk of thrombosis. Lupuslike inhibitors are mentioned because they commonly cause prolongations of the aPTT in children (Thiagarjan and Shapiro 1982). Specific investigation of either of these situations should be referred to a skilled coagulation specialty center.

SPECIFIC FACTOR ASSAYS

Specific factor assays are available for all known coagulation, fibrinolysis, and anticoagulation factors to quantitate their levels in plasma. These tests are not indicated unless a screening test result is abnormal. The only exceptions involve the patient with a history that is suspicious for von Willebrand's disease, factor XIII deficiency, or dysfibrinogenemia when the PT, aPTT, and fibrinogen activity measurement may not detect the decreased level or activity of von Willebrand's factor, factor XIII, or of the dysfunctional fibrinogen. Further testing may be justified by clinical suspicion based on the patient's history.

Tests for Disseminated Intravascular Coagulation

The usually available tests in most hospital laboratories for identification of DIC are semiquantitative fibrin/fibrinogen degradation product assays. Finding an increased amount of these degradation products provides evidence that either plasmin has circulated to lyse fibrin and fibrinogen or the patient's hepatic function is insufficient to clear the amounts of regularly produced degradation products. The D-dimer test is also a slide agglutination procedure that tests for the presence of two D subunits of fibrin crosslinked by factor XIII.

This test provides specific evidence that plasmin has digested fibrin clot and not fibrinogen. It is positive in patients with DIC, in patients resolving large intravascular clots, and in patients with hepatic insufficiency. Specific assays to demonstrate the presence of soluble fibrin monomer complexes or fibrinopeptides produced by the conversion of prothrombin to thrombin are available in specialized coagulation laboratories (Levi and Cate 1999).

INHERITED COAGULATION DISORDERS

Hemophilia A and B

Hemophilia is a congential bleeding disorder caused by the deficiency or absence of factor VIII activity for hemophilia A (classical hemophilia) and factor IX activity for hemophilia B (Christmas disease). These factors are zymogens that when activated, accelerate the activation of factor X, which results in the formation of thrombin. Deficiency of either of these factors causes hemorrhage because of delayed clot formation and an abnormally friable clot. The gene responsible for hemophilia A resides on the long arm of the X chromosome at Xq28, and the gene for hemophilia B resides near the terminus of the long arm of the X chromosome (Antonarakis 1998; Lillecrap 1998). This disorder predominantly affects males but can also cause bleeding in carrier females with low factor levels. It affects one in 5000 males (Soucie et al. 1998) and causes a spectrum of bleeding symptoms including hemarthrosis, deep muscle hemorrhages, easy bruising, oral bleeding, intracranial hemorrhage, and gastrointestinal and renal bleeding. Approximately one-third of all hemophilia cases represent new mutations, where there is no family history of a bleeding disorder (Forbes 1997).

The level of circulating factor VIII or IX determines the severity of the hemophilia. Table 20.1 gives the details of the factor levels associated with each level of severity. Moderate hemophiliacs have less frequent bleeding than severe hemophiliacs but still have joint

TABLE 20.1   Clinical Manifestations of Hemophilia by Severity

Factor Level	Severity	Bleeding Manifestations	Age at Diagnosis
1% or less	Severe	Spontaneous hemarthrosis and muscle bleeds, "target" joints, mucosal bleeding, CNS bleeding, GI bleeding, hematuria, trauma, or surgical induced bleeding	Infancy
2% to 5%	Moderate	Trauma-induced hemarthrosis and muscle bleeding, "target" joints, CNS bleeding, mucosal bleeding, GI bleeding, hematuria, trauma, or surgical-induced bleeding	Toddler to young child
More than 5%	Mild	Unusual, except with severe trauma or surgery	Late childhood to early adulthood

and muscle bleeding and can develop "target joints" as well as bleeding from other sites. Mild hemophiliacs may not be diagnosed until they are in their teens or early 20s, because of the lack of bleeding symptoms (Montgomery et al. 2003).

Treatment of Hemophilia A and B

Modern management of hemophilia started in the 1970s. Before that, no efficacious therapy was available, and the average life expectancy of a severe hemophilia patient was less than two decades. Cryoprecipitate provided the first therapeutic material that could provide sufficient factor VIII to treat or prevent bleeding in hemophilia A patients. However, use of cryoprecipitate tied the hemophilia patient to the hospital. In the mid-1970s, coagulation factor concentrates were developed for factors VIII and IX, which could be stored in a standard refrigerator. This development allowed the establishment of home infusion programs, so the hemophilia patient, his or her parents, or a home care nurse could do the infusion of the factor concentrates to control bleeding. Emergency room, clinic, or in-patient hospital time was markedly reduced. This self-infusion system, combined with multispecialty comprehensive hemophilia centers has decreased the mortality and morbidity from hemophilia and allowed the patient with hemophilia to better integrate into society (Mannucci and Tuddenham 2001).

Treating bleeding episodes in hemophilia is done by the prompt intravenous replacement of the deficient clotting factor to a sufficient level and then maintaining that replacement for the length of time needed for the overt bleeding to resolve or for the threat of bleeding to pass. The earlier bleeding can be treated, the less damage occurs to the involved joint or tissue. The type of bleeding episode dictates the necessary level of clotting factor replacement and the recommended doses and schedules of factor replacement are shown in Table 20.2. Prophylactic infusions of factor may be needed in central nervous system (CNS), GI, and retroperitoneal bleeding episodes after the overt bleeding has resolved, due to the risk of repeat bleeding at the same site (DiMichele 2001; Montgomery et al. 2003).

Surgery in Hemophilia

Management of surgery in patients with hemophilia should be done in specialty centers with experienced medical and nursing staff and with the necessary laboratory and pharmacy support. The patients must be screened for the presence of an inhibitor just before surgery, recovery and survival studies may need to be done to ensure that an appropriate dosing regimen is

used, and an adequate supply of the necessary clotting factor concentrate must be available before the procedure starts. On the morning of surgery, a 100% correction of clotting factor is given, and the factor level is maintained with scheduled infusions or a continuous infusion of clotting factor at greater than 50% to 60% for the next 5 to 7 days. After that, factor levels can be allowed to decrease to a 30% trough level for the period of time needed for postsurgical healing and rehabilitation to take place. Using a bolus-dosing regimen may be simpler than a continuous infusion of factor, but daily monitoring of trough factor levels is needed to ensure that adequate hemostatic levels are being maintained (Kobrinsky and Stegman 1997; Montgomery et al. 2003).

Continuous infusions of clotting factor have been used for surgical coverage and treatment of life-threatening hemorrhages in hemophilia A patients since the early 1980s. After the infusion of a 100% bolus dose, a 2 to 3 unit/kg/hour infusion of factor is given to maintain circulating levels of 50% to 60% (Hathaway et al. 1984; Montgomery et al. 2003). Advantages to using this over bolus dosing of factor include avoidance of the peaks and valleys of bolus dose factor replacement, up to a 30% reduction in the amount of factor needed to maintain a given factor level if the infusion extends for more than 3 to 4 days, and flexibility in scheduling the drawing of factor levels (Martinowitz et al. 1992). With the advent of low thrombogenicity factor IX concentrates, continuous infusion of factor has become possible for hemophilia B patients, although the 24-hour half-life of factor IX makes the bolus-dosing regimen easier to accomplish (Hoots et al. 2003). Previous prothrombin complex concentrates used for factor IX replacement had thrombosis risks, and their use is not recommended for surgery or other risky situations (Montgomery et al. 2003).

Prophylaxis in Hemophilia

Based on the experience of Swedish physicians in giving scheduled doses of factor VIII or IX concentrate to patients with moderate or severe hemophilia to prevent joint bleeding from occurring (Nilsson et al. 1992), the Medical and Scientific Advisory Council of the National Hemophilia Foundation recommended in 1994 that as optimal therapy, primary prophylaxis begin at age 1 to 2 years for all children with moderate or severe hemophilia. This was because of the proven ability of this treatment to prevent the long-term development of hemophilic arthropathy (DiMichele 2001). As practiced in North America, most centers institute prophylaxis as soon as joint bleeding occurs, and give 25 to 40 units/kg of factor VIII between three times per week to every other day. The dose and infusion sched-

TABLE 20.2 Guidelines for Factor Replacement Therapy in Hemophilia

Location of Bleed	Recommended Factor Dosing	Comments
Hemarthrosis, general	50% correction every 12 to 24 hours as needed.	RICE (rest, ice, compression, immobilization) may be used, in addition to factor. If target joint present, use 100% correction first day, 50% correction second day, and 50% correction on fourth day.
Intracapsular shoulder or hip	100% correction then 50% every 12 to 24 hours as needed.	Extremely painful bleeding, may require narcotic pain medications and up to 7 days of factor replacement.
Muscles or large hematoma	50% to 100% correction every 12 to 24 hours as needed. 30% correction daily with physical therapy to prevent rebleeding.	Check carefully for compartment syndrome on bleeding involving calf and forearm. Fasciotomy may be needed.
Psoas or other retroperitoneal location	100% correction then 50% every 12 hours till resolution, then 40% prophylaxis dosing for up to 2 weeks.	Significant blood loss possible, frequent recurrence of bleeding. Consider use of prophylaxis two or three times per week for several months to avoid recurrence.
CNS documented bleed	100% correction then 50% every 12 to 24 hours for 14 to 21 days. Consider 30% to 40% prophylaxis dosing for 2 to 3 months.	May need continuous infusion of factor, prophylactic use of anticonvulsants. Watch for repeat bleeding at 2 to 3 weeks from initial injury. Recurrence of of bleeding possible.
CNS preventive treatment following trauma	100% correction, then observation.	Factor must be infused before any evaluation. CT scan may or may not be needed, based on clinical situation.
Gastrointestinal	100% correction, then 30% to 50% daily till healing. Consider 30% to 40% prophylaxis dosing for 2 to 3 months based on severity of bleed.	Antifibrinolytic therapy for 3 to 5 days. Endoscopy may identify bleeding point and is recommended.
Genitourinary	100% correction, then 30% daily till resolution.	Evaluate for stones or urinary tract infection (UTI). Lesion usually not found. Prednisone 1 to 2 mg/kg/day for 5 to 7 days may help stop recurrent bleeding, if HIV-negative. Antifibrinolytic therapy is contraindicated.
Oral mucosa or epistaxis	50% to 100% correction depending on severity, then 30% to 50% daily to every other day till resolution. Antifibrinolytics for 5 to 7 days.	Antifibrinolytic therapy may be all that is needed to prevent recurrence, if bleeding has stopped. Can be given as mouthwash for oral bleeding. Avoid using prothrombin complex concentrates (PCCs). Activated prothrombin complex concentrate (APPCs) with antifibrinolytics.
Tongue, neck, or impending airway compromise	100% to 150% correction, then 50% correction every 12 hours or continuous infusion of factor to maintain levels at 100% for 24 hours, then 50% level till resolution of bleed.	If using bolus therapy, maintain trough level >50%. May need 7 to 14 days of therapy based on severity of bleeding.

ule are adjusted to maintain a nadir of greater than 1% factor activity. Hemophilia B patients are given a 30 to 60 unit/kg dose twice per week, due to the 24-hour half-life of factor IX in circulation. Barriers to the implementation of this treatment program include the cost of the factor, which may result in a fourfold increase in factor usage over on-demand therapy and the frequent need for central venous access devices in young children, with risks of infection and thrombosis (Aledort and Bohn 1996).

Inhibitors in Hemophilia

A complication of the treatment of hemophilia is the development of inhibitory antibodies (inhibitors) that block procoagulant function, which is one of the most serious complications of hemophilia treatment. Up to 46% of all hemophilia A patients develop some level of inhibitor at some point in their life, with 15% to 20% of severe hemophilia A patients developing high titer inhibitors (Briet et al. 1994). The low titer inhibitors, less than 5 Bethesda units in activity, are often transient, rarely have anamnestic responses to repeated exposure to factor, and can be treated with higher doses of factor. The high-titer inhibitors, more than 5 Bethesda units of inhibitor activity, rarely transient, have an anamnestic response to repeated exposure to factor, and bleeding cannot be treated with increased doses of factor. An additional complication exists for hemophilia B patients, who have a 5% rate of inhibitor development, and may exhibit anaphylaxis to factor IX infusions as a part of the development of the inhibitor (Warrier et al. 1997).

The treatment of hemophilia patients with inhibitors centers on two problems: first, to immediately control bleeding, and second, to eradicate the inhibitor, restoring normal response to regular factor infusions. Control of bleeding in inhibitor patients depends on the titer of the inhibitor. If it is a low-titer inhibitor, increasing the dose of factor for regular bleeding or giving a 100 unit/kg dose of factor VIII followed by a continuous infusion of factor VIII at 20 units/kg/hr frequently controls serious or life-threatening bleeding. Kasper has suggested that the low-titer inhibitor patients receive 40 units of factor VIII /kg body weight for each Bethesda unit of inhibitor (1989). If it is a high-titer inhibitor, then use of very high (100 to 200 units/kg/hour) continuous infusions of factor VIII may control the bleeding, but anamnestic responses may drive the inhibitor titer beyond control with factor VIII infusions. Then alternative treatment with porcine factor VIII concentrate, recombinant factor VIIa, prothrombin complex concentrates (PCCs), or activated prothrombin complex concentrates (APCCs) may be needed (Hoyer 1997; Montgomery et al. 2003).

The second part of inhibitor management is the long-term eradication of the inhibitor by use of immune tolerance regimens with scheduled infusions of factor VIII or IX to suppress production of the inhibitor and return to a normal recovery and half-life of infused factor. If tolerance can be achieved, then responsiveness to factor infusions returns, and bleeding episodes can be controlled with standard infusions of factor (Brackman and Gormsen 1977). Management of hemophilia patients with inhibitors should be done at a Comprehensive Hemophilia Treatment Center with staff who are experienced in managing these patients, as the costs of treatment are extremely high, and the morbidity and mortality associated with inhibitors is significant.

Types of Clotting Factor for Treating Hemophilia

Clotting factor concentrates come from two sources, either human plasma donations or production via recombinant methods from cell cultures. Currently produced human plasma-derived factor concentrates start with a pool of screened donor plasma that is processed to a combination of factor VIII and vWF called an intermediate purity factor concentrate, to a monoclonal immunoaffinity column purified factor called a high purity or monoclonal purified factor, or to a prothrombin complex concentrate that contains factor IX as well as other vitamin K-dependent proteins (Shapiro 2001; Mannuccci 2003).

Viral safety of the clotting factor concentrates has been a primary concern for the past two decades. The initial pooled, plasma-derived clotting factor concentrates for factor VIII and IX transmitted hepatitis B, hepatitis C, and human immunodeficiency virus (HIV). Over 50% of the U.S. hemophilia population became HIV-positive as a result of contaminated clotting factor concentrates, and the long-term effects of hepatitis C infection, including cirrhosis and hepatocellular carcinoma, are now the leading causes of liver transplant for hemophiliacs (Evatt and Hooper 1997; Gordon et al. 1998). With the development of virally inactivated, plasma-derived concentrates, and then recombinant-produced clotting factor concentrates, infection issues have receded but are not absent. Parvovirus can be transmitted by virally inactivated plasma-derived products, and there are still concerns about possible transmission of Creutzfeldt-Jakob disease (CJD) or "new variant" CJD by human or bovine protein-containing products, even though no cases have been associated with blood products or blood transfusions in humans (Evatt and Hooper 1997; Turner 1999).

Because of concerns about possible transmission of "new variant" CJD by transfusion of blood or blood products, efforts are being made to remove human or animal-derived proteins from the production of clotting factor concentrates. The first generation recombinant factor VIII product has human albumin as a stabilizer, while second generation factor VIII products do not use human albumin to stabilize the final preparation, although they have human and animal-derived material present in the manufacturing process. A third generation factor VIII product is being introduced, which uses very little human or animal-derived material at any stage in the production or stabilization of the concentrate. The present recombinant-produced factor VIIa and factor IX products do not have human albumin as a stabilizer in the final preparation (Mannucci 2003).

In guiding the choice of clotting factor concentrate to treat pediatric patients, a general recommendation is products providing the highest degree of safety and least associated risk should be used for all patients. Recombinant products are recommended for patients who have not been previously transfused with blood or treated with a plasma-derived factor concentrate. However, many older patients who have been treated with the virally inactivated, plasma-derived factor concentrates have remained on these products for many years without evidence of problems. Studies done by the Centers for Disease Control and Prevention (CDC) have not shown any evidence of transmission of infection by a clotting factor concentrate in the United States since 1987 (Evatt and Hooper 1997). High purity factor IX products, either monoclonally purified or recombinant, do not have the thrombosis risks that were seen with older and are the factors of choice for the treatment of hemophilia B patients. Because of the variability of dose response with the recombinant factor IX product, individual recovery studies may need to be done in the young child, in patients that do not demonstrate the expected factor level increment to factor infusions, and in pre-operative patient evaluations (Shapiro 2001). Table 20.3 lists a selection of the factor concentrates commercially available in the United States as of early 2003.

Desmopressin

For mild hemophilia A patients, desmopressin (DDAVP) offers an alternative to factor concentrate infusions for the control of bleeding symptoms. It is a synthetic analogue of vasopressin that when given in higher concentrations than used to treat diabetes insipidus, raises factor VIII and vWF levels, with few major side effects. The factor VIII and vWF levels are usually raised two to three times the baseline level, but specific testing on each patient is needed, as some fail to raise their factor levels in response to this medication (Mannucci et al. 1977). Dosing for intravenous use is 0.3 μg/kg/dose of DDAVP in 50 mL of normal saline and infused over at least 30 minutes. A nasal high-concentration DDAVP spray (Stimate, Aventis-Behring) is available and simplifies dosing and administration for some patients. Dosing of the nasal DDAVP is 150 μg/dose for patients weighing under 50 kg and 300 μg/dose for patients weighing 50 kg or more. This corresponds to one puff of desmopressin (Stimate) for the 150 μg dose and two puffs for the 300 μg dose. Factor VIII and vWF levels reach their peak within 30 to 60 minutes after administration, and the high levels of factor VIII and vWF last for 6 to 8 hours. Infusions can be repeated every 12 to 24 hours, but if they are given more frequently than every 48 hours, tachyphylaxis with a lessening of response to repeated dosing occurs. Side effects seen with DDAVP use include flushed face, headaches, nausea and vomiting, tachycardia, hypertension, and, rarely, hyponatremia due to the retention of free water. Because of the free water retention, DDAVP is not recommended for use in infants younger than 11 months, and older patients should be fluid restricted to satisfaction of thirst for 24 hours after its use. Seizures due to hyponatremia have been reported as a rare side effect of this medication (Montgomery et al. 2003).

Antifibrinolytics in the Treatment of Hemophilia

An adjunctive treatment to control bleeding from mucosal surfaces is antifibrinolytic therapy with either epsilon-aminocaproic acid (EACA) or with tranexamic acid. These have been found to be effective in controlling epistaxis, oral mucosal bleeding, and particularly bleeding from dental extractions. They may also be a significant adjunct to controlling excessive menstrual bleeding, which can be seen in symptomatic carrier females for hemophilia A or B or in von Willebrand's disease. Oral dosing for EACA is 100 mg/kg to 200 mg/kg initial dose with a maximum dose of 10 grams, followed by 50 to 100 mg/kg/dose with a maximum dose of 5 grams every 6 hours. The oral dose of tranexamic acid is 25 mg/kg every 6 to 8 hours. These medications are usually given for 7 to 10 days and can also be given as an oral mouthwash for localized control of bleeding (Stajcic 1985; Montgomery et al. 2003).

Dosage Calculation for Hemophilia A

To calculate the dose of factor VIII concentrate needed to treat specific bleeding, one unit of activity for factor VIII is defined as the amount of coagulant activity in 1 mL of fresh normal plasma. Infusion of one unit per kilogram of body weight of factor VIII increases the circulating factor VIII level by 2%. The following formula is used to estimate the number of units of factor to be given to obtain a specific factor level in circulation for hemophilia A patients:

$$\frac{(\% \text{ Activity desired} - \% \text{ Activity baseline}) \times \text{Weight (kg)}}{2} = \# \text{ of Units}$$

The percentage activity levels are expressed as a whole number, such as 100 for 100% factor activity level. An example of the calculation to get a 100% correction for a severe hemophilia A patient who weights 50 kg and has a baseline factor VIII activity of 1% is:

$$\frac{(100 - 1) \times 50\,\text{kg}}{2} = 2475 \text{ units}$$

Any calculation of dosing for factor products will need to be adjusted for the available vial sizes of factor concentrate. As a generalization, it is better to use the entire vial of factor than to waste material to give just the calculated number of units. Adjustment of the dose by +/– 10% will not affect efficacy.

Dosage Calculation for Hemophilia B

Calculation of factor dosing for hemophilia B patients is very similar to that for hemophilia A. The only difference is that one unit per kg of factor gives an increase in the circulating factor levels of 1% activity:

$$\frac{(\% \text{ Activity desired} - \% \text{ Activity baseline}) \times \text{Weight (kg)}}{1} = \# \text{ of Units}$$

TABLE 20.3   Factor Products Available in the United States

Product	Manufacturer	Preparation	Viral Attenuation/Inactivation	Clinical Use/Comments
Factor VIII/vWF Concentrates				
Humate-P	Aventis-Behring	Multiple precipitation	Pasteurization 60°C, 10 hours	Hemophilia A, von Willebrand's disease/albumin added
Koate DVI	Bayer	Multiple precipitation, size exclusion chromatography	TNBP/polysorbate 80, heat dry 80°C 72 hours	Hemophilia A, von Willebrand's disease/albumin added
Alphanate	Alpha	Heparin ligand, chromatography	TBNP/polysorbate 80, dry heat 80°C 72 hours	Hemophilia A, von Willebrand's disease/albumin added
Immunoaffinity Purified Factor VIII Concentrates (High Purity Product)				
Monoclate-P	Aventis-Behring	Monoclonal Ab affinity chromatography	Pasteurization 60°C, 10 hours	Hemophilia A/albumin added
Hemofil M AHF	Baxter	Monoclonal Ab affinity, ion exchange chromatography	TNBP/Triton X-100	Hemophilia A/albumin added
Monarc M	American Red Cross	Monoclonal Ab affinity, ion exchange chromatography	TNBP/Triton X-100	Hemophilia A/albumin added
Recombinant Factor VIII Concentrates				
Recombinate	Baxter	Recombinant, ion exchange, immunoaffinity chromatography	Viral attenuation through purification	Hemophilia A/full-length FVIII, albumin added
Refacto	Wyeth	Recombinant	TNBP/Triton X-100	Hemophilia A/B-domain-deleted FVIII, no albumin
Kogenate FS	Bayer	Recombinant, ion exchange, immunoaffinity chromatography	TNBP/Triton X-100	Hemophilia A/full-length FVIII, no albumin
Helixate FS	Aventis-Behring	Recombinant, ion exchange, immunoaffinity chromatography	TNBP/Triton X-100	Hemophilia A/full-length FVIII, no albumin
Factor IX- Prothrombin Complex Concentrates (PCC)				
Bebulin VH	Baxter	Ion exchange adsorption	Vapor heat 60°C 10 hours, 80°C 1 hour	Hemophilia B, heparin added
Proplex T	Baxter	Tricalcium phosphate adsorption, PEG fractionation	20% ethanol, dry heat 60°C 144 hours	Hemophilia B, heparin added
Profilnine SD	Alpha	DEAE cellulose absorption	TNBP/polysorbate 80	Hemophilia B, AT and heparin added, no albumin added
Coagulation Factor IX Concentrates				
Alpha-Nine SD	Alpha	Ion exchange, carbohydrate ligand chromatography	TNBP/polysorbate 80, nanofiltration	Hemophilia B, no albumin
Mononine	Aventis-Behring	Immunoaffinity chromatography	Sodium thiocyanate, ultrafiltration	Hemophilia B
Recombinant Factor IX Concentrates				
Benefix	Wyeth	Recombinant	Nanofiltration	Hemophilia B, no albumin added

TABLE 20.3—cont'd

Product	Manufacturer	Preparation	Viral Attenuation/ Inactivation	Clinical Use/Comments
Inhibitor Treatment Products				
Activated Prothrombin Complex Concentrates (APCCs)				
FEIBA VH	Baxter	Ethanol, DEAE-Sephadex, surface-activated PCC, batch controlled	Vapor heat 60°C 10 hours, 80°C 1 hour	Factors VIII and IX inhibitor bypassing agent-contains factor IX, no heparin
Autoplex T	NABI	TriCa PO$_4$ adsorption, PEG fractionation, surface-activated PCC	20% ethanol, dry heat 60°C 144 hours	Factors VIII and IX inhibitor bypassing agent-contains factor IX, heparin
Porcine Factor VIII Concentrate				
HYATE:C	IPSEN, Inc.	Cryoprecipitation Polyelectrolyte ion-exchange Chromatography	None: end product cell culture viral screen	Factor VIII inhibitor bypassing agent
Recombinant Factor VIIa Concentrate				
NovoSeven	NovoNordisk	Recombinant	None	Factors VIII and IX inhibitor bypassing agent, factor VII deficiency

Modified from Kasper and Costa e Silva, 2003.

Using the same information, a 50-kg patient with a baseline factor level of 1%, the following is an example of calculating the dose for a 100% correction:

$$\frac{(100 - 1) \times 50\,\text{kg}}{1} = 4950 \text{ units}$$

When calculating dosing for recombinant factor IX, due to differences in the volume of distribution, one unit per kilogram of body weight will only give an increase of 0.8% activity, and the equation is as follows:

$$\frac{(\% \text{ Activity desired} - \% \text{ Activity baseline}) \times \text{Weight (kg)}}{0.8} = \# \text{ of Units}$$

The example calculations are as follows, using the same baseline data:

$$\frac{(100 - 1) \times 50\,\text{kg}}{0.8} = 6188 \text{ units}$$

(rounded to next whole number)

Special attention has to be paid to the issue of dosing for pediatric patients with recombinant factor IX, as some patients get significantly less than a 0.8% increase for every unit per kilogram of body weight of factor that is infused (Shapiro 2001; Montgomery et al. 2003).

Dosing of Factor Concentrates in Inhibitor Patients

Dosing for control of bleeding in the inhibitor patient will initially be with increased doses of factor VIII or factor IX, as long as there is no anaphylaxis to the infusion of the factor. However, many high-titer inhibitor patients will fail to respond, and then porcine factor VIII, activated prothrombin complex concentrates (APCCs), or recombinant factor VIIa is used to control bleeding. Dosing for porcine factor VIII is similar to dosing for human factor VIII, and it can be used as long as the inhibitor does not crossreact with the porcine factor VIII molecule. An initial dose is often 100 to 150 units/kg, patients must be observed for allergic reactions, and a routine factor VIII assay can be used to monitor therapy (Hay et al. 1994). APCCs are usually used to control routine muscle and joint hemorrhages and are initially dosed at 75 units/kg of factor IX. If there is no response after two to three doses given every 12 hours, then it is unlikely that further dosing will be successful, and studies have shown only a 40% to 60% success rate in controlling hemarthrosis when APCCs are used (Lusher et al. 1980). APCCs contain factor IX, so they should not be used in a hemophilia B patient who has developed anaphylaxis along with his or her inhibitor (Mannucci 2003). Recombinant factor VIIa is initially dosed at 90 µg/kg every 2 hours till bleeding is controlled. Subsequent doses of factor VIIa can be increased to 180 µg/kg every 2 hours if the patient does not seem to be responding to the lower dose (Montgomery et al. 2003).

von Willebrand's Disease

von Willebrand's disease is a highly heterogenous, usually mild, autosomal dominant, inherited coagulation disorder. Laboratory abnormalities in vWF are found in roughly 1% of the world's population, with 10% of these patients having bleeding problems, making it the most prevalent inherited bleeding disorder (Sadler et al. 2000). vWF is a circulating plasma glycoprotein that is made up of high, intermediate, and low molecular weight multimers. The high molecular weight multimers are responsible for the vWF binding to glycoproteins Ib and IIb/IIIa on the surface of platelets, which is a critical step in the adhesion and aggregation of platelets to the injured vessel wall. A second role that vWF plays is to act as a carrier for the factor VIII molecule in circulation. By binding factor VIII to the vWF molecule, it protects the factor VIII from proteolytic degradation and helps to efficiently localize the factor VIII to the site of vascular injury.

Mucocutaneous bleeding is the hallmark of von Willebrand's disease, with patients showing symptoms that include easy bruising; epistaxis; menorrhagia; bleeding after dental extractions or ear, nose, throat (ENT) procedures; bleeding from minor wounds; postsurgical bleeding; postpartum bleeding; gastrointestinal bleeding; and rare joint hemarthrosis or central nervous system bleeding (Silwer 1973; Federici et al. 2002). With an autosomal dominant pattern of inheritance, both males and females with von Willebrand's disease can have bleeding symptoms. von Willebrand's disease has variable penetration of the gene, so that the offspring of a person with von Willebrand's disease can inherit the gene without having abnormal vWF levels and without having bleeding symptoms. This person can then pass the gene on to their offspring, who could have bleeding symptoms (Castaman et al. 1999).

Classification of von Willebrand's Disease

von Willebrand's disease is divided into three subtypes, with types 1 and 3 having quantitative defects in the vWF levels, and type 2 having qualitative defects in vWF structure and function (Sadler 1994). Details of the various types of von Willebrand's disease are summarized in Table 20.4. Bleeding symptoms are usually mild in type 1 and increase in severity in types 2 and 3. Type 3 gives severe bleeding symptoms that are similar

to hemophilia, including spontaneous bleeding and hemarthrosis (Sadler et al. 2000).

Diagnosis of von Willebrand's Disease

In order to make the diagnosis of von Willebrand's disease, there needs to be a combination of mucocutaneous bleeding symptoms in the patient and abnormal vWF laboratory tests combined with a family history of bleeding. The laboratory testing commonly involves serial determinations of plasma factor VIII (FVIII : C) level, vWF antigen (vWF : Agn) level, and ristocetin cofactor activity (vWF : RCo). The vWF : RCo is a measurement of the ability to agglutinate platelets in the presence of vWF and ristocetin. The vWF : RCo is used as a measurement of vWF activity. If the FVIIII : C level, the vWF : Agn, and the vWF : Rco are all low to unmeasurable, then the patient has type 3 von Willebrand's disease. If the vWF : Rco to vWF : Agn ratio is proportional, 0.7 to 1.2, then the patient has a normal vWF structure, and has either type 1 or type 2N von Willebrand's disease. If the vWF : Rco to vWF : Agn ratio is discordant, less than 0.7, then there is an abnormal vWF structure, and the patient has type 2A, type 2B, or type 2M von Willebrand's disease (Montgomery et al. 2003; Castaman et al. 2003).

To separate type 1 from type 2N von Willebrand's disease, the ratio of FVIII : C to vWF : Agn is used. If the ratio is greater than or equal to one, then it is concordant, and the patient has type 1 von Willebrand's disease. If the ratio is less than one, then suspect type 2N von Willebrand's disease, and a factor VIII binding assay is used to confirm the diagnosis. This measures the affinity of vWF for factor VIII, and will distinguish type 2N von Willebrand's disease from mild hemophilia A (Castaman et al. 2003). The PFA-100 platelet function analyzer is sensitive to all types of von Willebrand's disease except type 2N, where it gives normal results (Fressinaud et al. 1998; Dean et al. 2000).

To separate type 2A, type 2B, and type 2M von Willebrand's disease, von Willebrand multimer analysis and ristocetin-induced platelet agglutination (RIPA) at low concentrations of ristocetin are used. The von Willebrand's multimer assay is an agarose gel electrophoresis that separates low, intermediate, and high molecular weight multimers of vWF. The RIPA test is used to separate type 2B von Willebrand's disease from the other type 2 subtypes. Most von Willebrand's disease subtypes show a very low response in this assay when low concentrations of ristocetin are used. In type 2B von Willebrand's disease, there is a hyperresponsiveness to ristocetin, and the assay shows an elevated level of agglutination at low ristocetin concentrations (Sadler et al. 2000).

TABLE 20.4   Classification of von Willebrand's Disease

	Type 1	Type 2A	Type 2B	Type 2M	Type 2N	Type 3
Inheritance	Autosomal dominant	Autosomal dominant	Autosomal dominant	Autosomal dominant	Autosomal dominant or recessive	Autosomal recessive
Bleeding time	Normal or prolonged	Usually prolonged	Usually prolonged	Normal or prolonged	Normal	Prolonged
vWF:Agn	Reduced	Reduced	Reduced	Normal or reduced	Normal	Markedly reduced
vWF:RCo	Reduced	Reduced	Reduced	Reduced	Normal	Markedly reduced
FVIII:C	Normal or slightly reduced	Normal or slightly reduced	Normal or slightly reduced	Normal or slightly reduced	Reduced	Markedly reduced
RIPA at low concentrations of ristocetin	Absent	Absent	Increased	Absent	Absent	Absent
vWF multimer analysis	Normal	Absence of large and intermediate MW multimers	Absence of large MW multimers	Normal	Normal	Usually very low levels of multimers
vWF:RCo/ vWF:Agn	0.7–1.2	<0.7	<0.7	<0.7	0.7–1.2	NA
FVIII:C/ vWF:Agn	>1	>1	>1	>1	<1	NA
Therapy	Usually responsive to DDAVP	Variable response to DDAVP, FVIII/vWF factor concentrate if nonresponsive	DDAVP associated with thrombocytopenia, use FVIII/vWF factor concentrate	Variable response to DDAVP, FVIII/vWF factor concentrate if nonresponsive	DDAVP gives FVIII release with short survival, may need FVIII/vWF concentrate	No response to DDAVP, use FVIII/ vWF factor concentrate

FVIII:C = Factor VIII activity; vWF:Rco = von Willebrand factor ristocetin cofactor activity; vWF:Agn = von Willebrand factor antigen; RIPA = ristocetin-induced platelet agglutination; FVIII = Factor VIII.

A hallmark of von Willebrand's disease is the variability in factor levels and bleeding symptoms in the same patient from one time to the next. There can be variability in the severity of the bleeding symptoms between members of the same family (Abildgarrd et al. 1980). vWF levels are influenced by age, race, blood type, exercise, pregnancy, trauma, and inflammation. vWF levels are reduced in hypothyroidism and increased in hyperthyroidism. Combination birth control pills or estrogen therapy can elevate vWF levels. Because of this variability, serial sampling from a patient may be necessary before a diagnosis can be made, and two sets of abnormal laboratory tests should be found before a definitive diagnosis of von Willebrand's disease is made (Federici et al. 2002).

Treatment of von Willebrand's Disease

Treatment of patients with von Willebrand's disease has to address two objectives in order to control bleeding: the abnormal platelet adhesion and the low factor VIII levels. There are two treatments of choice, DDAVP and or transfusion with a vWF/factor VIII concentrate. Antifibrinolytic therapy is an important adjunct to either of these therapies for bleeding that involves mucous membranes. Estrogens raise plasma vWF levels, but in unpredictable amounts. Oral contraceptives may be useful in reducing the severity of menorrhagia in women with von Willebrand's disease (Castaman et al. 2003).

DDAVP use has been discussed in the section on hemophilia A, and its use in patients with von Willebrand's disease is the same, with the same doses, schedule, and side effects. It is the treatment of choice for type 1 von Willebrand's disease patients, but responsiveness to this material must be established for each patient before it is used. More than 90% of type 1 von Willebrand's disease patients will respond to DDAVP. Avoid use of the dilute form of intranasal DDAVP (100 µg/mL) that is intended for treatment of diabetes insipidus or enuresis. It is impossible to achieve hemostatic levels of DDAVP with this material. DDAVP is

TABLE 20.5   Guidelines for vWF/Factor VIII Concentrate Dosing in von Willebrand's Disease

Type of Bleeding	Dose (FVIII Units)	Number of Infusions	Duration of Therapy
Major surgery	40–50 Units/kg	Every 24 to 48 hours	Maintain FVIII >50% for at least 7 days
Minor surgery	20–40 Units/kg	Every 24 to 48 hours	Maintain FVIII >30% for at least 5–7 days
Dental extractions	20–40 Units/kg	Single	Maintain FVIII >30% for up to 6 hours, use antifibrinolytics
Spontaneous or posttraumatic bleed	20–40 Units/kg	Single	Usually one dose is sufficient, but may repeat as necessary till bleed resolves

contraindicated in type 2B von Willebrand's disease patients because of possible worsening of thrombocytopenia with increased levels of the patient's abnormal vWF. Type 2N patients have a good initial increase in factor VIII levels with DDAVP, but the increased levels are short-lived because of the lack of stabilization of the factor VIII molecule by vWF (Castaman et al. 2003; Montgomery et al. 2003).

Cryoprecipitate was initially used to treat patients with von Willebrand's disease and still represents a reasonable choice, but virucidal methods cannot be applied to cryoprecipitate, so it carries a small risk of transmission of blood borne infections. Because of this, virally inactivated vWF/factor VIII concentrates have been preferred for the treatment of von Willebrand's disease since the early 1990s for patients who are nonresponsive to DDAVP (Federici and Mannucci 1998). It is important to use the "intermediate purity" factor VIII concentrates, as the "high purity" monoclonal purified and recombinant factor VIII concentrates do not contain significant amounts of vWF. Treatment recommendations are summarized in Table 20.5 for patients nonresponsive to DDAVP.

With the infusions of the vWF/factor VIII concentrates, high postinfusion levels are consistently obtained, but factor VIII levels in patients rise higher than would be predicted over the next 24 hours, as the exogenous vWF stabilizes endogenously produced factor VIII. This can cause very high levels of factor VIII when multiple infusions of factor concentrate are given, such as in surgical situations. There is concern that sustained high levels of factor VIII can increase the risk of postoperative deep vein thrombosis (Mannucci 2002; Makris et al. 2002). Monitoring of factor VIII levels while patients are receiving multiple infusions of vWF/factor VIII concentrates and adjusting dose and schedule to avoid excessively high factor VIII levels is recommended.

Bleeding Nonresponsive to Standard Therapy

Infusion of exogenous vWF cannot replace vWF missing from the alpha granules in the platelets in some

variants of von Willebrand's disease. If bleeding is not controlled by vWF/factor VIII concentrates or cryoprecipitate alone, then DDAVP and/or platelet concentrates can be given to control the bleeding. Correction of the bleeding time helps to predict control of bleeding in these situations (Castillo et al. 1991). Rare patients with type 3 von Willebrand's disease develop anti-vWF alloantibodies, which not only render therapy with endogenous vWF ineffective, but also are associated with potentially life-threatening posttransfusion anaphylaxis. Recombinant factor VIII has been used in a high-dose continuous infusion to control bleeding during emergency surgery in a von Willebrand's disease patient with alloantibodies, because recombinant factor VIII does not contain vWF (Mannucci et al. 1987; Bergamaschini et al. 1995).

Rare Congenital Hemorrhagic Disorders

Specific information about the rare clotting factor deficiencies are provided in Table 20.6. All of the rare deficiencies listed are autosomal recessive in inheritance, with the exception of dysfibrinogenemia, which is autosomal dominant in inheritance. The autosomal recessive disorders have both heterozygous and homozygous presentations, where the bleeding manifestations in the heterozygous state are mild to asymptomatic and in the homozygous state are moderate to severe. The reader is referred to the review articles and textbook chapters in the references for more in-depth information on these disorders.

Factor XI Deficiency (Hemophilia C)

Hereditary factor XI deficiency is a mild, highly variable bleeding disorder that is often found in Jews of European descent, but can be found in other ethnic groups as well. Presentation is often for evaluation of a prolonged PTT or mild bleeding symptoms. The diagnosis is made by the measurement of factor XI activity on a fresh sample, as freezing may increase the factor XI activity in the sample (Pearson et al. 1981). Heterozygotes have factor XI levels of 25% to 60%, which

TABLE 20.6 Rare Hereditary Coagulation Factor Deficiencies

Factor Deficiency	Bleeding Symptoms (Homozygous Patients)	Treatment	Comments
Factor XI (hemophilia C)	Mild to moderate: epistaxis, easy bruising, bleed postdental or surgical procedures, menorrhagia	15–20 mL/kg FFP load, 7.5–10 mL/kg every 12–24 hours. (Factor XI concentrate available in Europe)	Maintain 40% to 60% factor level for hemostasis. Antifibrinolytics for dental procedures.
Factor XIII	Moderate to severe: umbilical stump bleeds, ICH, SQ, or muscle hematomas, bleeding postdental or surgical procedures	5–10 mL/kg FFP every 3–4 weeks. (Factor XIII concentrate available)	Prophylactic infusions to prevent bleeding. Antifibrinolytics for dental procedures.
Factor VII	Severe: umbilical stump, cephalohematomas, ICH, mucocutaneous bleeding, hemarthrosis, GI bleeding, menorrhagia, bleed postdental or surgical procedures, bleeding postpartum	10–15 mL/kg FFP every 4–6 hours. rFVIIa for severe hemorrhage or surgery.	rFVIIa dosing 20–30 µg/kg/dose, lower than in inhibitor therapy. Vitamin K therapy of no benefit.
Factor X	Moderate to severe: hemarthrosis, deep hematomas, menorrhagia, bleeding postsurgical procedure or trauma, neonatal ICH	20 mL/kg FFP load, 6 mL/kg every 12 hours for minor bleeding episodes. Prothrombin complex concentrates for surgery	Estrogens may help in reducing bleeding.
Factor V	Mild to moderately severe: epistaxis, hematomas, postsurgical or trauma, menorrhagia, postpartum, neonatal ICH, GI, and GU	20 mL/kg FFP load, 6 mL/kg every 12 hours. May need platelet transfusions in severe bleeding.	Hemarthrosis or deep muscle bleeds are rare.
Factor II (prothrombin)	Mild to moderate: epistaxis, hematomas, GI bleeding, menorrhagia, postsurgical or trauma, postpartum	10–20 mL/kg FFP load, 3 mL/kg every 12–24 hours. Prothrombin complex concentrates	Hemarthrosis are rare. 60 hour half-life, retreatment rarely needed.
Afibrinogenemia	Severe: ecchymosis, GI bleeding, hemarthrosis, bleeding postsurgery, posttrauma, or with dental procedures, severe menorrhagia, neonatal ICH, recurrent abortions, abruptio placenta	4 bags/10 kg (max dose 10 bags). Cryoprecipitate every other to every fourth day. (Fibrinogen concentrate available in Europe.)	Frequent problems with fetal loss in affected females. Consider prophylactic infusions every 3–4 days.
Hypofibrinogenemia	Mild: menorrhagia, postsurgical or posttraumatic, recurrent abortions, placental abruption	4 bags/10 kg (max dose 10 bags). Cryoprecipitate every other to every fourth day. (Fibrinogen concentrate available in Europe.)	Prophylaxis based on severity of bleeding symptoms.
Dysfibrinogenemia	Asymptomatic to mild: epistaxis, menorrhagia, bleeding postsurgical or posttrauma	4 bags/10 kg (max dose 10 bags) Cryoprecipitate every other to every fourth day. (Fibrinogen concentrate available in Europe.)	If fibrinogen activity is >50 mg/dL, fewer bleeding problems.
Alpha 2-Antiplasmin	Severe: umbilical cord stump, hemarthrosis, hematomas, epistaxis, posttraumatic, postsurgical or dental procedures, menorrhagia	Antifibrinolytics-epsilon aminocaproic acid or tranexamic acid	Differentiate from hemophilia
PAI-1	Severe: hemarthrosis, hematomas, menorrhagia, easy bruising, severe postsurgical or posttraumatic bleeding	Antifibrinolytics-epsilon aminocaproic acid or tranexamic acid	Differentiate from hemophilia

FFP = Fresh frozen plasma; ICH = intracranial hemorrhage; SQ = subcutaneous; rFVIIa = recombinant factor VIIa; GI = gastrointestinal; GU = genitourinary; PAI-1 = plasminogen activator inhibitor-1.

may overlap with the lower end of factor XI levels in normal individuals. Heterozygotes tend to have fewer bleeding problems than homozygotes, but the frequency and severity of bleeding disorders does not necessarily correlate with factor XI levels (Ragni et al. 1985). Patients with factor XI levels less than 15% or with a significant family history of bleeding should be treated before major surgery. Dental surgery can be done safely with only antifibrinolytic coverage (Berliner et al. 1992). Factor XI concentrates are available in the United Kingdom and France but have reported complications with thrombosis and DIC, even though the concentrates contain antithrombin or antithrombin and heparin (Bolton-Maggs et al. 1994; Mannucci et al. 1994).

Factor XIII Deficiency

A pattern of delayed bleeding is seen in homozygous factor XIII deficiency as clots repeatedly form and then break down 24 hours later. Diagnosis is made with a urea clot stability test, as standard hemostasis screening tests are normal (Lorand et al. 1980). There is often poor wound healing and delayed separation of the umbilical stump beyond 3 to 4 weeks. These patients have a very high rate of intracranial hemorrhage with little or no trauma, and this can be prevented with prophylactic infusions of fresh frozen plasma (FFP) or factor XIII concentrate (Abbondanzo et al. 1988). Females with severe factor XIII deficiency have spontaneous abortions if they become pregnant and can only carry to term if they are on prophylactic therapy throughout the pregnancy (Kobayashi et al. 1990). Very low levels of factor XIII, approximately 1% to 2%, are sufficient to provide hemostasis, and with a long half-life of 9 to 10 days, the every 3 to 4 week dosing intervals for the prophylaxis regimens are successful (Bauer 2003). Fibrogammin P (Aventis-Behring) is a virally inactivated plasma-derived factor XIII concentrate that is available in Europe and in clinical trials in the United States.

Factor VII Deficiency

Presentation of homozygous patients is often in infancy with bleeding symptoms, including intracranial hemorrhage, suggestive of hemophilia, but with a prolonged PT and a normal aPTT. Diagnosis is made by measurement of factor VII activity. Heterozygous patients are usually asymptomatic, but factor VII levels are poorly predictive of bleeding risk (Ragni et al. 1981). Patients with less than 1% factor VII activity have bleeding comparable to a severe hemophilia patient. The half-life of factor VII is short, 3 to 6 hours, so plasma infusions may control a single bleeding episode with factor VII levels of 15% but not provide sufficient coverage for severe hemorrhagic events or surgery. Recombinant factor VIIa (NovoNordisk) has been used successfully for control of bleeding (Mariani et al. 1999), and there is a virally inactivated factor VII concentrate (Baxter-Immuno) available in Europe and in the United States as part of clinical trials.

Factor X Deficiency

Clinical presentation is with mucocutaneous or post-traumatic bleeding combined with a prolonged PT and aPTT but normal fibrinogen. The diagnosis is made by measurement of factor X activity. There are variants where only the PT or the PTT is prolonged. The Russell's viper venom (RVV) time is a test where factor X is directly activated by the venom. It is usually prolonged in patients with severe deficiency but may be normal in some variants (Cooper et al. 1997). Patients with less than 1% activity have severe bleeding symptoms; if the factor level is greater than 10%, bleeding symptoms are mild, and factor levels of 20% to 30% are sufficient to prevent bleeding. The use of PCCs has been recommended in place of FFP because of the viral inactivation that is part of their production, however, this must be balanced against the possible risk of thrombotic complications. PCCs are recommended for use in complicated surgical procedures where higher levels of factor X are required (Blanchette et al. 1999). Successful prophylactic treatment of a severe patient with recurrent hemarthrosis has been reported using twice per week infusions of PCCs (Kouides and Kulzer 2001).

Factor V Deficiency

Presentation of these patients is with a mild to moderately severe bleeding disorder combined with a prolonged PT and PTT. Occasional patients will have prolonged bleeding times, due to low factor V content in platelets. Specific factor assays for factor V confirm the diagnosis, and assays for factor VIII should be done because combined deficiencies of factor V and VIII can occur (Ginsburg et al. 1998). The bleeding symptoms can be variable, with some patients having marked bleeding and others with the same factor V level having few symptoms. Menarche is frequently associated with severe menorrhagia (Mammen 1983). Hemostasis is achieved with factor V levels above 10% to 20%.

Factor II (Prothrombin) Deficiency

Prothrombin deficiency may be due to a lack of production of factor II (hypoprothrombinemia), due to a qualitative abnormality in the prothrombin produced (dysprothrombinemia) or to a combination of both disorders. It usually presents as a mild to moderate bleeding disorder with prolongation of both the PT and aPTT. The diagnosis is made using a combination of factor II activity and antigen measurements. Care must be taken to look for evidence of a lupus anticoagulant with an acquired antiprothrombin inhibitor, vitamin K deficiency, and liver disease (Blanchette et al. 1999). Bleeding symptoms in the neonate can be confused with hemorrhagic disease of the newborn. Factor II levels of 20% to 25% are sufficient to control bleeding. Use of PCCs has been recommended due to their viral inactivation, but this must be balanced against the risk PCCs have of thrombotic complications. Monitoring of factor II levels during replacement therapy is recommended to avoid excessive treatment, given the 3-day half-life of prothrombin (Roberts and White 2001).

Afibrinogenemia and Hypofibrinogenemia

Afibrinogenemia and hypofibrinogenemia are the homozygous and heterozygous presentations of decreased fibrinogen synthesis (al-Mondhiry and Ehmann 1994). The clinical presentation for afibrinogenemia patients is with a bleeding disorder comparable to moderate or severe hemophilia. Prolongations of the PT, PTT, TCT, bleeding time, and very low fibrinogen levels by both functional and immunological assays are seen. These patients may present in the neonatal period due to the trauma of delivery. Treatment with FFP or preferably cryoprecipitate to an initial fibrinogen level of 80 to 100 mg/dL controls bleeding symptoms, and levels are maintained above 50 to 60 mg/dL unitl bleeding resolves. Recurrent spontaneous intracranial hemorrhage has been reported in patients with afibrinogenemia (Henselmans et al. 1999). Prophylactic infusions every 3 to 4 days have been recommended for patients with severe symptoms (Rodriguez et al. 1988). Hypofibrinogenemia is often asymptomatic if the plasma fibrinogen level is greater than 50 mg/dL, but with lower plasma fibrinogen levels, patients have mild bleeding symptoms (al-Mondhiry and Ehmann 1994; Blanchette et al. 1999).

Dysfibrinogenemia

Dysfibrinogenemia is the secretion of a functionally abnormal fibrinogen, which has an autosomal dominant inheritance. Patients are either asymptomatic, hemorrhage, or have arterial or venous thrombosis. Poor wound healing has been reported in some patients. Laboratory evaluation shows a combination of usually normal PT and aPTT, prolonged TCT, and/or reptilase time with a normal or low fibrinogen activity measurement and a normal to elevated antigenetic fibrinogen. Fibrinogen levels must be significantly lower with the functional assay than with antigenic assay. When the TCT is prolonged, patients are more likely to have hemorrhagic symptoms. Some patients with thrombotic problems have a shortened TCT. Approximately half of dysfibrinogenemia patients with functional fibrinogen

levels of less than 50 mg/dL have hemorrhagic symptoms (McDonagh 2001; Bauer 2003).

Alpha 2-antiplasmin

Severe deficiency of alpha 2-antiplasmin is associated with a severe bleeding disorder caused by excessive fibrinolysis. The usual hemostatic screening tests are normal, and the euglobulin clot lysis time, a measure of fibrinolytic activity, is usually shortened. Testing for alpha 2-antiplasmin is done with functional and antigenic assays. The heterozygous presentation of this disorder is asymptomatic or has mild bleeding with menorrhagia, easy bruising, postdental or postsurgical bleeding. FFP contains alpha 2-antiplasmin, but most bleeding episodes can be controlled with the antifibrinolytics epsilon aminocaproic acid or tranexamic acid (Griffin et al. 1993).

Plasminogen Activator Inhibitor-1 (PAI-1)

The clinical presentation is of severe bleeding in patients with severe PAI-1 deficiency caused by a lack of regulation of plasminogen activity and excessive fibrinolysis. Regular hemostatic tests are normal, and the euglobulin clot lysis time can be abnormally short. Specific testing for PAI-1 with activity and antigenic tests is available. While FFP contains PAI-1, the recommended therapy for bleeding episodes is with the antifibrinolytics epsilon aminocaproic acid or tranexamic acid (Dieval et al. 1991).

Multiple Factor Deficiencies and Nonhemorrhagic Deficiencies

Most often multiple factor deficiencies are due to acquired disease states, such as liver failure or vitamin K deficiency. There are two congenital multiple factor deficiencies reported, one for factors V and VIII and the other for factors II, VII, IX, and X. The combined factors V and VIII deficiency is an autosomal recessive disorder, where the homozygotes have epistaxis, menorrhagia, postsurgical or dental bleeding, and postpartum hemorrhage, while the heterozygotes have mild bleeding symptoms. Hemarthrosis is rare (Peyvandi et al. 1998). Treatment is with DDAVP (to increase factor VIII) and/or FFP (to replace factor V), depending on the severity of the bleed.

The combined deficiencies of factors II, VII, IX, and X have been associated with pseudoxanthoma elasticum as well as the phenotype of warfarin embryopathy. The bleeding disorder can vary from mild to severe. Vitamin K will temporarily increase factor levels in some patients and has been used prophylactically to prevent bleeding. FFP has been used to control hemor-rhage, if the patient is nonresponsive to vitamin K (Brenner et al. 1998).

Deficiencies of factor XII, prekallikrein, and high molecular weight kininogen have all been shown to prolong the aPTT, sometimes markedly, while not having any accompanying bleeding symptoms. This is often found as a result of pre-operative screening test-ing in patients without bleeding symptoms. Some indi-viduals with factor XII deficiency have a mild thrombotic tendency, attributed to reduced plasma fibrinolytic activity, but epidemiological studies have questioned this finding (Bauer 2003).

ACQUIRED HEMORRHAGIC DISORDERS

As tertiary care pediatrics has improved, more criti-cally ill children are being cured, but acquired hemor-rhagic problems have become a significant problem for tertiary care pediatrics. This section provides a brief overview of those problems with a focus on diagnosis and treatment. For more extensive discussions of the specific topics, the reader is referred to the articles and review chapters in the references.

Disseminated Intravascular Coagulation

DIC is a very heterogeneous syndrome with endoge-nous activation of thrombin and plasmin due to multi-ple causes and associated with a significant mortality rate. Fibrin deposition in the microvasculature and consumption of coagulation factors results in both end-organ dysfunction and a bleeding diathesis (Levi and Cate 1999). The consensus definition of DIC from the Scientific Subcommittee on DIC of the Interna-tional Society on Thrombosis and Haemostasis (ISTH) is "an acquired syndrome characterized by the intravas-cular activation of coagulation with loss of localization arising from different causes. It can originate from and cause damage to the microvasculature, which, if suffi-ciently severe, can cause organ dysfunction" (Taylor et al. 2001). It is a secondary event, with a partial listing of causes in pediatrics in Box 20.1. Chronic as well as acute DIC has been described, with the chronic forms seen in the examples of malignancies and giant heman-giomas (Kasabach-Merritt syndrome). Presentation of overt DIC is most often with bleeding from surgical or venipuncture sites, bruising, and purpura. However, with the intensive monitoring done in pediatric inten-sive care units, DIC may be a laboratory diagnosis that is not accompanied by bleeding or microvascular thrombosis. Purpura fulminans is a very severe form of

---

BOX 20.1   Pediatric Causes of Disseminated Intravascular Coagulation

Bacterial infections—meningococcemia, streptococcus, gram-negative bacteria
Viral infections—varicella, hepatitis, CMV
Rickettsial infections—Rocky Mountain spotted fever
Systemic fungal infections
Malignancy—acute promyelocytic leukemia, disseminated neuroblastoma
Trauma—head injury, burns
Profound shock or asphyxia
Intravascular hemolysis—ABO incompatible transfusion reaction
Hemolytic uremic syndrome
Severe collagen vascular disease
Kasabach-Merritt syndrome
Liver disease

DIC that occurs with superficial purpura followed by bullae and hemorrhagic skin necrosis, with acquired deficiencies of proteins C and S (Smith et al. 1999). It is most often seen in meningococcal sepsis but can also be seen in streptococcal infections and chickenpox and is similar in appearance to the neonate with severe protein C or protein S deficiency.

Acute DIC usually occurs with the laboratory findings of thrombocytopenia; prolonged PT, aPTT, and TCT; low fibrinogen; positive D-dimers; fibrin degradation products; positive fibrin monomers; and a peripheral smear compatible with a microangiopathic hemolytic anemia, but no single test can reliably diag-nose DIC. Coagulation factors V and VIII, antithrom-bin, protein C, protein S, and alpha 2-antiplasmin are usually low due to consumption. Chronic DIC is a com-pensated disorder, with maintenance of fibrinogen and factor levels, platelet count, and normal PT and aPTT. Specific tests of thrombin activation (thrombin-antithrombin complexes, prothrombin fragment 1 + 2, and fibrinopeptide A) and plasmin generation ($B\beta$ 15–42 peptide) can be used to demonstrate low-grade DIC but are usually used as research tools (Levi and Cate 1999).

Treatment of DIC depends on resolution of the underlying disorder. In chronic or nonovert DIC, therapy to support the hemostatic system may not be needed. In overt DIC, transfusion of plasma products may help to control hemorrhage while the underlying disorder is addressed. There is significant controversy over management strategies, but there is general agree-ment on the transfusion of FFP for broad-spectrum clot-ting and inhibitor replacement, cryoprecipitate for fibrinogen and factor VIII replacement, and platelets for thrombocytopenia. It is reasonable to maintain the platelet count at more than 50,000/μl and the fibrinogen concentration at more than 100 mg/dL (Chalmers and Gibson 1999). The evidence to support the use of antithrombin and protein C concentrates is growing at present and controlled trials are needed but may be very difficult to perform (Smith et al. 1999; Messori et al. 2002; Lee and Downey 2000). Activated protein C has shown the ability to reduce the mortality rate for patients with severe sepsis, but with an increased risk of bleeding (Bernard et al. 2001). The use of heparin to suppress the consumption of coagulation factors continues but remains controversial. In situations of chronic DIC or with large vessel thrombosis, systemic heparin is generally used. Use of low dose heparin (10 units/kg/hour) or low molecular weight heparin in patients with chronic DIC unlikely to quickly resolve may offer an important adjunct to the control of ongoing consumption of coagulation factors (Corrigan 1977; Oguma et al. 1990). DIC in the newborn can also been treated with exchange transfusion (Monagle et al. 2003).

Liver Disease

The hemostatic defects of liver disease are combinations of the following pathophysiology: lack of production of most coagulation and fibrinolytic factors due to impaired hepatic synthesis, reduced clearance of activated coagulation factors, activation of the fibrinolytic system, thrombocytopenia due to splenic sequestration and other mechanisms, qualitative defects in platelet function, chronic low-grade consumptive coagulopathy, loss of hemostatic proteins into ascitic fluid and concurrent vitamin K deficiency (Mammen 1994; Kang 2000). Hemorrhagic problems in liver failure are variable and include superficial bleeding, mucocutaneous bleeding, and hemorrhage with invasive procedures such as liver biopsy. Bleeding from gastric and esophageal varices, into the abdomen, and into the CNS can be life-threatening. Specific problems occur with the placement of peritoneal-venous shunts, where the development of DIC is very common secondary to the infusion of procoagulants in the ascitic fluid into the bloodstream (Gleysteen et al. 1990). Liver transplant surgery can produce extraordinary amounts of bleeding, due to severe portal hypertension, previous surgery, a brisk rise in tissue plasminogen activator and fall in plasminogen activator inhibitor-1 levels during the anhepatic and reperfusion phases of surgery, the reciprocal fall in fibrinogen that occurs during the anhepatic stage, and a coagulopathy with release of a heparin-like substance immediately after reperfusion of the liver (Porte et al. 1989; Harper et al. 1989).

Standard coagulation screening tests are used to assess possible hemostatic defects in hepatic failure patients with bleeding or requiring invasive procedures. Initial prolongation is usually seen in the PT, due to the short half-life of factor VII, followed by prolongation of both the PT and aPTT as declines in other vitamin K-dependent clotting factors and the nonvitamin K-dependent factors occur with worsening hepatic synthesis. The degree of prolongation of the PT correlates with the degree of hepatocellular damage. Factor VIII and vWF levels are increased as hepatic disease worsens, unless concomitant DIC occurs, when factor VIII, vWF, and antithrombin are consumed. Fibrinogen production is maintained by the liver until late in the course of the disease, resulting in normal or elevated fibrinogen levels until end stage liver disease is reached. In advanced liver failure, a wide spectrum of disorders is seen, with prolongation of the PT/INR, PTT, TCT, low fibrinogen levels, shortened euglobulin clot lysis times secondary to increased fibrinolysis, elevating levels of D-dimers, and FDPs due to both DIC and decreased hepatic clearance, and increased bleeding times (Joist and George 2001).

Therapy of hepatic-induced coagulopathy is mostly supportive, with replacement therapy given to stop bleeding or to prepare the patient for invasive procedures. This is based on the results of ongoing monitoring of the hemostatic system using the PT/INR, aPTT, TCT, fibrinogen, and D-dimer level. Oral or intravenous vitamin K_1 should be started early in the management of liver disease, avoiding intramuscular injections due to the risk of hematomas. Mild prolongations of the PT (for example, 3 to 4 seconds or INR <1.5), with normal aPTT and fibrinogen levels do not require correction before invasive procedures (McVay and Toy 1990).

As the status of the hemostatic system worsens, with prolongation of the PT to an INR >1.5, FFP can be used to give a broad-spectrum replacement of both coagulation factors and inhibitors normally present in circulation. In practice, as liver function decreases, such large and repeated infusions of FFP are needed to correct the INR to <1.5 that volume overload occurs. Exchange transfusion can be used to correct a patient with severe coagulopathy before surgery or an invasive procedure, but the short half-life of factor VII (approximately 6 hours) requires additional infusions of FFP every 6 to 12 hours to maintain hemostasis. Cryoprecipitate can be used to replace low levels of fibrinogen without excessive volume requirements. Prothrombin complex concentrates are not recommended due to increased risk of thromboembolic complications and DIC (Joist and George 2001). Factor VIIa has been used in place of FFP for correction of the PT before procedures, using a single dose of 40 μg/kg (Shami et al. 2003). Advantages include more consistent correction of the PT than with FFP, very small volumes of infused material, and the infectious safety of a recombinant product. Concerns continue about thrombotic risks, and there is a need for further studies to define optimal dosing, safety, and efficacy. Platelet transfusions are indicated to stop bleeding if the platelet count is less than 50,000/μl and DDAVP can be used to shorten the bleeding time. Heparin may make bleeding worse and antifibrinolytic agents may promote thrombotic complications. In the situation of excessive fibrinolysis, replacement of alpha 2-antiplasmin is given with FFP and antifibrinolytics are used, even with their increased risk of thrombosis (Bovill 2001).

Renal Disorders

End-stage renal disease causes bleeding from mucosal surfaces secondary to uremia-induced platelet dysfunction. Bleeding times are often extended beyond 20 minutes, and it is common for end-stage renal patients to have mild thrombocytopenia, but with platelet counts of greater than 100,000/μL. This cannot explain the degree of prolongation of the bleeding time (Enkoyan et al. 1969). Use of DDAVP or cryoprecipitate to raise vWF levels can correct the bleeding time (Mannuchi et al. 1983). Effective dialysis and maintenance of the hematocrit between 30% and 40%, either with transfusion of PRBCs or with erythropoietin, will also correct the bleeding time (Akizawa et al. 1991). Renal transplantation that corrects the uremia resolves the platelet dysfunction (Monagle and Andrew 2003). Conjugated estrogens can be given orally, subcutaneously, or intravenously and have been reported to shorten the bleeding time and decrease bleeding in patients with renal failure (Livio et al. 1986). The main cause of hemorrhagic complications in patients on dialysis therapy is excessive anticoagulation therapy with heparin.

Vitamin K Deficiency

Vitamin K deficiency in healthy children is rare, because intestinal bacteria produce vitamin K_2. If broad-spectrum antibiotics reduce vitamin K_2 production, a normal diet provides sufficient vitamin K. However, the combination of inadequate vitamin K intake with prolonged use of broad-spectrum antibiotics can produce vitamin K deficiency. Groups of children at high risk for this include breast-fed newborns, particularly if the parents have forgone neonatal vitamin K prophylaxis (Sutor 1995), chronically ill children with inadequate intake of vitamin K, or with disorders that interfere with the adsorption of vitamin K, and children with poor nutrition who are receiving prolonged courses of antibiotics. These include children with cystic fibrosis, biliary atresia, celiac disease, Crohn's disease, and obstructive jaundice. The final group of children who have to be considered are those on warfarin or who have ingested long-acting coumarin "superwarfarin" found in rodenticides (Greeff et al. 1987). Second- and third-generation cephalosporins have a weak vitamin K antagonism that is a risk only if other predisposing factors are present (Lipsky 1988).

Clinical bleeding due to vitamin K deficiency is relatively infrequent, given the frequent laboratory evidence of this deficiency in children with predisposing conditions. If bleeding does occur, it is usually mild to moderate, such as bruising or bleeding from venipunctures. Internal bleeding is rare although CNS hemorrhage can occur. The laboratory diagnosis is made with an isolated prolonged PT in early deficiency. As the deficiency increases, prolongation of both the PT and aPTT occurs while there are normal fibrinogen levels and TCT. A 1 : 1 mix of patient and normal plasma should correct any prolongations of the PT or the aPTT and rule out the presence of heparin or other inhibitor. Demonstration of low levels of the vitamin K-dependent clotting factors (usually factors VII and IX) with normal levels of factors V and VIII rules out liver disease and consumptive coagulopathy, confirming the diagnosis (Monagle and Andrew 2003).

Treatment of vitamin K deficiency should be dictated by the clinical situation. Intravenous vitamin K should be given only in settings of high risk or where other routes are not feasible, due to the possible anaphylactic reactions to intravenous vitamin K. If oral absorption is normal, then oral vitamin K replacement can be used, although correction of the PT is slower than with other routes. Subcutaneous dosing of vitamin K is safe and effective. Correction of the coagulation abnormality takes place within 6 to 8 hours with oral replacement and 2 to 6 hours with subcutaneous administration. If acute bleeding is occurring or the risk of bleeding is high, then an infusion of 10 to 15 mL/kg of FFP can help to rapidly correct the PT. In life-threatening hemorrhage, such as an intracranial hemorrhage, both PCCs and recombinant factor VIIa have been used to stop bleeding. Dosing of vitamin K is 1 to 5 mg subcutaneously, depending on size, for asymptomatic patients with mild deficiency; 2 to 10 mg subcutaneously if there is bleeding; and with life-threatening bleeding, 5 to 20 mg should be given systemically in addition to rapid intravenous factor replacement (Monagle and Andrew 2003). If there has been ingestion of a superwarfarin rodenticide, then vitamin K replacement may need to continue for a prolonged period of time (Greeff et al. 1987). Prophylactic vitamin K supplementation is given to children on total parenteral nutrition and is recommended for patients with severe noncholestatic or cholestatic liver disease, major small bowel resection for intestinal complications, pulmonary disease necessitating long-term use of antibiotics, and for patients with pancreatic insufficiency. Oral administration of 10 mg of daily oral vitamin K_1 should be given to pregnant females on oral anticonvulsant therapy for 2 weeks before birth, to help prevent overt vitamin K deficiency in their infants at birth (Morrow and Craig 2003).

Acquired Inhibitors of Coagulation

Specific inhibitors of single coagulation factors are well-recognized complications of hemophilia treatment and are discussed in the section on hemophilia. Acquired inhibitors to von Willebrand's disease are described in children with Wilms' tumor and autoimmune disorders (Scott et al. 1981; Michiels et al. 2001). Resolution of the inhibitor came after treatment of the Wilms' tumor, but this entity is important, as surgery may need to occur before resolution of the tumor. A therapeutic approach is to give DDAVP and then cryoprecipitate. If this fails, then intravenous immunoglobulin, prednisolone, or platelet transfusions can be used (Michiels et al. 2001). Acquired inhibitors to factor VIII have been reported in nonhemophiliac children who have severe recurrent bleeding. Some have had recent viral infections or penicillin exposure, but others have no clear etiology. Treatment to control bleeding is the same as for hemophilia A patients with inhibitors, with the addition of intravenous immunoglobulin, immunosuppressive agents, or extracorporeal immunoadsorption (Brodeur et al. 1980; Green and Lechner 1981; Grunewald et al. 2001). Acquired antibodies to factor V develop in some patients after exposure to bovine thrombin as part of fibrin glue, most often used during cardiac surgery. Others develop following surgery in which no bovine thrombin was used, after exposure to aminoglycosides, in patients with an underlying factor V deficiency, and as an idiopathic process. Bleeding problems are variable, ranging from none to significant hemorrhages (Muntean et al. 1994; Knobl and Lechner 1998). While many of these antibodies appear to be transient, re-exposure to the same sealant product can result in the reappearance of the antibody. Treatment has been with platelet transfusions, FFP, plasmapheresis, intravenous immunoglobulin, and recombinant factor VIIa (Chalmers and Gibson 1999). Hemorrhagic lupus anticoagulant syndrome has been described in the clinical setting of a previously well child who develops bleeding symptoms of variable severity following a viral infection. Laboratory studies show a prolonged aPTT that does not correct on a 1:1 mixing study, with evidence of a lupus anticoagulant on specific testing and low prothrombin levels. The antibody usually clears spontaneously in three months and usually does not require treatment. Steroids have been used in some cases, and if there is significant bleeding, bypassing agents are used in the same way as in hemophilia patients with inhibitors (Becton and Stine 1997; Bernini et al. 1993).

References

Abbondanzo SL, Gottenberg JE, Loftus RS, et al. 1988. Intracranial hemorrhage in congenital deficiency of factor XIII. *Am J Pediatr Hematol Oncol* 10:65–68.

Abildgaard CF, Suzuki Z, Harrison J, et al. 1980. Serial studies in von Willebrand's disease: variability vs "variants." *Blood* 56:712–716.

Akizawa T, Kinugasa E, Kitaoka T, Koshikawa S. 1991. Effects of recombinant human erythropoietin and correction of anemia on platelet function in hemodialysis patients. *Nephron* 58:400–406.

al-Mondhiry H and Ehmann WC. 1994. Congenital afibrinogenemia. *Am J Hematol* 46:343–347.

Aledort LM and Bohn RL. 1996. Prophylaxis and continuous infusion for hemophilia: can we afford it? *Blood Coagul Fibrinolysis* 7(Suppl 1):S35–S37.

Andrew M, Paes B, Johnston M. 1990. Development of the hemostatic system in the neonate and young infant. *Am J Pediatr Hematol Oncol* 12:95–104.

Antonarakis SE. 1998. Molecular genetics of coagulation factor VIII gene and hemophilia A. *Haemophilia* 4(Supp 2):1–11.

Bauer KA. 2003. Rare hereditary coagulation factor abnormalities. In *Nathan and Oski's Hematology of infancy and childhood*. 6th ed. Nathan DG, Orkin SH, Ginsburg D, Look AT, eds. Philadelphia, PA: WB Saunders.

Becton DL and Stine KC. 1997. Transient lupus anticoagulants associated with hemorrhage rather than thrombosis: the hemorrhagic lupus anticoagulant syndrome. *Thromb Res* 130:998–1000.

Bergamaschini L, Mannucci PM, Federici AB, et al. 1995. Posttransfusion anaphylactic reactions in a patient with severe von Willebrand disease: role of complement and alloantibodies to von Willebrand factor. *J Lab Clin Med* 125:348–355.

Berliner S, Horowitz I, Martinowitz U, et al. 1992. Dental Surgery in patients with severe factor XI deficiency without plasma replacement. *Blood Coagul Fibrinol* 3(4):465–468.

Bernard GR, Vincent JL, Laterre PF, et al. 2001. Efficacy and safety of recombinant human activated protein C for severe sepsis. *N Engl J Med* 344:699–709.

Bernini JC, Buchanan GR, Ashcraft J. 1993. Hypoprothrombinemia and severe hemorrhage associated with a lupus anticoagulant. *J Pediatr* 123:937–939.

Blanchette VS, Dean J, Lillecrap D. 1999. Rare congential hemorrhagic disorders. In *Pediatric hematology*. 2nd ed. Lilleyman J, Hann I, Blanchette VS, eds. Portland, OR: Book News, Inc.

Bolton-Maggs PHB, Colvin BT, Satchi G, et al. 1994. Thrombogenic potential of factor XI concentrate. *Lancet* 334:748–749.

Bolton-Maggs PHB. 2000. Factor XI deficiency and its management. *Haemophilia* 6(Suppl 1):100–109.

Bovill EG. 2001. Liver diseases. In *Disorders of hemostasis and thrombosis, a clinical guide*. Goodnight SH and Hathaway WE, eds. New York: McGraw-Hill.

Brenner B, Sanchez-Vega B, Wu SM, et al. 1998. A missense mutation in gamma-glutamyl carboxylase gene causes combined deficiency of all vitamin K-dependent blood coagulation factors. *Blood* 92:4554–4559.

Briet E, Rosendaal FR, Kreuz W, et al. 1994. High titer inhibitors in severe hemophilia A. A meta-analysis based on eight long-term follow-up studies concerning inhibitors associated with crude or intermediate purity factor VIII products. *Thromb Haemost* 72: 162–164.

Brodeur GM, O'Neill PJ, Wilimas JA. 1980. Acquired inhibitors of coagulation in nonhemophiliac children. *J Pediatr* 96 Pt I:439–441.

Castaman G, Eikenboom JC, Bertina RM, et al. 1999. Inconsistency of association between type 1 von Willebrand disease phenotype and genotype in families identified in an epidemiological investigation. *Thromb Haemost* 82:1065–1070.

Castaman G, Federici AB, Rodeghiero F, Mannucci PM. 2003. Von Willebrand's disease in the year 2003: towards the complete identification of gene defects for correct diagnosis and treatment. *Haematologica* 88:94–108.

Castillo R, Monteagudo J, Escolar G, et al. 1991. Hemostatic effect of normal platelet transfusion in severe von Willebrand disease patients. *Blood* 77:1901–1905.

Chalmers EA, Gibson BES. 1999. Acquired disorders of hemostasis during childhood. In *Pediatric hematology*. 2nd ed. Lilleman J, Hann I, Blanchette V, eds. London: Churchill Livingston.

Close HL, Kryzer TC, Nowlin JH, Alving BA. 1994. Hemostatic assessment of patients before tonsillectomy: a prospective study. *Otolaryngol Head Neck Surg* 111:733–738.

Colman RW, Clowes AW, George JN, et al. 2001. Overview of hemostasis. In *Hemostasis and thrombosis, basic principles and clinical practice*. Colman RW, Hirsch J, Marder VJ, Clowes AW, George JN, eds. Philadelphia, PA: Lippincott, Williams & Wilkins.

Cooper DN, Millar DS, Wacey A, et al. 1997. Inherited factor X deficiency: molecular genetics and pathophysiology. *Thromb Haemostas* 78:161–172.

Corrigan JJ, Jr. 1977. Heparin therapy in bacterial sepsis. *J Pediatr* 91:695–700.

Dean JA, Blanchette VS, Carcao MD, et al. 2000. von Willebrand disease in a pediatric-based population—comparison of type 1 diagnostic criteria and use of the PFA-100 and a von Willebrand factor/collagen-binding assay. *Thromb Haemost* 84:401–409.

DeLoughery TG. 2001. Oral anticoagulants. In *Disorders of hemostasis and thrombosis, a clinical* guide. Goodnight SH and Hathaway WE, eds. New York: McGraw-Hill.

Dieval J, Nguyen G, Gross S, et al. 1991. A lifelong bleeding disorder associated with a deficiency of plasminogen activator inhibitor type 1. *Blood* 77:528–532.

DiMichele MD. 2001. Hemophilia A FVIII Deficiency. In *Disorders of hemostasis and thrombosis, a clinical guide*. Goodnight SH and Hathaway WE, eds. New York: McGraw-Hill.

Enkoyan G, Wacksman S, Glueck H, et al. 1969. Platelet function in renal failure. *N Engl J Med* 280:677–681.

Esmon CT. 2003. Blood coagulation. In *Nathan and Oski's hematology of infancy and* childhood. 6th ed. Nathan DG, Orkin SH, Ginsburg D, Look AT, eds. Philadelphia, PA: WB Saunders.

Evatt BL and Hooper WC. 1997. Infectious complications of blood products. In *Hemophilia*. Forbes CD, Aledort L, Madhok R, eds. London: Chapman & Hall.

Federici AB, Castaman G, Mannucci PM. 2002. Guidelines for the diagnosis and management of von Willebrand disease in Italy. Italian Association of Hemophilia Centers AICE. *Haemophilia* 8:6007–6621.

Federici AB and Mannucci PM. 1998. Optimizing therapy with factor VIII/von Willebrand factor concentrates in von Willebrand disease. *Haemophilia* 4(Suppl 3):7–10.

Forbes CD. 1997. The early history of hemophilia. In *Hemophilia*. Forbes CD, Aledort L, Madhok R, eds. London: Chapman & Hall.

Fressinaud E, Veyradier A, Truchaud F, et al. 1998. Screening for von Willebrand disease with a new analyzer using high shear stress: a study of 60 cases. *Blood* 91:1325–1331.

George JN and Shattil SJ. 1991. The clinical importance of acquired abnormalities of platelet function. *N Engl J Med* 324:27–39.

Ginsburg D, Nichols WC, Zivelin A, et al. 1998. Combined factors V and VIII deficiency—the solution. *Haemophilia* 4:677–682.

Gleysteen JJ, Hussey CV, Heckman MG. 1990. The cause of coagulopathy after peritoneovenous shunt for malignant ascites. *Arch Surg* 125:474–477.

Goodnight SH and Hathaway WE. 2001. Mechanisms of hemostasis and thrombosis. In *Disorders of hemostasis and thrombosis, a clinical guide*. Goodnight SH and Hathaway WE, eds. New York: McGraw-Hill.

Gordon FH, Mistry PK, Sabin CA, et al. 1998. Outcome of orthotopic liver transplantation in patients with hemophilia. *Gut* 42:744–749.

Greaves M and Preston FE. 2001. Approach to the bleeding patient. In *Hemostasis and thrombosis, basic principles and clinical practice*. 4th ed. Colman RW, Hirsch J, Marder VJ, Clowes AW, George JN, eds. Philadelphia, PA: Lippincott, Williams & Wilkins.

Greeff MC, Mashile O, MacDougall LG. 1987. "Superwarfarin" bromodialone. poisoning in two children resulting in prolonged anticoagulation. *Lancet* 28570:1269.

Green D, Lechner K. 1981. A survey of 215 nonhemophilic patients with inhibitors to factor VIII. *Thromb Haemostas* 45: 200–203.

Griffin GC, Mammen EF, Sokol RJ, et al. 1993. Alpha 2-antiplasmin deficiency. An overlooked cause of hemorrhage. *Am J Pediatr Hematol Oncol* 15:328–330.

Grunewald M, Beneke H, Guthner C, et al. 2001. Acquired haemophilia: experiences with a standardized approach. *Haemophilia* 7:164–169.

Harker LA and Slichter SJ. 1972. The bleeding time as a screening test for evaluation of platelet function. *N Engl J Med* 287:155–159.

Harper PL, Luddington RJ, Jennings I, et al. 1989. Coagulation changes following hepatic revascularization during liver transplantation. *Transplantation* 48:603–607.

Hathaway WE, Christian MJ, Clarke SL, et al. 1984. Comparison of continuous and intermittent factor VIII concentrate therapy in hemophilia A. *Am J Hematol* 17:85–88.

Hay CR, Lozier JN, Lee CA, et al. 1994. Porcine factor VIII therapy in patients with congenital hemophilia and inhibitor: efficacy, patient selection, and side effects. *Semin Hematol* 31(2 Suppl 4):20–25.

Henselmans JM, Meijer K, Haaxma R, et al. 1999. Recurrent spontaneous intracerebral hemorrhage in a congenitally afibrinogenemic patient: diagnostic pitfalls and therapeutic options. *Stroke* 30: 2479–2482.

Hoots WK, Leissinger C, Stabler S, et al. 2003. Continuous intravenous infusion of a plasma-derived factor IX concentrate Mononine in haemophilia B. *Haemophilia* 9:164–172.

Hoyer LW. 1997. Inhibitors in hemophilia. In *Hemophilia*. Forbes CD, Aledort L, Madhok R, eds. London:Chapman & Hall.

Joist JH, George JN. 2001. Hemostatic abnormalities in liver and renal disease. In *Hemostasis and thrombosis: basic principles and clinical practice*. Colman RW, Hirsh J, Marder VJ, Clowes AW, George JN, eds. Philadelphia, PA: Lippincott, Williams & Wilkins.

Kang Y. 2000. Coagulopathies in hepatic disease. *Liver Transplant* 64 Suppl 1:S72–S75.

Kasper CK. 1989. Treatment of factor VIII inhibitors. In *Progress in hemostasis and thrombosis*. 9th ed. Collier BS, ed. Philadelphia, PA: WB Saunders.

Kasper CK and Costa e Silva M. 2003. *Monograph: registry of clot-*

ting factor concentrates. 4th ed. Montreal: World Federation of Hemophilia.

Knobl P and Lechner K. 1998. Acquired factor V inhibitors. *Baillieres Clin Haematol* 11:305–318.

Kobayashi T, Terao T, Kojima T, et al. 1990. Congenital factor XIII deficiency with treatment of factor XIII concentrate and normal vaginal delivery. *Gynecol Obstet Invest* 29:235–238.

Kobrinsky NL and Stegman DA. 1997. Management of hemophilia during surgery. In *Hemophilia.* Forbes CD, Aledort LM, Madhok R, eds. London: Chapman & Hall.

Kouides PA and Kulzer L. 2001. Prophylactic treatment of severe factor X deficiency with prothrombin complex concentrate. *Haemophilia* 7:220–223.

Kundu SK, Heilmann EJ, Sio R, et al. 1995. Description of an in vitro platelet function analyzer PFA–100. *Semin Thromb Hemost* 21: 106–112.

Lee WL and Downey GP. 2000. Coagulation inhibitors in sepsis and disseminated intravascular coagulation. *Intensive Care Med* 26: 1701–1706.

Levi M and Ten Cate H. 1999. Disseminated intravascular coagulation. *N Engl J Med* 341:586–592.

Lillecrap D. 1998. The molecular basis of haemophilia B. *Haemopilia* 4:350–357.

Lipsky JJ. 1988. Review: antibiotic-associated hypoprothrombinemia. *J Antimicrob Chemother* 21:281–300.

Livio M, Mannucci PM, Vigano G, et al. 1986. Conjugated estrogens for the management of bleeding associated with renal failure. *N Engl J Med* 315:731–735.

Lorand L, Losowsky MS, Miloszewski KJM. 1980. Human factor XIII: Fibrin-stabilizing factor. *Prog Hemost Thromb* 5:245–290.

Lusher JM. 2003. Clinical and laboratory approach to the bleeding patient. In *Nathan and Oski's hematology of infancy and childhood.* 6th ed. Nathan DG, Orkin SA, Ginsburg D, Look AT, eds. Philadelphia, PA: WB Saunders.

Makris M, Colvin B, Gupta V, et al. 2002. Venous thrombosis following the use of intermediate purity FVIII concentrate to treat patients with von Willebrand's disease. *Thromb Haemost* 88:387–388.

Mammen EF. 1994. Coagulation defects in liver disease. *Med Clin North Am* 78:545–554.

Mammen EF. 1983. Factor V deficiency. *Semin Thromb Hemosta* 9: 17–18.

Mammen EF, Comp PC, Gosselin R, et al. 1998. PFA-100 system: a new method for assessment of platelet dysfunction. *Semin Thromb Hemost* 24:195–202.

Mannucci PM. 2003. Hemophilia: treatment options in the twenty-first century. *J Thromb Haemost* 1:1349–1355.

Mannucci PM. 2002. Venous thromboembolism in von Willebrand disease. *Thromb Haemost* 88:378–379.

Mannucci PM, Bauer KA, Santagostino E, et al. 1994. Activation of the coagulation cascade after infusion of a factor XI concentrate in congenitally deficient patients. *Blood* 84:1314–1319.

Mannuci PM, Remuzzi G, Pusineri F, et al. 1983. Deamino-8-D-argenine vasopressin shortens the bleeding time in uremia. *N Engl J Med* 315:8–12.

Mannucci PM, Ruggeri ZM, Pareti FL, et al. 1977. 1-Deamino-8-d-arginine vasopressin: a new pharmacological approach to the management of haemophilia and von Willebrands' diseases. *Lancet* 18017:869–872.

Mannucci PM, Tamaro G, Narchi G, et al. 1987. Life-threatening reaction to factor VIII concentrate in a patient with severe von Willebrand disease and alloantibodies to von Willebrand factor. *Eur J Haemotol* 39:567–570.

Mannucci PM and Tuddenham EGD. 2001. The hemophilias—from royal genes to gene therapy. *N Engl J Med* 344:1773–1779.

Mariani G, Testa MG, Di Paolantonio T, et al. 1999. Use of recombinant, activated factor VII in the treatment of congenital factor VII deficiencies. *Vox Sang* 77:131–136.

Martinowitz U, Schulman S, Gitel S, et al. 1992. Adjusted dose continuous infusion of factor VIII in patients with hemophilia. *Br J Haemat* 82:729–734.

McDonagh J. 2001. Dysfibrinogenemia and other disorders of fibrinogen structure and function. In *Hemostasis and thrombosis: basic principles and clinical practice.* 4th ed. Colman RW, Hirsh J, Marder VJ, Clowes AW, George JN, eds. Philadelphia, PA: Lippincott Williams & Wilkins.

McVay PA and Toy PTCY. 1990. Lack of increased bleeding after liver biopsy in patients with mild hemostatic abnormalities. *Am J Clin Pathol* 94:747–753.

Michiels JJ, Budde U, van der Planken M, et al. 2001. Acquired von Willebrand syndromes: clinical features, aetiology, pathophysiology, classification, and management. *Best Pract Res Clin Haematol* 14:401–436.

Messori A, Vacca F, Vaiani M, et al. 2002. Antithrombin III in patients admitted to intensive care units: a multicenter observational study. *Crit Care* 6:447–451.

Monagle P and Andrew M. 2003. Acquired disorders of hemostasis. In *Nathan and Oski's hematology of infancy and childhood.* 6th ed. Nathan DG, Orkin SH, Ginsburg D, Look AT, eds. Philadelphia, PA: WB Saunders.

Montgomery RR, Gill JC, Scott JP. 2003. Hemophilia and von Willebrand disease. In *Nathan and Oski's hematology of infancy and Childhood.* 6th ed. Nathan DG, Orkin SH, Ginsburg D, Look AT, eds. Philadelphia, PA: WB Saunders.

Morrow JI and Craig JJ. 2003. Anti-epileptic drugs in pregnancy: current safety and other issues. *Expert Opin Pharmacother* 4: 445–456.

Muntean W, Zenz W, Finding K, et al. 1994. Inhibitor to factor V after exposure to fibrin sealant during cardiac surgery in a two-year-old child. *Acta Paediatr* 83:84–87.

Nilsson IM, Berntorp E, Lofqvist T, et al. 1992. Twenty-five years' experience of prophylactic treatment in severe haemophilia A and B. *J Intern Med* 232:25–32.

Oguma Y, Sakuragawa N, Maki M, et al. 1990. Treatment of disseminated intravascular coagulation with low molecular weight heparin. Research Group of FR-860 on DIC in Japan. *Semin Thromb Hemost* 16 Suppl:34–40.

Pearson RW and Triplett DA. 1981. Factor XI assay results in the CAP survey (1981). *Am J Clin Pathol* 78(4 Suppl):615–620.

Peyvandi F, Tuddenham EG, Akhtari AM, et al. 1998. Bleeding symptoms in 27 Iranian patients with the combined deficiency of factor V and factor VIII. *Br J Haematol* 100, 773–776.

Porte RJ, Bontempo FA, Knot EAR, et al. 1989. Hemostasis in liver transplantation. *Gastroenterology* 97:488–501.

Ragni MV, Lewis JH, Spero JA, et al. 1981. Factor VII deficiency. *Am J Hematol* 10:79–88.

Ragni MV, Sinha D, Seaman F, et al. 1985. Comparison of bleeding tendency, factor XI coagulant activity, and factor XI antigen in 25 factor XI–deficient kindreds. *Blood* 65:719–724.

Rappaport SI. 1983. Preoperative hemostatic evaluation: which tests, if any? *Blood* 61:229–231.

Roberts HR and White GC. 2001. Inherited disorders of prothrombin conversion. In *Hemostasis and thrombosis: basic principles and clinical practice.* 4th ed. Colman RW, Hirsh J, Marder VJ, Clowes AW, George JN, eds. Philadelphia, PA: Lippincott, Williams & Wilkins.

Rodgers RP and Levin J. 1990. A critical reappraisal of the bleeding time. *Semin Thromb Hemost* 16:1–20.

Rodriguez RC, Buchanan GR, Clanton MS. 1988. Prophylactic cryoprecipitate in congenital afibrinogenemia. *Clin Peditr* 27:543–545.

Rohrer MJ, Michelotti MC, Nahrwold DL. 1988. A prospective evaluation of the efficacy of preoperative coagulation testing. *Ann Surg* 208:554–557.

Sadler JE. 1994. A revised classification of von Willebrand's disease for the Subcommittee on von Willebrand factor of the Scientific and Standardization Subcommittee of the International Society on Thrombosis and Haemostasis. *Thromb Haemost* 71:520–523.

Sadler JE, Mannucci PM, Berntorp E, et al. 2000. Impact, diagnosis and treatment of von Willebrand disease. *Thromb Haemost* 84:160–174.

Scott J, Montgomery R, Tubergen D, et al. 1981. Acquired von Willebrand's disease in association with Wilm's tumor: regression following treatment. *Blood* 58:665–669.

Shami VM, Caldwell SH, Hespenheide EE, et al. 2003. Recombinant activated factor VII for coagulopathy in fulminant hepatic failure compared with conventional therapy. *Liver Transpl* 9:138–143.

Shapiro AD. 2001. Coagulation factor concentrates. In *Disorders of hemostasis and thrombosis, a clinical guide.* Goodnight SH and Hathaway WE, eds. New York: McGraw–Hill.

Silwer J. 1973. von Willebrand's disease in Sweden. *Acta Paediatr Scand Suppl* 238:1–159.

Smith OP and White B. 1999. Infectious purpura fulminans: diagnosis and treatment. *Br J Haematol* 104:202–207.

Soucie JM, Evatt B, Jackson D. 1998. Occurrence of hemophilia in the United States. The Hemophilia Surveillance System Project Investigators. *Am J Hematol* 59:288–294.

Sramek A, Eikenboom JCJ, Briet E, et al. 1995. Usefulness of patient interview in bleeding disorders. *Arch Intern Med* 155:1409–1415.

Stajcic Z. 1985. The combined local/systemic use of antifibrinolytics in hemophiliacs undergoing dental extractions. *Int J Oral Surg* 14:339–345.

Sutor AH. 1995. Vitamin K deficiency bleeding in infants and children. *Semin Thromb Hemostas* 21:317–329.

Taylor FB, Toh CH, Hoots WK, et al. 2001. Towards definition, clinical and laboratory criteria, and scoring system for disseminated intravascular coagulation. *Thromb Haemostas* 86:1327–1330.

Thiagarjan P and Shapiro SS. 1982. Lupus anticoagulants. *Progr Hemost Thromb* 5:263–285.

Thompson JM and Poller L. 1985. The activated partial thromboplastin time. In *Blood coagulation and haemostasis: a practical guide.* Thompson JM, ed. Edinburgh: Churchill Livingston.

Turner M. 1999. The impact of new-variant Creutzfeldt-Jakob disease on transfusion practice. Review. *Brit J Haematol* 106:842–850.

Warrier I, Ewenstein BM, Koerper MA, et al. 1997. Factor IX: inhibitors and anaphylaxis in hemophilia B. *J Pediatr Hematol Oncol* 19:23–27.

Zwack GC, Derkay CS. 1997. The utility of preoperative hemostatic assessment in adenotonsillectomy. *Int J Pediatr Otorhinolaryngol* 39:67–76.

21

Transfusion of the Patient with Autoimmune Hemolysis

KAREN E. KING, MD

INTRODUCTION

The autoimmune hemolytic anemias (AIHAs) are characterized by the following two features: (1) decreased red cell survival and (2) the presence of an autoantibody directed against red cell antigens.

The AIHAs may be either primary (idiopathic) or secondary. Secondary AIHA is seen in association with autoimmune diseases, lymphoproliferative syndromes, or infections (viral or mycoplasma). In children, AIHA has been seen following vaccinations (Seltsam 2000) (Johnson et al. 2002).

The overall incidence of AIHA is estimated to be one to three cases per 100,000 per year (Gehrs 2002). In children, AIHA has been seen in all age groups. Infants and toddlers are more likely to have AIHA following an infection. Teenagers with AIHA are more likely to have an underlying disease.

CLINICAL PRESENTATION AND LABORATORY FINDINGS

Clinical Presentation

Children with AIHA often present with signs and symptoms of anemia. The clinical severity will differ based on the time course of the disease. These children may report fatigue, exercise intolerance, dyspnea on exertion, and dizziness. On physical examination, they may have pallor and jaundice. Fever may be present. Cardiac examination may reveal tachycardia and a systolic flow murmur associated with the high output anemic state. If the onset of anemia was gradual, patients may appear to have compensated and they may have relatively mild symptoms even in the presence of severe anemia.

Laboratory Findings

Evaluation of the peripheral smear is critical to the diagnosis of AIHA. The peripheral smear will reveal microspherocytosis, polychromatophilia, and reticulocytosis. The microspherocytes result from the splenic ingestion of portions of the red cell, which may have had immunoglobulin or complement attached. Polychromatophilia and reticulocytosis are a result of increased bone marrow activity and compensation for the anemia.

Laboratory tests may reveal elevated total bilirubin and increased serum lactate dehydrogenase. Decreased haptoglobin may also be seen, but haptoglobin can be less informative since it is also an acute phase reactant and young children do not synthesize haptoglobin well. Increased reticulocytes are usually present; however, reticulocytopenia has been seen (Greenberg et al. 1980), and this phenomenon may be prolonged in some cases requiring transfusion support. Patients with a brisk intravascular hemolytic process may have evidence of hemoglobinuria and hemoglobinemia.

CLASSIFICATION

The AIHAs are classified based on the characteristics of the causative antibody, including its immunoglobulin class and thermal reactivity as shown in Table 21.1. On the basis of the in vivo and in vitro characteristics of the causative autoantibody, AIHA can be subdivided

TABLE 21.1   Autoimmune Hemolytic Anemia: Classification and Serologic Features

	Warm AIHA	Cold Agglutinin Syndrome	Paroxysmal Cold Hemoglobinuria
Direct antiglobulin test	IgG, IgG, and C3	C3 only	C3 only
Immunoglobulin class	IgG (sometimes, IgA, rarely IgM)	IgM	IgG
Eluate	IgG	Nonreactive	Nonreactive
Serum	IgG agglutinating RBCs at the antihuman globulin phase (panagglutinin)	IgM agglutinating antibody, usually titers >1000, reacting at 30°C in albumin	IgG biphasic hemolysin (Donath-Landsteiner antibody)
Specificity	Rh specificity common	Anti-I, anti-i	Anti-P

into those associated with warm antibodies, reacting optimally at 37°C, and those associated with cold antibodies, reacting optimally at 0–5°C. The AIHAs associated with cold antibodies can be further subdivided into the more common cold agglutinin syndrome and the rare paroxysmal cold hemoglobinuria. Typically, warm AIHA is due to IgG autoantibodies, whereas cold agglutinin syndrome is due to IgM autoantibodies. Paroxysmal cold hemoglobinuria is due to an IgG, biphasic hemolysin which is described in more detail below. Numerous drugs are capable of inducing an immune hemolytic anemia, clinically indistinguishable from warm AIHA. Consequently, drug-induced hemolytic anemia is included as a separate type of AIHA.

In a large study including 865 cases of AIHA over an 18-year time period, Sokol et al. (1981) found 41% warm AIHA, 32% cold agglutinin syndrome, 7% mixed-type AIHA, 18% drug-induced hemolytic anemia, and 2% paroxysmal cold hemoglobinuria.

WARM AUTOIMMUNE HEMOLYTIC ANEMIA

Serologic Findings

Warm AIHA is generally due to IgG autoantibodies, which are optimally reactive at 37°C. The serologic hallmark of the AIHAs is the finding of a positive direct antiglobulin test. In warm AIHA, the direct antiglobulin test usually identifies IgG with or without complement coating of the red cells. Eluate studies can further elucidate the nature and specificity of the red cell bound IgG. Serum studies including the indirect antiglobulin test using either untreated or enzyme-treated allogeneic red cells at 20°C and 37°C will demonstrate antibody present in the serum. Often, the serum and eluate reveal a panagglutinin with reactivity against all cells tested.

Selected cell panel studies may reveal a specificity of the antibody. Often autoantibodies implicated in warm AIHA have broad Rh specificity, only showing negative reactivity with D- or Rh null red cells. Relative specificity for an Rh blood group antigen can also be seen and must be distinguished from an alloantibody. This often requires sophisticated adsorption tests. Specificity for other blood group antigens occurs less commonly.

Treatment

The therapeutic approach to warm AIHA depends, in part, on whether the disease is primary or secondary. The endpoint of therapy is not the resolution of the direct antiglobulin test. Patients may have positive direct antiglobulin tests without hemolysis or anemia.

Corticosteroids

For patients with either idiopathic or secondary forms of AIHA, corticosteroids are the initial therapy of choice. The use of steroids can result in therapeutic effect within 24 to 48 hours of initiation. The clinical response to corticosteroids is felt to result primarily from their effect on tissue macrophages, which become less efficient in clearing IgG or C3-coated red cells (Fries et al. 1983). A significant decrease in antibody production may not be seen for several weeks. Approximately 80% of patients will respond to corticosteroid therapy (Collins and Newland 1992).

Intravenous Immunoglobulin G

Since intravenous IgG (IVIG) has been useful in the therapy for immune-mediated thrombocytopenia, its efficacy in AIHA was considered (Bussel et al. 1983; Warrier et al. 1997). Unfortunately, it is not as effective in treating AIHA. Only about one-third of patients with warm AIHA will respond to IVIG therapy, although children tend to respond somewhat better than adults (Bussel et al. 1986; Flores et al. 1993). Consequently, IVIG should be considered as adjunctive therapy for warm AIHA.

Splenectomy

The efficacy of splenectomy in the treatment of warm AIHA is well established. Coon (1985) found an excellent response in 64% of patients and an improved status in an additional 21% of patients. In children, the clinical option of a splenectomy needs to be considered carefully. Postsplenectomy, children are at risk of overwhelming sepsis due to encapsulated bacterial organisms. If splenectomy is performed, children should receive immunization with a polyvalent vaccine against *Streptococcus pneumoniae* as prophylactic treatment. Additionally, children should receive lifelong penicillin prophylaxis. Penicillin allergic children can receive erythromycin. Patients and their families should be instructed to seek medical attention whenever the patient experiences a significant fever, since this may be the first indication of bacterial sepsis.

Immunosuppressive Therapy

Several immunosuppressive agents have been reported to be successful in the treatment of warm AIHA. Unfortunately, the majority of these reports have been small series or case reports. Consequently, the use of these immunosuppressive therapies should be considered only for patients who fail conventional treatment (Petz 2001).

Azathioprine

Azathioprine can be used in the treatment of warm AIHA, however, adverse effects can be limiting. Although gastrointestinal intolerance can be problematic, evidence of bone marrow suppression should be closely monitored. Since this is a cytotoxic agent, there is a potential risk of neoplasia and prolonged administration should be avoided.

Cyclosporine

Review of the literature reveals numerous case reports of successful use of cyclosporine in the treatment of warm AIHA (Petz 2001). Unfortunately, there are also case reports of failures with this therapy. When cyclosporine is used, patients need to be followed for potential nephrotoxicity.

Cyclophosphamide

Recently, high dose cyclophosphamide followed by granulocyte colony-stimulating factor was used with success in a small number of refractory patients; one child was included in this report (Moyo et al. 2002).

Rituximab

This genetically engineered chimeric murine/human monoclonal anti-CD20 antibody targets B-cell precursors and mature B cells. Plasma cells do not carry the CD20 antigen. Zecca et al. (2003) reported their prospective evaluation of the use of rituximab in treating 15 children with refractory warm AIHA. With a median follow-up of 13 months, 13 patients (87%) responded to treatment, and two patients failed to improve. Quartier et al. (2001) reported successful treatment of five pediatric patients using rituximab. They noted that although useful in the treatment of AIHA, rituximab therapy was associated with prolonged B-cell deficiency and hypogammaglobulinemia.

Additional Forms of Therapy

Danazol

Danazol is an attenuated androgen, which has proven to be helpful in some cases of warm AIHA. Ahn (1990) reported 28 patients treated with danazol and 77% achieved excellent or good responses with therapy. The mechanism of its action in this clinical setting is unknown.

Plasmapheresis

The usefulness of plasmapheresis in the long-term treatment of warm AIHA is limited. Plasmapheresis can, however, be used in extreme cases as a temporizing measure while waiting to achieve the therapeutic effect of an immunosuppressive agent.

COLD AGGLUTININ SYNDROME

Cold agglutinin syndrome (CAS) can occur in either an acute or chronic form. The acute form of this disease is often secondary to a lymphoproliferative syndrome or *Mycoplasma pneumoniae* infection. Chronic CAS is generally seen in elderly patients and not seen in children.

Serologic Findings

CAS is the AIHA associated with cold reactive, IgM autoantibodies. In CAS, the direct antiglobulin test reveals complement coating of the red cells. Since the autoantibody is an IgM antibody, it will not be detected by routine direct antiglobulin tests, and consequently, eluate studies are rarely indicated. Serum evaluation should include a cold agglutinin titer and thermal amplitude studies; these studies are helpful in determining and confirming the clinical significance of the cold autoantibody. In general, the cold autoantibodies that clinically cause CAS have a high titer (>1000) and/or serum studies using 30% bovine albumin that show reactivity at 30°C (Garratty et al. 1975).

Although it is only of academic interest, the specificity of the antibody is most often directed against I antigen. Less often, specificity for i antigen is found, usually in patients with infectious mononucleosis. Rarely, other antigen specificities have been reported.

Treatment

The mainstay of treatment for patients with CAS is avoidance of the cold. Acute hemolytic events can be prevented if patients keep themselves warm by maintaining a high temperature of their environment and by wearing additional clothing when going outdoors. Through diligent avoidance of the cold, many patients with CAS can be managed without requiring transfusion. Patients with mild, compensated anemia are often not treated for long periods of time or may require only episodic transfusions. If transfusion is required, the use of a blood warmer is generally recommended.

For patients with more severe, partially compensated idiopathic CAS, medical therapy is generally unsatisfactory. Steroids are not efficacious in most patients with CAS. Similarly, splenectomy is not usually effective, presumably because the liver is the dominant site of sequestration of red cells heavily sensitized with C3. Chlorambucil has been used with some success, but the associated bone marrow suppression can be limiting. As in the treatment of warm AIHA, plasmapheresis may have a role as a temporizing measure. If plasmapheresis is performed in this setting, special attention may be necessary to maintain core body temperature during the exchange process at 37°C. Potentially, an in-line blood warmer will suffice, but some patients may require more elaborate efforts.

MIXED-TYPE AUTOIMMUNE HEMOLYTIC ANEMIA

Occasionally, unusual patients may not be classifiable into either warm AIHA or CAS. These patients appear to have a combination of both warm and cold AIHA, or mixed-type AIHA, representing approximately 7% of all cases of AIHA. Serologically, these patients have both a warm reactive IgG autoantibody and an IgM autoantibody with a lower titer than is usually seen in CAS, but a high thermal amplitude. These patients may present with severe hemolysis; however, they frequently have a rapid response to corticosteroid therapy. It is important to accurately identify these patients, since a patient with mixed-type AIHA who is mistakenly diagnosed with CAS would not receive corticosteroids.

DRUG-INDUCED HEMOLYTIC ANEMIA

An acquired hemolytic anemia can develop in association with drugs. In investigating the possibility of a drug-induced hemolytic anemia, one must consider not only the prescription medications that the patient is taking, but also over-the-counter medications and even possible chemical exposures. Clinically, these patients may present with severe hemolysis and life-threatening anemia (Garratty et al. 1992). Although the specific use of various drugs changes over time as newer drugs become available, certainly the cephalosporins are often implicated in drug-induced hemolytic anemia and should be considered a potential culprit when evaluating a patient with this entity (Arndt et al. 1999).

Proposed Mechanisms

Traditionally, three mechanisms have been proposed to explain the development of autoantibodies in association with drugs, but it is important to note that very little evidence supports these proposed mechanisms (Brecher 2002). Additionally, some patients have been found to have drug-induced hemolytic anemia with features suggestive of multiple mechanisms (Habibi 1985). Consequently, a unifying theory has been suggested (Garratty 1994).

Drug Adsorption

In this mechanism, the drug binds to the surface of the red cell. An antibody directed against the drug is formed. The antibody-antigen interaction occurs at the surface of the red cell, leading to destruction of the red cell. In these cases, the antibody is generally a warm reactive IgG antibody. The direct antiglobulin test is positive due to IgG. Complement coating of the red cells may or may not be found. The prototypic drug implicated in this mechanism is penicillin.

Immune Complex

This proposed mechanism suggests that unbound drug stimulates the production of drug-specific antibody, which may be IgG, IgM, or both. The antibody and drug form immune complexes, which are capable of binding to red cells. It is unclear if this interaction between the immune complexes and the red cells is specific or nonspecific. The direct antigobulin test is generally positive, revealing complement coating of the red cells. The prototypic drug demonstrating this mechanism is quinidine.

Clinically, patients with this form of drug-induced hemolytic anemia may present with acute intravascular hemolysis, hemoglobinemia, and hemoglobinuria. Renal failure may also occur.

Drug Independent Autoantibodies

In this situation, the drug induces a warm AIHA with serologic similarities to idiopathic warm AIHA. The direct antiglobulin test is generally positive for IgG. A panagglutinin can be found in both the serum and the eluate. The autoantibody may even show relative specificity for Rh blood group antigens, and it may be detected for prolonged periods after the drug has been discontinued. The prototypic drug is alpha-methyldopa.

Unifying Theory

Following the identification of patients with drug-induced hemolytic anemias who had features of more than one of the proposed mechanisms, the concept of a unifying theory has evolved (Habibi 1985; Mueller-Eckhardt 1990; Salama 1992; Garratty 1994). Since drugs are capable of binding to red cells, either loosely or firmly, the combination of drug and red cell membrane invoke an antibody response. Depending on the specific interaction of drug and red cell membrane, the antibody may be directed against any of the following: predominantly red cell membrane, predominantly drug, or a part of both membrane and drug. Consequently, antibodies directed against all of these may be identified, explaining the patients who appear to have multiple traditional mechanisms simultaneously.

Treatment

In treating drug-induced hemolytic anemia, the implicated drug should be discontinued immediately. Although there is no supportive data available, some advocate the empiric use of steroid therapy. Supportive therapy is indicated in more severe cases, and transfusion therapy may be required. In cases of drug independent autoantibody formation, prolonged transfusion support may be necessary, since the autoantibody can persist well after the drug is discontinued.

LESS COMMON TYPES OF AUTOIMMUNE HEMOLYTIC ANEMIA

Paroxysmal Cold Hemoglobinuria

Paroxysmal cold hemoglobinuria (PCH) is an uncommon type of AIHA, accounting for approximately 2% of all cases of AIHA. PCH is more commonly seen in children as compared to adults. One study found no cases of PCH in 531 adults with AIHA, but reported 22 cases of PCH in 68 (32%) children with AIHA (Gottsche et al. 1990). PCH often occurs in association with infections. Although syphilis was the most commonly associated infection in the past, currently PCH is most often preceded by a viral infection, often an upper respiratory infection.

The causative autoantibody is a biphasic, IgG antibody known as the Donath-Landsteiner (DL) antibody. The DL antibody is capable of binding to red cells in the cold causing the irreversible binding of C3 and C4. At warmer temperatures, the antibody dissociates from the red cell, and complement activation leads to hemolysis. Consequently, routine direct antiglobulin tests reveal the presence of complement (C3) coating of the red cells; IgG is not found coating the red cells, and eluate studies are generally negative. Although other specificities have been demonstrated, the DL antibody often has specificity for the P antigen. A DL test demonstrates the biphasic nature of the hemolysin in vitro. With the addition of normal serum as a source of fresh complement, the patient's serum and test red cells expressing P antigen are placed in the cold and then at 37°C; a positive test results in the final demonstration of hemolysis.

In the setting of severe hemolysis, transfusion therapy may be required. Since the DL antibody does not usually react above 4°C, it does not interfere with routine compatibility testing. Some have suggested that the transfusion of red blood cells with p phenotype is indicated; however, red blood cells with this phenotype are uncommon and difficult to obtain under urgent circumstances and may only be available through rare donor registries. One may consider the transfusion of p red blood cells when a patient with PCH fails to respond to the transfusion of routine donor blood.

Autoimmune Hemolytic Anemia Due to Warm IgM Antibodies

Salama and Mueller-Eckhardt (1987) reported 15 children who clinically presented with evidence of warm AIHA. Serological evaluation of these unusual patients revealed the presence of a noncomplement-binding IgM autoantibody. Based on these very unusual findings, the authors suggested that this may represent a new type of warm AIHA.

In contrast, hemolytic anemia has been rarely seen in association with warm-reacting IgM autoantibodies, which are capable of activating complement and agglutinating in vivo. Although most of the cases reported in the literature are adult patients, it has been reported to occur in children (Friedmann et al. 1998; Nowak-Wegrzyn et al. 2001). Clinical findings include marked intravascular agglutination leading to ischemia and tissue infarction, in addition to severe hemolysis. Numerous therapies have been attempted including whole blood exchange, immunosuppressive therapy, and cytotoxic therapy. Unfortunately, these therapeutic interventions are usually unsuccessful, and this type of autoimmune hemolytic anemia is virtually uniformly fatal.

Autoimmune Hemolytic Anemia Due to IgA Antibodies

In about 14% of patients with warm AIHA, IgA class antibodies are present, usually in association with IgG and/or IgM antibodies. Cases solely attributed to IgA autoantibody are considered rare (0.1% to 0.2%) (Janvier et al. 2002). Pediatric cases of AIHA due to IgA alone have been reported. This entity should be considered when a patient clinically presents with signs and symptoms of AIHA, and the direct antiglobulin test is negative for IgG.

Autoimmune Hemolytic Anemia with a Negative DAT

Patients with DAT-negative warm AIHA are uncommon. These patients have all of the clinical and hematological features of warm AIHA, except that they have a negative DAT and no detectable autoantibodies in their serum. In these cases, more sensitive testing, such as the enzyme-linked antiglobulin test or radiolabeled antiglobulin tests, may be necessary to identify red cell-bound IgG.

TRANSFUSION THERAPY IN AUTOIMMUNE HEMOLYTIC ANEMIA

Indications for Transfusion

Specific treatments for each type of AIHAs have been discussed. Red blood cell transfusion is a significant component of the supportive care for all patients with AIHA. Although many patients will present with mild to moderate anemia that does not require urgent transfusion, an occasional patient presents with life-threatening anemia, and the need for red blood cell transfusions is emergent.

Careful communication between the clinician and the transfusion service is required, since transfusion may be required before the serologic evaluation is completed. Even after thorough serologic evaluation, the optimal blood for transfusion may be serologically incompatible. The clinician must understand that in some cases, serologically incompatible blood is safe for transfusion and can be expected to have in vivo survival comparable to the patient's own red cells. Reluctance to transfuse these patients because of serologic incompatibility or an incomplete workup can be devastating. Patients with severe anemia may appear to be hemodynamically stable, but these patients have life-threatening anemia and should be transfused immediately regardless of the status of the serologic evaluation or compatibility.

Selection of Blood for Transfusion

Transfusion management is complicated for patients with AIHA, due to their serologic complexities. Although patients with AIHA can present with an ABO discrepancy, ABO typing can usually be performed following removal of IgG autoantibody in warm AIHA or warm washing red cells to remove IgM autoantibody in CAS. In an emergency or if results are not clear-cut, Group O donor red cells can be used. Rh typing can also be difficult if the patient's cells are heavily coated with autoantibody. Low protein, monoclonal reagents are available for Rh typing in the setting of immunoglobulin-coated red cells.

The most pressing problem in a patient with previous pregnancy or transfusions is detecting and identifying alloantibody, which may be hidden or masked by the autoantibody. The exclusion of underlying clinically significant alloantibodies is time consuming, and the clinical situation may not allow for completion of these studies before transfusion is needed. In the untransfused patient, autologous adsorption studies are required to investigate the presence of underlying alloantibody. If the patient has been transfused in the preceding 3 months, more complex, differential allogeneic adsorption studies are required.

In the untransfused patient, determination of the red cell phenotype is invaluable. Techniques and procedures are available to dissociate autoantibody so that phenotyping can be performed. Once a red cell phenotype has been determined, this phenotype can guide the exclusion of alloantibodies by indicating which antigen specificities are at risk of eliciting an alloantibody. Shirey et al. (2002) has shown that when available, phenotypically matched red cells are an efficient method for providing safe red cells for transfusion in this setting. Additionally, patients with AIHA are at risk for requiring chronic transfusion therapy; knowing the red cell phenotype of the patient is helpful in identifying alloantibodies if they develop over time.

Many of these serologic complexities may be avoidable in the future with the development of hemoglobin-based oxygen carriers (Mullon et al. 2000).

Transfusion Management

Once the decision has been reached to begin transfusions, several additional steps can enhance patient safety. In some cases, transfusions of small aliquots of blood may be sufficient to provide relief of symptoms while avoiding the complications of fluid overload. The use of leukocyte-reduced red blood cells is helpful in the prevention of febrile, nonhemolytic transfusion reactions whose initial presentation can suggest a hemolytic transfusion reaction with more severe consequences (Lumadue et al. 1996). In the case of transfusion for patients with CAS, blood warmers are often suggested, but there are limited data to support or refute this recommendation.

References

Ahn YS. 1990. Efficacy of danazol in hematologic disorders. *Acta Haematol* 84:122–129.

Arndt PA, Leger RM, and Garratty G. 1999. Serology of antibodies to second- and third-generation cephalosporins associated with immune hemolytic anemia and/or positive direct antiglobulin tests. *Transfusion* 39:1239–1246.

Brecher ME. 2002. *Technical manual*. 14th ed. Bethesda, MD: AABB.

Bussel BJ, Cunningham-Rundles C, and Abraham C. 1986. Intravenous treatment of autoimmune hemolytic anemia with very high dose gammaglobulin. *Vox Sang* 51:264–269.

Bussel JB, Schulman I, Hilgartner MW, and Barandun S. 1983. Intravenous use of gammaglobulin in the treatment of chronic immune thrombocytopenic purpura as a means to defer splenectomy. *J Pediatr* 103:651–654.

Collins PW and Newland AC. 1992. Treatment modalities of autoimmune blood disorders. *Semin Hematol* 29:64–74.

Coon WW. 1985. Splenectomy in the treatment of hemolytic anemia. *Arch Surg* 120:625–628.

Flores G, Cunningham-Rundles C, Newland AC, and Bussel JB. 1993. Efficacy of intravenous immunoglobulin in the treatment of autoimmune hemolytic anemia: results of 73 patients. *Am J Hematol* 44:237–242.

Friedmann AM, King KE, Shirey RS, et al. 1998. Fatal autoimmune hemolytic anemia in a child due to warm-reactive immunoglobulin M antibody. *J Pediatr Hematol Oncol* 20:502–505.

Fries LF, Brickman CM, and Frank MM. 1983. Monocyte receptors for the Fc portion of IgG increase in number in autoimmune hemolytic anemia and other hemolytic states and are decreased by glucocorticoid therapy. *J Immunol* 131:1240–1245.

Garratty G. 1994. Review: immune hemolytic anemia and/or positive direct antiglobulin tests caused by drugs. *Immunohematology* 10:41–50.

Garratty G, Nance S, Lloyd M, et al. 1992. Fatal immune hemolytic anemia due to cefotetan. *Transfusion* 32:269–271.

Garratty G, Petz LD, and Hoops JK. 1975. The correlation of cold agglutinin titrations in saline and albumin with haemolytic anemia. *Br J Haematol* 35:587–595.

Gehrs BC and Friedberg RC. 2002. Autoimmune hemolytic anemia. *Am J Hematol* 69:258–271.

Gottsche B, Salama A, and Mueller-Eckhardt C. 1990. Donath-Landsteiner autoimmune hemolytic anemia in children: a study of 22 cases. *Vox Sang* 58:281–286.

Greenberg J, Curtis-Cohen M, et al. 1980. Prolonged reticulocytopenia in autoimmune hemolytic anemia of childhood. *J Pediatr* 97:784–786.

Habibi B. 1985. Drug induced red blood cell autoantibodies codeveloped with drug specific antibodies causing haemolytic anaemias. *Br J Haematol* 61:139–143.

Janvier D, Sellami F, Missud F, et al. 2002. Severe autoimmune hemolytic anemia caused by a warm IgA autoantibody directed against the third loop of band 3 (RBC anion-exchange protein 1). *Transfusion* 42:1547–1552.

Johnson ST, McFarland JG, Kelly KJ, et al. 2002. Transfusion support with RBCs from an Mk homozygote in a case of autoimmune hemolytic anemia following diphtheria-pertussis-tetanus vaccination. *Transfusion* 42:567–571.

Lumadue JA, Shirey RS, Kickler TS, and Ness PM. 1996. Leukocyte reduction of red cells when transfusing patients with autoimmune hemolytic anemia: a strategy to decrease the incidence of confounding transfusion reactions. *Immunohematology* 12:84–86.

Moyo VM, Smith D, Brodsky I, Crilley P, Jones RJ, and Brodsky RA. 2002. High-dose cyclophosphamide for refractory autoimmune hemolytic anemia. *Blood* 100:704–706.

Mueller-Eckhardt C and Salama A. 1990. Drug-induced immune cytopenias: a unifying pathogenic concept with special emphasis on the role of drug metabolites. *Trans Med Rev* 4:69–77.

Mullon J, Giacoppe G, Clagett C, et al. 2000. Transfusions of polymerized bovine hemoglobin in a patient with severe autoimmune hemolytic anemia. *N Engl J Med* 342:1638–1643.

Nowak-Wegrzyn A, King KE, Shirey RS, et al. 2001. Fatal warm autoimmune hemolytic anemia resulting from IgM autoagglutinins in an infant with severe combined immunodeficiency. *J Pediatr Hematol Oncol* 23:250–252.

Petz LD. 2001. Treatment of autoimmune hemolytic anemias. *Curr Opin Hematol* 8:411–416.

Quartier P, Brethon B, Philippet P, et al. 2001. Treatment of childhood autoimmune haemolytic anaemia with rituximab. *Lancet* 358:1511–1513.

Salama A and Mueller-Eckhardt C. 1992. Immune-mediated blood cell dyscrasias related to drug. *Sem Hematol* 29:54–63.

Salama A and Mueller-Eckhardt C. 1987. Autoimmune haemolytic anaemia in childhood associated with non-complement binding IgM autoantibodies. *Br J Haematol* 65:67–71.

Seltsam A, Shukry-Schulz S, and Salama A. 2000. Vaccination-associated immune hemolytic anemia in two children. *Transfusion* 40:907–909.

Shirey RS, Boyd JS, Parwani AV, Tanz WS, Ness PM, and King KE. 2002. Prophylactic antigen-matched donor blood for patients with warm autoantibodies: an algorithm for transfusion management. *Transfusion* 42:1435–1441.

Sokol RJ, Hewitt S, and Stamps BK. 1981. Autoimmune haemolysis: an 18 year study of 865 cases referred to a regional transfusion center. *Br Med J* 282:2023–2027.

Warrier I, Bussel JB, Valdez L, Barbosa J, Beardsley DS, and the Low-Dose IVIG Study Group. 1997. Safety and efficacy of low-dose intravenous immune globulin (IVIG) treatment for infants and children with immune thrombocytopenic purpura. *J Pediatr Hematol Oncol* 19:197–201.

Zecca M, Nobili B, Ramenghi U, Perrotta S, et al. 2003. Rituximab for the treatment of refractory autoimmune hemolytic anemia in children. *Blood* 101:3857–3861.

Platelet Transfusions in the Infant and Child

MATTHEW SAXONHOUSE, MD, WILLIAM SLAYTON, MD,
AND MARTHA C. SOLA, MD

INTRODUCTION

Many advances have been made to ensure the safety of platelet transfusions, including intensive donor screening, infectious disease marker testing, and increased use of leukocyte reduction techniques. However, platelet transfusions still carry known and unknown risks and may represent a significant expense during a patient's hospital stay. Certain disease states require special components, thus further increasing costs. Because of these issues, it is important for the clinician to order and administer the correct product for his or her patient to maximize effectiveness and minimize complications. In this chapter, we will review the platelet products that can be used for neonatal and pediatric patients, followed by an overview of issues specific to transfusion in children and in neonates.

PLATELET PRODUCTS AND SPECIAL PREPARATIONS

Product Description

The two types of platelet components available for use in most hospital settings are whole blood-derived platelet concentrates, referred to as "random-donor platelets," and apheresis-derived platelets, referred to as "single-donor platelets" or "platelet pheresis." Both types have similar hemostatic properties. However, the preparation, contents, and possibility for adverse reactions differ between products.

Whole Blood-Derived Platelet Concentrates

Whole blood-derived platelet concentrates, or random-donor platelets, are prepared from a single donated unit of whole blood and contain approximately 50 mL of volume. Although platelet content varies, each concentrate contains about 10×10^9 platelets per 10 mL, with the remaining volume consisting of donor plasma and preservative-anticoagulant solution. In the adult setting, three to eight platelet concentrates are pooled into a single product that contains 3 to 6×10^{11} platelets. As a result, the recipient is exposed to multiple donors. In contrast, since standard neonatal and pediatric dosing is 10 to 15 mL/kg of a platelet suspension, a single random-donor platelet unit is usually sufficient to provide a platelet transfusion to neonates and small infants. Thus, they are exposed to only one blood donor and receive approximately 10×10^9 platelets/kg body weight, a dose expected to increase the blood platelet count to $>100 \times 10^9$/L posttransfusion (Andrew et al. 1993; Strauss 2000a). However, each concentrate contains approximately 10^8 leukocytes, which increases the risk for leukocyte-mediated reactions.

Apheresis-Derived Platelets

Apheresis-derived platelets are prepared from a single donor through a semi-automated apheresis process. In this process, the donor is connected to a blood cell separator that removes whole blood from the donor, isolates platelets into a separate container, and then returns platelet-depleted blood to the donor. To prevent coagulation, ACD-A or a similar citrate anticoagulant is used. The process lasts 1 to 2 hours and results in 3 to 8×10^{11} platelets suspended in 200 to 300 mL (Roback 2001). In addition to the increased number of platelets, this component differs from random-donor platelet units in that it only contains 10^4 to 10^6 leukocytes per suspension, which is <1% of the leukocytes found in a platelet concentrate pool and satisfies Food and Drug Administration (FDA) requirements for leukocyte reduction. Thus, the risk of leukocyte-mediated adverse reactions is lower when using this component. Apheresis-derived platelets also contain up to 0.5 mL of RBCs and can be both crossmatched (on rare occasions, when visibly pink) and human leukocyte antigen (HLA)-matched when necessary.

Storage and Bacterial Contamination

Both whole blood-derived and apheresis-derived platelet units are stored at 20°C to 24°C with continuous gentle agitation. With adequate oxygenation, bicarbonate buffers the lactate produced to maintain a stable pH for 1 week (Andrew et al. 1993; Strauss 2000a). However, current practice is to limit the storage period to 5 days, due to the concern that longer storage periods could allow the proliferation of contaminating bacteria (Klein et al. 1997). In fact, postplatelet-transfusion sepsis is the second most common cause of death associated with transfusion. The risk for bacterial contamination is 1:1000 to 1:2000 platelet units, which translates into 2000 to 4000 bags being contaminated and results in 333 to 1000 clinical cases of sepsis each year in the United States (Brecher et al. 2001). Neonates and infants, in particular, are at increased risk for serious infection if a contaminated unit is received, due to the immaturity of their immune system.

Special Preparations

When ordering platelet transfusions, it is important to meet the specific requirements of the recipient in order to prevent adverse reactions. However, the more options selected, the longer the preparation time and the higher the cost.

Cytomegalovirus (CMV)-Seronegative Components

Transfusion-transmitted CMV (TT-CMV) infection is a process by which CMV is transmitted from a blood component to a CMV-seronegative recipient. While TT-CMV is infrequent in immunocompetent transfusion recipients (Preiksaitis et al. 1988; Wilhelm et al. 1986), it is a significant cause of morbidity and mortality in immunocompromised patients. Before the installation

of preventive measures, including screening for human immunodeficiency virus (HIV) and hepatitis, 13% to 37% of immunocompromised transfusion recipients contracted CMV from unscreened blood components (Bowden et al. 1986; Bowden et al. 1991; Miller et al. 1991; Yeager et al. 1981).

One effective way to diminish the acquisition of CMV from platelet transfusions is to use CMV-antibody-negative blood products (Lee et al. 1992; Miller et al. 1991). However, the incidence of CMV infection following transfusion with CMV-seronegative blood components is still 1% to 4% (Lang et al. 1977; Miller et al. 1991; Strauss 1995). Reasons for this include an intrinsic false-negative rate of the test for antibody to CMV, a low antibody titer, or transient viremia quenching circulating antibody (Strauss 2000a). The major limitation with obtaining CMV-antibody-negative blood is that <50% of the blood donor population is CMV-antibody-negative. Because of this limitation, alternative options have been explored to provide CMV-safe blood products. Leukocyte reduction (see next section) has been reported to offer safety comparable to that of blood products supplied by negative donors.

Leukocyte Reduction

Leukocyte reduction is accomplished by passing platelet components through a polyester-based filtration device or by apheresis technology. The filtration process removes 99.9% of leukocytes, resulting in 1 to 5×10^6 residual leukocytes per concentrated platelet pool (Roback 2001). The purpose of leukocyte reduction is to reduce complications associated with platelet transfusions. These complications are (1) febrile, non-hemolytic transfusion reactions (FNHTR), (2) alloimmunization to leukocyte antigens, (3) transmission of infectious organisms associated with leukocytes (such as CMV), and (4) immune modulation. Many of these complications occur frequently in older children and adults, and in those instances leukocyte reduction of platelet components before transfusion is warranted. However, transfusion reactions are relatively rare in infants and neonates. In fact, the major indication for transfusing leukocyte-reduced platelets in neonates is to diminish the risk of TT-CMV.

The prevention of CMV by leukocyte reduction has been extensively studied. Although the precise dose of white blood cells (WBCs) that transmits CMV is unknown, evidence suggests that transfusions with less than 1 to 5×10^6 WBCs/unit are unlikely to be infectious (Strauss 1993; Gilbert et al. 1989; Eisenfeld et al. 1992). In fact, the American Association of Blood Banks (AABB) has stated that seronegative and leukocyte-reduced units are equivalent for prevention of TT-CMV. However, others have disagreed with this statement, arguing that the data are inconclusive to declare leukocyte reduction and CMV-antibody-negative blood products to be equivalent in preventing TT-CMV (Zwicky et al. 1999; Blajchman et al. 2001).

Leukocyte reduction is an expensive technology and represents the largest incremental cost in the history of blood component preparation (Dzik et al. 2002). Still, several countries have adopted practices of universal leukocyte reduction before transfusion of their blood components (Roseff et al. 2002). The benefits and cost-effectiveness of that approach are still under debate. A recent large-scale, randomized controlled trial found no beneficial effect of universal leukocyte reduction versus selective leukocyte reduction for patients with clear indications on in-hospital mortality and length of stay, although patients who received leukocyte-reduced blood did have a lower incidence of febrile reactions (Dzik et al. 2002).

Gamma Irradiation

Transfusion-associated graft-versus-host disease (TA-GVHD) is a severe disease resulting from the transfusion of immunocompetent T lymphocytes present in platelet concentrates into an immunocompromised host, although it can rarely also occur in immunocompetent hosts. The mortality of TA-GVHD is approximately 90%, and it rarely responds to treatment. However, exposure of blood components to a minimum of 2500 cGy of gamma irradiation before transfusion effectively prevents GVHD. Gamma irradiation of 3000 cGy has no adverse effect on platelet quality in WBC-reduced single-donor apheresis platelet concentrates (Zimmermann et al. 2001). All patients receiving immunosuppressive therapy and all patients with a suspected or confirmed immunodeficiency should receive irradiated platelets. For example, patients with conal-truncal abnormalities should receive irradiated products until they can be evaluated for DiGeorge syndrome, which is associated with thymic abnormalities and T-cell immunodeficiency. In addition, blood products obtained from family members, HLA-matched products, and blood products destined for intrauterine or exchange transfusions also should be irradiated. The need for irradiated blood products in *all* neonates is controversial (Strauss 2000b), although it has been argued that the immaturity of their immune system predisposes them to develop TA-GVHD and that the existence of a primary immunodeficiency may be unrecognized during the neonatal period. This controversy is discussed in the section on platelet transfusions in neonates.

Washing

Allergic reactions to platelet transfusions rarely occur in neonates and young children. However, they still can occur, and monitoring during transfusions is indicated. In the older pediatric population as in adults, when repeated allergic reactions occur despite adequate premedication, platelet components may be spun down in a centrifuge and resuspended in saline (Vesilind et al. 1988; Vo et al. 2001). This removes the plasma proteins that are thought to initiate the allergic reaction. We do not recommend routine washing of platelets, however, because it can lead to loss and functional impairment of platelets. An important exception is when maternal platelets are transfused to neonates with alloimmune thrombocytopenia. In this circumstance, the platelets have to be washed or plasma-reduced to remove the pathological antiplatelet antibodies present in the mother's plasma.

Volume Reduction

The preparation of volume-reduced platelets by centrifugation can be performed in most blood banks. The problem with this process is that up to half of the platelets in the original unit may be lost during the volume reduction (Simon and Sierra 1984). Increased processing of platelet components also increases clumping and dysfunction. Therefore, this procedure is not done frequently, and we do not recommend it routinely for neonates, since most neonates are able to tolerate 10 to 15 mL/kg, and this dose results in an adequate platelet increment of up to 100×10^9/L.

Fresh, Matched Platelets

Increased storage time of platelet units decreases posttransfusion platelet increments. Accordingly, physicians can request fresh ABO-matched platelets, but this should be limited to situations in which the patient has had prior transfusions and is now refractory. This should be tried in patients with platelet refractoriness before ordering crossmatched or HLA-matched platelet components.

Crossmatched Platelets

Crossmatched platelet components can be used to support thrombocytopenic patients who have become refractory to wholeblood-derived platelet units. The process of platelet crossmatching is similar to packed red blood cell (PRBC) crossmatching. The patient's serum is screened for the presence of alloantibodies against platelet components. In one study, the use of crossmatched platelet components improved the platelet count significantly in 50% of patients who were refractory to whole blood-derived platelet units (Gelb et al. 1997).

HLA-Matched Platelets

This process is reserved for the refractory patient on whom the previously mentioned measures have failed. Most antibodies that cause platelet refractoriness are directed against HLA class I antigens. Providing a platelet transfusion with an HLA type that closely matches the recipient's can therefore improve platelet increments. In fact, up to 90% of refractory patients will benefit from an HLA-matched product.

ABO and Rh Compatibility

ABO

When selecting platelet units for transfusion, it is desirable for the neonate/infant and the platelet donor to be the same ABO blood group. It is particularly important to minimize giving repeated transfusions of Group O platelets to Group A or B recipients, as large quantities of passive anti-A or anti-B can lead to hemolysis (Pierce et al. 1985). Volume reduction of donated O platelets can reduce the amount of passive antibodies, but it increases platelet loss, clumping, and dysfunction and is therefore not routinely recommended. Conversely, small amounts of A and B antigen found on platelets may lead to poor platelet increments in recipients with high titers of anti-A and/or anti-B isoagglutinins (Carr et al. 1990; Lee and Schiffer 1989), and transfusion of ABO-mismatched platelets may accelerate the development of platelet refractoriness (Carr et al. 1990). For that reason, patients who are likely to require long-term platelet support should receive leukocyte-reduced, ABO-identical platelets, in order to delay the development of platelet alloimmunization.

Rh

Even though most platelet components contain less than 1 mL of RBCs, the development of anti-Rh antibodies can occur. This can affect future transfusions or future pregnancies in female patients who are Rh-negative. The transfusion of Rh-positive platelet products to Rh-negative recipients is therefore discouraged.

Ordering and Administration

When administering platelets, a 170 μm filter should be interposed between the unit and recipient to prevent the infusion of microaggregates and fibrin clots.

Platelets are usually administered over 30 to 60 minutes. During the transfusion, and especially within the initial 15 minutes, it is important to monitor for signs of bacterial infection, fever, allergy, and anaphylactic reactions. Mild reactions can be treated by slowing or stopping the infusion and administering diphenhydramine. For more severe reactions (hypotension, tachycardia, tachypnea, and/or apnea), the transfusion should be stopped and supportive care provided. The remaining product should be sent to the blood bank for analysis.

PLATELET TRANSFUSIONS IN CHILDREN

In this section, we describe current pediatric platelet transfusion practices. However, most of these practices have been developed with reference to adult studies, as few randomized clinical trials have been conducted in children (Luban 2002).

Childhood Conditions Leading to Thrombocytopenia

A number of conditions lead to significant thrombocytopenia in children (Table 22.1). Infection is typically associated with mild to moderate thrombocytopenia (platelets 50,000 to 150,000/microliter), but rarely causes profound thrombocytopenia leading to hemorrhage (platelets 0 to 50,000), unless it is accompanied by disseminated intravascular coagulation (DIC) (van Gorp et al. 1999). A variety of pharmaceuticals can also cause mild to moderate thrombocytopenia and, rarely, profound thrombocytopenia (Blackburn et al. 1998; Finsterer et al. 2001; Gesundheit et al. 2002; Patnode et al. 2000; Yamreudeewong et al. 2002). Autoimmunity, specifically immune thrombocytopenic purpura, is a common cause of profound thrombocytopenia, occurring in 1/10,000 children per year (Lilleyman 1983). Marrow infiltrating malignancies can cause thrombocytopenia but are usually associated with a deficiency in a second cell lineage (Calpin et al. 1998; Jubelirer et al. 2002; Klaassen et al. 2001). These disorders have an incidence in the order of 5 to 20/1,000,000 children per year. Finally, congenital bone marrow failure syndromes, such as Fanconi's anemia (Alter 1992; Alter 1996), amegakaryocytic thrombocytopenia (Auerbach et al. 1997; Lonial et al. 1999; Manoharan et al. 1989), and Wiskott-Aldrich syndrome (Balduini et al. 2002; Nonoyama, 2001) are very rare causes of profound thrombocytopenia in children and have an incidence of approximately 1/1,000,000 children per year or less. Platelet transfusions are also indicated to treat hemorrhage in patients with disorders of platelet function,

TABLE 22.1 Disorders Associated with Thrombocytopenia in Children

Infections

Viral infections
Bacterial infections
Fungal infections
Disseminated intravascular coagulation

Pharmaceuticals

Antibiotics
Antihistamines
Antimalarials
Antiseizure
Antineoplastic

Immune

Immune thrombocytopenic purpura
Systemic lupus erythematosus
Evans's syndrome
Posttransfusion purpura
Heparin-induced thrombocytopenia

Neoplasm

Leukemia
Lymphoma
Neuroblastoma
Rhabdomyosarcoma
Ewing's sarcoma
Medulloblastoma

Congenital

Thrombocytopenia, absent radii
Fanconi's anemia
Amegakaryocytic thrombocytopenia
Wiskott-Aldrich syndrome
Felty's syndrome
May-Heglin anomaly
Sebastian syndrome
Bernard-Soulier Syndrome
Alport's syndrome

TABLE 22.2 Indications and Contraindications for Platelet Transfusion in Children

Platelet Transfusions in Children Indicated

Platelet count <10,000 and decreased platelet production, without other risk factors for bleeding (see Factors that Increase Bleeding Risk).
Platelet count <50,000 with active bleeding or planned procedure in patient with decreased platelet production.
Platelet count <100,000/microliter with hemorrhage, DIC, or other coagulation problem.
Bleeding with qualitative platelet defect at any platelet count.
Cardiovascular bypass surgery with excessive bleeding at any platelet count.

Platelet Transfusions Contraindicated

Immune thrombocytopenic purpura, unless there is profound bleeding.
Hemolytic uremic syndrome.
Thrombotic thrombocytopenic purpura.

Factors that Increase Bleeding Risk and Warrant Increased Platelet Threshold

Fever.
Active bleeding.
Surgery.
Clotting abnormalities.

such as Glanzmann's thrombasthenia, Bernard-Soulier syndrome, and platelet-type von Willebrand disease (Abzug et al. 1991; Bennett et al. 1992; Vuckovic 1996).

Platelet Transfusions

Transfusion Triggers

The decision to transfuse platelets usually falls into one of two categories: treatment of hemorrhage or prophylaxis (Table 22.2). The need to transfuse platelets in patients with severe thrombocytopenia and hemorrhage is evident. In contrast, the use of platelet transfusions to prevent hemorrhage is more controversial, particularly in patients who are not currently bleeding and have no

other condition that predisposes them to bleeding. Many studies have demonstrated a decrease in major hemorrhage in thrombocytopenic patients who receive prophylactic platelet transfusions, but an effect on mortality has not been demonstrated (Higby et al. 1974; Murphy et al. 1982; Roy et al. 1973). Most adult oncologists, nevertheless, use platelet transfusions prophylactically (Pisciotto et al. 1995). Historically, pediatric and adult oncologists have used a transfusion trigger level of 20×10^9/L (Feusner 1984). However, two recent studies in adults compared a threshold of 10 versus 20 $\times 10^9$/L and showed no therapeutic difference and a large cost-savings associated with the lower threshold (Lawrence et al. 2001; Wandt et al. 1998). As a result of this work, many pediatric oncologists have altered their approach, and now use a platelet transfusion threshold of 10×10^9/L in a selected subset of children who have no other risks of bleeding.

In patients with chronic thrombocytopenia who do not have other underlying problems that predispose them to hemorrhage, many clinicians withhold prophylactic platelet transfusions and transfuse only with bleeding. Conversely, most patients who undergo treatment for malignancies receive platelet transfusions when their platelet counts drop below a certain level.

Co-incident medical conditions can substantially increase the risk of hemorrhage in thrombocytopenic patients and necessitate a higher trigger for platelet transfusion. These conditions include any problem where a small hemorrhage could cause a permanent disability, such as intracranial or retinal bleeding (Iyori et al. 2000; Medeiros et al. 1998). Certain medicines, such as aspirin (Van Hecken et al. 2002) and nonsteroidal anti-inflammatory agents (Nishizawa et al. 1981; Ragni et al. 1992), hinder platelet function and should be avoided in patients with thrombocytopenia. Medical conditions such as uremia also lead to platelet dysfunction, which, in combination with thrombocytopenia, significantly increases the risk of a major hemorrhage (Noris et al. 1999). Finally, conditions associated with a decrease in clotting factors, such as DIC and hepatic disease, can lead to hemorrhage at higher platelet counts, and many centers target a higher platelet count for these conditions (Rebulla 2001).

Major surgery can also lead to hemorrhage and is an indication for targeting higher platelet levels. Studies have supported levels between 30 and 70×10^9 platelets/L (Murphy et al. 1969; Simpson, 1987). In contrast, minor procedures such as bone marrow aspirates are well tolerated without major bleeding at very low platelet levels. A report suggests that lumbar punctures can be performed with platelet counts below 10×10^9/L in the absence of other factors affecting the coagulation system (Howard et al. 2000). However, reports of spinal hematomas due to lumbar puncture in patients with thrombocytopenia have led some to transfuse platelets immediately before lumbar puncture in patients with platelet counts below 20×10^9/L (Blade et al. 1983; Edelson et al. 1974).

Platelet Product

A child's weight or body surface area primarily determines the dose of platelets, although it should also take into account the starting platelet count. A given dose of platelets tends to give a better platelet increment, and platelets survive longer in children than in adults (Davis et al. 1999). The standard dose of platelets is 10 mL/kg of a platelet suspension, regardless of whether whole blood-derived or apheresis-derived platelet units are used as the source. Alternatively, one whole-blood derived platelet unit per 10 kg of body weight can be used in older children. This dose typically increases the platelet counts by 40 to 60×10^9/L (Herman et al. 1987). Most centers cap adult doses at 6 to 8 whole-blood derived platelet units (random donor platelets), or 1 apheresis unit. Platelet counts should be tested between 15 minutes and 1 hour posttransfusion to determine the platelet increment, which is the difference between the pre- and posttransfusion platelet count.

Poor Response to Platelet Transfusion

Platelet refractoriness is defined as the consistent inability to increase platelet counts following platelet transfusion. The success of platelet transfusion has classically been measured using the corrected platelet count index (CCI). The CCI is calculated using the following formula:

$$CCI = \frac{(\text{Posttransfusion Platelet Count} - \text{Pretransfusion Platelet}) \times \text{Body Surface Area}}{\text{Number of Platelets Transfused} \times 10 - 11}$$

Refractoriness is defined as a CCI <5000 after two sequential transfusions, where at least one transfusion used fresh platelets (<48 hours old) (Schiffer et al. 1986).

A number of factors can lead to a poor platelet increment following transfusion. These include recipient factors, such as sepsis, DIC, fever, and certain pharmaceuticals. Box 22.1 lists factors that can lead to poor platelet recovery. There are also a number of maneuvers that can improve platelet recovery following transfusion. These are listed in Box 22.2. Patients who have had frequent platelet transfusions can develop antibodies to HLA and platelet-specific antigens (Duquesnoy et al. 1977; Roy et al. 1996). These antibodies can lead to rapid destruction of transfused platelets, which can be life-threatening in a hemorrhaging patient. Platelet refractoriness is a major problem in patients with chronic thrombocytopenia, such as patients undergoing treat-

---

BOX 22.1 Factors Leading to Decreased Platelet Increment

Age of component
Washing platelets
Fever, infection
Disseminated intravascular coagulation
Graft-versus-host disease
Hepatosplenomegaly
Alloantibodies/autoantibodies
Massive blood loss
Hemolytic uremic syndrome/thrombotic thrombocytopenic purpura (platelet transfusions discouraged/contraindicated)
Necrotizing enterocolitis
Medications (amphotericin, vancomycin, ciprofloxacin)

BOX 22.2 Approaches to Platelet Refractoriness

Transfuse ABO identical platelets
Use fresh platelets
HLA-matched platelets
Crossmatched platelets

ment for acute myelogenous leukemia, with aplastic anemia, or with other congenital thrombocytopenia syndromes. The sensitization in most cases is due to antibodies against HLA-I, which are expressed on platelets and contaminating leukocytes. Both leukocyte filtration and ultraviolet irradiation of platelet products have been shown to decrease the development of antiplatelet antibodies (TRAPSG 1997).

Conditions Where Platelet Transfusion Should Be Avoided or Limited

Because of the risk of alloimmunization and platelet refractoriness, patients with congenital or chronic forms of thrombocytopenia should be given platelet transfusions sparingly. At one time, the number of transfusions was correlated with the success of engraftment following bone marrow transplant in patients with aplastic anemia, but that seems to be less the case with more modern conditioning regimens (McCann et al. 1994). In immune thrombocytopenic purpura, platelet transfusions are not effective, unless they are combined with a treatment that will decrease the immune destruction of platelets. Transfusions are therefore generally avoided unless the patient has a life-threatening bleed, in which case they are combined with treatments such as prednisone, intravenous immunoglobulin, or (in the presence of an intracranial bleed) splenectomy at the time of clot evacuation (Medeiros et al. 1998). Disorders such as thrombotic thrombocytopenic purpura that are associated with a consumptive coagulopathy are thought to be exacerbated by platelet transfusion, which therefore should be avoided (Gordon et al. 1987).

PLATELET TRANSFUSIONS IN NEONATES

Neonatal Thrombocytopenia

The definition of thrombocytopenia for neonates at any gestational or postconceptional age is the same than for adults, namely a platelet count of $<150 \times 10^9$/L (Andrew et al. 1984). However, the significance of platelet counts between 100 and 150×10^9/L in neonates is unclear, because a relatively high incidence of counts in this range has been found in otherwise healthy neonates (Aballi et al. 1968). Most neonatologists choose to closely monitor these neonates and to repeat their platelet counts before starting a full evaluation. Platelet counts $<100 \times 10^9$/L should be considered definitely abnormal at any age and warrant evaluation.

Thrombocytopenia is a frequent problem among ill neonates, affecting 18% to 35% of all patients admitted

to the neonatal intensive care unit (NICU) (Castle et al. 1986; Mehta et al. 1980; Oren et al. 1994). Although multiple clinical conditions have been associated with thrombocytopenia (Table 22.3), the mechanisms responsible for thrombocytopenia in neonates, as in adults, are only three: decreased platelet production, increased platelet destruction, platelet sequestration (mostly secondary to hypersplenism), or a combination of these processes. In many affected neonates, the

TABLE 22.3 Pathophysiological Classification of Neonatal Thrombocytopenia

Immune-Mediated Thrombocytopenia

Neonatal alloimmune thrombocytopenia (NAIT)
Autoimmune thrombocytopenia

Infections

Bacterial infections
Viral infections
Fungal infections

Genetic Conditions

Thrombocytopenia-absent radius (TAR) syndrome
Congenital amegakaryocytic thrombocytopenia
   Fanconi's anemia
   Congenital hypoplastic thrombocytopenia with microcephaly
Congenital thrombocytopenia, Robin's syndrome, agenesis of the corpus callosum, distinctive facies, and developmental delay
Familial macrothrombocytopenias
   Bernard-Soulier syndrome
   May-Hegglin anomaly
   Paris-Trousseau thrombocytopenia
   X-linked recessive thrombocytopenia
Chromosomal abnormalities
   Trisomies 13, 18, and 21 and Turner's syndrome
Other genetic disorders associated with neonatal thrombocytopenia
   Wiskott-Aldrich syndrome
   Noonan's syndrome
   Alport's syndrome
Inherited metabolic disorders (methylmalonic acidemia, ketoglycinemia, isovalericacidemia, holocarboxylase synthetase deficiency)

Disseminated Intravascular Coagulation
Miscellaneous Causes

Pregnancy-induced hypertension
Intrauterine growth restriction
Necrotizing enterocolitis
Thrombosis
Perinatal asphyxia

Medication-Induced Thrombocytopenia

Heparin
Vancomycin
Medications administered to the mother (such as, quinine)

responsible mechanism remains unclear despite extensive evaluation.

In regard to the clinical impact of neonatal thrombocytopenia, the bleeding time has been shown to be significantly prolonged in most neonates with a platelet count $<100 \times 10^9$/L (Andrew et al. 1987). Whether this translates into an increased risk of clinically significant bleeding is a subject of debate. While some studies have reported a higher incidence of severe intraventricular hemorrhage (IVH) and serious neurological morbidity in thrombocytopenic compared to nonthrombocytopenic infants (Andrew et al. 1987), others have reported a lack of correlation between thrombocytopenia and IVH (Lupton et al. 1988). Two recent studies, which focused exclusively on neonates who received one or more platelet transfusions, reported a tenfold or higher increase in mortality among NICU patients who received platelet transfusions, compared to untransfused neonates (Del Vecchio et al. 2001; Garcia et al. 2001). However, none of those neonates died as a consequence of uncontrollable hemorrhage, but rather as a consequence of the severe underlying disease processes that also lead to thrombocytopenia, suggesting that severe thrombocytopenia in neonates should be considered a marker of overall disease severity.

Platelet Transfusions

Transfusion Triggers

Several studies have highlighted the significant disparities that exist worldwide in the use of platelet transfusions to treat thrombocytopenic neonates (Del Vecchio et al. 2001; Garcia et al. 2001; Murray et al. 2002; Strauss et al. 1993). Although multiple factors probably contribute to these disparities, they mostly reflect the lack of evidence-based guidelines for platelet transfusions in neonates. In the only prospective controlled randomized trial performed on this subject, Andrew et al. (1993) randomly assigned thrombocytopenic premature infants to either a group in whom platelet counts were maintained above 150×10^9/L, or to a group that only received platelet transfusions for clinical indications or for a platelet count $<50 \times 10^9$/L. Most platelet transfusions in the latter group were administered at a platelet count of approximately 60×10^9/L. These investigators found no differences in frequency or severity of intracranial hemorrhages between the two groups, and thus concluded that maintaining platelet counts in the normal range ($>150 \times 10^9$/L) provided no evident benefit, and that nonbleeding premature infants with platelet counts $>60 \times 10^9$/L should not receive prophylactic platelet transfusions. Since all neonates with platelet counts $<50 \times 10^9$/L were automatically transfused, however, this study did not address the issue of whether lower platelet counts could be tolerated in clinically stable infants without an increased risk of hemorrhage.

In an attempt to answer this question, Murray et al. (2002) performed a retrospective review of their use of platelet transfusions among neonates with platelet counts $<50 \times 10^9$/L (53 over 3 years). Overall, they transfused 51% of these neonates: all infants with a platelet count $<30 \times 10^9$/L and neonates with a platelet count between 30 and 50×10^9/L who had a previous hemorrhage (severe IVH), or were clinically unstable. These authors did not observe any major hemorrhage in this group of severely thrombocytopenic neonates, regardless of whether platelet transfusions were given or withheld. Five of the 53 neonates studied died, but none as a result of hemorrhage. They concluded that a prophylactic platelet transfusion trigger threshold of $<30 \times 10^9$/L probably represents a safe practice *for clinically stable* NICU patients.

In the absence of other studies to provide evidence-based practice parameters, numerous experts and consensus groups have published guidelines for the administration of platelet transfusions to neonates. However, these represent best-guess estimates and consensus practices rather than solid, evidence-derived recommendations. In the case of bleeding neonates, experts agree that platelet transfusions are indicated for platelet counts <50 to 100×10^9/L. However, the majority of platelet transfusions are not given to treat bleeding, but rather are given prophylactically to decrease the risk of hemorrhage associated with a platelet count below a certain threshold. Establishing this threshold has been particularly difficult in neonates because of the large number of variables—in addition to the platelet count—that play a role in determining the risk of severe hemorrhage in any individual neonate. Among these are the gestational age (premature neonates have a higher risk of bleeding), the days of life (most intraventricular hemorrhages occur in the first week), the degree of illness, the mechanism of thrombocytopenia (decreased platelet production versus increased platelet destruction), the presence of other coagulation abnormalities (that is, DIC), and the use of medications that compromise hemostasis. Recognizing the importance of these factors, experts have formulated guidelines that attempt to incorporate most of them (Blanchette et al. 1991; Blanchette et al. 1995; Calhoun et al. 2000; Murray 2002; Roberts et al. 1999; Strauss 2000a; Voak et al. 1994). These guidelines are summarized in Table 22.4.

Regardless of differences in the exact threshold for transfusion, there is overall agreement that neonates of any gestational or postconceptional age should be transfused at higher platelet counts than children or adults, because of multiple developmental abnormalities in platelet function (Israels et al. 1990; Israels et al. 1997; Rajasekhar et al. 1994; Stuart 1979; Stuart et al. 1984)

TABLE 22.4 Summary of Platelet Transfusion Triggers Recommended for Neonates (Platelet counts ×10⁹/L)

Author	Nonbleeding Sick Preterm	Nonbleeding Stable Preterm	Nonbleeding Term	Before Invasive Procedure	Active Bleeding
Blanchette et al. 1991*	<100	<50	<20	<50 if failure of production <100 if DIC	<50 if failure of production <100 if DIC
Voak et al. 1994†	<50	<30	<30	Not addressed	Not addressed
Blanchette et al. 1995	<50	<30	<20 if stable <30 if sick	<50 for minor procedure <100 for major surgery	<50 in all cases <100 if DIC Any platelet count if functional disorder
Roberts et al. 1999	<50 if DIC <100 if falling rapidly	<50	<30	<100	<100 if major organ bleeding <50 if minor bleeding
Calhoun et al. 2000	<50	<25	Same as preterm	<50	Not addressed
Strauss 2000a	<100	<20	<20	<50	<100
Murray 2002	<50	<30	<30	<50	<100
Roseff et al. 2002	<100	<50	<30	<50 if failure of production <100 if DIC	<50 if stable <100 if sick

*Guidelines from the Pediatric Hemotherapy Committee of the American Association of Blood Banks for the conduct of pediatric blood transfusion audits.
†Guidelines from the British Committee for Standards in Haematology Blood Transfusion Task Force.

and clotting factors (Andrew et al. 1990; Andrew et al. 1988) that potentiate their risk of bleeding. Experts also concur in recommending that sick premature neonates (although the definition of "sick" is variable) should be transfused at higher platelet counts (50 to 100 × 10⁹/L) than stable or term neonates (20 to 50 × 10⁹/L).

Platelet Product

Neonates qualifying for a platelet transfusion should receive 10 to 15 mL/kg of a CMV-safe, ABO-matched standard platelet suspension, prepared either from a fresh unit of whole blood or by platelet pheresis. Volume reduction is not routinely recommended, because a transfusion of 10 mL/kg of any platelet suspension provides sufficient platelets to increase the platelet count by >50 × 10⁹/L (Andrew et al. 1993; Strauss 2000b), and because of the risks of platelet loss, clumping, and dysfunction associated with the additional handling. The only exception are neonates in whom rigorous volume restriction is essential (that is, neonates with renal disease leading to oliguria or anuria, or neonates in congestive heart failure). Platelets should be transfused as rapidly as the infant's condition allows, usually over 1 hour, but definitely within 2 hours.

Although there is some controversy regarding the importance of transfusion-transmitted CMV infection in neonates, it has become standard practice to transfuse neonates only with blood components with low risk of transmitting CMV (Calhoun et al. 2000; Strauss et al. 1993). There is a definite indication in the case of low birth weight infants (<1500 g) and infants born to a seronegative mother or to a mother whose CMV status is unknown (Roseff et al. 2002; Wong and Luban 2002). In most institutions, however, the practice has been to transfuse all infants with blood products obtained from donors without detectable antibodies to CMV (CMV-negative) (Strauss et al. 1993), an approach considered the standard of care for patients at risk of severe CMV infection. Alternatively, leukoreduction by any method capable of reducing the WBC count to less than 1 to 5 × 10⁶/unit has been shown to effectively reduce the risk of CMV infection to levels comparable to those achieved with the use of seronegative blood components (Strauss 2000b).

Neonates are also considered to be at increased risk for transfusion-associated graft-versus-host disease (TA-GVHD). However, the degree of risk, and therefore the need to provide irradiated blood components to all neonates, has been the subject of debate. A large review of cases of TA-GVHD among infants (Ohto et al. 1996; Sanders et al. 1990; Strauss 2000b) described 73 infants who developed this complication. The great majority of infants had recognized risk factors for which irradiated cellular blood components are clearly indicated: underlying immunodeficiency disease, intrauterine or exchange transfusion, or transfusion from a relative. Only 6 of the 73 infants did not fit into one of these categories (Strauss 2000b). Based on the fact that an underlying primary immunodeficiency disorder may not be apparent in the neonatal period, however, many institutions have chosen to routinely irradiate all blood products for neonates. Other centers have adopted a more selective approach and only irradiate blood products administered to neonates with a specific indication (as previously discussed). In this regard, a survey performed in 1989 among members of the AABB revealed a spectrum of practices, which varied between routinely providing irradiated blood products for all neonates (13% of hospitals), for some neonates (46% of hospitals), or to none (41% of all hospitals) (Strauss et al. 1993). It is likely that these relative proportions have changed over the last decade, but the diversity of practices probably persists.

Special Conditions in Neonatology

Neonatal Alloimmune Thrombocytopenia

The classic presentation of neonatal alloimmune thrombocytopenia (NAIT) is that of an otherwise healthy, usually term or near-term neonate who is found to have severe thrombocytopenia. The pathogenesis of this type of thrombocytopenia resembles that of erythroblastosis fetalis. When incompatibility between parental platelet antigens exist, the mother can become sensitized to an antigen expressed on the fetal platelets. If a maternal antibody is formed and crosses the placenta, it can bind to the fetal platelets, which are then removed from the circulation by the fetal reticuloendothelial system. Unlike neonatal Rh disease, however, the first pregnancy can produce an affected fetus, and there is no routine prenatal screening test available.

There are multiple antigens expressed on platelets, including class I HLA antigens, ABO antigens, and several biallelic platelet-specific antigens. The nomenclature for these platelet-specific antigens is rather confusing. In an attempt to solve this problem, a simplified system was recently implemented. According to this nomenclature, each human platelet antigen (HPA) was assigned a number (HPA-1 to -11), and the different allelic forms were distinguished by an "a" or "b" suffix (Blanchette et al. 2000). An "a" suffix indicates the more common and "b" the more rare allele. The frequency with which a particular antigen is responsible for cases of NAIT in a specific population depends on the relative frequency of that antigen in that population. For example, approximately 70% of Caucasian individuals have the genotype HPA-1a/HPA-1a (*PLA-1 positive*), while only 2.9% of the population is homozygous for the HPA-1b antigen (*PLA-1 negative*) (George et al. 1995). For that reason, HPA-1a (or PLA-1) accounts for 80% to 90% of the cases of NAIT in the Caucasian pop-

ulation (Homans 1996). In Asian subjects, the HPA-4 (or *Yuk/Pen*) system is most frequently implicated in the pathogenesis of NAIT (George et al. 1995). Cases mediated by antibodies against HPA-5 (Kaplan et al. 1991), HPA-3 (von dem Borne et al. 1980), HPA-2 (Kroll et al. 1994), and HPA-6 (Westman et al. 1997) have also been described, although with much smaller frequency.

Since intracranial hemorrhage has been reported in 10% to 15% of cases of NAIT (Blanchette 1988), it is extremely important that effective therapy be provided as soon as possible to infants with suspected or confirmed NAIT and severe thrombocytopenia (platelet counts $<30 \times 10^9/L$). The most effective therapy for these neonates is the transfusion of antigen-negative compatible platelets, and the most reliable source of these platelets is the mother. The process of obtaining maternal platelets should therefore be initiated as soon as the diagnosis of severe NAIT appears likely. If maternal platelets are used, however, they have to be irradiated (for prevention of GVHD) and washed or plasma-depleted to remove pathogenic maternal alloantibodies present in the plasma. If maternal platelets are unavailable, transfusion with HPA-1a and -5b antigen-negative platelets has been recommended in Caucasian populations, in which these two antigen systems are responsible for >95% of the cases of NAIT (Blanchette et al. 2000). This approach is only possible when large centralized blood services have registries of HPA-1a and -5b-negative blood donors who donate regularly, and thus maintain platelet concentrates for this particular use readily available (Win et al. 1997).

While awaiting compatible platelets, or in cases where those are not available, the use of high dose intravenous immunoglobulin (IVIG) or a trial of random donor platelets is indicated. Since 98% of Caucasian donors are HPA-1a positive, the absence of a response to random-donor platelets strongly supports the diagnosis of NAIT in this population. However, a very small percentage of neonates will have a response to random platelets, either due to the unlikely scenario of receiving HPA-1- negative random platelets, or because of the presence of weak HPA-1a antibodies, or because of NAIT due to platelet alloantibodies against antigens other than HPA-1a. For example, random-donor platelets would be expected to be effective in approximately 80% of cases of HPA-5b-mediated NAIT (Blanchette et al. 2000). In regard to IVIG, it has been reported to be beneficial in some cases of NAITP (Massey et al. 1987; Sidiropoulos et al. 1984). The recommended total dose is 2 g/kg administered either as 0.4 g/kg daily for 5 days, or 1 g/kg daily for 2 consecutive days (Blanchette et al. 2000; Calhoun et al. 2000).

The Neonate on Extracorporeal Membrane Oxygenation

Extracorporeal membrane oxygenation (ECMO) is a mainstay therapy of modern neonatology used to treat patients with respiratory and/or cardiac failure unresponsive to maximal medical therapy. The platelet count falls immediately after a neonate is placed on ECMO, presumably due to a combination of hemodilution by the additional blood in the circuit and accelerated platelet destruction secondary to activation and adherence of platelets to the circuit tubing or the membrane oxygenator.

Multiple platelet transfusions are invariably given to neonates on ECMO, averaging 1.3 platelet transfusions per day (Chevuru et al. 2002). In fact, platelet transfusions account for the majority of donor exposures among neonates on ECMO (Chevuru et al. 2002; Rosenberg et al. 1994; Zavadil et al. 1998). To some degree, this is due to the fact that most centers transfuse neonates on ECMO to maintain platelet counts above 100 or even 110×10^9/L (Chevuru et al. 2002). This higher platelet transfusion threshold has been recommended based on the high risk of bleeding in these patients resulting from a combination of severe illness (Dela Cruz et al. 1997), thrombocytopenia, platelet dysfunction (Cheung et al. 2000; Robinson et al. 1993), and systemic heparinization while on ECMO. In an attempt to decrease the number of donor exposures, some investigators have lowered their threshold for platelet transfusions to 80×10^9/L (Minifee et al. 1990) and even 50×10^9/L (Bjerke et al. 1992), and others have recommended the use of aliquoted platelet pheresis from either a single or a limited number of donors (Rosenberg et al. 1994). These interventions, particularly the latter one, have resulted in a significant reduction in platelet-related donor exposures among neonates on ECMO, but the safety of a lower platelet transfusion threshold has not been studied.

Medications that Affect Platelet Function

Neonates admitted to NICUs are frequently exposed to medications thought to induce platelet dysfunction. Indomethacin and inhaled nitric oxide, in particular, are medications routinely used in the NICU setting that have been extensively studied. Indomethacin, which impairs platelet function by reversibly inhibiting cyclooxygenase, is frequently administered to preterm neonates to close a patent ductus arteriosus (Van Overmeire et al. 2001) or to prevent severe intraventricular hemorrhage (Hanigan et al. 1988; Ment 1988; Ment et al. 1985; Ment et al. 1994). Several authors have reported decreased platelet aggregation (Friedman

et al. 1978) and significant prolongations (>twofold) of the bleeding time in treated neonates (Andrew et al. 1989; Corazza et al. 1984; Sola et al. 2001b), although it is not clear whether this prolongation is associated with a significantly increased risk of severe hemorrhage (Maher et al. 1985; Ment et al. 1994).

Studies looking at the effects of inhaled nitric oxide have been mostly performed in adults and have led to inconclusive results. While some have reported decreased platelet function in response to nitric oxide (Gries et al. 1998; Gries et al. 2000; Samama et al. 1995), others have observed no effect (Albert et al. 1999a; Christou et al. 1998). Furthermore, some investigators have described a prolongation in the bleeding time in the absence of changes in platelet function (Albert et al. 1999b; Albert et al. 1996; George et al. 1998), raising the possibility that the effects of nitric oxide could be exerted at the level of the platelet-endothelial interface.

Although the clinical significance of these effects on platelet function is unclear, it is prudent to take these medications, as well as the presence of any other clotting defects, into consideration when trying to determine the risk of hemorrhage in a thrombocytopenic neonate. Indomethacin, in particular, is almost always administered to preterm neonates during the first week of life, at a time when platelet dysfunction has been well documented (Tanindi et al. 1995) and when the risk of intraventricular hemorrhage is highest. We therefore recommend that, under those circumstances, and particularly in unstable neonates, platelets be transfused at a higher threshold ($<100 \times 10^9$/L).

REACTIONS TO PLATELET TRANSFUSIONS

Infection

During the 1980s, the number of patients receiving PRBC transfusions decreased significantly, mainly due to the risk of transmitting HIV and hepatitis B (Surgenor et al. 1990). Simultaneously, however, the use of platelet transfusions increased substantially. While volunteer donor programs and infection screening techniques have significantly reduced the risks of transmitting infection, a certain risk of transmitting life-threatening or debilitating infections remains.

Recently, attention has been focused on the risk of life-threatening septicemia due to contaminating bacteria from platelet units. Bacteria are more likely to contaminate platelet products because they are maintained for several days at room temperature (Brecher et al. 2000; Buchholz et al. 1971; Kuehnert et al. 2001; Liu et al. 1999; Pink et al. 1993) The length of storage of platelets in the United States is limited to 5 days because of this risk. In a study performed at Johns Hopkins University, reactions to platelet transfusions were evaluated during the 1987–1998 time period. During this period, the center had switched from using random donor platelets approximately 50% of the time to using almost exclusively apheresis platelet units. During this study period, there was a threefold decrease in the rate of transmission of bacterial infection, from 1/4818 to 1/15,098 transfusions (Ness et al. 2001). This experience has led some hospitals to use only apheresis platelets in order to decrease the sepsis risk. Beginning in March 2004, all blood banks in the United States will be required to monitor for bacterial contamination of platelet products in some way.

Viruses that are known to taint the blood supply in the United States include CMV and parvovirus, but less commonly hepatitis B and C viruses, HIV (Dodd 1995), and newer agents such as the organism causing West Nile encephalitis (CDC 2002). The risk of contracting these infections is increased by repeated transfusion in patients with chronic thrombocytopenia. Other theoretical risks include the emergence of new infectious agents that we are currently unable to screen for, similar to the situation in the early 1980s, when a number of patients were infected by HIV before the virus was identified.

Allergic Reactions

Plasma proteins are the cause of allergic reactions following transfusion. Because plasma is a major component of the platelet product, allergic reactions are a significant cause of morbidity following platelet transfusion (Kluter et al. 1999), particularly in patients with IgA deficiency (Sloand et al. 1990). These reactions can be treated in most cases with antihistamines and corticosteroids, but they can be harder to treat in the setting of IgA deficiency. The plasma proteins can be removed by washing platelets, which involves centrifuging them down and resuspending them in a solution that does not contain plasma proteins (Silvergleid et al. 1977; Sloand et al. 1990). However, this can decrease the number and function of the platelets in the product. Washed platelets should therefore be reserved for patients who have repeated allergic reactions and who fail to respond to pretreatment with antihistamine and corticosteroids.

Fever

Fever can be the result of contaminating bacteria, ABO incompatibility, or contaminating leukocytes. While sepsis and hemolysis should be considered in a patient who develops fever during or shortly after a platelet transfusion, nonhemolytic febrile transfusion reactions are much more common. Leukocytes that contaminate platelet products are the major cause of fever following transfusion, due to the release of inflammatory cytokines (Heddle et al. 1994; Muylle et al. 1996). For that reason, prestorage filtration of the platelet products can reduce the incidence of fever (Couban et al. 2002).

Graft-Versus-Host Disease

GVHD is the result of contaminating T lymphocytes that are present in platelet concentrates. GVHD usually results from the transfusion of immunocompetent T cells into an immunocompromised host (Ohto et al. 1996; Schroeder 2002; Williamson 1998), although it can also occur in immunocompetent hosts (Ohto et al. 1996). GVHD presents between 8 to 10 days following transfusion and is characterized by rash, diarrhea, elevated hepatic transaminases, and hyperbilirubinemia. TA-GVHD also commonly causes pancytopenia. This syndrome is associated with very high mortality (>90%) and can be prevented by irradiation of platelets before transfusion. Specific indications for the use of irradiated blood products are listed on p. 255 under Gamma-Irradiation.

Transfusion-Related Lung Injury

Transfusion-related lung injury (TRALI) should be a consideration in any patient who develops respiratory distress and hypoxemia during or shortly after a platelet transfusion. TRALI typically occurs 1 to 6 hours after transfusion of platelets and is due to the presence of antibodies to leukocytes that are contained in the donor plasma. These antibodies cause agglutination of neutrophils within the pulmonary vasculature and subsequently inflammation. Patients classically present with hypotension, fever, and pulmonary infiltrates associated with severe hypoxemia (Kao et al. 2003; Silliman et al. 2003; Silliman et al. 1997).

Posttransfusion Purpura

Posttransfusion purpura is a rare disorder that occurs in patients who have previously been transfused with blood products and in HPA-1a-negative women who have been exposed to HPA-1a during pregnancy. Posttransfusion purpura is due to alloantibodies to platelet-specific glycoproteins, most commonly to HPA-1a (Valentin et al. 1990). This results in the destruction not only of the transfused platelets, but also of the patient's own platelets. The resulting thrombocytopenia typically lasts between 5 and 10 days. This disorder typically occurs in adults but has been seen in teenagers. Treatment with IVIG can help in patients with hemorrhage. Some authors have also recommended transfusions with washed, HPA-1a-negative platelets, though their use is controversial. (Brecher et al. 1990).

POTENTIAL ALTERNATIVES TO PLATELET TRANSFUSIONS

Recombinant thrombopoietin (rTpo) and recombinant IL-11 (rIL-11) are hematopoietic growth factors with significant thrombopoietic activity that hold promise as alternatives to platelet transfusions. While IL-11 is already FDA-approved for the treatment of thrombocytopenia, rTpo is still undergoing clinical trials. Neither one of these factors has been used in the neonatal setting to this date.

Tpo is the most potent known stimulator of platelet production (Kaushansky et al. 1995), and studies have shown that it can ameliorate chemotherapy-induced thrombocytopenia (Basser et al. 1997; Basser et al. 1996; Fanucchi et al. 1997; Vadhan-Raj et al. 1997) and certain cases of refractory immune thrombocytopenic purpura (Nomura et al. 2002). In vitro and in vivo animal studies have shown that megakaryocyte progenitors of neonates are more sensitive to rTpo than progenitors from adults (Sola et al. 2000a; Sola et al. 2000b). However, it takes 5 days from the start of therapy to see a rise in the platelet counts (Sola et al. 2000a). For that reason, it is evident that only selected neonates with severe and persistent thrombocytopenia and with an element of insufficient platelet production would be candidates for this therapy (Sola et al. 2001a). In addition, concerns remain about potential toxicities, including thrombocytosis, increased platelet function, thrombosis, and the development of neutralizing antibodies. Clearly, more studies are necessary to identify neonates that would potentially benefit from rTpo administration, and the use of this medication should be restricted to well-controlled clinical trials with appropriate parental consent.

Recombinant IL-11 has been FDA-approved for the prevention of severe thrombocytopenia following myelosuppressive chemotherapy for nonmyeloid malignancies (Isaacs et al. 1997). However, its administration has been associated with sometimes significant side effects, mostly related to fluid retention (edema, pleural effusions) and atrial arrythmias (Smith 2000; Bussel et al. 2001). There is limited clinical data in pediatric populations (Braithwaite et al. 2002; Goldman et al. 2001), and its use has never been reported in neonates. As with thrombopoietin, the use of this hematopoietic growth factor in the pediatric population should be restricted to controlled clinical trial settings.

References

Aballi AJ, Puapondh Y, et al. 1968. Platelet counts in thriving premature infants. *Pediatrics* 42:685–689.

Abzug MJ and Levin MJ. 1991. Neonatal adenovirus infection: four patients and review of the literature. *Pediatrics* 87:890–896.

Albert J, Norman M, et al. 1999a. Inhaled nitric oxide does not influence bleeding time or platelet function in healthy volunteers. *Eur J Clin Invest* 29:953–959.

Albert J, Wallen NH, et al. 1999b. Neither endogenous nor inhaled nitric oxide influencesthe function of circulating platelets in healthy volunteers. *Clin Sci (Lond)* 97:345–353.

Albert J, Wallen NH, et al. 1996. Effects of inhaled nitric oxide compared with aspirin on platelet function in vivo in healthy subjects. *Clin Sci (Lond)* 91:225–231.

Alter BP. 1996. Aplastic anemia, pediatric aspects. *Oncologist* 1:361–366.

Alter BP. 1992. Arm anomalies and bone marrow failure may go hand in hand. *J Hand Surg [Am]* 17:566–571.

Andrew M, Castle V, et al. 1989. Modified bleeding time in the infant. *Am J Hematol* 30:190–191.

Andrew M, Castle V, et al. 1987. Clinical impact of neonatal thrombocytopenia. *J Pediatr* 110:457–464.

Andrew M and Kelton J. 1984. Neonatal thrombocytopenia. *Clin Perinatol* 11:359–391.

Andrew M, Paes B, et al. 1990. Development of the hemostatic system in the neonate and young infant. *Am J Pediatr Hematol Oncol* 12:95–104.

Andrew M, Paes B, et al. 1988. Development of the human coagulation system in the healthy premature infant. *Blood* 72:1651–1657.

Andrew M, Vegh P, et al. 1993. A randomized, controlled trial of platelet transfusions in thrombocytopenic premature infants. *J Pediatr* 123:285–291.

Auerbach AD, Verlander PC, et al. 1997. New molecular diagnostic tests for two congenital forms of anemia. *J Clin Lab Anal* 11:17–22.

Balduini CL, Iolascon A, et al. 2002. Inherited thrombocytopenias: from genes to therapy. *Haematologica* 87:860–880.

Basser RL, Rasko JE, et al. 1997. Randomized, blinded, placebo–controlled phase I trial of pegylated recombinant human megakaryocyte growth and development factor with filgrastim after dose–intensive chemotherapy in patients with advanced cancer. *Blood* 89:3118–3128.

Basser RL, Rasko JE, et al. 1996. Thrombopoietic effects of pegylated recombinant human megakaryocyte growth and development factor (PEG–rHuMGDF) in patients with advanced cancer. *Lancet* 348:1279–1281.

Bennett JS and Kolodziej MA. 1992. Disorders of platelet function. *Dis Mon* 38:577–631.

Bjerke HS, Kelly RE, et al. 1992. Decreasing transfusion exposure risk during extracorporeal membrane oxygenation (ECMO). *Transfus Med* 2:43–49.

Blackburn SC, Oliart AD, et al. 1998. Antiepileptics and blood dyscrasias: a cohort study. *Pharmacotherapy* 18:1277–1283.

Blade J, Gaston F, et al. 1983. Spinal subarachnoid hematoma after lumbar puncture causing reversible paraplegia in acute leukemia. Case report. *J Neurosurg* 58:438–439.

Blajchman MA, Goldman M, et al. 2001. Proceedings of a consensus conference: prevention of posttransfusion CMV in the era of universal leukoreduction. *Transfus Med Rev* 15:1–20.

Blanchette VS. 1988. Neonatal alloimmune thrombocytopenia: a clinical perspective. *Curr Stud Hematol Blood Transfus* 54:112–126.

Blanchette VS, Hume HA, et al. 1991. Guidelines for auditing pediatric blood transfusion practices. *Am J Dis Child* 145:787–796.

Blanchette VS, Johnson J, et al. 2000. The management of alloimmune neonatal thrombocytopenia. *Baillieres Best Pract Res Clin Haematol* 13:365–390.

Blanchette VS, Kuhne T, et al. 1995. Platelet transfusion therapy in newborn infants. *Transfus Med Rev* 9:215–230.

Bowden RA, Sayers M, et al. 1986. Cytomegalovirus immune globulin and seronegative blood products to prevent primary cytomegalovirus infection after marrow transplantation. *N Engl J Med* 314:1006–1010.

Bowden RA, Slichter SJ, et al. 1991. Use of leukocyte-depleted platelets and cytomegalovirus-seronegative red blood cells for prevention of primary cytomegalovirus infection after marrow transplant. *Blood* 78:246–250.

Braithwaite K, Abu-Ghosh A, et al. 2002. Treatment of severe thrombocytopenia with IL-11 in children with Wiskott-Aldrich syndrome. *J Pediatr Hematol Oncol* 24:323–326.

Brecher ME, Holland PV, et al. 2000. Growth of bacteria in inoculated platelets: implications for bacteria detection and the extension of platelet storage. *Transfusion* 40:1308–1312.

Brecher ME, Means N, et al. 2001. Evaluation of an automated culture system for detecting bacterial contamination of platelets: an analysis with 15 contaminating organisms. *Transfusion* 41:477–482.

Brecher ME, Moore SB, et al. 1990. Posttransfusion purpura: the therapeutic value of PlA1-negative platelets. *Transfusion* 30:433–435.

Buchholz DH, Young VM, et al. 1971. Bacterial proliferation in platelet products stored at room temperature. Transfusion-induced Enterobacter sepsis. *N Engl J Med* 285:429–433.

Bussel JB, Mukherjee R, et al. 2001. A pilot study of rhuIL-11 treatment of refractory ITP. *Am J Hematol* 66:172–177.

Calhoun DA, Christensen RD, et al. 2000. Consistent approaches to procedures and practices in neonatal hematology. *Clin Perinatol* 27:733–753.

Calpin C, Dick P, et al. 1998. Is bone marrow aspiration needed in acute childhood idiopathic thrombocytopenic purpura to rule out leukemia? *Arch Pediatr Adolesc Med* 152:345–347.

Carr R, Hutton JL, et al. 1990. Transfusion of ABO-ismatched platelets lead to early platelet refractoriness. *Br J Haematol* 75:408–413.

Castle V, Andrew M, et al. 1986. Frequency and mechanism of neonatal thrombocytopenia. *J Pediatr* 108:749–755.

Centers for Disease Control and Prevention. 2002. Investigations of West Nile virus infections in recipients of blood transfusions. *JAMA* 288:2535–2536.

Cheung PY, Sawick G, et al. 2000. The mechanisms of platelet dysfunction during extracorporeal membrane oxygenation in critically ill neonates. *Crit Care Med* 28:2584–2590.

Chevuru SC, Sola MC, et al. 2002. Multicenter analysis of platelet transfusion usage among neonates on extracorporeal membrane oxygenation. *Pediatrics* 109:e89.

Christou H, Magnani B, et al. 1998. Inhaled nitric oxide does not affect adenosine 5′ diphosphate–dependent platelet activation in infants with persistent pulmonary hypertension of the newborn. *Pediatrics* 102:1390–1393.

Corazza MS, Davis RF, et al. 1984. Prolonged bleeding time in preterm infants receiving indomethacin for patent ductus arteriosus. *J Pediatr* 105:292–296.

Couban S, Carruthers J, et al. 2002. Platelet transfusions in children: results of a randomized, prospective, crossover trial of plasma removal and a prospective audit of WBC reduction. *Transfusion* 42:753–758.

Davis KB, Slichter SJ, et al. 1999. Corrected count increment and percent platelet recovery as measures of posttransfusion platelet response: problems and a solution. *Transfusion* 39:586–592.

Del Vecchio A, Sola MC, et al. 2001. Platelet transfusions in the neonatal intensive care unit: factors predicting which patients will require multiple transfusions. *Transfusion* 41:803–808.

Dela Cruz TV, Stewart DL, et al. 1997. Risk factors for intracranial hemorrhage in the extracorporeal membrane oxygenation patient. *J Perinatol* 17:18–23.

Dodd RY. 1995. Viral contamination of blood components and approaches for reduction of infectivity. *Immunol Invest* 24:25–48.

Duquesnoy RJ, Filip DJ, et al. 1977. Successful transfusion of platelets "mismatched" for HLA antigens to alloimmunized thrombocytopenic patients. *Am J Hematol* 2:219–226.

Dzik WH, Anderson JK, et al. 2002. A prospective, randomized clinical trial of universal WBC reduction. *Transfusion* 42:1114–1122.

Edelson RN, Chernik NL, et al. 1974. Spinal subdural hematomas complicating lumbar puncture. *Arch Neurol* 31:134–137.

Eisenfeld L, Silver H, et al. 1992. Prevention of transfusion-associated cytomegalovirus infection in neonatal patients by the removal of white cells from blood. *Transfusion* 32:205–209.

Fanucchi M, Glaspy J, et al. 1997. Effects of polyethylene glycol-conjugated recombinant human megakaryocyte growth and development factor on platelet counts after chemotherapy for lung cancer. *N Engl J Med* 336:404–409.

Feusner J. 1984. Supportive care for children with cancer. Guidelines of the Childrens Cancer Study Group. The use of platelet transfusions. *Am J Pediatr Hematol Oncol* 6:255–260.

Finsterer J, Pelzl G, et al. 2001. Severe, isolated thrombocytopenia under polytherapy with carbamazepine and valproate. *Psychiatry Clin Neurosci* 55:423–426.

Friedman Z, Whitman V, et al. 1978. Indomethacin disposition and indomethacin-induced platelet dysfunction in premature infants. *J Clin Pharmacol* 18:272–279.

Garcia MG, Duenas E, et al. 2001. Epidemiologic and outcome studies of patients who received platelet transfusions in the neonatal intensive care unit. *J Perinatol* 21:415–420.

Gelb AB and Leavitt AD. 1997. Crossmatch-compatible platelets improve corrected count increments in patients who are refractory to randomly selected platelets. *Transfusion* 37:624–630.

George D and Bussel JB. 1995. Neonatal thrombocytopenia. *Semin Thromb Hemost* 21:276–293.

George TN, Johnson KJ, et al. 1998. The effect of inhaled nitric oxide therapy on bleeding time and platelet aggregation in neonates. *J Pediatr* 132:731–734.

Gesundheit B, Kirby M, et al. 2002. Thrombocytopenia and megakaryocyte dysplasia: an adverse effect of valproic acid treatment. *J Pediatr Hematol Oncol* 24:589–590.

Gilbert GL, Hayes K, et al. 1989. Prevention of transfusion–acquired cytomegalovirus infection in infants by blood filtration to remove leucocytes. Neonatal Cytomegalovirus Infection Study Group. *Lancet* 1:1228–1231.

Goldman SC, Bracho F, et al. 2001. Feasibility study of IL–11 and granulocyte colony stimulating factor after myelosuppressive chemotherapy to mobilize peripheral blood stem cells from heavily pretreated patients. *J Pediatr Hematol Oncol* 23:300–305.

Gordon LI, Kwaan HC, et al. 1987. Deleterious effects of platelet transfusions and recovery thrombocytosis in patients with thrombotic microangiopathy. *Semin Hematol* 24:194–201.

Gries A, Bode C, et al. 1998. Inhaled nitric oxide inhibits human platelet aggregation, P selectin expression, and fibrinogen binding in vitro and in vivo. *Circulation* 97:1481–1487.

Gries A, Herr A, et al. 2000. Randomized, placebo-controlled, blinded and cross-matched study on the antiplatelet effect of inhaled nitric oxide in healthy volunteers. *Thromb Haemost* 83:309–315.

Hanigan WC, Kennedy G, et al. 1988. Administration of indomethacin for the prevention of periventricular-intraventricular hemorrhage in high-risk neonates. *J Pediatr* 112:941–947.

Heddle NM, Klama L, et al. 1994. The role of the plasma from platelet concentrates in transfusion reactions. *N Engl J Med* 331:625–628.

Herman JH and Kamel HT. 1987. Platelet transfusion. Current techniques, remaining problems, and future prospects. *Am J Pediatr Hematol Oncol* 9:272–286.

Higby DJ, Cohen E, et al. 1974. The prophylactic treatment of thrombocytopenic leukemic patients with platelets: a double blind study. *Transfusion* 14:440–446.

Homans A. 1996. Thrombocytopenia in the neonate. *Pediatr Clin North Am* 43:737–756.

Howard SC, Gajjar A, et al. 2000. Safety of lumbar puncture for children with acute lymphoblastic leukemia and thrombocytopenia. *JAMA* 284:2222–2224.

Isaacs C, Robert NJ, et al. 1997. Randomized placebo-controlled study of recombinant human interleukin-11 to prevent chemotherapy-induced thrombocytopenia in patients with breast cancer receiving dose-intensive cyclophosphamide and doxorubicin. *J Clin Oncol* 15:3368–3377.

Israels SJ, Daniels M, et al. 1990. Deficient collagen-induced activation in the newborn platelet. *Pediatr Res* 27:337–343.

Israels SJ, Odaibo FS, et al. 1997. Deficient thromboxane synthesis and response in platelets from premature infants. *Pediatr Res* 41:218–223.

Iyori H, Bessho F, et al. 2000. Intracranial hemorrhage in children with immune thrombocytopenic purpura. Japanese Study Group on childhood ITP. *Ann Hematol* 79:691–695.

Jubelirer SJ and Harpold R. 2002. The role of the bone marrow examination in the diagnosis of immune thrombocytopenic purpura: case series and literature review. *Clin Appl Thromb Hemost* 8:73–76.

Kao GS, Wood IG, et al. 2003. Investigations into the role of anti–HLA class II antibodies in TRALI. *Transfusion* 43:185–191.

Kaplan C, Morel-Kopp MC, et al. 1991. HPA-b (Br(a)) neonatal alloimmunethrombocytopenia: clinical and immunological analysis of 39 cases. *Br J Haematol* 78:425–429.

Kaushansky K, Broudy VC, et al. 1995. Thrombopoietin, the Mp1 ligand, is essential for full megakaryocyte development. *Proc Natl Acad Sci U S A* 92:3234–3238.

Klaassen RJ, Doyle JJ, et al. 2001. Initial bone marrow aspiration in childhood idiopathic thrombocytopenia: decision analysis. *J Pediatr Hematol Oncol* 23:511–518.

Klein HG, Dodd RY, et al. 1997. Current status of microbial contamination of blood components: summary of a conference. *Transfusion* 37:95–101.

Kluter H, Bubel S, et al. 1999. Febrile and allergic transfusion reactions after the transfusion of white cell-poor platelet preparations. *Transfusion* 39:1179–1184.

Kroll H, Muntean W, et al. 1994. Anti Ko(a) as a cause of neonatal alloimmune thrombocytopenia. *Beitr Infusionsther Transfusionsmed* 32:244–246.

Kuehnert MJ, Roth VR, et al. 2001. Transfusion-transmitted bacterial infection in the United States, 1998 through 2000. *Transfusion* 41:1493–1499.

Lang DJ, Ebert PA, et al. 1977. Reduction of postperfusion cytomegalovirus infections following the use of leukocyte depleted blood. *Transfusion* 17:397–405.

Lawrence JB, Yomtovian RA, et al. 2001. Lowering the prophylactic platelet transfusion threshold: a prospective analysis. *Leuk Lymphoma* 41:67–76.

Lee EJ and Schiffer CA. 1989. ABO compatibility can influence the results of platelet transfusion. Results of a randomized trial. *Transfusion* 29:384–389.

Lee PI, Chang MH, et al. 1992. Transfusion-acquired cytomegalovirus infection in children in a hyperendemic area. *J Med Virol* 36:49–53.

Lilleyman JS. 1983. Management of childhood idiopathic thrombocytopenic purpura. *Br J Haematol* 54:11–14.

Liu HW, Yuen KY, et al. 1999. Reduction of platelet transfusion-associated sepsis by short-term bacterial culture. *Vox Sang* 77:1–5.

Lonial S, Bilodeau PA, et al. 1999. Acquired amegakaryocytic thrombocytopenia treated with allogeneic BMT: a case report and review of the literature. *Bone Marrow Transplant* 24:1337–1341.

Luban NL. 2002. Not just a small adult. *Transfusion* 42:666–668.

Lupton BA, Hill A, et al. 1988. Reduced platelet count as a risk factor for intraventricular hemorrhage. *Am J Dis Child* 142:1222–1224.

Maher P, Lane B, et al. 1985. Does indomethacin cause extension of intracranial hemorrhages: a preliminary study. *Pediatrics* 75:497–500.

Manoharan A, Williams NT, et al. 1989. Acquired amegakaryocytic thrombocytopenia: report of a case and review of literature. *Q J Med* 70:243–252.

Massey GV, McWilliams NB, et al. 1987. Intravenous immunoglobulin in treatment of neonatal isoimmune thrombocytopenia. *J Pediatr* 111:133–135.

McCann SR, Bacigalupo A, et al. 1994. Graft rejection and second bone marrow transplants for acquired aplastic anaemia: a report from the Aplastic Anaemia Working Party of the European Bone Marrow Transplant Group. *Bone Marrow Transplant* 13:233–237.

Medeiros D and Buchanan GR. 1998. Major hemorrhage in children with idiopathic thrombocytopenic purpura: immediate response to therapy and long-term outcome. *J Pediatr* 133:334–339.

Mehta P, Vasa R, et al. 1980. Thrombocytopenia in the high-risk infant. *J Pediatr* 97:791–794.

Ment LR. 1988. Intraventricular hemorrhage of the preterm neonate: prevention studies. *Mead Johnson Symp Perinat Dev Med*: 19–26.

Ment LR, Duncan CC, et al. 1985. Randomized indomethacin trial for prevention of intraventricular hemorrhage in very low birth weight infants. *J Pediatr* 107:937–943.

Ment LR, Oh W, et al. 1994. Low-dose indomethacin therapy and extension of intraventricular hemorrhage: a multicenter randomized trial. *J Pediatr* 124:951–955.

Miller WJ, McCullough J, et al. 1991. Prevention of cytomegalovirus infection following bone marrow transplantaiton: a randomized trial of blood product screening. *Bone Marrow Transplant* 7:277–234.

Minifee PK, Daeschner CW, et al. 1990. Decreasing blood donor exposure in neonates on extracorporeal membrane oxygenation. *J Pediatr Surg* 25:38–42.

Murphy S and Gardner FH. 1969. Effect of storage temperature on maintenance of platelet viability—deleterious effect of refrigerated storage. *N Engl J Med* 280:1094–1098.

Murphy S, Litwin S, et al. 1982. Indications for platelet transfusion in children with acute leukemia. *Am J Hematol* 12:347–356.

Murray NA. 2002. Evaluation and treatment of thrombocytopenia in the neonatal intensive care unit. *Acta Paediatr Suppl* 91:74–81.

Murray NA, Howarth LJ, et al. 2002. Platelet transfusion in the management of severe thrombocytopenia in neonatal intensive care unit patients. *Transfus Med* 12:35–41.

Muylle L, Wouters E, et al. 1996. Febrile reactions to platelet transfusion: the effect of increased interleukin 6 levels in concentrates prepared by the platelet-rich plasma method. *Transfusion* 36:886–890.

Ness P, Braine H, et al. 2001. Single-donor platelets reduce the risk of septic platelet transfusion reactions. *Transfusion* 41:857–861.

Nishizawa EE and Wynalda DJ. 1981. Inhibitory effect of ibuprofen (Motrin) on platelet function. *Thromb Res* 21:347–356.

Nomura S, Dan K, et al. 2002. Effects of pegylated recombinant human megakaryocyte growth and development factor in patients with idiopathic thrombocytopenic purpura. *Blood* 100:728–730.

Nonoyama S. 2001. Wiskott-Aldrich syndrome (role of WASP). *J Med Dent Sci* 48:1–6.

Noris M and Remuzzi G. 1999. Uremic bleeding: closing the circle after 30 years of controversies? *Blood* 94:2569–2574.

Ohto H and Anderson KC. 1996. Posttransfusion graft-versus-host disease in Japanese newborns. *Transfusion* 36:117–123.

Oren H, Irken G, et al. 1994. Assessment of clinical impact and predisposing factors for neonatal thrombocytopenia. *Indian J Pediatr* 61:551–558.

Patnode NM and Gandhi PJ. 2000. Drug-induced thrombocytopenia in the coronary care unit. *J Thromb Thrombolysis* 10:155–167.

Pierce RN, Reich LM, et al. 1985. Hemolysis following platelet transfusions from ABO incompatible donors. *Transfusion* 25:60–62.

Pink JM, MacCallum S, et al. 1993. Platelet transfusion-related sepsis. *Aust N Z J Med* 23:717.

Pisciotto PT, Benson K, et al. 1995. Prophylactic versus therapeutic platelet transfusion practices in hematology and/or oncology patients. *Transfusion* 35:498–502.

Preiksaitis JK, Brown L, et al. 1988. Transfusion-acquired cytomegalovirus infection in neonates. A prospective study. *Transfusion* 28:205–209.

Ragni MV, Miller BJ, et al. 1992. Bleeding tendency, platelet function, and pharmacokinetics of ibuprofen and zidovudine in HIV(+) hemophilic men. *Am J Hematol* 40:176–182.

Rajasekhar D, Kestin AS, et al. 1994. Neonatal platelets are less reactive than adult platelets to physiological agonists in whole blood. *Thromb Haemost* 72:957–963.

Rebulla P. 2001. Platelet transfusion trigger in difficult patients. *Transfus Clin Biol* 8:249–254.

Roback JD. 2001. Platelets and related products. In *Handbook of transfusion medicine*. San Diego: Academic Press.

Roberts IA and Murray NA. 1999. Management of thrombocytopenia in neonates. *Br J Haematol* 105:864–870.

Robinson TM, Kickler TS, et al. 1993. Effect of extracorporeal membrane oxygenation on platelets in newborns. *Crit Care Med* 21:1029–1034.

Roseff SD, Luban NL, et al. 2002. Guidelines for assessing appropriateness of pediatric transfusion. *Transfusion* 42:1398–1413.

Rosenberg EM, Chambers LA, et al. 1994. A program to limit donor exposures to neonates undergoing extracorporeal membrane oxygenation. *Pediatrics* 94:341–346.

Roy AJ, Jaffe N, et al. 1973. Prophylactic platelet transfusions in children with acute leukemia: a dose response study. *Transfusion* 13:283–290.

Roy V and Verfaillie CM. 1996. Refractory thrombocytopenia due to anti-PLA1 antibodies following autologous peripheral blood stem cell transplantation: case report and review of literature. *Bone Marrow Transplant* 17:115–117.

Samama CM, Diaby M, et al. 1995. Inhibition of platelet aggregation by inhaled nitric oxide in patients with acute respiratory distress syndrome. *Anesthesiology* 83:56–65.

Sanders MR and Graeber JE. 1990. Posttransfusion graft-versus-host disease in infancy. *J Pediatr* 117:159–163.

Schiffer CA, Lee EJ, et al. 1986. Clinical evaluation of platelet concentrates stored for one to five days. *Blood* 67:1591–1594.

Schroeder ML. 2002. Transfusion-associated graft-versus-host disease. *Br J Haematol* 117:275–287.

Sidiropoulos D and Straume B. 1984. The treatment of neonatal isoimmune thrombocytopenia with intravenous immunoglobin (IgG i.v.). *Blut* 48:383–386.

Silliman CC, Boshkov LK, et al. 2003. Transfusion-related acute lung injury: epidemiology and a prospective analysis of etiologic factors. *Blood* 101:454–462.

Silliman CC, Paterson AJ, et al. 1997. The association of biologically active lipids with the development of transfusion-related acute lung injury: a retrospective study. *Transfusion* 37:719–726.

Silvergleid AJ, Hafleigh EB, et al. 1977. Clinical value of washed-platelet concentrates in patients with non-hemolytic transfusion reactions. *Transfusion* 17:33–37.

Simon TL and Sierra ER. 1984. Concentration of platelet units into small volumes. *Transfusion* 24:173–175.

Simpson MB. 1987. Prospective-concurrent audits and medical consultation for platelet transfusions. *Transfusion* 27:192–195.

Sloand EM, Fox S, et al. 1990. Preparation of IgA-deficient platelets. *Transfusion* 30:322–326.

Smith JW, II. 2000. Tolerability and side-effect profile of rhIL-11. *Oncology (Huntingt)* 14:41–47.

Sola MC, Christensen RD, et al. 2000a. Pharmacokinetics, pharmacodynamics, and safety of administering pegylated recombinant megakaryocyte growth and development factor to newborn rhesus monkeys. *Pediatr Res* 47:208–214.

Sola MC, Du Y, et al. 2000b. Dose-response relationship of megakaryocyte progenitors from the bone marrow of thrombocytopenic and nonthrombocytopenic neonates to recombinant thrombopoietin. *Br J Haematol* 110:449–453.

Sola MC, Dame C, et al. 2001a. Toward a rational use of recombinant thrombopoietin in the neonatal intensive care unit. *J Pediatr Hematol Oncol* 23:179–184.

Sola MC, del Vecchio A, et al. 2001b. The relationship between hematocrit and bleeding time in very low birth weight infants during the first week of life. *J Perinatol* 21:368–371.

Strauss RG. 2000a. Blood banking and transfusion issues in perinatal medicine. In *Hematologic problems of the* neonate. R. Christensen, ed. Philadelphia: W.B. Saunders.

Strauss RG. 2000b. Data-driven blood banking practices for neonatal RBC transfusions. *Transfusion* 40:1528–1540.

Strauss RG. 1999. Leukocyte reduction to prevent transfusion-transmitted cytomegalovirus infections. *Pediatr Transplant* 3 Suppl 1:19–22.

Strauss RG. 1995. Red blood cell transfusion practices in the neonate. *Clin Perinatol*. 22:641–655.

Strauss RG. 1993. Selection of white cell-reduced blood components for transfusions during early infancy. *Transfusion* 33:352–357.

Strauss RG, Levy GJ, et al. 1993. National survey of neonatal transfusion practices: II. Blood component therapy. *Pediatrics* 91:530–536.

Stuart MJ. 1979. Platelet function in the neonate. *Am J Pediatr Hematol Oncol* 1:227–234.

Stuart MJ, Dusse J, et al. 1984. Differences in thromboxane production between neonatal and adult platelets in response to arachidonic acid and epinephrine. *Pediatr Res* 18823–18826.

Surgenor DM, Wallace EL, Hao SH, et al. 1990. Collection and transfusion of blood in the United States, 1982–1988. *N Engl J Med* 322:1646–1651.

Tanindi S, Kurekci A, et al. 1995. The normalization period of platelet aggregation in newborns. *Thromb Res* 80:57–62.

TRAPSG. 1997. Leukocyte reduction and ultraviolet B irradiation of platelets to prevent alloimmunization and refractoriness to platelet transfusions. The Trial to Reduce Alloimmunization to Platelets Study Group. *N Engl J Med* 337:1861–1869.

Vadhan-Raj S, Murray LJ, et al. 1997. Stimulation of megakaryocyte and platelet production by a single dose of recombinant human thrombopoietin in patients with cancer. *Ann Intern Med* 126: 673–681.

Valentin N, Vergracht A, et al. 1990. HLA-DRw52a is involved in alloimmunization against PL-A1 antigen. *Hum Immunol* 27:73–79.

van Gorp EC, Suharti C, et al. 1999. Review: infectious diseases and coagulation disorders. *J Infect Dis* 180:176–186.

Van Hecken A, Juliano ML, et al. 2002. Effects of enteric-coated, low-dose aspirin on parameters of platelet function. *Aliment Pharmacol Ther* 16:1683–1688.

Van Overmeire B, Van de Broek H, et al. 2001. Early versus late indomethacin treatment for patent ductus arteriosus in premature infants with respiratory distress syndrome. *J Pediatr* 138:205–211.

Vesilind GW, Simpson MB, et al. 1988. Evaluation of a centrifugal blood cell processor for washing platelet concentrates. *Transfusion* 28:46–51.

Vo TD, Cowles J, et al. 2001. Platelet washing to prevent recurrent febrile reactions to leucocyte-reduced transfusions. *Transfus Med* 11:45–47.

Voak D, Cann R, et al. 1994. Guidelines for administration of blood products: transfusion of infants and neonates. British Committee for Standards in Haematology Blood Transfusion Task Force. *Transfus Med* 4:63–69.

von dem Borne AE, von Riesz E, et al. 1980. Baka, a new platelet-specific antigen involved in neonatal allo-immune thrombocytopenia. *Vox Sang* 39:113–120.

Vuckovic SA. 1996. Glanzmann's thrombasthenia revisited. *J Emerg Med* 14:299–303.

Wandt H, Frank M, et al. 1998. Safety and cost effectiveness of a $10 \times 10(9)/L$ trigger for prophylactic platelet transfusions compared with the traditional $20 \times 10(9)/L$ trigger: a prospective comparative trial in 105 patients with acute myeloid leukemia. *Blood* 91:3601–3606.

Westman P, Hashemi-Tavoularis S, et al. 1997. Maternal DRB1*1501, DQA1*0102, DQB1*0602 haplotype in fetomaternal alloimmunization against human platelet alloantigen HPA-6b (GPIIIa-Gln489). *Tissue Antigens* 50:113–118.

Wilhelm JA, Matter L, et al. 1986. The risk of transmitting cytomegalovirus to patients receiving blood transfusions. *J Infect Dis* 154:169–171.

Williamson LM. 1998. Transfusion associated graft versus host disease and its prevention. *Heart* 80:211–212.

Win N, Ouwehand WH, et al. 1997. Provision of platelets for severe neonatal alloimmune thrombocytopenia. *Br J Haematol* 97: 930–932.

Wong EC and Luban N. 2002. Cytomegalovirus and parvovirus transmission by transfusion. In *Rossi's principles of transfusion medicine*. DW Simon, TL Snyder EL, Stowell CP, Strauss RG, ed. Philadelphia: Lippincott, Williams & Wilkins.

Yamreudeewong W, Fosnocht BJ, et al. 2002. Severe thrombocytopenia possibly associated with TMP/SMX therapy. *Ann Pharmacother* 36:78–82.

Yeager AS, Grumet FC, et al. 1981. Prevention of transfusion-acquired cytomegalovirus infections in newborn infants. *J Pediatr* 98:281–287.

Zavadil DP, Stammers AH, et al. 1998. Hematological abnormalities in neonatal patients treated with extracorporeal membrane oxygenation (ECMO). *J Extra Corpor Technol* 30:83–90.

Zimmermann R, Schmidt S, et al. 2001. Effect of gamma radiation on the in vitro aggregability of WBC-reduced apheresis platelets. *Transfusion* 41:236–242.

Zwicky C, Tissot JD, et al. 1999. Prevention of posttransfusion cytomegalovirus infection: recommendations for clinical practice. *Schweiz Med Wochenschr* 129:1061–1066.

Bone Marrow-Derived Stem Cells

GRACE S. KAO, MD, AND STEVEN R. SLOAN, MD, PhD

INTRODUCTION

Bone marrow (BM) is one source of hematopoietic stem cells. BM-derived stem cell products contain stem cells and hematopoietic progenitor cells that are committed to particular hematopoietic lineages. Hence, the term "hematopoietic progenitor cells" (HPC) is more accurate than "stem cells" and will be used in this chapter. HPCs are crucial for hematological reconstitution in patients who undergo high-dose chemotherapy and myeloablation to treat their underlying diseases.

There are two major sources of human BM HPCs. Autologous BM transplants are performed using BM HPCs donated by the person who is also the recipient of the transplant. Allogeneic BM transplants use BM HPCs donated by a person other than the recipient. In the past few years, peripheral blood-derived HPCs have become the preferred HPC source for both autologous and allogeneic hematopoietic stem cell transplantation in adult patients. This shift from BM-derived to peripheral blood-derived HPCs for transplantation has not been observed as much in the pediatric setting. However, the majority of pediatric HPCs used for autologous hematopoietic stem cell transplantations are collected from peripheral blood (see Chapter 24), while the majority of allogeneic hematopoietic stem cell transplants conducted in pediatric patients still use HPCs harvested from BM, according to data collected in Italy from 1995 to 1998.

BM cell transplants are still relatively new when compared with transfusions of other commonly used blood products and derivatives. Procedures and practices for collection and preparation vary between insti-

tutions. Various organizations, including the Food and Drug Administration (FDA), Foundation for the Accreditation of Cell Therapy (FACT), and the American Association of Blood Banks (AABB), have either recently adopted standards or are planning to do so in the near future. Most of these standards focus on the quality of the environment, human resources, and procedures for HPC collection and processing and do not specify product content or usage.

PRODUCT DESCRIPTION

Cellular Constituents

In addition to HPCs, BM-derived stem collections contain all of the cellular components that are normally present in BM. Mature and immature RBCs (RBCs), lymphocytes, myeloid cells, and platelets are present in BM-derived HPCs. Some of these cells, especially T lymphocytes, can contribute to graft-versus-host diseases (GVHD) and/or a graft-versus-tumor effect. At the same time, these T lymphocytes are crucial for hematological engraftment. In addition, other cells may contaminate the BM. Of particular concern is the fact that malignant cells may contaminate an autologous BM harvest for oncology patients. Therefore, BM products may undergo additional processing to remove unwanted cells or to preferentially select for HPCs.

Characterization of HPC Cell Content

Traditionally, the quantity of useful cells in a BM-HPC product was determined by measuring the number of mononuclear cells. The number of mononuclear cells may roughly correlate with the number of HPCs and stem cells. Stem cells comprise about 1 in 100,000 of the BM mononuclear cells, but this proportion can vary significantly between donors. In general BM-HPCs collected for autologous transplants should contain approximately 1 to 3×10^8 mononuclear cells/kg recipient body weight, and BM-HPCs collected for allogeneic transplants should contain approximately 2 to 4×10^8 mononuclear cells/kg recipient body weight, with some variation for different clinical situations. For example, guidelines might indicate that with BM-HPCs for human leukocyte antigen (HLA)-identical sibling allogeneic transplants for aplastic anemia should contain at least 3×10^8 cells per kg, BM-HPCs for unrelated donors should also contain 3×10^8 cells per kg, and BM-HPCs for HLA-identical allotransplants for leukemia, hemoglobinophathies, or inborn errors of metabolism should contain at least 2×10^8 cells per kg (Atkinson 1998). The National Marrow Donor Program (NMDP) in the United States requires 2.0×10^8 mononuclear cells per kg for unrelated allogeneic BM-HPC transplants. Most institutions use additional means to enumerate the HPCs in BM products, such as granulocyte-macrophage colony forming unit assays (CFU-GM) or CD34+ cell counts by flow cytometry. CD34 is a glycophosphoprotein expressed on most hematopoietic progenitor cell surfaces. CD34+ cells comprise 1% to 5% of peripheral blood mononuclear cells following mobilization. Many cells in addition to stem cells also express the CD34+ antigen (Siena et al. 1993). Studies have shown that approximately 0.1 to 1×10^4 CFU-GM per kg patient body weight are needed to ensure timely engraftment. However, CFU-GM assays usually provide retrospective data only since they take two weeks to complete. Using standard dual-color CD34/CD33 direct immunofluorescence assay, the enumeration of CD34+ cells in the BM product was found to have positive correlation with CFU-GM assay (Siena et al. 1991). Hence, most institutions primarily rely on CD34+ cell counts to quantify HPCs with 2 to 5×10^6 CD34+ cells per kg patient body weight usually being required for timely engraftment. These values must be determined by each institution and may depend on disease and whether the cells are intended for an autologous transplant, a related allogeneic transplant, or an unrelated allogeneic transplant. These assays are also performed for peripheral blood-derived HPC (PB-HPC), and the target number of CD34+ cells are similar. The assays are described in Chapter 24.

Unprocessed BM-HPCs include everything that was collected from the donor's BM and the anticoagulant. This product will be in a volume that ranges from 500 mL to 2000 mL.

Anticoagulants, Additives

BM-HPCs contain plasma, anticoagulant, and additional buffered solution. The anticoagulant may be heparin, acid-citrate-dextrose (ACD), citrate-phosphate-dextrose (CPD), or CPD-adenine (CPDA-1). Acid citrate dextrose should be used if the marrow will be kept in liquid form for extended times for long-distance transport or for other reasons (Gee and Lee 1998). BM-HPCs collected at some institutions contain tissue culture media that contains electrolytes and buffers and may contain vitamins and/or minerals. Another practice is the addition of a buffered electrolyte solution instead of tissue culture media. Tissue culture media provides conditions for long-term cell growth but is not needed for short-term storage of BM-HPCs. Because tissue culture media is not currently approved for human use in the United States, FDA-approved solutions such as Normosol or other infusion grade solutions are more commonly used.

Human plasma contaminates collections of BM-HPCs. While this plasma is not usually specifically removed from the final BM-HPC product, the plasma may be removed during other processing such as CD34+ selection. Additional processing may result in the addition of other additives. For example, some products depleted of red cells may contain hetastarch or additional albumin.

Labeling

Labeling requirements are determined by accrediting organizations such as the AABB and the FACT and regulatory agencies such as the FDA. The label should contain the name and address of the processing facility unless it is an unrelated allogeneic transplant, the unit's unique numeric or alphanumeric identifier, the date and time (and time zone if applicable) of collection, the volume of the product, the ABO and Rh type of the donor, the type and volume of additives such as anticoagulants and cryprotectants, a biohazard label, the expiration date, manipulation method(s), and the recommended storage temperature range. In addition the label must contain the phrases "Hematopoietic Progenitor Cells, Marrow," "Warning: This Product May Transmit Infectious Agents," and "Properly Identify Intended Recipient." An HPC intended for an autologous transplant must contain the phrase "For Autologous Use Only" on the label. An HPC intended for an allogeneic transplant must contain the phrase "For Use By Intended Recipient Only." Units from unrelated donors must be labeled with the donor registry name and unique donor registry number. The most recent results of infectious disease tests must be on the label or must be communicated to transplant facility by some other means. A biohazard label must be attached to the product if infectious disease tests indicate that the donor is likely to have been infected with human immunodeficiency virus (HIV), hepatitis B, or hepatitis C.

COLLECTION AND STORAGE

Donor Evaluation

HLA Compatibility

The HLA types of potential allogeneic BM donors are determined so that the patient will receive an optimal match (for reviews see Little et al. 1998). The major HLA system encodes for class I (A, B, and C) and class II (DR, DP, and DQ) cell surface molecules that are important in the immune response. There are multiple alleles for each HLA locus resulting in significant diversity in the population. Patients transplanted with donor HPCs that are HLA-incompatible with the patient are more likely to develop GVHD and more likely to have delayed immune reconstitution. However, HLA-incompatible transplants are more likely to produce graft-versus-leukemia (GVL) effects than are HLA-compatible transplants.

Because of the risks associated with GVHD and delayed immunological recovery, transplant donors are usually selected to maximize HLA compatibility. This is accomplished by attempting to identify people who share identical HLA-A, HLA-B, and HLA-DR genes with the patient. Some institutions may also attempt to match HLA-C genes. Matching HLA-DP and HLA-DQ genes is not important. Up to one mismatched allele may be acceptable. The specific mismatch is important. Two alleles may present only minor mismatches if they are in the same crossreactive group (CREG). Other mismatched alleles may cause severe GVHD. Minor histocompatibility antigens (mHag) can also contribute to GVHD. These antigens are not normally measured and used for identifying donors. Transplants from relatives pose a lower risk from minor histocompatibility antigens incompatibility when compared to transplants from unrelated donors, but mHag can cause GVHD in all transplants, including transplants from identical siblings. Traditionally sera that react against specific HLA antigens have been used to type patients and potential donors. For a variety of reasons, serological techniques are error prone and fail to discriminate between some related alleles. DNA-based techniques that are based on the polymerase chain reaction (PCR) have been advancing rapidly in recent years. These techniques provide greater precision and discrimination between related alleles.

Family members are often the most likely histocompatible donors. There is a 97% to 99% chance that an identical twin sibling is HLA identical. (This percentage is not 100% because of the relatively high frequency of chromosomal recombination near the HLA genes.) Theoretically, there is a 25% chance that any one nonidentical sibling of a patient is HLA identical with the patient. Certain HLA genes are inherited preferentially, however, and there is a 35% that a child with leukemia and any one sibling are HLA identical (Chan et al. 1982). Other family members may also be histocompatible.

If no compatible family members are identified, then national donor registries can be searched. The National Donor Marrow Program (NDMP) in the United States is the largest registry, and they can request registries in other nations to search for donors if none are identified in the NDMP. The registries contain HLA typing information that is usually based on serological typing and often only includes HLA-A, HLA-B, and sometimes HLA-DR typing information. Potential donors then need to be typed using PCR-based methods at all alleles that the transplant center normally considers.

Medical History

Once a potential donor is identified, his or her health is assessed by history, physical, and laboratory tests. This evaluation is performed in part to determine if the donor is at high risk for donating BM. Up to 0.4% of BM donors develop life-threatening complications, however, this may be reduced to 0.1 to 0.2% if older donors, donors with cardiovascular disease, and obese donors are excluded (Buckner et al. 1994). FACT also requires a pregnancy test for all female donors of childbearing potential. Donor risk behavior and health is also evaluated to determine the likelihood of infectious disease transmission to the patient. Standards do not specify specific exclusion criteria. However, infectious disease tests routinely performed for whole blood donation should be performed on BM donors no more than 30 days before donation. In addition, FACT requires cytomegalovirus (CMV) testing of donors. Individuals excluded for whole blood donation may be acceptable BM donors. Bone marrow transplant programs determine the acceptability of donors on a case by case basis if there is risk of infectious disease transmission.

Potential autologous BM transplant candidates are screened using criteria similar to that described in Chapter 24 for peripheral blood-derived stem cells. In addition, BM fibrosis can make BM donation difficult or impossible.

Collection Procedure

BM is collected in the operating room from anesthetized donors. The donor usually receives general anesthesia but may receive spinal or epidural anesthesia. Syringes are used to aspirate BM from the posterior iliac crests and pelvic rim. In some patients, BM is also aspirated from the anterior iliac crests and sternum (Thomas and Storb 1970). The harvested marrow is placed into a sterile container with an anticoagulant (usually heparin) and tissue culture media or buffered normal saline solution. After collection, the marrow is filtered through stainless steel mesh screens to remove clots, bone fragments, fat, and fibrin and transferred to a blood bag or other sterile container. The most serious risks to the donor are related to anesthesia. Most donors experience fatigue, pain at the collection site, and lower back pain that lasts for a few weeks (Stroncek et al. 1993). Acetaminophen with codeine or a similar minor narcotic for a few days following the harvest usually alleviates the pain.

Laboratory Evaluation of Bone Marrow Harvest

Various quality control measurements may be made from aliquots of the product. Many institutions measure percentage cell viability in addition to mononuclear cell concentrations and/or CD34+ cell concentrations. Culture of BM product for bacterial and fungal contamination is a recommended standard. In addition, some institutions perform colony forming unit assays (see Chapter 24). In most cases, the results of cultures and CFU assays will not be available until several days after the transplant, but these assays can provide data for the quality of the laboratory's processing and may help in the treatment and diagnosis of a patient whose engraftment is delayed or who is septic. Repeat ABO typing is usually performed on the BM product to ensure correct donor-recipient identification and appropriate cellular manipulation procedures to be conducted, such as RBC and plasma depletion to prevent hemolytic transfusion reactions.

Storage and Transport

Allogeneic BM is usually stored for two to 36 hours at room temperature (not to exceed 37°C or drop below 2°C). While storage time should be minimized, HPCs that have been stored at 4°C to 25°C for 24 to 36 hours have been successfully transplanted (Lasky et al. 1986). If the donor and patient are at the same institution, then the marrow is usually harvested and transplanted on the same day. If the donor and patient are in different loca-

tions, then the marrow is usually shipped from the donor's location to the recipient's location within hours after the harvest. A courier from the patient's institution will often personally transport the cells to ensure that the cells are kept at an acceptable temperature.

Cryopreservation

As mentioned previously, harvested BM products are often contaminated with fat and bony spicules, a significant volume of RBCs, tissue culture media, or buffered normal saline solution with anticoagulants such as heparin. Harvested BMs are washed and red cell-depleted to get rid of debris before cryopreservation. Similar to PB-HPCs, the BM-HPCs are cryopreserved in freezing media containing 5% to 10% of DMSO using a rate-controlled freezer and placed in vapor phased liquid nitrogen for storage (see Chapter 24). Allogeneic BM-HPCs are usually not cryopreserved for fear of damaging the cells and reducing the chances of engraftment. However, allogeneic BM-HPCs have been successfully transplanted following cryopreservation.

SPECIAL PREPARATION AND PROCESSING

Red Blood Cell/Plasma Depletion

Allogeneic BM-HPCs may contain RBCs or plasma that is incompatible with the patient and these are indications for red cell or plasma depletion. When RBC depletion is necessary, the laboratory performing the manipulation will determine a satisfactory level of depletion, which is usually about 90% depletion of the RBCs. In addition, RBC-depleted products may contain hetastarch or albumin depending on the procedure that was used. Plasma depletion is less labor intensive as compared to red cell depletion. When plasma depletion is indicated, the harvested BM product is spun down, plasma was extracted with plasma extractor, and the BM-HPCs were then suspended in 5% human serum albumin before patient infusion.

Cell Selection

Harvested BM product is extremely heterogeneous. Beside immature HPCs that are crucial for hematological reconstitutions, other mature hematopoietic or non-hematopoietic cells can also be present. Techniques used to isolate specific type of cells may potentially improve the specificity of BM-HPC therapy. Currently, T cells, CD34+ cells, and malignant cells are the targets of these cell selection techniques.

Subsets of allogeneic donor T cells are thought to cause GVHD. Studies that deplete CD8+ T cells from BM product showed significant reduction of GVHD in allogeneic transplantation (Giralt et al. 1995; Alyea et al. 1998). Therefore, allogeneic BM-HPCs may be depleted of T cells to reduce the chances of severe GVHD. Reduced graft-versus-cancer effects were observed in patients who did not present with GVHD, and increased relapse of diseases was seen in patients who received T cell-depleted product. As more clinical trials involving T cell depletion have been conducted in allogeneic BM-HPC transplant patients who are at high risk for GVHD, increased complications, such as delayed or lack of engraftment and presentation of opportunistic diseases, such as CMV reactivation and posttransplant lymphoproliferative disorders, were observed (Broers et al. 2000; Chiang et al. 2001).

There are several techniques that may be used to deplete allogeneic BM-HPC harvests of T cells. These techniques are identical to those used to deplete T cells from PB-HPC (see Chapter 24). Counterflow centrifugation elutriation separates cells based on their size and density. Elutriation is used to separate lymphocytes from most of the other cells. CD34+ hematopoietic progenitor cells vary in size and density. While the smaller more primitive CD34+ cells segregate with the lymphocytes, at least half of the CD34+ cells can be separated from the lymphocytes (Gee and Lee 1998). The lymphocyte-depleted cell population is transplanted. Because the transplanted product may lack the most immature small CD34+ cells, long-term engraftment may be at risk. For this reason, some institutions may select CD34+ cells from the lymphocyte fraction and add these back to the lymphocyte-depleted cells (Gee and Lee 1998). Lectin separation and E-rosette depletion are sometimes used to remove T cells from BM. With improvement in magnetic bead technology, negative selection using nickel beads attached to anti-CD8 antibodies have been used in clinical trials to reduce amount of T cells in harvested BM products.

T cell-depleted products may contain additional solutions, antibodies, or other chemicals depending on the T cell depletion method. BM-HPCs that have been depleted of T cells by CD34+ selection contain fewer red cells, platelets, and white cells of all types although a large proportion of the CD34+ cells is retained. CD34+-selected products are enriched for HPCs and are usually contained in a small volume.

For autologous BM transplant patients, the malignant cell contamination poses the most concern. At present, most of autologous HPC transplant patients, both adult and pediatric, received peripheral blood rather than BM-derived HPCs. Hence, the techniques used to purge or negatively select malignant tumor cells will be discussed further in Chapter 24.

Genetic Engineering

Gene transfer into repopulating hematopoietic cells to correct or regulate blood cell related disease would create many therapeutic opportunities. The progress of gene therapy in human has been slow but progressive. Most recently, a retrovirus vector containing the γc chain has been successfully used to transduce autologous CD34+ cells from BM for pediatric patients with severe combined immunodeficiency disease (SCID) (Hacein-Bey-Abina et al. 2002). HLA-identical allogeneic HPC transplantations have been used successfully for SCID patients to reconstitute their T-cell immune function. However, all of them depend on lifelong immunoglobulin infusion due to lack of B cell reconstitution (Haddad et al. 1998). These patients who underwent genetically altered autologous CD34+ HPC transplantation appeared to have both T- and B-cell immune reconstitution after transplantation. Hence, the gene therapy was viewed as an improvement over HLA-identical allogenetic HPC transplantations. Unfortunately, some of these patients were found to have leukemia-like disease after long-term follow up. The cause of this complication is most likely due to random insertion of the retrovial vector in the host genome(Hacein-Bey-Abina et al. 2003).

Beside SCID patients, progress has also been made for diseases such as sickle cell disease and thalassemia. Lentiviral vectors containing a globin gene with introns linked to regulatory elements from the locus control region (LCR) have been used successfully to transduced BM-derived CD34+ cells, and autologous transplantation of globin gene-modified CD34+ cells have corrected the thalassemia and sickle cell anemia phenotypes in murine models (Walters et al. 2002).

ABO and Rh Compatibility

Selection of BM-HPC donors for transplantation is made primarily based on HLA types. RBC incompatibility between donor and recipient is thus considered secondarily. ABO and Rh types of the donor and recipient should be determined in advance of allogeneic transplants to anticipate management of incompatibilities.

While FACT standards require that a test for the ABO group and Rh type be performed on a donor specimen collected at the time of the BM harvest, this is not a requirement of all accreditation agencies. However, the ABO and Rh type of the donor must be included on the BM-HPC label to help ensure that the patient is transplanted with the correct BM-HPC.

Approach to Major Red Cell Incompatibility

A major blood group incompatibility occurs when naturally occurring antibodies (anti-A, anti-B, anti-A,B) in the patient's plasma react with donor RBCs. In addition, tests of the patient's serum and donor RBCs may reveal the presence of other unexpected clinically significant antibodies against RBC antigens on donor RBCs (for example, Kell, Duffy, Kidd, and so on). If clinically significant antibodies react against donor RBCs by routine pretransfusion testing, then the red cells should be removed from the BM. Acceptable methods for red cell depletion use hydroxyethyl starch (HES) sedimentation or automated cell separators. In addition, some cell selection methods that are designed to purify the stem cell population, such as methods that select for CD34+ cells, also remove RBCs. HES sedimentation, automated cell separators, and repeated sedimentation and dilution with compatible RBCs remove most, but not all, of the incompatible RBCs. Hence, these procedures mitigate but do not eliminate the risk of hemolytic transfusion reactions associated with BM-HPC transplants.

Hydroxyethyl Starch Sedimentation

HES sedimentation had been commonly used to remove incompatible RBCs from BM harvests (Dinsmore et al. 1983). HES added to BM causes the RBCs to form rouleaux and accelerates their sedimentation. The sedimented RBCs are drained from the bottom of the bag leaving the rest of the BM harvest including the HPCs in the original bag. Laboratories often find it difficult to remove enough RBCs while retaining most of the white blood cells (WBCs) using HES sedimentation (Atkinson 1998). This method results in infusion of HES to the patient, which does not have significant side effects.

Cell Separator

Many transplant laboratories now use a cell separator to remove RBCs from BM harvests. Cell separators use the same principles as apheresis machines to separate the various components in blood and selectively remove specified components (Chapter 30). During this processing albumin, saline, and ACD are often added to the product. (In addition to removing RBCs from the BM, some transplant programs remove isoagglutinins from the patient by plasmapharesis,

although few studies indicate that this is necessary [Atkinson 1998]).

Sedimentation and Dilution

Repeated sedimentation and dilution with compatible RBCs is a lengthy process because of the time required for RBCs to sediment in the absence of HES. This technique is not used by many processing laboratories, but some laboratories have diluted some products with compatible RBCs in conjunction with HES sedimentation or automated cell separators.

Approach to Minor ABO Red Cell Incompatibility

The ABO group of the BM transplant donor and recipient may have a minor ABO incompatibility in which plasma from the donor contains antibodies against ABO antigens on the recipient's RBCs. Additionally, the donor's plasma may contain unexpected antibodies against other RBC antigens that are present on donor RBCs (for example, Kell, Duffy, Kid, and so on). These antibodies would be discovered when performing antibody screens on the donor's blood. If donor plasma contains clinically significant antibodies that react with the patient's RBCs, then one should consider removing plasma from the BM harvest to reduce the chance of clinically significant hemolysis of the patient's RBCs (Lasky et al. 1983). Plasma can be removed by plasma extractor following centrifuging the BM collection. Alternatively, a cell separator can be used to remove plasma (see Chapter 29).

Red Cell Crossmatches

Many institutions crossmatch donor RBCs with patient plasma and donor plasma with patient RBCs, even though current accreditation standards do not explicitly require this testing in all cases. This can serve as a check to verify whether there is a need for red cell or plasma depletion of the BM product. If the cells are incompatible, then the product should be labeled as being incompatible. If a major incompatibility exists, this product may be safely administered following red cell depletion; if a minor incompatibility exists, this product may be safely administered. The person performing the transplant should verify that the product has been depleted of red blood or plasma cells by checking the label and visually examining the product. Rh blood group mismatches do not impair engraftment, reduce patient survival, or increase the risk of GVHD. Some Rh-positive donors who have received Rh-negative BM have developed anti-D antibodies. While these patients' RBCs may undergo increased hemolysis, Rh-negative RBCs should repopulate their circulatory systems as the donor BMs engraft.

Ordering and Administration

When considering an allogeneic BM-HPC transplant, a donor with a compatible HLA type must be identified. Family members, especially siblings, are usually the best potential donors and their HLA types are determined. If no donor is identified in the patient's family, searches for potential donors can be made through national and international BM transplant registries. High-resolution HLA types of potential donors are determined to identify histocompatible donors. If a compatible donor is identified, his or her health status is assessed. Autologous BM-HPCs and autologous PB-HPCs are ordered and administered the same way (see Chapter 24).

If an allogeneic BM transplant is planned, efforts must be coordinated between the processing laboratory, the physicians harvesting the BM, the patient, and the donor. This is made even more complex if the donation is made at a location that is different from the patient's location. In those cases, the BM is harvested at a hospital that serves as the collection facility and a courier, often a member of the transplant team from the patient's hospital, transports the BM to the patient's hospital.

Before administration of the marrow, proper identification of the patient and product is critical to ensure that the patient is receiving the correct BM-HPCs. Cells are usually infused at least 24 hours after completion of chemotherapy to prevent the cytotoxic effects of chemotherapy from damaging the infused cells. Cells can be immediately infused following radiotherapy, however. The patient should be well hydrated before infusion. Oxygen and anti-anaphylaxis treatment such as epinephrine should be available. Allogeneic BM-HPCs that had been cryopreserved are administered using procedures that are similar to those used for cryopreserved PB-HPCs (see Chapter 24). Cells are administered intravenously, usually through a central venous catheter. Cells may be administered rapidly through an intravenous push, or, unlike cells that have been cryopreserved with DMSO, cells that have not been cryopreserved may be infused over several hours. It may be advisable to initiate the infusion slowly to observe for any adverse reactions and then accelerate the infusion rate. The patient should be closely monitored and vital signs should be taken periodically during the infusion as for any blood product in view of risks for allergic, ana-phylactic, hemolytic or febrile nonhemolytic transfusion reaction.

Indications

HPC are usually administered to patients whose own hematopoietic system is defective. Although a disease may directly cause the BM defect, toxic cancer treatment is the more frequent cause of the BM damage. Hematopoietic stem cell transplants have been used for immunodeficiencies, autoimmune diseases, and genetic disorders.

Autologous

Autologous BM-HPC transplants are indicated for all clinical cases in which autologous stem cell transplants are indicated. However, with improved method for PB-HPC mobilization, autologous BM-HPCs are now used infrequently even in pediatric settings. According to the nationwide pediatric BM transplant registry in Italy, there was a rapid shift in the source of autologous HPCs employed for transplantation starting in 1997, where greater than 70% are of PB-HPCs (Pession et al. 2000). Autologous BM harvests are now reserved for situations when mobilization of peripheral HPCs was impossible or inadequate. For pediatric patients, they are often due to the complications related to peripheral access placement needed for PB-HPC collection. These complications include central line-related deep venous thrombosis and line infection, which can lead to inadequate number of HPCs collected. As the result, additional BM harvests may become necessary in order to obtain sufficient number of HPCs needed for adequate hematopoietic reconstitution after myeloablation from high dose chemotherapy or radiation.

Allogeneic

In the past few years, the number of allogeneic HPC transplantation conducted in the United States and European pediatric population has approached or even surpassed the number of autologous HPC transplantations. Despite the dramatic shift to the use of PB-HPCs in autologous settings, more than 90% of pediatric patients continue to receive BM-HPCs for allogeneic HPC transplantation. However, the main indication for allogeneic HPC transplantation in childhood remained to be lymphomyeloproliferative disorders (66%) followed by nonmalignant diseases (33%) such as hemoglobinopathy, immunodeficiency, and metabolic disorders. In particular, leukemia was the main indication for allogeneic transplants (Pession et al. 2000).

Allogeneic HPC transplantation can potentially stabilize or reverse some of the complications associated with sickle cell disease. However, these procedures remained experimental and transplantation-related complication remained extremely high in this population. The best candidates for HLA-identical sibling-related allograft transplantation are children younger than 16 years of age with severe vaso-occlusive disease, stroke, and acute chest syndrome. The recent advances in nonmyeloablative transplant may induce sufficient mixed hematopoietic chimerism to treat sickle cell disease-related complication and reduce transplant-related toxicities (Steinberg and Brugnara 2003). BM-HPCs probably will remain the main source of HPCs for allogeneic transplantation in sickle cell patients because the majority of sibling donors for these patients have sickle cell traits and the growth factors-induced sickle cell crisis have been reported in sickle cell trait PB-HPC donors during cell mobilization (Adler et al. 2001; Wei and Grigg 2001).

Contraindications

Donor Issues

The physical condition of the donor may make BM collection and associated anesthesia especially dangerous. The importance of these risks when deciding whether to harvest BM from a prospective donor depends on whether the BM-HPCs are intended to be used for an autologous or an allogeneic transplant. Only minimal risk is acceptable when harvesting BM from a healthy allogeneic donor while some risk may be acceptable when harvesting BM for an autologous transplant.

In addition to the general health of the donor, a prospective autologous donor's BM needs to be evaluated. Marrow fibrosis in the autologous donor is a contraindication because BM harvests are impossible in some patients with BM fibrosis. In addition, autologous transplants for malignancy are usually not considered in patients who have evidence of cancer in their BM by bone imaging studies and biopsy. This evaluation can miss minimal disease that may be important. For example, immunohistochemistry studies have shown micrometastates of breast cancer cells in 17% to 60% of marrow harvested from patients who were not thought to have BM metastasis by conventional techniques. Unfortunately PB-HPCs may not offer an advantage; studies have shown 10% to 78% of PB-HPCs can be contaminated by malignant cells. Furthermore, some studies have shown that micrometastases correlate with poor outcome, though it is unknown if the poor outcomes are due to transplanting malignant cells

or to the more advanced stage of the disease in these patients. Regardless, most transplant programs currently collect autologous BM only from those patients who have healthy BM as determined by conventional techniques.

Allogeneic BM harvests from healthy donors are contraindicated when there is significant risk to the donor associated with the collection and anesthesia. Donors who are obese, older, or have cardiovascular or pulmonary disease are at increased risk (Buckner et al. 1994). In addition, allogeneic donors who test positive for infectious diseases such as hepatitis B, hepatitis C, or CMV can pose increased risk to recipients, and the transplant physicians must decide whether to transplant marrow from these donors. Most transplant physicians will not transplant BM from donors who test positive for HIV.

Recipient Issues

BM transplantation is a potentially dangerous treatment that is contraindicated in some patients who are at especially high risk. The risk-benefit analysis must consider the fact that the risk associated with transplant depends on the relationship between the donor and the patient. Allogeneic BM-HPC transplants from unrelated donors are the riskiest BM transplants, and autologous BM-HPC transplants are the safest BM-HPC transplants. Allogeneic BM-HPC transplants from HLA-identical siblings are of intermediate risk. Each transplant program must establish its own guidelines. The age of the recipient is a major risk factor for allogeneic transplants, other risk factors often used as contraindications may include organ dysfunction as indicated by serum creatinine >2.8 mg/dL, serum bilirubin >2.4 mg/100 mL, PaO_2 <70 mm Hg, a left ventricular ejection fraction <50%, or active infection, or a Karnofsky performance score <70%. However, patients with renal failure have been successfully transplanted (Mehta and Singhal 1998).

EXPECTED RESPONSE

BM-HPCs should reconstitute the hematopoietic system. Engraftment, measured as a neutrophil count $\geq 0.5 \times 10^9$/L, usually occurs between eight and 30 days posttransplant. Platelet ($\geq 20 \times 10^9$/L) and red cell (reticulocyte >1.55%) engraftment usually follows neutrophil recovery. Mean time for neutrophil and platelet engraftment following autologous BM-HPC transplant ranges from 11 to 14 days and 17 to 23 days, respectively (Schmitz et al. 1996). Engraftment kinetics depend on the condition of the supporting BM stroma, the dose of HPCs infused, the underlying disease, and posttransplant GVHD prophylaxis treatment. Autologous stem cell transplants engraft more rapidly than allogeneic stem cell transplants. Growth factors administered after the transplant can speed engraftment of neutrophils (Gisselbrecht et al. 1994; Stahel et al. 1994). Some experimental protocols involve the use of nonmyeloablative treatments followed by allogeneic stem cell treatments. With these therapies, no period of severe neutropenia or thrombocytopenia normally occurs.

In the long term, the patient's hematopoietic system should be completely replaced by the donor's hematopoietic system. For patients with a history of leukemia who are transplanted, failure to completely and permanently replace the patient's hematopoietic system indicates a higher chance of disease relapse. This replacement is measured by "chimerism analysis." Chimerism analysis determines the phenotype and/or genotype of the hematopoietic cells in the transplant recipient. The blood type and the HLA type of the patient should change to the donor's types. If there are no HLA differences, microssatellite DNA markers can be used for chimerism analysis.

Potential Adverse Effects

Acute Reactions During Product Infusion

Acute adverse reactions to allogeneic BM transplant infusions include hemolytic reactions, allergic reactions (mild or anaphylactic), reactions to rapid volume changes, febrile nonhemolytic reactions, fluid overload reactions, and sepsis or endotoxic shock. These reactions are associated with the same signs and symptoms described in Chapters 26–28. Management of these reactions differs than management of identical reactions that can occur with transfusions of more traditional blood components; however, BM-HPCs are usually irreplaceable. For this reason, the patient is normally treated for signs and symptoms of the reaction as described in Chapters 26–28, but the infusion is continued if possible. In some cases the infusion may be temporarily halted, but the infusion should be restarted as soon as the patient can tolerate it. If the patient is not being prophylactically treated with antimicrobial therapy, then a febrile reaction could be an indication to commence such therapy.

Signs and symptoms such as flank pain, hypotension, hematuria, and dyspnea should be investigated. Some of these signs and symptoms could be due to a hemolytic transfusion reaction, or due to reactions to DMSO and lysed RBCs contained in cryopreserved products. Whenever a suspected hemolytic transfusion reaction occurs, the label on the BM-HPC should be rechecked immediately to confirm that the correct product is being infused. A DAT can be performed on the patient's RBCs. Additional tests may include retyping RBCs from the patient and BM-HPCs, performing antibody screens on serum or plasma from the patient and the BM-HPCs (or BM donor), and performing cross-matches between donor RBCs and the patient's serum and the patient's RBCs and donor's plasma or serum. These tests will provide additional evidence concerning the correct identity of the BM-HPCs and the patient and may suggest that the patient's or donor's RBCs are undergoing hemolysis. In these cases, further depletion of RBCs or plasma in the BM-HPC, or plasmapheresis of the patient may be warranted.

Chronic Adverse Effects

Graft-versus-host Disease

GVHD is a potentially serious adverse reaction of allogeneic BM-HPC transplants. T lymphocytes derived from allogeneic donor BM can recognize the patient's cells as foreign and react against those cells. The skin, gastrointestinal tract, and liver are the principal targets of this reaction. By definition, acute GVHD occurs within the 100 days of the transplant but usually occurs around the time of BM engraftment. Risk factors for development of GVHD include unrelated donors, HLA-mismatched donors, multiparous female donors, and older patients. Cutaneous symptoms can include erythema, a macular/papular rash, bullous lesions, and epidermal necrosis. Liver manifestations can include increased conjugated bilirubin and/or transaminases, hepatomegaly, and right upper quadrant tenderness. Gastrointestinal manifestations include diarrhea, nausea, vomiting, and cramping. Chronic GVHD, which arises more than 100 days posttransplant, resembles collagen vascular diseases with multiple systems affected including the skin, mouth, eyes, sinuses, gastrointestinal tract, liver, lungs, vagina, muscle, nervous system, urological system, hematopoietic system, and lymphoid system (Atkinson 1990).

Infectious Disease

Allogeneic transplants may also transmit the same infectious diseases that can potentially be transmitted by blood transfusions.

Alternative HPC Sources

The main alternatives to transplants of BM-HPCs are PB-HPCs and umbilical cord blood-derived HPC. These are described in Chapters 12 and 13. PB-HPCs engraft more quickly than BM-HPCs resulting in decreased times of neutropenia and thrombocytopenia (Schmitz et al. 1996). For this reason, almost all autologous hematopoietic stem cell transplants are collected from peripheral blood.

Although PB-HPCs offer several short-term advantages over BM-HPCs, allogeneic BM-HPC transplants are still performed because of possible increased risk of GVHD associated with PB-HPC transplants. PB-HPCs contain nearly 10 times more T lymphocytes than BM-HPCs (Dreger et al. 1994), and T cells are the principal mediators of GVHD. Several studies with limited numbers of patients suggest that there is no increased incidence of severe acute GVHD in patients who receive allogeneic PB-HPC (Hagglund et al. 1998). However, in contrast, some studies have shown an increased incidence and severity of chronic GVHD following PB-HPC transplants (Scott et al. 1998). These studies have followed a limited number of patients for a limited time. Furthermore, drug therapy or T cell depletion of PBSC may overcome this problem. Currently, the possible increased risk of chronic GVHD associated with PBSC transplants must be taken into account when considering this alternative to BM-HPCs.

Umbilical cord blood-derived HPCs have primarily been used for allogeneic (unrelated and sibling) transplants of pediatric patients. In the future, many more patients may be candidates for transplants of umbilical cord blood-derived HPC. Chapter 25 contains more information regarding the potential advantages and disadvantages of umbilical cord blood.

References

Adler BK, Salzman DE, Carabasi MH, et al. 2001. Fatal sickle cell crisis after granulocyte colony-stimulating factor administration. *Blood* 97:3313–3314.

Alyea EP, Soiffer RJ, Canning C, et al. 1998. Toxicity and Efficacy of Defined Doses of CD4+ Donor Lymphocytes for Treatment of Relapse After Allogeneic Bone Marrow Transplant. *Blood* 91:3671–3680.

Atkinson K. 1998. *The BMT data book: a manual for BM and blood stem cell transplantation.* Cambridge England; New York, Cambridge University Press.

Atkinson K. 1990. Chronic graft-versus-host disease. *Bone Marrow Transplant* 5:69–82.

Broers AEC, van der Holt R, van Esser JWJ, et al. 2000. Increased transplant-related morbidity and mortality in CMV-seropositive patients despite highly effective prevention of CMV disease after allogeneic T-cell-depleted stem cell transplantation. *Blood* 95:2240–2245.

Buckner CD, Petersen FB, and Bolonese BA. 1994. Bone Marrow Donors. In *Bone marrow transplantation.* ED Thomas, SJ Forman and KG Blume, eds. Boston: Blackwell Scientific Publications.

Chan KW, Pollack MS, Braun D, Jr., et al. 1982. Distribution of HLA genotypes in families of patients with acute leukemia. Implications for transplantation. *Transplantation* 33:613–615.

Chiang K, Hazlett L, Godder K, et al. 2001. Epstein-Barr virus-associated B cell lymphoproliferative disorder following mismatched

related T cell-depleted BM transplantation. *Bone Marrow Transplantation* 28:1117–1123.

Dinsmore RE, Reich LM, Kapoor N, et al. 1983. ABH incompatible BM transplantation: removal of erythrocytes by starch sedimentation. *Br J Haematol* 54:441–449.

Dreger P, Haferlach T, Eckstein V, et al. 1994. G-CSF-mobilized peripheral blood progenitor cells for allogeneic transplantation: safety, kinetics of mobilization, and composition of the graft. *Br J Haematol* 87:609–613.

Gee AP and Lee C. 1998. T-cell deplation of allogeneic stem-cell grafts. In *The clinical practice of stem-cell transplantation.* J Barrett and JG Treleaven, eds. Oxford: Isis Medical Media Ltd. 2.

Giralt S, Hester J, Huh Y, et al. 1995. CD8-depleted donor lymphocyte infusion as treatment for relapsed chronic myelogenous leukemia after allogeneic BM transplantation. *Blood* 86:4337–4343.

Gisselbrecht C, Prentice HG, Bacigalupo A, et al. 1994. Placebo-controlled phase III trial of lenograstim in bone-marrow transplantation [published erratum appears in Lancet 1994 Mar 26;343(8900):804]. *Lancet* 343:696–700.

Hacein-Bey-Abina S, Le Deist F, Carlier F, et al. 2002. Sustained correction of X-linked severe combined immunodeficiency by ex vivo gene therapy. *N Engl J Med* 346:1185–1193.

Hacein-Bey-Abina S, von Kalle C, Schmidt M, et al. 2003. A serious adverse event after successful gene therapy for X-linked severe combined immunodeficiency. *N Engl J Med* 348:255–256.

Haddad E, Landais P, Friedrich W, et al. 1998. Long-term immune reconstitution and outcome after HLA-nonidentical T-cell-depleted bone marrow transplantation for severe combined immunodeficiency: a European retrospective study of 116 patients. *Blood* 91:3646–3653.

Hagglund H, Ringden O, Remberger M, et al. 1998. Faster neutrophil and platelet engraftment, but no differences in acute GVHD or survival, using peripheral blood stem cells from related and unrelated donors, compared to BM. *Bone Marrow Transplant* 22:131–136.

Lasky LC, McCullough J, and Zanjani ED. 1986. Liquid storage of unseparated human BM. Evaluation of hematopoietic progenitors by clonal assay. *Transfusion* 26:331–334.

Lasky LC, Warkentin PI, Kersey JH, et al. 1983. Hemotherapy in patients undergoing blood group incompatible BM transplantation. *Transfusion* 23:277–285.

Little AM, Marsh SG, and Madrigal JA. 1998. Current methodologies of human leukocyte antigen typing utilized for BM donor selection. *Current Opinion in Hematology* 5:419–428.

Mehta J and Singhal S. 1998. Pretransplant evaluation of the patient and donor. In *The clinical practice of stem-cell transplantation.* J Barrett and J Treleaven. Oxford: Isis Medical Media.

Pession A, Rondelli R, Paolucci P, et al. 2000. Hematopoietic stem cell transplantation in childhood: report from the BM transplantation group of the Associazione Italiana Ematologia Oncologia Pediatrica (AIEOP). *Haematologic* 85:638–646.

Schmitz N, Bacigalupo A, Labopin M, et al. 1996. Transplantation of peripheral blood progenitor cells from HLA-identical sibling donors. European Group for Blood and Marrow Transplantation (EBMT). *Br J Haematol* 95:715–723.

Schmitz N, Linch DC, Dreger P, et al. 1996. Randomised trial of filgrastim-mobilised peripheral blood progenitor cell transplantation versus autologous bone-marrow transplantation in lymphoma patients [see comments] [published erratum appears in Lancet 1996 Mar 30;347(9005):914]. *Lancet* 347:353–357.

Scott MA, Gandhi MK, Jestice HK, et al. 1998. A trend towards an increased incidence of chronic graft-versus-host disease following allogeneic peripheral blood progenitor cell transplantation: a case controlled study. *Bone Marrow Transplant* 22:273–276.

Siena S, Bregni M, and Gianni AM. 1993. Estimation of peripheral blood CD34+ cells for autologous transplantation in cancer patients [letter; comment]. *Exp Hematol* 21:203–205.

Stahel RA, Jost LM, Cerny T, et al. 1994. Randomized study of recombinant human granulocyte colony-stimulating factor after high-dose chemotherapy and autologous BM transplantation for high-risk lymphoid malignancies. *J Clin Oncol* 12:1931–1938.

Steinberg MH and Brugnara C. 2003. Pathophysiological-based aproaches to treatment of sickle cell disease. *Annu. Rev. Medicine* 54:89–112.

Stroncek DF, Holland PV, Bartch G, et al. 1993. Experiences of the first 493 unrelated marrow donors in the National Marrow Donor Program. *Blood* 81:1940–1946.

Thomas ED and Storb R. 1970. Technique for human marrow grafting. *Blood* 36:507–515.

Walters MC, Nienhuis AW, and Vichinsky E. 2002. Novel therapeutic approaches in sickle cell disease. Hematology 2002:10–34.

Wei A and Grigg A. 2001. Granulocyte colony-stimulating factor-induced sickle cell crisis and multiorgan dysfunction in a patient with compound heterozygous sickle cell/beta + thalassemia. *Blood* 97:3998–3999.

Peripheral Blood Stem Cells

GRACE S. KAO, MD, AND STEVEN R. SLOAN, MD, PhD

INTRODUCTION

Peripheral blood (PB) as a stem cell source was introduced in 1981. It is now known that hematopoietic stem cells traffic constantly between extravascular marrow spaces and PB. Therefore, the quality of stem cells is not thought to be different between bone marrow (BM) and PB stem cell pools (Korbling and Anderlini 2001). Both BM- and PB-derived stem cell products contain stem cells and hematopoietic progenitor cells that are committed to particular hematopoietic lineages. The term "hematopoietic progenitor cells" (HPCs) is more accurate than "stem cells" and will be used in this chapter. Committed and partially differentiated HPCs are probably responsible for the initial circulating leukocytes and platelet recovery, called short-term engraftment, following a HPC transplant. However, the cells that contribute to long-term multilineage reconstitution of the hematopoietic system are the pluripotent stem cells that have the capacity for self-renewal. BM- and PB-derived allografts differ in their reconstitutive and immunogenic characteristics, which seem to be based on the proportion of early pluripotent and self-renewing pluripotent stem cells, to lineage-committed late progenitor cells, and on the number of accessory cells, particularly T-cell subsets, contained in the HPC product.

Similar to BM-derived HPCs, PB-derived HPCs can be used in autologous and allogeneic settings. In autologous PB-HPC transplants, HPCs are donated by the person who also is the recipient and the HPC products are usually cryopreserved before the patient receiving myeloablative chemoradiotherapy. For allogeneic HPC transplants, the HPCs are donated by a person other than the recipient. For adult patients, PB-derived HPC

products have became the major stem cell source for both allogeneic and autologous HPC transplantation. For pediatric patients, the majority of autologous HPCs are derived from PB while only a minority of allogeneic HPCs are conducted using PB-HPCs. Because of the patient size, more pediatric than adult patients receive allogeneic HPCs collected from umbilical cord blood (see Chapter 25).

Despite more than a decade of experience using HPC for the reconstitution of lymphohematopoietic function after myeloablative treatments, the collection and mobilization of PB-derived HPCs in pediatric patients remained relative new, especially in allogeneic settings. As a result, the practice of PB-HPC collection can vary between institutions. Regulatory agencies and accrediting organizations have recently adopted, or are still in the process of developing, standards and regulations for PB-HPC transplant programs. The same agencies that regulate and accredit BM transplants will also regulate and accredit PB-HPC transplants (see Chapter 23).

PRODUCT DESCRIPTION

Cellular Constituents

In addition to progenitor cells, PB-HPC collections contain other hematopoietic cells. Although the PB-HPC collection procedure is designed to enrich for MNCs, all PB-HPC collections contain granulocytes, erythrocytes, and platelets. Certain methods of collection can remove more platelets than others. Lymphocytes and immature myeloid cells are also present in

PB-HPC harvests. Compared to BM-derived HPCs, PB HPC products contain a three- to fourfold higher number of CD34$^+$ cells and an approximately tenfold higher total number of lymphoid subsets when mobilized with growth factor such as granulocyte-colony stimulating factor (G-CSF) (Ottinger et al. 1996; Korbling and Anderlini 2001). Like BM-HPCs, malignant cells may potentially contaminate autologous PB-HPC collections for oncology patients. The autologous malignant cells and allogeneic lymphoid subsets can sometimes pose harm to the patients. Hence, PB-HPCs may be processed to remove unwanted cells.

Characterization of HPC Content

Various assays can be used to assess the number and types of cellular constituents in a PB-HPC product. Before increased accessibility and standardization of CD34+ cell counts, total MNC counts were used to estimate the stem cell concentration in the collection. Now, CD34+ cell counts are used as an indirect measurement of pluripotent stem cells and HPCs. Biological growth assays such as the colony-forming unit (CFU) assay and the long-term culture-initiating cell (LTCIC) assay may also be performed to measure both the quantity and quality of HPCs in a PB-HPC product. Other methods have been developed to measure the most immature progenitor cells, but these assays are complex and not performed in most clinical transplant laboratories. Because CFU assays take weeks to produce results, such assays are not useful for determining whether enough cells have been collected from a donor before the transplant. Because total MNC concentrations and/or CD34+ cell concentrations are measured within hours or days of the collection, they are frequently used to measure the cell dose obtained after each peripheral collection and determine if repeat collections are need to reach a targeted cell value.

Characterization of Mononuclear Cell Content

The mononuclear cell (MNC) count is one measure that may help determine whether sufficient cells are present in the PB-HPC collection to result in a timely engraftment of the patient's hematopoietic system. Though some studies have suggested that the MNC counts correlate with engraftment; this correlation has not been seen by others (Roberts et al. 1993). However, MNC cell counts are easily performed on automated hematology analyzers or hemocytometers, and several centers still use these results. MNC counts remained crucial for centers that use dual-platform assay to determine the absolute CD34+ counts because the enumer-

ation of CD34+ cells are based on percentage of total MNCs (Gratama et al. 1999).

The target dose of MNCs usually ranges from 2 to 6×10^8 cells/kg, but each transplant program must establish its own guidelines. Target doses may depend upon the source of the cells (for example, unrelated allogeneic transplants usually require higher doses than autologous transplants), or on the patient's diagnosis. Most laboratories cryopreserve cells at a concentration of 2×10^7 to 8×10^8 MNCs/mL.

Characterization of CD34+ Cell Content

The number of CD34+ cells in the PB-HPC product is the most widely used measurement to predict whether sufficient cells are present for timely engraftment of the transplanted cells. CD34 is a cell surface protein that is expressed on most stem cells and many other immature hematopoietic cells. CD34+ cells comprise 1% to 5% of PB MNCs following mobilization. CD34+ cell counts are determined by flow cytometry. Two flow cytometry techniques are used to determine CD34+ cell counts in PB-HPC products. Most techniques for CD34+ cell enumeration are dual-platform assays, where the percentage of CD34+ cells is determined flow cytometrically, and the percent the white blood cell (WBC) count is determined on a hematology cell analyzer. Recently, so-called single-platform assay have been developed, in which the absolute number of CD34+ cells is directly derived from a single-flow cytometric measurement by incorporating a known number of fluorescent counting beads in the flow cytometric assay (Gratama et al. 1999). The ratio between the number of beads and CD34+ cells counted allows direct calculation of the absolute CD34+ cell numbers. These procedures require technical expertise and judgment, and results between laboratories may not correlate well. However, these methods have now standardized and should help improve inter-laboratory reproducibility (Keeney et al. 1998). Single-platform assays are less likely to produce variability between laboratories because they avoid the need for a second instrument.

Several, but not all studies, suggest that a minimum number of CD34+ cells must be transplanted to ensure rapid engraftment. The target number of CD34+ cells depends on whether autologous cells, allogeneic cells from a related donor, or allogeneic cells from an unrelated donor are transplanted. The target number of cells to transplant must be determined by each transplant program but studies suggest that that a dose of 2 to 5×10^6 CD34+ cells/kg is adequate to ensure trilineage engraftment in a timely fashion (Weaver et al. 1995).

Characterization of CFU Content

CFU assays can be used as an indirect measure of hematopoietic progenitor cells. This method identifies and counts hematopoietic progenitor cells based on their ability to proliferate and give rise to more mature hematopoietic cells. Cells from the PB-HPC product are cultured in a semisolid media, and the types of cell colonies that grow from individual immature hematopoietic cells are identified and counted. CFU culture conditions are not completely standardized, and interpretation of results is somewhat subjective. Not surprisingly, results from CFU assays can vary significantly between laboratories. While some studies have shown a correlation between CFU assay results and engraftment speed, other studies have revealed no such correlation (To et al. 1992). Each transplant program determines whether it will strive for a specific minimum CFU dose. In most cases CFU assays cannot be used to determine whether additional PB-HPC collections are necessary for any individual patient because colonies cannot be scored until approximately two weeks after plating the cells. However, CFU assays may be used for data analysis and monitoring quality of various aspects of the transplant program. Other methods have been developed to measure the most immature progenitor cells, but these assays are complex and are not performed in most clinical transplant laboratories.

Anticoagulants and Additives

PB-HPC products contain plasma, anticoagulant, and additional buffered solution. In most cases acid-citrate-dextrose formula A (ACD-A) or a similar anticoagulant is added. Human plasma is present in all PB-HPC products whether or not the product is cryopreserved. Cryopreserved PB-HPC products usually contain 10% to 20% of a protein solution such as plasma or human albumin and a cryoprotectant solution consisting of 10% dimethylsulfoxide (DMSO) or 5% DMSO, 6% hydroxyethyl starch, and albumin (Stiff 1991). Cryopreserved PB-HPCs usually contain a buffered electrolyte solution such as Normosol or other infusion grade solutions. Some institutions use tissue culture media that also contains vitamins and/or minerals, but its use is discouraged because tissue culture media is not currently approved for infusion to humans in the United States. Red cell-depleted products may contain additional albumin.

Labeling

PB-HPCs are labeled in the same manner as BM-HPCs (see Chapter 18), except that they contain the phrase "Hematopoietic Progenitor Cells, Apheresis" instead of "Hematopoietic Progenitor Cells, Marrow."

COLLECTION

Donor Evaluation and Preparation

Allogeneic donors undergo the same screening (history and infectious disease testing) as for BM donation to ensure the HPC products are safe (see Chapter 23). The pediatric donor screening process is slightly different from adult donors because the majority of pediatric donors are minors and the medical history of the donors are often conducted through their parents or guardians. Some of the screening questions concern sensitive issues such as sexual and drug histories. Older teenagers should be asked these questions in private. While no specific guidelines have been formulated for medical screening of pediatric donors, children as young as 10 to 11 years can often be asked sensitive screening questions in private with the consent of their parent or guardian. Many parents appreciate the fact that a donor screening questionnaire administered by a trained health care provider can be an educational experience for their child.

Like all apheresis procedures, both autologous and allogeneic PB-HPC donors will need to present with good venous access for either short-term or long-term blood collection. Peripheral venous accesses are usually used to collect PB-HPCs from older allogeneic donors. Central venous accesses are usually not needed during allogeneic donation from older donors because sufficient number of HPCs can easily be collected in one to two days from growth factor-mobilized donors. However, central venous or femoral line catheters are usually needed for donors younger than 10 years and sometimes needed from older donors because of inadequate peripheral venous access. In addition, some young donors cannot comply with collection using peripheral venous access because each collection lasts at least three to four hours and the donors need to be relatively still throughout the entire procedure. The placement of central venous access increases the risk of donation because of increased bleeding, infection, and development of deep venous thrombosis associated with line placement. For autologous donors, central venous accesses were often placed for PB-HPC collection because the same venous accesses are also needed for chemotherapy administration and transfusion support during and after transplantation. For pediatric allogeneic and autologous donors, additional problems can arise during collection procedures. These donors are also subjected to the same risks of anticoagulants and volume shifts that are normally associated with apheresis procedures (see Chapter 29). With close monitoring and modifications of the collection procedure, the donor can remain isovolumetric throughout the procedure. Studies have shown that PB-HPC collection is safe in pediatric donors with very small blood volume.

More recently, the use of pediatric donors for adult HPC transplantation has been considered as these donors appear to have improved PB HPC mobilization using growth factors. Unlike BM-derived HPCs, the number and volume of HPC collections do not depend on the body size of the donor, and the collection can always be repeated if insufficient number of cells are collected.

Relative contraindications to the collection procedure include hemodynamic instability, symptomatic anemia, evidence of active infections, and recent ingestion of angiotensin converting enzyme (ACE) inhibitors in the donor.

Mobilization

Mobilization is the increased shift of pluripotent stem cells, hematopoietic progenitor cells, and mature and immature hematopoietic cells from BM to the blood. This is usually accomplished by administration of chemotherapy or growth factors/cytokines. The donor is usually treated with cytokines and/or chemotherapy before collection. Mobilization treatment causes an increased peripheral white blood cell (WBC) count that predominantly consists of myeloid cells at all stages of development. HPCs are the most critical cells to mobilize, and the number of HPCs correlates with the number of CD34+ cells. While CD34+ cells normally represent approximately 0.1% of PB MNCs, mobilization can increase this proportion to more than 1% (Stadtmauer et al. 1995).

Donors are treated with mobilization drugs daily for approximately for four to seven days before collection by apheresis. Mobilization kinetics vary significantly between donors and can depend on the mobilization regimen and previous treatments the donor has received. Though some institutions use the peripheral WBC count to determine when to begin apheresis, the PB CD34+ cell count is probably a better indicator to determine the optimal day to begin peripheral cell collection (Stadtmauer et al. 1995). The target cell counts vary. Examples include protocols that commence leukapheresis when the WBC $\geq 8.0 \times 10^9$ cells/L, others when the WBC $\geq 1.0 \times 10^9$ cells/L, and others when the CD34+ cell count reaches 10 cells/μL (Haas et al. 1994). If cell counts reveal that the patient's HPCs are mobilizing poorly with G-CSF, mobilization may improve with G/GM-CSF. Growth factors, such as G-CSF, can cause unpleasant side effects such as bone

pain, fever, and malaise, which are readily treated with acetaminophen and/or mild narcotics.

Autologous Transplants

HPCs from autologous donors can be mobilized with cytotoxic chemotherapeutic agents and/or growth factors. The choice of drugs will depend on a variety of factors, including the type of tumor and prior exposure to chemotherapeutic drugs. A small portion of autologous donors are poor mobilizers (Goldman et al. 2001). Despite appropriate cytotoxic chemotherapy and growth factors, these donors do not have enough CD34+ cells in the PB. Some of them do respond to increased amounts of the same or additional type of growth factors. These patients tend to have significant disease in the BM and often underwent multiple chemotherapy treatments in the past before mobilization. It is also unclear if the quality of the graft obtained by these patients is the same as ones from donors who are easily mobilized.

Allogeneic Transplants

While HPCs can be mobilized with cytotoxic chemotherpeutic agents and/or growth factors, chemotherapy is not usually used to mobilize cells from an allogeneic donor because of risks associated with administering chemotherapy to healthy individuals. Though G-CSF is the most widely used growth factor used for mobilization, other growth factors including GM-CSF, IL-1, IL-3, IL-8, IL-11, and SCF also mobilize HPCs (Mauch et al. 1995). Cytokine combinations like G-CSF + SCF, G-CSF + GM-CSF, or IL-11 + SCF may improve mobilization, and some institutions use growth factor combinations or a growth factor/chemotherapy combination to mobilize HPCs (Mauch et al. 1995).

Young allogeneic HPC donors are usually human leukoctye antigen (HLA)-matched siblings of pediatric patients who need to undergo transplantations. PB collections of HPCs following three to five days of G-CSF mobilization have been successfully performed in several centers (Watanabe et al. 2000; Benito et al. 2001). Despite the lack of short-term G-CSF-related complications in these donors, the long-term effects of G-CSF on these young, healthy donors are still unknown. Hence, the use of cytokines in young, healthy donors becomes somewhat of a dilemma. As the result, BM-derived HPCs remained the more popular source of stem cells for transplantation in pediatric settings.

HLA Compatibility-Allogeneic Donors

HLA matching is the principal means of choosing an allogeneic HPC donor. Issues concerning HLA matching for allogeneic PB-HPC transplants and allogeneic BM-HPC transplants are identical and are discussed in Chapter 23.

Collection Procedure

PB-HPCs are collected by leukapheresis. The MNCs are collected using some of the same machines that can be used for therapeutic apheresis (see Chapters 29–31). The leukapheresis procedure is conducted to specifically remove MNCs. Donors are subjected to the same risks associated with anticoagulants and volume shifts that are normally associated with apheresis. Citrate anticoagulants and heparin are the most common anticoagulants administered to the donors during leukapheresis. Citrate prevents blood coagulation by chelating the calcium needed to activate calcium-dependent coagulation factors. As a side effect, donors with lower body weight, especially the pediatric population, are more likely to experience symptoms of hypocalcemia during cell collection. The symptoms can often be eliminated or prevented by calcium administration. In some PB-HPC collection centers, intravenous calcium gluconate is given during leukapheresis to prevent hypocalcemia. Citrate anticoagulant is often preferred over heparin for allogeneic donors because it is rapidly metabolized and never produces systemic anticoagulation in healthy donors without liver diseases. Heparin may enhance systemic anticoagulation more than is expected because of the additional effect of clotting factor dilution by the nonplasma replacement solutions. Thus, intrapheresis monitoring of clotting time is crucial to prevent any bleeding-related complications when heparin is used.

Laboratory Evaluation

Laboratory evaluation includes quantitation of HPCs in the product and assessment of the sterility of the product. Cellular contents are quantified indirectly by measuring MNCs and CD34+ cells, as described in earlier. In addition, a microscopic examination of trypan-blue-stained cells is often performed to determine cellular viability. Sterility is usually determined by culturing products for bacteria and fungi. Proliferation assays such as CFU measurements may also be performed to retrospectively measure HPC content before and after special manipulations.

Storage and Transport

Autologous products are usually cryopreserved before storage. At the end of cryopreservation, cells are maintained in liquid nitrogen or in −80°C freezers. Many centers store cells in liquid nitrogen tanks because cells may be stored longer than −80°C freezers,

but freezers are more likely to fail. Liquid nitrogen tanks can fail as well, and all freezers and liquid nitrogen tanks that contain HPCs should be regularly monitored and have alarm systems. Frozen cells can be stored for years. Cells stored for five years in liquid nitrogen and cells stored at −80°C for over two years have been successfully transplanted (Rowley 1992).

Hepatitis viruses and fungi can contaminate liquid nitrogen storage tanks and contaminate the HPCs stored in the tanks, and thus in turn infect transplant recipients (Hernandez-Navarro et al. 1995). Double bags and/or storage in the vapor phase of the liquid nitrogen tank may help prevent these contamination problems, but this has not been proven.

PB-HPCs should be administered soon after being thawed. Cryopreserved autologous PB-HPCs are usually transplanted to a patient who is being treated in the institution that is storing the cells. In these cases the PB-HPCs need only be transported from the storage area to the patient treatment area. Cells may be transported in liquid nitrogen and thawed at the bedside, or they may be thawed immediately before transport to the patient's room. If cryopreserved cells need to be transported to another institution, then they are transported frozen in liquid nitrogen or packed in dry ice.

Allogeneic PB-HPCs are not usually cryopreserved and are stored and transported the same as BM-HPCs.

SPECIAL PREPARATION

Most of the special preparation and processing that is performed on PB-HPCs is designed to purify the hematopoietic progenitor cells. While PB-HPCs collected by apheresis contain many different cells, the cells responsible for reconstituting the hematopoietic system are the hematopoietic progenitor cells. Autologous PB cells collected from a cancer patient may contain tumor cells that could contribute to relapse of disease, and allogeneic T cells in PB grafts may cause GVHD. Additionally, allogeneic PB-HPCs may contain incompatible RBCs that could be hemolyzed by antibodies in the patient's plasma and/or may contain plasma with antibodies that can hemolyze the patient's RBCs. Thus, transplant laboratories perform a variety of procedures to purity the PB-HPCs. The processing may involve depletion of RBCs, lymphocytes, or malignant cells, or positive selection for HPCs expressing CD34 surface antigens. These processed products are in reduced volumes and usually have different cell populations when compared to minimally processed HPCs.

RBC and Plasma Depletion

PB-HPCs do not usually need RBC or plasma depletion because the products are not contaminated with large volumes of RBCs. When major ABO incompatibility between the allogeneic donor and recipient occur, the volume of RBCs in the collection is first assessed. When the number of RBCs in HPC products is more than the RBC volume limitation (usually between 15 to 20 mL) set by the cell processing center, RBC depletion may be performed to prevent hemolytic reactions. Minor ABO incompatibility or presence of significant alloantibodies in allogeneic donors may necessitate depletion of plasma from PB-HPCs intended for transplantation. The indications for depleting PB-HPCs of RBCs or plasma and the techniques used to deplete them are identical to the indications and techniques that apply to BM-HPCs.

Selection

Positive Selection for CD34+ Cells

Most stem cells and many committed progenitor cells express the CD34 antigen, which can be used as a marker for selection. Because CD34+ cells in the HPC graft is an indirect measure of hematopoietic progenitor cells and possible pluripotent stem cells that are important for both short- and long-term hematopoietic reconstitution after BM myeloablation, isolation of CD34+ cells for infusion can potentially eliminate the contamination of CD34-malignant cells in autologous PB-HPC grafts and reduce the number of donor T cells in allogeneic PB-HPC collections that can induce GVHD while ensuring appropriate hematological engraftment.

CD34+ selections are available from many hematopoietic cell processing laboratories. While methods vary, the first step often involves separation of the MNCs from the erythrocytes and polymorphonuclear cells by density gradient centrifugation. Monoclonal anti-CD34 antibodies bound to a solid phase matrix are then used to selectively adsorb the CD34+ cells. The bound cells are then eluted from the solid matrix. Unfortunately, most of the positively selected CD34+ cells infused to the patients remain bound to the antibodies that are often derived from murine origin. As the result, these procedures are not FDA-approved and remain experimental. The Isolex 300i system is designed and currently FDA-approved for the processing of autologous PB progenitor cells (PBPCs) under a closed system (Prince et al. 2002). Using this system, the CD34+ cells from the PB-HPCs are first coated with murine anti-CD34 monoclonal antibody. Sensitized cells are then rosetted with paramagnetic, polystyrene beads with affinity purified sheep anti-mouse IgG covalently bound to the surface. Rosetted CD34+ cells are magnetically retained, while nontargeted cells are washed away to become the negative fraction. The CD34+ cells are then separated from antibodies/beads complexes by releasing peptides—in this case the positively selected product containing CD34+ cells without the antibodies. Several studies have shown that CD34+ selected cells can reconstitute hematopoiesis in the BM. A dose of at least 1.2×10^6 CD34+ cells selected cells per kilogram appears necessary for rapid platelet recovery, and some institutions transplant at double or triple that amount (Shpall et al. 1997). With a sufficient dose, the BM will be reconstituted with CD34+ selected cells as rapidly as with total PB-HPC transplants. This is an area of intensive investigation and minimum cell dosages have not been determined.

Selection of CD34+ cells would not benefit autologous transplants for patients with tumors that express CD34. CD34 is normally expressed on capillary endothelium and stromal cell precursors in the BM, and tumors derived from these cells frequently express CD34 (Fina et al. 1990). Specifically, CD34 is often expressed on vascular tumor cells including angiosarcoma, hepatic hemangioendothelioma, and Kaposi's sarcoma (Fina et al. 1990). Additionally, about 40% of acute myeloid leukemias, 65% of pre-B acute lymphoblastic leukemias, 1% to 5% of acute T cell lymphoid leukemia express CD34 (Borowitz et al. 1990). CD34 antigen has also been reported to be expressed on a few cases of squamous cell lung carcinoma, neuroblastoma, and Ewing's sarcoma.

CD34+ selected cells are usually suspended in a small volume of buffered normal saline solution and contain few if any RBCs, platelets, or mature leukocytes but may contain additional proteins that were added during processing. If the cells are frozen, 10% DMSO and at least 10% to 20% protein solution (plasma or albumin) will usually be included in the final infused product.

Negative Selection

Most of the negative selection techniques used in PB-HPC grafts are aimed at reducing the number of donor T cells that often cause GVHD during allogeneic transplantaions. Several methods are used to specifically deplete T lymphocytes from HPC products. Methods that use the soybean agglutinin (SBA) remove most mature blood cells. Counterflow centrifugal elutriation, a method that removes most lymphocytes, is more frequently used to deplete T cells from BM and is described in Chapter 11. Some methods specifically remove T cells. These include methods that are based on the binding affinity between T cells and sheep RBCs and other methods that use antibodies that specifically bind to T cells.

SBA binds to N-acetyl-D-galactosamine, which is expressed on all mature blood cells, and is used to remove those cells. SBA added to the HPC collection will agglutinate all but the progenitor cells. The agglutinated cells are then removed on a 5% bovine serum albumin gradient (Collins et al. 1992). Alternatively SBA-coated plastic can be used to selectively remove all but the progenitor cells by panning (Lebkowski et al. 1994). T-lymphocytes rosette around sheep RBCs, and this can be used to selectively remove T cells. The rosettes sediment or can be removed by density-gradient centrifugation (Collins et al. 1992).

Other methods of T-cell removal utilize monoclonal antibodies that recognize antigens that are specifically expressed on lymphocytes, T cells, or a subset of T cells. A variety of antibodies have been used, including the "CAMPATH" series of rat monoclonal antibodies that recognize DCw52, an antibody that recognizes the T-cell receptor heterodimer. Other antibodies that specifically recognize CD2, 3, 4, 5, 6, or 8 have been used as well (Gee et al. 1989). These antibodies can be used to separate cell populations or to lyse the cells targeted by the antibodies. To separate cells targeted by the antibody, the antibody must be physically bound to a solid matrix such as paramagnetic microspheres, paramagnetic nanoparticles, or relatively large plastic surfaces used for panning (Gee et al. 1989).

Purging

Though CD34+ cell selection is the most common way to remove tumor cells, many other techniques can be used to purge tumor cells from autologous HPC collections. Most of these techniques are investigational and exact approaches vary depending on the particular research protocol and the type of malignancy. Many of the techniques begin with a physical purification technique such as density-gradient centrifugation or counterflow centrifugal elutriation. Additional techniques to purge tumor cells include methods that use heat, cell culture, cytotoxic effector cells, cytotoxic drugs, molecular biology-based molecules, and antibodies.

A variety of cytotoxic drugs have been used to purge tumor cells. The most widely used drugs are 4HC and mafosfamide. Purging with these drugs reduces the risk of disease relapse in patients with acute myeloid leukemia (AML) (Gorin et al. 1990; Gorin et al. 1991). While these purging protocols can destroy committed progenitor cells and prolong engraftment times, pluripotent progenitors survive the purging protocols (Rowley and Davis 1991; Douay et al. 1995). A variety of other drugs have been suggested. Some of these are photoactive drugs that are added to the HPC, which are then exposed to fluorescent light.

Molecular biology-based agents include antisense DNA oligonucleotides and ribozymes that are designed to inhibit expression of specific genes that promote cancer cell growth, though few. Only a few clinical trials using these approaches have been reported. RNAi may be explored in the future as another method of inhibiting expression of genes responsible for a variety of diseases.

Antibodies can lyse the target cells either by activating complement that is added to the cells or by utilizing complement present in the plasma that is collected with the HPCs. Alternatively, some investigators have conjugated a toxin, such as ricin, directly to the monoclonal antibody (Uckun and Myers 1993).

Cryopreservation

Unlike most allogeneic PB-HPC collections, most autologous PB-HPC collections are cryopreserved. Cryopreservation is designed to preserve the MNCs for extended periods of time. Cryopreservation is not designed to alter the cell population in the PB-HPC collection and therefore is not considered a manipulation of the product by accrediting agencies such as the Foundation for the Accreditation of Cellular Therapy (FACT). The cell solution that was collected by apheresis is centrifuged, the plasma is removed, and the white cells are resuspended in a cryopreservation solution at a concentration determined by the laboratory that usually ranges between 2×10^7 to 8×10^8 cells/mL. Two techniques commonly used, although with some variations, have been described.

Commonly, the cryopreservation solution is designed so that the final product contains 10% DMSO and at least 10% to 20% human protein solution such as plasma or albumin (Rowley 1992; Burger et al. 1996). The remainder of the acellular volume consists of a buffered saline solution such as tissue culture media, Normosol, or other infusion grade solutions. Tissue culture media is not approved for human infusion in the United States, and its use is discouraged for this purpose. After resuspension, the cell solution is dispensed into freezing bags. The bags are chilled in a controlled rate freezer −1°C to −2°C/min until they reach about −50°C and then −5°C/min until they reach about −90°C. The bags are then transferred to a liquid nitrogen storage tank. Cells stored for five years in liquid nitrogen have been successfully transplanted (Rowley 1992).

In another common technique, the cryopreservation solution is designed so that the final product contains albumin, 5% DMSO + 6% hydroxyethylstarch (Stiff 1991). With this technique, the freezing bags are placed horizontally in a −80°C freezer for freezing and storage.

Thawing

Cryopreserved cells must be thawed before transplantation. Cells are thawed rapidly by immersing the product in a 37°C water bath. Rapid thawing can lead to bag breakage, and various strategies have been used to minimize this problem. Bags that have a low breakage rate should be chosen, and some centers double-bag the HPCs. Products exposed to water baths due to bag breakage have been safely infused.

Cell death due to thawing can be minimized by infusing cells as soon as possible. Thus cells are usually thawed immediately before infusion in the same room with the patient and are often infused rapidly. Each bag of cells is sequentially thawed and infused, minimizing waste in the event of a patient reaction. Another approach is to thaw the cells in the laboratory and then remove the DMSO by centrifugation. While this method avoids the unpleasant side effects of DMSO, the cells are exposed to DMSO for a longer period of time. In this case, all of the cells to be infused are thawed simultaneously making it difficult to temporarily discontinue the transplant if the patient experiences an adverse reaction.

ABO and Rh Compatibility

ABO and Rh typing should be performed on the donor and patient. In addition, FACT requires that donors be typed at the time of collection as described in Chapter 23. The PB-HPC label should contain the ABO and Rh type. This helps ensure that the HPCs will be infused to the intended recipient. Crossmatches should be performed on allogeneic transplants, and incompatible products should be depleted of RBCs or plasma using the same criteria and essentially the same techniques described in Chapter 23 and previously in this chapter. If red cells are depleted, they should be depleted before cryopreserving the cells. If plasma in the PB-HPCs is incompatible with the patient's RBCs, the incompatible plasma can be removed and compatible plasma can be added during the cryopreservation processing procedures.

In some allogeneic transplants, the donor's plasma contains isohemagglutinins against the recipient's RBCs. This occurs when the donor is group O and the recipient is group B, or when the donor is group B and the recipient is group A. Although these are known as "minor" ABO incompatibilities, their effect can be serious. While a minor ABO incompatibility does not usually cause immediate significant hemolysis, massive delayed hemolysis can occur (Salmon et al. 1999). This hemolysis, which is caused by stimulation and proliferation of donor-derived B lymphocytes, usually develops five to 16 days following the transplant and can be more severe than analogous reactions seen with BM-HPC transplants. Some drugs such as methotrexate that are used as prophylaxis for GVHD inhibit B lymphocytes and reduce the chances of severe hemolysis. However, other anti-GVHD drugs such as cyclosporine and FK506 do not inhibit B lymphocytes and do not prevent severe hemolysis. Although massive hemolysis occurs in no more than 10% to 20% of susceptible patients, it can be abrupt, severe, and fatal. The direct antiglobulin test (DAT) will usually be positive before the onset of significant hemolysis and can be used to identify patients at risk for massive hemolysis, but not all patients with a positive DAT will develop massive hemolysis. Treatment consists of empiric use of corticosteroids, hydration, and transfusion with blood products that are compatible with both the blood group of the donor and the original blood group of the recipient. Severe cases may be additionally treated with methotrexate and RBC exchange transfusion.

ORDERING AND ADMINISTRATION

The process for selection of the donor (for example, HLA matching) is the same as that for BM-HPC discussed in Chapter 23. PB-HPCs are infused intravenously using essentially the same methods used for transplantation of BM-HPCs. The patient should be well hydrated before infusion and is often premedicated with acetaminophen, an antihistamine, and a corticosteroid. Oxygen and an anti-anaphylaxis treatment such as epinephrine should be available. HPCs are administered intravenously, usually through a central venous catheter, without any filters; gamma irradiation of HPCs is contraindicated. The patient should be closely monitored, and vital signs should be taken periodically during the infusion (as for any blood product infusion) due to the risks of allergic, anaphylactic, hemolytic, or febrile nonhemolytic transfusion reactions.

Indications

Autologous Transplants

The majority of autologous HPC transplantations are performed in pediatric patients with lymphomyeloproliferative disorders, while approximately 25% are for pediatric patients with solid tumors/malignancies. Many patients with lymphomyeloproliferative disorders, such as acute lymphoblastic leukemia (ALL), AML, chronic myeloblastic leukemia (CML), myelodysplastic syndromer (MDS), and non-hodgkin lymphoma (NHL), undergo allogeneic HPC transplants if HLA-matched donors are available. However, autologous HPC grafts are the product of choice for pediatric patients with solid tumors who require high dose chemotherapy followed by HPC rescue (Horowitz and Rowlings 1997). These solid tumor diseases included Ewing's sarcoma, Wilms' tumor, and neuroblastoma. Neuroblastoma patients represent the majority of solid tumor patients who undergo autologous transplants.

Solid tumor patients who need high dose chemotherapy or radiation and HPC rescue are usually those who are at high risk for recurrence or progression of disease but demonstrate responsiveness after conventional chemotherapy. In the past few years, there has been a significant switch from BM to PB as the main source of HPCs for autologous transplantation. In 2002, almost all autologous HPC transplants performed at The Children's Hospital in Boston were conducted using PB-HPCs.

Allogeneic Transplants

Allogeneic HPC transplants are mainly indicated for pediatric patients with hematological diseases or congenital immune deficiencies. The hematological diseases include malignancies of the hematopoietic system such as AML, CML, and ALL, or patients with severe complications associated with hemoglobinopathies like sickle cell diseases and thalassemia. Unlike autologous HPC transplantation, most pediatric allogeneic HPC transplants are still performed with BM-derived HPCs due to ethical issues relating to the use of cytokines to mobilize young healthy donors. Hence, limited data are available for allogeneic PB-HPC transplantations in pediatric population.

Contraindications

The contraindications for PB-HPC transplants are similar to the ones for BM-HPC transplants. However, PB-HPC donors do not undergo general anesthesia during collection, and thus this risk is eliminated. As with BM-HPC transplants, the risk-benefit analysis depends on the relationship between the donor and the patient. Allogeneic PB-HPC transplants from unrelated donors have the highest risks, while autologous PB-HPC transplants are the safest. Allogeneic PB-HPC transplants from HLA-identical siblings are of intermediate risk. Transplant protocols are designed to consider the age, organ function, and Karnofsky performance score of the patient.

Expected Response

PB-HPCs usually reconstitute the hematopoietic system within a few weeks to a month. PB neutrophil counts should rise to $\geq 0.5 \times 10^9$/liter between eight and 30 days posttransplant, with a mean time of 11 to 17 days. PB platelet counts should rise to $\geq 20 \times 10^9$/liter between six and 140 days posttransplant, with a mean time of nine to 31 days (Russell and Miflin 1998). The time to engraftment times depend on the number of CD34+ cells infused, the condition of the supporting BM stroma, and the underlying disease. Autologous transplants engraft more rapidly than allogeneic transplants. Administration of a myeloid growth factor (G-CSF or GM-CSF) shortens the time of neutropenia but not thrombocytopenia (Russell and Miflin 1998). With allogeneic transplants, the hematopoietic system should be eventually completely replaced by the donor's hematopoietic system. This can be measured by chimerism analysis (see Chapter 23).

POTENTIAL ADVERSE EFFECTS

Acute Adverse Reactions

Some acute adverse reactions are specifically associated with cryopreserved products. DMSO in cryopreserved products usually causes an unpleasant taste and odor. In addition headache, flushing, nausea, vomiting, cramping, diarrhea, bradycardia, hypertension, hypotension, and dyspnea can occur (Zambelli et al. 1998). Other reactions that have been reported include anaphylactic-type reactions, neurological complications, and signs and symptoms of leukostasis. The severity of these adverse reactions varies significantly between patients, but higher DMSO doses are more likely to result in more severe reactions. The LD_{50} for dogs receiving DMSO PO is >0 g/kg, and some institutions infuse no more that 1 g/kg/day into human patients. This corresponds to about 500 mL of cells per day in a 70-kg patient. Other institutions continue the transplant as long as the patient tolerates it. Furthermore, lysis of RBCs during cryopreservation and thawing can lead to hemoglobinuria and hemoglobinemia. Because the HPCs are usually irreplaceable, the infusion is usually continued. Severe reactions should be investigated as described in Chapter 23.

Acute reactions associated with the infusion of a large volume of HPCs can occur. The signs and symptoms are identical to volume overload associated with transfusions. CD34+ selected cells have a small total volume, and acute reactions of any kind are uncommon.

Substantial delayed hemolysis can develop five to 16 days after an allogeneic PB-HPC transplant complicated by a minor ABO incompatibility (Salmon et al. 1999). Massive hemolysis can be abrupt, severe, and fatal. Treatment consists of corticosteroids, hydration to maintain adequate renal blood flow, and transfusion with blood products that are compatible with both the blood group of the donor and the original blood group of the recipient. Severe cases may also be treated with methotrexate and RBC exchange transfusions.

Chronic Adverse Reactions

Like BM-HPC transplants, allogeneic PB-HPC transplants can cause GVHD. Because the absolute T-cell number is higher in unmanipulated PBSC allografts than in BM allografts by approximately one log, a higher incidence of acute GVHD might be expected in PB-HPC recipients. Despite more T cells in the PB-HPC products, immunomodulatory effects of in vivo cytokine treatment, such as G-CSF, and cell-to-cell interaction in the apheresis product conceivably could reduce the incidence of acute GVHD after allogeneic PBSCT. Thus, the cumulative incidence of acute GVHD was found to be statistically no different whether using PB-HPCs or BM-HPCs for hematopoietic reconstitution. The probability of chronic GVHD developing varies by report; four of nine studies show a significantly higher probability of chronic GVHD after allogeneic PB-HPC transplants than after allogeneic BM-HPC transplants (Korbling and Anderlini 2001). GVHD continued to be the allogeneic transplant-related complication causing the most mortality and morbidity. Attempts to reduce this risk have focused on techniques to reduce the concentration of T cells in PB-HPC collections by positive selection for HPCs or negative selection and removal of T cells. Clinical trials are needed to substantiate whether reduction in GVHD results from this manipulation.

Chronic adverse reactions that occur with allogeneic BM-HPC transplants can also occur with allogeneic PB-HPC transplants. Compared with BM-HPC transplants, allogeneic PB-HPC transplants have a higher risk of severe chronic GVHD, though this may not be true for transplants of PB-HPCs that have been depleted of T cells (Storek et al. 1997). Allogeneic PB-HPC transplants can also potentially transmit the same infectious diseases transmissible by other types of blood transfusions.

ALTERNATIVE HPC SOURCES

The main alternatives to PB-HPC transplants are BM-HPC transplants and umbilical cord blood transplants. These are described in Chapters 23 and 25, respectively. PB-HPCs engraft more quickly than BM-HPCs, resulting in a decreased interval of neutropenia and thrombocytopenia following PB-HPC transplants (Schmitz et al. 1996). For this reason, almost all autologous HPC transplants are collected from PB.

BM-HPC or PB-HPC can be used for allogeneic transplants of HPCs. PB-HPCs provide faster engraftment but have a higher risk of chronic GVHD. To avoid the potential increased risk of chronic GVHD, BM-HPCs may be chosen. Because pediatric allogeneic PB-HPC transplants are relatively new, the data concerning the long-term outcomes of these transplants are limited compared to BM-HPC transplants. Umbilical cord blood-derived stem cells are also used for allogeneic transplants of pediatric patients.

References

Benito A, Gonzalez-Vicent M, Garcia F, et al. 2001. Allogeneic PB stem cell transplantation (PBSCT) from HLA-identical sibling donors in children with hematological diseases: a single center pilot study. *BM Transplantation* 28:537–543.

Borowitz MJ, Shuster JJ, Civin CI, et al. 1990. Prognostic significance of CD34 expression in childhood B-precursor acute lymphocytic leukemia: a Pediatric Oncology Group study. *J Clin Oncol* 8:1389–1398.

Burger SR, Fautsch SK, Stroncek DF, et al. 1996. Concentration of citrate anticoagulant in PB progenitor cell collections [see comments]. *Transfusion* 36:798–801.

Collins NH, Kernan NA, Bleau SA, et al. 1992. T-cell depletion of allogeneic human BM grafts by soybean lectin agglutination and either sheep red blood cell rosetting or adherence on the CD5/CD8 CELLector. BM Processing and Purging: A Practical Guide. A. P. Gee. Boca Raton, CRC Press:201–212.

Douay L, Giarratana MC, Labopin M, et al. 1995. Characterization of late and early hematopoietic progenitor/stem cell sensitivity to mafosfamide. *BM Transplant* 15:769–775.

Fina L, Molgaard HV, Robertson D, et al. 1990. Expression of the CD34 gene in vascular endothelial cells. *Blood* 75:2417–2426.

Gee AP, Mansour V, and Weiler M. 1989. T-cell depletion of human BM. *J Immunogenet* 16:103–115.

Goldman SC, Bracho F, Davenport V, et al. 2001. Feasibility study of IL-11 and granulocyte colony-stimulating factor after myelosuppressive chemotherapy to mobilize PB stem cells from heavily pretreated patients. *Journal of Pediatric Hematology/Oncology* 231:300–305.

Gorin NC, Aegerter P, Auvert B, et al. 1990. Autologous BM transplantation for acute myelocytic leukemia in first remission: a European survey of the role of marrow purging. *Blood* 75:1606–1614.

Gorin NC, Labopin M, Meloni G, et al. 1991. Autologous BM transplantation for acute myeloblastic leukemia in Europe: further evidence of the role of marrow purging by mafosfamide. European Co-operative Group for BM Transplantation (EBMT). *Leukemia* 5:896–904.

Gratama JW, Braakman E, Kraan J, et al. 1999. Comparison of single and dual-platform assay formats for CD34+ haematopoietic progenitor cell enumeration. *Clin Lab Haem* 21:337–346.

Haas R, Mohle R, Fruhauf S, et al. 1994. Patient characteristics associated with successful mobilizing and autografting of PB progenitor cells in malignant lymphoma. *Blood* 83:3787–3794.

Hernandez-Navarro F, Ojeda E, Arrieta R, et al. 1995. Single-centre experience of PB stem cell transplantation using cryopreservation by immersion in a methanol bath. *BM Transplant* 16:71–77.

Horowitz MM and Rowlings PA. 1997. An update from the International BM Transplant Registry and the Autologous Blood and Marrow Transplant Registry on current activity in hematopoietic stem cell transplantation. *Current Opinion in Hematology* 4:395–400.

Keeney M, Chin-Yee I, Weir K, et al. 1998. Single platform flow cytometric absolute CD34+ cell counts based on the ISHAGE guidelines. International Society of Hematotherapy and Graft Engineering. *Cytometry* 34:61–70.

Korbling M and Anderlini P. 2001. PB stem cell versus BM allotransplantation: does the source of hematopoietic stem cells matter? *Blood* 98:2900–2908.

Lebkowski JLS and Harvey M. 1994. Isolation and culture of human CD34+ hematopoietic stem cells using AIS CELLectors. *Hematopoietic stem cells—the Mulhouse* manual. E. Wunder, H. Sovalat, PR Henon, and S Serker. Dayton:Alpha Med Press.

Mauch P, Lamont C, Neben TY, et al. 1995. Hematopoietic stem cells in the blood after stem cell factor and interleukin-11 administration: evidence for different mechanisms of mobilization. *Blood* 86:4674–4680.

Ottinger H, Beelen D, Scheulen B, et al. 1996. Improved immune reconstitution after allotransplantation of PB stem cells instead of BM. *Blood* 88:2775–2779.

Prince H, Bashford J, A3 DW, et al. 2002. Isolex 300i CD34-selected cells to support multiple cycles of high-dose therapy. *Cytotherapy* 4:137–145.

Roberts MM, To LB, Gillis D, et al. 1993. Immune reconstitution following PB stem cell transplantation, autologous BM transplantation and allogeneic BM transplantation. *BM Transplant* 12:469–475.

Rowley SD and Davis JM. 1991. The use of 4-HC in autologous purging. *BM processing and purging: a practical* guide. A. P. Gee. Boca Raton: CRC Press.

Rowley SD. 1992. Hematopoietic stem cell cryopreservation: a review of current techniques. *J Hematother* 1:233–250.

Russell NH and Miflin G. 1998. PB stem cells for autologous and allogeneic transplantation. *The clinical practice of stem-cell* transplantation. J. Barrett and J. Treleaven. Oxford: Isis Medical Media.

Salmon JP, Michaux S, Hermanne JP, et al. 1999. Delayed massive immune hemolysis mediated by minor ABO incompatibility after allogeneic PB progenitor cell transplantation. *Transfusion* 39:827.

Schmitz N, Linch DC, Dreger P, et al. 1996. Randomised trial of filgrastim-mobilised PB progenitor cell transplantation versus autologous bone-marrow transplantation in lymphoma patients [see comments] [published erratum appears in Lancet 1996 Mar 30; 347(9005):914]. *Lancet* 347:353–357.

Shpall EJ, LeMaistre CF, Holland K, et al. 1997. A prospective randomized trial of buffy coat versus CD34-selected autologous BM support in high-risk breast cancer patients receiving high-dose chemotherapy. *Blood* 90:4313–4320.

Stadtmauer EA, Schneider CJ, and Silberstein LE. 1995. PB progenitor cell generation and harvesting. *Seminars in Oncology* 22:291–300.

Stiff PJ. 1991. Simplified BM cryopreservation using dimethysulfoxide and hydroxyethylstarch as cryoprotectants. *BM processing and purging: a practical* guide. A. P. Gee. Boca Raton: CRC Press.

Storek J, Gooley T, Siadak M, et al. 1997. Allogeneic PB stem cell transplantation may be associated with a high risk of chronic graft-versus-host disease [see comments]. *Blood* 90:4705–4709.

To LB, Roberts MM, Haylock DN, et al. 1992. Comparison of haematological recovery times and supportive care requirements of autologous recovery phase PB stem cell transplants, autologous BM transplants and allogeneic BM transplants. *BM Transplant* 9:277–284.

Uckun FM and Myers DE. 1993. Allograft and autograft purging using immunotoxins in clinical BM transplantation for hematologic malignancies. *J Hematother* 2:155–163.

Watanabe T, Kajiume T, Abe T, et al. 2000. Allogeneic PB stem cell transplantation in children with hematologic malignancies from HLA-matched siblings. *Medical & Pediatric Oncology.* 34:171–176.

Weaver CH, Hazelton B, Birch R, et al. 1995. An analysis of engraftment kinetics as a function of the CD34 content of PB progenitor cell collections in 692 patients after the administration of myeloablative chemotherapy. *Blood* 86:3961–3969.

Zambelli A, Poggi G, Da Prada G, et al. 1998. Clinical toxicity of cryopreserved circulating progenitor cells infusion. *Anticancer Res* 18:4705–4708.

25

Umbilical Cord Blood Stem Cells

LAURA C. BOWMAN, MD, MICHAEL A. BRIONES, DO, AND ANN E. HAIGHT, MD

INTRODUCTION

Bone marrow transplantation with allogeneic hematopoietic stem cells has been successful therapy for a variety of malignant and nonmalignant diseases and has demonstrated great promise in recent decades. However, wider use of this therapy has been limited by the paucity of suitable human leukocyte antigen (HLA)-matched donors, lengthy volunteer bone marrow procurement process, and the morbidity and mortality due to severe graft-versus-host disease (GVHD), especially when unrelated donors are used. Use of umbilical cord blood (UCB) as the graft source is one strategy to address these limitations. The first report of using human UCB for transplantation in patients with malignant diseases was published by Ende et al. in 1972, though there were questions of true marrow reconstitution in these patients. The first related-donor UCB transplantation was reported in 1989 (Gluckman et al. 1989), and the first unrelated-donor UCB transplantation in 1993. To date, UCB transplantation has been used to treat a wide variety of malignant and nonmalignant disorders (Table 25.1). When compared to other graft sources, UCB has distinct advantages and disadvantages described later in this chapter.

COLLECTION, STORAGE, AND HANDLING

The first cord blood bank was established at the Indiana University School of Medicine by Broxmeyer et al., and the subsequent first UCB transplant was performed using units from this bank (Gluckman et al. 1989; Broxmeyer et al. 1989). Public cord blood banks have been established in New York, Los Angeles, Durham (North Carolina), as well as others in the United States. Eurocord is an international registry operating on behalf of the European Blood and Marrow Transplant Group and participation is open to European and nonEuropean centers conducting UCB transplants. Moreover, there are private, for-profit cord blood banks that have been established to provide autologous and sibling-directed cord blood transplantation to be used in the future.

Collection Process

After delivery of the infant, the cord is ligated before clamping of the umbilicus to prevent crushing of the tissue. Then, the umbilicus is cut between the clamp and the ligation, and the cord is wiped with alcohol, followed with betadine to ensure sterility of the harvested blood (Bertolini et al. 1995; Wagner et al. 1992; Harris et al. 1994). Collection of human UCB before the third stage of labor is ended, while there are still uterine contractions, permits the recovery of an additional 80 to 160 mL of blood. The procedure of cord blood collection is simple and does not endanger the mother or fetus. Yet, in order to collect the highest amount of sterile cord blood, expertise, patience, and attention to detail are required of the obstetrical team. Some of the methods of UCB collection are described in Table 25.2. The risk of contamination of the harvested blood (either with maternal blood or with microorganisms) is reduced when a closed system is used, thus the closed or semi-closed systems should be used. Collection of cord blood

TABLE 25.1   Diseases Treated with Umbilical Cord
            Blood Transplantation

Malignant Disorders	Nonmalignant Disorders
Acute leukemias (lymphoid and myeloid)	Bone marrow failure syndromes
Chronic myelogenous leukemia	Severe aplastic anemia
Myelodysplastic syndromes	Fanconi's anemia
Non-Hodgkins lymphoma	Blackfan-Diamond syndrome
Neuroblastoma	Dyskeratosis congenita
	Kostmann's syndrome
	Chronic granulomatous disease
	Hemoglobinopathies
	Sickle cell anemia
	Thalassemia
	Immunodeficiency syndromes
	Severe combined immunodef.
	Leukocyte adhesion defect
	Wiskott-Aldrich syndrome
	Chronic granulomatous disease
	Inborn errors of metabolism
	Storage diseases
	Histiocytic disorders

TABLE 25.2   Methods of Umbilical Cord Blood Collection

Method	Depiction
Open technique Syringe/flush/drain	While the placenta is still in utero, the blood is aspirated from the umbilcal vein using 60 mL syringes. It is then followed by injection of heparin and saline into the delivered placenta and withdrawl of additional product by a syringe. Then the end of the cord is cut and the blood is drained in a sterile container.
Closed technique Blood bag	The placenta is delivered and the umbilical vein is cannulated using a stanard blood donor set.
Semiclosed technique Syringe	While the placenta is still in utero, the blood is aspirated from the umbilcal vein using 60 mL syringes.
Syringe/Flush	The placenta is still in utero, the blood is aspirated from the umbilcal vein using 60 mL syringes. It is then followed by injection of heparin and saline into the delived placenta and withdrawl of additional product by a syringe.
Syringe/Flush/ Blood Bag	Same as the syringe/flush method with the addition of the umbilical vein being cannulated with a standard blood donor set.

by the "syringe only" method, as is standardly used in clinical practice, results in volumes of an average of 75 mL. However, several other techniques have routinely resulted in the collection of volumes of placental/UCB that are several times greater (Bertolini et al. 1995; Wagner et al. 1992; Harris et al. 1994; Broxmeyer et al. 1991). The collection of blood is performed by first withdrawing as much blood as possible by syringe while the placenta is still in utero, followed by heparin flush and cannulation of the delivered placenta with a standard blood donor set (blood bag) and has resulted in the greatest amounts of blood being obtained (range 120 to 220 mL, mean 163) (Wagner et al. 1992). The volume of umbilical cord blood that can be harvested is 42 to 240 mL (median 103 mL). Needle aspiration of the placental vein after placental delivery can produce an additional 8 to 85 mL (median 31 mL) (Harris et al. 1994). With respect to volume of cord blood collected when comparing cesarean section versus vaginal deliveries, cesarean may allow significantly more cord blood volume collected and CD34+ cells compared to vaginal deliveries (Yamada et al. 2000). Both ACD (acid, citrate, dextrose) or CPD (citrate, phosphate, dextrose) can be used as anticoagulants. CPD is preferable as it is less affected by variations in the volume of the collected blood. The average number of mononuclear cells per mL of blood is similar in all of these methods. Thus, when more blood is harvested, more cells are potentially available for transplantation.

Separation and Cryopreservation

Separation and isolation of the mononuclear cell population is usually performed before cryopreservation. UCB may be cryopreserved without additional processing; however, this retains a large volume of cells not involved in engraftment, such as granulocytes and red blood cells (RBCs). Thawed granulocytes and RBCs are predominately lysed, which may cause adverse effects in the recipient as well as risk of ABO incompatibility. There are two general methods of separating cord blood into mononuclear cells. The first method uses standard density gradient separation using Ficoll-Hypaque, in which the visible mononuclear cell interface is harvested, washed, and counted. With the second method, the cord blood cells are centrifuged over standard Ficoll-Hypaque density gradients followed by plasma being pipetted off, and the interface as well as the entire gradient down to the RBC pellet is collected, then the cells are diluted and centrifuged to a pellet. The cells are then centrifuged again over a Ficoll-Hypaque density gradient and harvested. This method was found to completely deplete the samples of mature as well as nucleated RBCs and results in maximal mononuclear cell recovery compared with the single Ficoll-Hypaque density gradient separation (Harris et al. 1994).

In general UCB mononuclear cells are cryopreserved using an automated, microprocessor-controlled cell freezer. Cells are stored in flat plastic bags designed to maximize heat transfer and to withstand the cryoreservation and thawing procedures. Small volume cryovials and bags of the product are also cryopreserved and used for testing. The cells are resuspended in freezer media at densities of 5 to 50×10^6 cells/mL in one to two mL cryovials for example. Equal volumes of cryopreservative solution containing dimethyl sulfoxide (DMSO), which is a cryoprotectant, are added. The final concentration of DMSO is between 5% and 10%. Programmed or rate-controlled freezing is the standard for clinical cell cryopreservation. This procedure is as follows: 1°C/min cooling down to −4°C followed by a rapid drop to −40°C, a 1°C/min drop to −45°C, and a 10°C/min drop to −90°C. Another method that does not utilize rate-controlled freezing (Hernandez-Navarro et al. 1998) is one in which the DMSO-treated cells are cooled on ice and placed into a methanol bath and then transferred to a −80°C freezer for two hours. With either procedures the cells are then stored in a liquid nitrogen freezer in the liquid phase or the liquid/gas interphase. The maximum storage time of cryopreserved cells is unknown, but there have been reports as long as 10 years demonstrating minimal effect on cell viability, cellular composition of the UCB, and progenitor/stem cell capacity (Broxmeyer and Cooper 1997).

Histocompatibility and Infectious Agent Testing

Cord blood units undergo HLA typing using serological technique for class I antigens and molecular, DNA-based HLA typing for class II alleles. The cord blood unit is also tested for the volume of the product, nucleated cell count, percent CD34+ cells, ABO-Rh type, and bacterial cultures. Infectious disease testing of the cord blood is performed in accordance with the American Association of Blood Banks (AABB), American Society of Blood and Marrow Transplantation (ASBMT) and specifically includes hepatitis B and C, human immunodeficiency virus (HIV), human T-cell lymphotrophic virus (HTLV), cytomegalovirus (CMV), and syphilis.

Thawing and Infusion Procedure

Cryopreserved UCB units are thawed before infusion in a 37°C water bath with mild shaking. It is important not to bend the bag during the thawing process, thus preventing possible breakage of the bag secondary to the plastic being brittle from the freezing. Recovery of viable cells from cryopreserved products can be significantly improved by diluting to restore the osmolarity of the suspension and removing the supernatant containing DMSO (Rubinstein et al. 1998). The cells are washed and resuspended in an infusion solution. The infusion product is routinely tested for white blood cell (WBC) count, total nucleated cell count, and cell recovery, as well as viability of the cells, ABO/Rh type, bacterial and fungal cultures. Cells should be infused no longer than 60 minutes from thawing and washing of the cells. The product is infused by slow intravenous push or by gravity. Leukodepletion filters should not be used as these would remove the cells. DMSO commonly causes nausea, histamine release, and cardiac effects including hypertension, bradycardia, and arrythmias; thus patients are closely monitored and routinely premedicated before the transfusion with antihistamine, antipyretic, and antiemetic. Alkalinization, mannitol, and furosemide may also be used to minimize the renal impact of DMSO-induced hemolysis.

Ex Vivo Expansion of Cord Blood Stem Cells

A limiting factor of cord blood transplantation is the small numbers of cord blood stem cells that can be obtained from each collection and the dose-limiting number of cells. This limits the use of UCB as a graft source in larger children and adults. Thus, the potential for ex vivo expansion of UCB-derived stem cells to increase graft size is of significant importance. Ex vivo expansion may permit a single cord blood unit to be expanded and used for multiple transplant recipients and may help reduce the duration of cytopenias. Several studies have shown that primitive cord blood cell can be expanded (Moore and Hopkins 1994; Koller et al. 1998; McNiece et al. 2000; Lewis et al. 2001). These studies, however, point to a greater expansion of the more mature progenitor cells than the most immature or stem cells in this primative population. There is lack of a quantitative assay to measure these stem cells which makes ex vivo expansion that much more difficult. Currently, there are laboratories using ex vivo expanded and nongrowth factor-manipulated cells in the clinical setting.

CLINICAL APPLICATIONS

Advantages of UCB Transplantation

Without physical risk to the donor and with low rates of contamination by herpes family viruses, UCB provides a rich source of hematopoietic progenitor cells with great proliferative capacity in a small volume, allowing for successful engraftment with an average of one-log fewer total nucleated cells (TNC) than with the use of bone marrow. Wider HLA disparity is tolerable without increases in GVHD, such that 0-, 1-, or 2-antigen mismatches are commonly accepted.

This flexibility, combined with the public banking of donated UCB boasting a broader racial and ethnic representation and lack of the donor attrition that complicates unrelated volunteer donation, serve to significantly expand the unrelated donor pool in a very meaningful way. Worldwide, there are currently approximately 70,000 cryopreserved HLA-A, -B, and -DRB1 typed units available for UCB transplantation (Wagner 2003). Further enhancing rates of successful procurement is the ease of international transfer. Such exchange is facilitated by Netcord, an international cooperative group of UCB banks, which uses detailed banking standards.

When compared to other unrelated graft sources, procurement time is dramatically compressed, from a median of four months down to a matter of weeks. Such time-savings is crucial to patients with high-risk hematological malignancies whose remission and clinical status is often tenuous, as well as children with inherited diseases that may be rapidly progressive.

Disadvantages of UCB Transplantation

There are several important disadvantages of UCB. Successful engraftment with UCB depends importantly on cell dose (TNC per kilogram of recipient body weight), generally limiting UCB transplantation to smaller patients. Unless the UCB graft is from a family member, additional donor cells cannot be obtained for treatment of graft failure, relapse, posttransplant lymphoproliferative disease (PTLD), or other complications. This is an important consideration for chronic myelogenous leukemia, where donor lymphocyte infusion (DLI) is a well-established treatment for posttransplant relapse.

Other disadvantages include the uncertain existence of a graft-versus-leukemia (GVL) effect and the relatively limited data available on the durability of the UCB graft given the relative youth of the UCB transplantation field.

Clinical Trials of UCB Transplantation

Engraftment

Despite intial concern about engraftment potential of UCB when compared to marrow or PBSC, successful engraftment occurs in approximately 80% to 90% of unrelated UCB recipients (Rocha 2000; Rocha 2001; Rubinstein 1998). Though favorable, these rates are lower than the 98% matched sibling and 90% to 96% unrelated marrow donor engraftment rates found in comparative studies (Rocha 2000; Rocha 2001). Engraftment does occur more slowly than with other graft sources for pediatric patients despite growth factor support, with a median 26 to 27 days to neutrophil engraftment (Rocha 2000; Thomson 2000) and more significant delays in platelet transfusion independence (75 days to untranfused count of 50,000) (Thomson 2000).

The single most important factor impacting on time to engraftment is the total nucleated cell content relative to recipient body weight (Gluckman 1997; Locatelli 1999; Rubinstein 1998; Rocha 2001; Wagner 2002). Though degree of HLA-mismatch impacts significantly as well (Rubinstein 1998; Gluckman 1997), a higher cell dose has the potential to at least partially overcome the negative impact of HLA-mismatch (Wagner 2002). For cord selection, minimum threshold doses of 1.5×10^7 TNC/kg or 1.7×10^5 CD 34+ cells/kg has been suggested (Wagner 2002; Gluckman 2001), though higher doses are optimal.

Though UCB transplant preparative regimens may vary, current common standard regimens include stopped here (1) hyperfractionated total body irradiation, high-dose cyclophosphamide, and equine antithymocyte globulin (ATG), and (2) busulfan, melphalan, and ATG, with cyclosporine A and prednisone or low-dose methotrexate for GVHD prophylaxis.

GVHD

Eurocord and International Bone Marrow Transplant Registry (IBMTR) pediatric studies have demonstrated lower incidences of acute and chronic GVHD using UCB when retrospectively compared to marrow grafts for both related and unrelated donors when adjusting for other factors affecting GVHD risks (Rocha 2000; Rocha 2001). This phenomenon may be in part a result of reduced numbers of T cells in the UCB graft but is likely also due to dampened alloreactive potential of UCB progenitor cells.

Epstein-Barr virus (EBV)-associated PTLD

The naivete of neonatal lympocytes and the low dose of infused donor T cells in UCB transplantation raises theoretical concern for EBV-associated PTLD. However, a recent study of 272 UCB transplantation recipients found a 2% incidence of PTLD, which is comparable to unmanipulated, unrelated marrow donor transplantation and lower than that seen in T cell-depleted marrow transplants (Barker 2001).

Survival

Though prospective studies are ongoing, comparative studies of stem cell sources in children are primarily retrospective to date and difficult to control for the potential bias of variables that may impact on choice of graft source. The most important factor impacting on survival appears to be CD34+ cell dose. A 2002 study of 102 UCB transplant recipients concluded that CD34+ cell dose should be used to select UCB grafts when multiple potential UCB donor units exist with an HLA disparity of two or less antigens (Wagner 2002Blood). Not all banked UCB units have CD34+ data available, however, so this factor may not be uniformly used in graft selection.

Family and Autologous UCB

The private UCB industry has grown considerably in recent years, providing collection and cryopreservation of UCB units for families who pay a fee for storage, as well as a fee for release of the product. While there are a number of theoretical uses for autologous UCB, there are few current practical indications, such that the cost-to-benefit ratio does not favor private banking for families who do not have an identified member with a disease amenable to transplantation. Of note is the 1997 development of an NHLBI-funded sibling UCB bank at the Children's Hospital Oakland, which has now stored over 500 units for families with hemoglobinopathies, immunodeficiencies, malignancies, and inborn errors of metabolism at no cost to the families (Reed 2003).

References

Barker JN, Davies SM, DeFor T, et al. 2001. Survival after transplantation of unrelated donor umbilical cord blood is comparable to that of human leukocyte antigen-matched unrelated donor bone marrow: results of a matched-pair analysis. *Blood* 97:2957–2961.

Barker JN, Martin PL, Coad JE, DeFor T, et al. 2001. Low incidence of Epstein-Barr virus-associted posttransplatnation lymphoproliferative disorders in 272 unrelated-donor umbilical cord blood transplant recipients. *Biol Blood Marrow Transplant* 7:395–399.

Bertolini F, Lazzuri L, Lacri E, et al. 1995. Comparative study of different procedures for the collection and banking of umbilical cord blood. *J Hematother* 4(1):29–36.

Broxmeyer HE, Douglas GW, Hungar G, et al. 1989. Human umbilical cord blood as a potential source of transplantable hematopoietic stem/progenitor cells. *Proc. Natl Acad Sci Amer USA* 86:3828–3832.

Broxmeyer HE and Cooper S. 1997. High efficiency recovery of immature haematopoietic progenitor cells with extensive proliferative capacity from human cord blood cryopreserved for 10 years. *Clin Exp Immunol* 107(1):45–53.

Broxmeyer HE, Kurtzburg J, Gluckman E, et al. 1991. Umbilical cord blood hematopoietic stem and repopulating cells in humans clinical transplantation. *Blood cells* 17:313–329.

Ende M and Emile N. 1972. Hematopoietic transplantation by mean of fetal (cord): a new method. *Virginia Med Mar* E9:276.

Gluckman E. 2001. Hematopoietic stem-cell transplants using umbilical-cord blood. *N Engl J Med* 344:1860–1861.

Gluckman E, Broxmeyer HE, Auerbach SO, et al. 1989. Hematopoietic reconstitution in a patient with Fanconi's Anemia by mean of umbilical cord blood from an HLA identical siblings. *New Engl J Med* 381:174–178.

Gluckman E, Rocha V, Boyer-Chammard A, et al. 1997. Outcome of cord-blood transplantation from related and unrelated donors. *N Engl J Med* 337:373–381.

Harris DT, Schumacher MJ, Rychlik S, et al. 1994. Collection, separation and cryopreservation of umbilical cord blood for use in transplantation bone marrow. *Transplantation* 13:135–143.

Hernandez-Navarro F, Ojuda E, Arieta R, et al. 1998. Hematopoietic cell transplantation using plasma and DMSO without HES, with non-programmed freezing by immersion in a methanol bath. Results in 2/3 cases. *Bone Marrow Transplantation* 21(5):571–617.

Koller MR, Manchel I, Maher RJ, et al. 1998. Clinical scale human umbilical cord blood cell expansion in a novel automated perfusion culture system. *Bone Marrow Transplant* 21(7):653–663.

Lewis ID, Almeida-Porada G, Du J, et al. 2001. Umbilical cord blood cell capable of engrafting in primary, secondary, and tertiary xenogeneic hosts are preserved after ex vivo culture in a noncontact system. *Blood* 97:3441–3449.

Locatelli F, Rocha V, Chastang C, et al. 1999. Factors associated with outcome after cord blood transplantation in children with acute leumekia. Eurocord-Cord Blood Transplant Group. *Blood* 93:3662–3671.

McNiece I, Kubegov D, Kerzic P, et al. 2000. Increased expansion and differentiation of cord blood products using a two-step expansion culture. *Exp Hematol* 28:1181–1186.

Moore MAS and Hopkins I. 1994. Ex vivo expansions of cord blood derived stem cells and progenitors. *Blood Cells* 20:468–481.

Reed W, Smith R, Dekovic F, et al. 2003. Comprehensive banking of sibling donor cord blood for children with malignant and nonmalignant disease. *Blood* 101:351–357.

Rocha V, Cornish J, Sievers EL, et al. 2001. Comparison of outcomes of unrelated bone marrow and umbilical cord blood transplants in children with acute leukemia. *Blood* 97:2962–2971.

Rocha V, Wagner JE Jr, Sobocinski KA, et al. 2000. Graft-versus-host disease in children who have received a cord-blood or bone marrow transplant from an HLA-identical sibling. Eurocord and International Bone Marrow Transplant Registry Working Committee on Alternative Donor and Stem Cell Sources. *N Engl J Med* 342:1846–1854.

Rubinstein P, Carrier C, Scaradavou A, et al. 1998. Outcomes among 562 recipients of placental-blood transplants from unrelated donors. *N Engl J Med* 339:1565–1577.

Thomson BG, Robertson KA, Gowan D, et al. 2000. Analysis of engraftment, graft-versus-host disease, and immune recovery following unrelated donor cord blood transplantation. *Blood* 96:2703–2711.

Wagner JE, Barker JN, DeFor TE, Baker S, et al. 2003. Transplantation of unrelated donor umbilical cord blood in 102 patients with malignant and nonmalignant diseases; influence of CD34 cell dose and HLA disparity on treatment-related mortality and survival. *Blood* 100:1611–1618.

Wagner WE, Broxmeyer HE, and Cooper S. 1992. Umbilical cord and placental blood hematopoietic stem cells collection cryo preservation and storage. *J Hematotner* 1:167–174.

Yamada T, Okamoto Y, Kasamatsu H, et al. 2000. Factors affecting the volume of umbilical cord blood collections. *Acta Obstet Gynecol Scand* 19:830–833.

C H A P T E R

26

Transfusion Reactions

ANNE F. EDER, MD, PhD

INTRODUCTION

Adverse reactions to blood component transfusion range from brief episodes of fever to life-threatening hemolysis. The clinical challenge lies in recognizing these reactions, especially since the early signs and symptoms, such as fever and chills, may herald either benign febrile reactions or potentially lethal ABO incompatibility. Consequently, all transfusions should be carefully monitored, and adverse reactions to blood components should be appropriately investigated. By definition, acute reactions occur during transfusion or within 24 hours; delayed reactions, after at least 24 hours.

The further classification of reactions to blood components among children is the same as in adults, but important differences between these patient populations emerge with respect to the underlying predisposition, clinical presentation, and frequency of transfusion reactions. All transfusion recipients regardless of age may experience immune-mediated hemolytic reactions, febrile nonhemolytic transfusion reactions, or allergic reactions. Newborn infants, however, are much less likely than children and adults to experience acute allergic or febrile reactions and rarely become alloimmunized to red cells because of their immature immune system; but they are more susceptible to developing cytopenias from passively transferred antibodies in blood components. Young children also appear less likely than adults to have acute febrile or allergic reactions to blood component therapy, and the risk of alloimmunization is inversely related to the age at which children receive their first red cell transfusion. All transfusion recipients are at risk of acute immunologic

complications from incompatible transfusion, but the clinical presentation of acute hemolytic transfusion reactions may differ in infants and children compared to adults.

ACUTE HEMOLYTIC TRANSFUSION REACTIONS

Acute hemolytic transfusion reactions (AHTRs) are caused by the immune-mediated destruction of transfused red cells, with the most serious reactions occurring in the setting of inadvertent ABO incompatible red cell transfusion. Red cells from group A, B, or AB individuals are rapidly destroyed upon transfusion to individuals who lack these carbohydrate antigens and express the corresponding "naturally-occurring" ABO isohemagglutinins, anti-A, anti-B, or anti-A,B. ABO incompatible transfusions account for more than half of all transfusion-related deaths and usually result from the failure to properly identify the intended transfusion recipient, either at the time of initial phlebotomy for pretransfusion testing or before administering the red cell unit (Linden et al. 2000; Williamson et al. 1999).

On occasion, IgM antibodies other than anti-A and anti-B or complement-fixing IgG alloantibodies in the recipient cause AHTRs, such as anti-P[k] and anti-Vel and rarely, Lewis (anti-Le[a]), Kidd (anti-Jk[a], anti-Jk[b]), and Kell (anti-K1) specificities (Ried and Lomas-Francis 1997). Alternatively, acute hemolysis can result from passively-transferred antibodies in plasma or plasma derivatives that cause destruction of the recipient's red cells or other nonimmune mechanisms.

Incidence

The incidence of ABO incompatible transfusions is not known but has been estimated to occur in one in 38,000 to one in 70,000 red cell transfusions (Linden et al. 2000; Williamson et al. 1999). These estimates are derived from voluntary or mandated reporting of transfusion errors or surveillance strategies and may underestimate the frequency of ABO incompatibility due to a failure to either recognize or report the transfusion reaction. By all current measures, the risk of receiving an ABO-incompatible unit (1:38,000 to 70,000) exceeds the risk of receiving a unit potentially infectious for HIV or HCV (less than 1:1,000,000) (Dodd et al. 2002). AHTRs are the leading cause of transfusion-related mortality, implicated in at least 18 deaths per year in the United States (Sazama 1990). In contrast, only three transfusion-transmitted HIV infections have been documented in the first three years following implementation of nucleic acid testing.

Pathophysiology

Anti-A and anti-B antibodies occur without prior red cell exposure in individuals who lack the corresponding carbohydrate antigens. In group A and B individuals, these antibodies are predominantly IgM; whereas, IgG accounts for a greater proportion of the anti-A and anti-B reactivity than IgM in group O individuals (Brecher 2002). Upon exposure to incompatible red cells, anti-A and anti-B antibodies immediately bind to the corresponding red cell antigen, forming immune complexes and initiating complement activation, coagulation cascades, and neuroendocrine and systemic inflammatory responses. Complement activation triggers the interrelated coagulation and kinin cascades via Hageman factor and fixes the terminal pore complex, C5–9, on the red cell membrane, causing osmotic lysis of circulating red cells. This intravascular hemolysis produces the hallmark features of ABO incompatible transfusion, hemoglobinemia and hemoglobinuria. The released complement fragments, C3a and C5a, are anaphylotoxins that activate white cells to release histamine and other vasoactive amines and inflammatory cytokines, particularly interleukin (IL)-1, IL-6, IL-8, and tumor necrosis factor alpha (TNF-α). The actions of anaphylotoxins, inflammatory cytokines, histamines, bradykinin, and vasoactive amines produce fever, hypotension, wheezing, chest pain, nausea, and vomiting. Disseminated intravascular coagulation (DIC) may result from the activation of the coagulation and fibrinolytic systems. Intravascular hemolysis and hypotension induce a compensatory sympathetic nervous system response, leading to renal, splanchnic, and cutaneous vasoconstriction, which may ultimately result in shock and circulatory collapse. Impairment of renal function results from renal ischemia, hypotension, antigen-antibody complex deposition, and thrombosis in renal vessels, and these physiological insults may ultimately culminate in acute tubular necrosis and renal failure.

Diagnosis

Clinical Evaluation

In the event of an acute transfusion reaction, the transfusion should be immediately discontinued, and the intravenous (IV) line maintained for vigorous hydration and fluid resuscitation, if necessary (Table 26.1). The patient should be evaluated, monitored, and given immediate supportive care. If the patient received more than one red cell unit, all units are suspect not just the unit infusing at the time the symptoms were noted. The bedside check of the patient's identifying information and unit labeling should be repeated, and

TABLE 26.1   Initial Approach to Acute Transfusion Reactions

Clinical Service

Stop the transfusion and maintain intravenous access with normal saline (0.9% sodium chloride).
Evaluate the patient for evidence of a severe reaction and carefully monitor symptoms, vital signs, and urine output.
Provide supportive medical care to the patient, if necessary.
Repeat bedside check of the patient's identification wrist band against component labels. Return unit(s) to blood bank along with patient's posttransfusion blood samples and completed transfusion record.

Blood Bank

Clerical check of unit, requisition, computer records.
DAT (purple-top, EDTA tube) on posttransfusion specimen.
Visual inspection of posttransfusion plasma for hemolysis.
Additional testing on the pre- and posttransfusion specimens, if acute hemolysis is suspected:
  Repeat ABO and Rh testing
  Repeat crossmatch with implicated unit(s)
  Repeat antibody screening tests
  Repeat DAT and antibody screen on additional specimens obtained at intervals after the transfusion reaction

Clinical Service and Blood Bank

Review results of blood bank evaluation before subsequent transfusions.
Investigate the possible risk to other patients who may have been affected by patient identification error.

the units should be returned to the blood bank for further investigation.

Fever, defined as a 1°C (1.8°F) increase above normal body temperature, is the most common sign of an acute hemolytic transfusion reaction and may be accompanied by chills; pain at the infusion site or in the chest, abdomen, or lower back; dyspnea; and hypotension. The patient may become intensely uncomfortable and extremely apprehensive. Children may be less able to verbalize their feelings and explain their symptoms but may suddenly develop overwhelming anxiety and complain of discomfort or pain. Infants and anesthetized patients cannot report symptoms, and the initial signs of an AHTR may be generalized oozing from intravenous sites, uncontrollable bleeding, or laboratory signs of DIC, such as hypofibrinogenemia, thrombocytopenia, and the presence of fibrin degradation products and D-dimers. Circulatory collapse and renal failure may ensue.

The severity of the reaction to ABO-incompatible transfusion is extremely variable and usually reflects the rate and volume administered. Severe symptoms may be evident after transfusing only 10 to 15 mL incompatible red cells, and as little as 30 mL has been fatal (Sazama 1990). Nearly half (47%) of the recipients of incompatible red cells, however, suffer no ill effects even after receiving a full unit; 41% result in AHTRs, and only 2% result in death (Linden et al. 2000). The risk of mortality increases if more than one incompatible unit is transfused (Brecher 2002).

Laboratory Evaluation

In all acute transfusion reactions, the blood bank should check for clerical errors, visible hemolysis, and serologic incompatibility as soon as possible after stopping the transfusion (Table 26.1). Additional laboratory testing may be required for further clinical investigation of hemolysis.

Clerical Check

The blood bank must first rule out medical error, such as the possibility that a patient received a unit of blood intended for a different recipient. If the clerical check implicates a patient identification or issuing error, the possibility that other patients are at risk due to a blood component switch or specimen labeling mix-up must be addressed immediately. For example, samples from new patients received in the blood bank from the same location in the hospital (for example, the emergency room) or at the same approximate time should be redrawn, blood components issued at the same time or to the same location in the hospital should be recalled, and blood components stored in refrigerators outside the blood bank should be immediately inventoried.

After the acute crisis, the circumstances surrounding the erroneous transfusion should be investigated in a root cause analysis, and corrective action taken as appropriate.

Visual Inspection

Visual inspection of the patient's plasma after transfusion and comparison to a pretransfusion specimen is a sensitive method to detect intravascular hemolysis because destruction of as little as 5 mL of transfused red cells results in reddish discoloration indicative of hemoglobinemia. A second posttransfusion specimen may be required to rule out the possibility of artifactual hemolysis resulting from poor collection technique. Myoglobinemia, seen in severe trauma, burns, or other muscle injury, may also cause pink or red-tinged plasma unrelated to acute hemolysis. Icteric plasma (hyperbilirubinemia) either suggests a hemolytic process that has been ongoing for several hours or may be a sign of coincidental liver disease or unrelated chronic hemolysis due to congenital hemolytic anemias.

Direct Antiglobulin Test

The direct antiglobulin test (DAT; direct Coombs' test) detects antibody or complement binding to circulating red cells. Pineda reported positive DATs in 41 of 47 (89%) hemolytic transfusion reactions (Pineda et al. 1978). Negative DAT results in this setting likely reflect complete destruction and clearance of transfused, incompatible cells and are accompanied by telltale hemoglobinemia or hemoglobinuria in most cases of ABO incompatibility. Additional serum laboratory findings consistent with intravascular hemolysis include increased lactate dehydrogenase, decreased haptoglobin, and increased bilirubin, several hours into the reaction.

If clerical discrepancies are found or any of the tests suggest serologic incompatibility, the ABO/Rh typing and antibody screen of the patient should be determined on a new specimen. The retained segments of the transfused units should be crossmatched against the new, posttransfusion specimen. If the investigation rules out the possibility of an incompatible transfusion, alternative etiologies of acute intravascular hemolysis should be explored.

Differential Diagnosis

Apart from incompatible blood transfusion, acute intravascular hemolysis may be caused by other immune and nonimmune mechanisms (Table 26.2). Nonimmune hemolysis results from physical damage to transfused red cells due to improper handling or administration. Exposure to excessive temperatures may

TABLE 26.2. Differential Diagnosis of Acute
Intravascular Hemolysis

Immune-Mediated

Acute hemolytic transfusion reaction with incompatible blood
    transfusion
Autoimmune hemolytic anemia
T-activation
Bystander hemolysis in sickle cell disease patients
Congenital hemolytic anemia, for example, glucose 6-phosphate
    dehydrogenase deficiency

Nonimmune

Osmotic lysis, due to addition of the following solutions to the unit
    or infused in the same intravenous line:
    Hypotonic solutions
    Calcium-containing solutions
    Drugs
Mechanical destruction:
    Artificial heart valves
    Extracorporeal circulation (ECMO, cardiac bypass, apheresis,
        dialysis)
    Mechanical pumps
    Small-bore IV catheters (≥24 gauge), rapid transfusion
Thermal damage:
    Malfunctioning blood warmers
    Inappropriate use of radiant heaters in incubators or microwaves
Bacterial contamination of the unit
Infections

occur with malfunctioning blood warmers, or the improper warming of blood with the use of the radiant incubator heaters or microwaves. Incompatible IV fluids, such as hypotonic saline, 5% dextrose in water (D5W), water, or calcium-containing (Ringer's lactated) saline, or drugs, added either directly to the unit or infused through the same IV line, may cause osmotic lysis or aggregation and subsequent physical destruction of transfused red cells. Mechanical hemolysis may be caused by rapid transfusion through small bore needles (less than 24 gauge), by roller pumps, such as those used in cardiac bypass surgery, or by pressure infusion cuffs. In premature infants, blood can be transfused safely through 24-gauge catheters if transfused slowly, but more rapid administration through these lines may produce hemolysis. Incorrect preparation of frozen red cells and inadequate deglycerolization may cause the cells to lyse after transfusion. Finally, bacterial growth in a red cell unit may cause hemolysis of the unit before transfusion.

Clinical conditions associated with acute intravascular hemolysis that may confound investigation of a possible AHTR include underlying immune-mediated hemolysis as occurs with autoimmune hemolytic anemia, congenital hemolytic anemia, and drug-induced

hemolytic anemia. In sickle cell patients, sickle cell crises and potentially fatal posttransfusion hemolysis may develop after red cell transfusion. The cephalosporin antibiotics, in particular ceftriaxone and cefotetan, can cause immune-mediated hemolytic anemia, which may be clinically indistinguishable from an AHTR (Brecher 2002). Intravascular hemolysis may occur in patients with red cell T activation and polyagglutination, independent of transfusion or subsequent to receiving plasma-containing blood components. T activation has been associated with necrotizing enterocolitis in infants and severe infections in children, most notably with *Clostridia* bacteremia, *Streptococcus pneumoniae*, and influenza (Eder and Manno 2001). Artificial heart valves may also produce mechanical hemolysis of red cells and result in acute hemoglobinemia and hemoglobinuria.

Management

The initial steps for managing an acute transfusion reaction include immediately discontinuing the transfusion and maintaining IV access with normal saline infusion and performing the clinical evaluation and laboratory investigation described in the previous section (Table 26.1).

In the event of a confirmed ABO incompatibility or other acute hemolytic transfusion reaction, the primary concern is to promote adequate renal blood flow, maintain blood pressure, and provide additional supportive care, as needed (Table 26.3). The adequacy of renal perfusion should be monitored by measurement of urine output to maintain urine flow rates well above 1 mL/kg/hr in children (>100 mL/hr in adults) for at least 18 to 24 hours. This is usually accomplished with IV normal saline, but diuretics (for example, furosemide) may be necessary. IV furosemide (40 to 80 mg/dose for an adult or 1 to 2 mg/kg/dose for a child) promotes diuresis and improves blood flow to the renal cortex (Brecher 2002). The balance between maintaining adequate hydration and avoiding fluid overload may be difficult in patients with preexisting cardiac and renal insufficiency. If oliguria develops (urine output less than 1 mL/kg/hr), renal tubular necrosis has likely occurred and further fluid administration may be harmful.

Hypotension may be treated with dopamine in low doses (1 to 5 µg/kg/min), as a means to increase cardiac output and dilate the renal vasculature (Brecher 2002). Higher doses of dopamine or other pressors that decrease renal blood flow should be avoided. Supportive treatment of DIC is directed at preventing serious hemorrhagic and thrombotic complications. Patients with active bleeding may require administration of platelets, FFP, and cryoprecipitated AHF, but patients

TABLE 26.3.   Acute Reactions to Blood Components

Acute Hemolytic Transfusion Reaction (AHTR)

Clinical Presentation	Treatment	Prevention
Fever Chills, rigors Apprehensiveness Pain in the lower back, flanks, chest, along infusion vein Hypotension Bleeding Disseminated intravascular coagulation Renal failure	Supportive treatment of hypotension, maintain adequate renal perfusion, which may require the following interventions: Intravenous colloid or crystalloid (e.g., 10–20 mL/kg normal saline (0.9% NaCl) Furosemide (IV) to maintain urine flow 30–100 mL/hr or greater (>1 mL/kg/hr) Children: 1–2 mg/kg/dose Adults: 40–80 mg/dose Low dose dopamine (1–5 µg/kg/min) for hypotension FFP*, platelets, cryoprecipitated AHF*, for DIC with active bleeding	Adherence to proper patient identification procedures at the time of blood collection for pretransfusion testing and before transfusion Implementation of technological innovations to reduce the probability of human error, such as bar-coded blood component and patient identification systems
Laboratory findings: Hemoglobinemia Hemoglobinuria Positive direct antiglobulin test		

Febrile Nonhemolytic Transfusion Reaction (FNHTR)

Clinical Presentation	Treatment	Prevention
Fever (rise in temperature >1°C) Chills Rigors Malaise Headache Nausea/vomiting	Acetaminophen (orally, 10–15 mg/kg/dose) Supportive care, as needed	Acetaminophen (orally, 10–15 mg/kg/dose) 30–60 min before transfusion Prestorage leukocyte-reduced components Washed blood components, for recurrent FNHTRs

Allergic Transfusion Reactions

Clinical Presentation	Treatment	Prevention

Cutaneous
  Urticaria
  Pruritis
  Flushing
  Facial edema
  Angioedema
Respiratory
  Wheezing
  Stridor
  Dyspnea
  Cough
Cardiovascular
  Hypotension
  Loss of consciousness
Gastrointestinal
  Nausea
  Vomitting
  Abdominal cramps
	Antihistamines
	Diphenhydramine, orally or IV
	  Children: 1–1.5 mg/kg/dose
	  Adults: 25–50 mg/dose
	Inhaled beta-2 agonists for bronchospasm
	  Nebulized albuterol:
	    0.05 mL/kg of 0.5% solution (max. dose, 1 mL diluted in 1–2 mL normal saline)
	
	Severe allergic reactions may require additional, aggressive intervention, as described for anaphylaxis, below.
		Antihistamines, at least one hour before transfusion:
		Diphenhydramine, orally or IV
		  Children: 1–1.5 mg/kg/dose
		  Adults: 25–50 mg/dose
		For moderate to severe reactions, corticosteroids, 6–12 hours before transfusion, such as one of the following regimens:
		  Methylprednisolone IV 1 mg/kg/dose
		  Hydrocortisone IV 1 mg/kg/dose
		  Prednisone orally 1 mg/kg/dose
		Severe allergic reactions may require aggressive prophylaxis, as described for anaphylaxis, below.

TABLE 26.3—cont'd

Anaphylaxis (Pawlowski 1998)

Clinical Presentation	Treatment	Prevention
Immediate respiratory distress	Anaphylaxis is a medical emergency and may require aggressive resuscitation, including but not limited to administration of the following medications:	Avoid transfusion
Upper airway obstruction		Autologous blood donation for elective procedures
Lower airway bronchospasm	Epinephrine (1 : 1000) subcutaneous 0.01 mg/kg/ dose (0.01 mL/kg) (Max. dose 0.4 mg [0.4 mL])	IgA deficient or washed blood components
Wheezing	Diphenhydramine IV or IM 1 mg/kg/dose (Max. dose 50 mg)	Aggressive premedication regimens with corticosteroids, H1 blockers, on the day before procedure and continued through the transfusion
Retractions		
Shortness of breath	Methylprednisolone IV 1–2 mg/kg/dose (Max. dose, 125 mg)	
Hypotension	Albuterol, nebulizer 0.01–0.05 mL/kg of 0.5% solution (Max. dose 1 mL)	
Weak pulse		
Circulatory collapse	Supplemental oxygen	
Loss of consciousness	Additional supportive care, as needed	

*FFP = Fresh frozen plasma, AHF = Antihemophilic factor.

Note: Doses are provided as a general guide for children >1 year or adults as indicated. Age-appropriate dosing is required for infants, and doses should be modified as appropriate for children with significant comorbidities (for example, renal failure, cardiac disease). Consultation with appropriate pharmacy or medical specialists is recommended before implementing a treatment approach at individual institutions.

without active bleeding likely do not benefit from blood component therapy. Anticoagulation with heparin is controversial and may aggravate hemorrhage in patients with acute DIC who are actively bleeding but may be considered if serious thrombotic complications arise.

Administration of additional units of red cells before completion of the transfusion reaction investigation requires clinical judgement. In general, transfusion should be avoided until the cause of the reaction is determined or an AHTR is ruled out. If a patient is actively bleeding and patient identification issues are unresolved, uncrossmatched group O-negative red cells may be the safest option.

Prevention

Prevention is directed at efforts to eradicate the potential for patient misidentification and other technical errors. In one study, more than half of the errors occurred in the patient care area, with the single most frequent event being the administration of a unit of blood to the wrong patient followed by specimen collection and sample labeling errors (Linden 2000). In response to errors or near misses, a facility must have a defined process to identify the need for corrective or preventive action. Analysis of near misses may identify problematic processes before an adverse event results. A system to track transfusion-related events in hospitals and facilitate causal analysis is the Medical Event Reporting System-Transfusion Medicine (Kaplan et al.

1998). A "root cause" analysis establishes the primary cause of an adverse outcome and evaluates all facets of a process, including equipment, materials, methods, environment, and human factors. Such a comprehensive analysis is intended to identify failures in the system that predispose to human error, not to place blame on an individual.

Quality improvement initiatives aim to redesign processes to minimize the occurrence of human error, especially in routine or repetitive tasks. Retraining of individuals involved in the error may be an element of a corrective action plan, but a formal system analysis should be performed to identify, and rectify, flawed processes. The analysis should include those individuals closest to the error, such as the nurses, phlebotomists, and technologists responsible for administering the transfusion or performing the procedure, as well as medical directors, technical supervisors, charge nurses, physicians, and administrative staff members with a more global understanding of institutional organizational relationships.

Technological innovations afford the greatest opportunities to decrease the potential for human error and prevent transfusion-related fatalities. A new federal regulation, proposed in March, 2003, aims to improve patient safety by requiring blood components and drugs to have bar codes on their labels and identifying patients by bar-coded wrist bands (FDA 2003). Bar coding is part of a computerized system that enables health care professionals to check whether they are giving the right drug, or in the case of blood components, the right unit, to the right patient at the right time. Although some transfusion services have already implemented these systems, this new federal regulation will expedite its more widespread utilization.

FEBRILE NONHEMOLYTIC TRANSFUSION REACTIONS

A febrile nonhemolytic transfusion reaction (FNHTR) is defined as a temperature increase of greater than 1°C (1.8°F) associated with a transfusion that cannot be attributed to other causes (Brecher 2002). The reaction may occur either during or within one to four hours following the transfusion, and associated symptoms may include chills or rigors and, rarely, nausea and vomiting. Although self-limiting, FNHTRs may be extremely uncomfortable and frightening for a patient, especially if accompanied by rigors. Because fever is an early and nonspecific sign of several different types of transfusion reactions, a diagnosis of FHNTR cannot be made until other causes, in particular acute hemolytic and septic transfusion reactions, are excluded. Among hospitalized patients, FNHTRs may be difficult to distinguish from fevers temporally associated with transfusion but due to concurrent illness, indwelling catheters, or medication.

Incidence

Overall, FNHTRs complicate 0.5% to 2.0% of transfusions (Brecher 2002). The risk of FNHTRs depends on the blood component and method of preparation as well as patient-related factors, such as age, parity, prior transfusion history, and underlying illness. FNHTRs are more frequent with transfusion of platelets than red cells, occurring in 1% to 38% of platelet transfusions as opposed to 0.5% to 6% of red cell transfusions (Brecher 2002). The wide range observed in reported rates of FNHTRs with platelet transfusions primarily reflects the method of preparation and leukocyte content of the units. Pooled random-donor platelet units are more often implicated in FNHTRs than single-donor apheresis platelet units; leukocyte reduction decreases the incidence of FNHTRs with both platelet preparations (Miller and Aubuchon 1999). Individuals who have been previously transfused or pregnant are at higher risk of FNHTRs than those with no such prior history. In contrast, newborn infants only rarely demonstrate FNHTRs, and the risk of febrile reactions among other pediatric transfusion recipients is lower than among adults. In a study of pediatric patients, 5% to 12% of platelet transfusions were associated with FNHTRs; whereas, comparable studies in adults suggest rates of 18% to 38% (Couban et al. 2002).

Pathophysiology

Fever, the quintessential feature of FNHTRs, is triggered by the action of pyrogenic cytokines (for example, IL-1, IL-6, TNF-α) on the anterior hypothalamus, inducing production of prostaglandin E2 by cells in the thermoregulatory center. Exposure to cytokines with transfusion may occur as a result of the activation of donor leukocytes by preformed antibodies in the recipient, activation of recipient leukocytes and endothelial cells by transfused donor leukocytes or plasma constituents, or by the passive transfer of cytokines that accumulated in the unit during storage. The association between prior alloimmunization of the patient to HLA or granulocyte antigens through transfusion or pregnancy and FNHTRs supports a causal role for the pathogenic antibodies against transfused leukocytes in the reactions. Prestorage and poststorage leukocyte reduction likely blocks most of these reactions, by removing the inciting leukocytes in the donor unit. Alternatively, FNHTRs may be caused by production of cytokines or other biologically active substances by leukocytes in the donor unit that accumulate with storage. Stored platelet units have been shown to contain cytokines such as IL1β, IL-6, IL-8, and TNF, and transfusion of the plasma supernatant of the component can elicit FNHTRs independent of the cellular portion (Heddle et al. 1994). HLA- or granulocyte antibodies (leukagglutinins) in the donor unit may stimulate endogenous cytokine production by recipient monocytes. Another biologic mediator, CD40 ligand (CD154), in donor units has also been associated with febrile responses to platelets likely through the stimulation of the recipient's endothelial cells to produce prostaglandin E2 (Phipps et al. 2001). Poststorage leukoreduction will not prevent these types of reactions, but prestorage leukoreduction may reduce their occurrence.

Diagnosis

FNHTR is a diagnosis of exclusion. Fever should not be attributed to a FNHTR without appropriate investigation because it may be the first sign of acute hemolysis due to an incompatible unit or sepsis due to a bacterially contaminated unit. Other causes of fever must also be eliminated before FNHTR is diagnosed, such as concurrent infection, underlying illness, or coincidental drug reaction.

Clinical Evaluation

The common pathway to evaluate all acute reactions to blood components is first aimed at eliminating acute hemolysis as a possible explanation for the observed symptoms (see Table 26.1). The transfusion should be discontinued, and the patient should be carefully monitored for development of additional symptoms. Additional investigation should address the patient's condition as well as any recently administered medications in the 24 hours before the transfusion. The patient's temperature chart, hospital course in the days leading up to the transfusion, status of indwelling catheters, as well as any recently drawn blood cultures should be reviewed to determine if the fever was likely coincidental to the transfusion. The patient's transfusion history, including any previous adverse reactions to blood components, should be investigated to determine if preventive measures should be implemented for recurrent FNHTRs.

Laboratory Evaluation

Laboratory evaluation of FNHTRs should proceed as described for acute hemolytic transfusion reactions (see Table 26.1). Although various cytokines in blood components have been correlated with the risk of FNHTRs, no routine laboratory tests are currently available to identify specific plasma constituents that may be responsible for individual reactions.

Management

In the event of an acute transfusion reaction, the transfusion should be discontinued and investigated as described previously (see Table 26.1). The fever associated with FNHTRs is self-limiting and usually resolves within one to two hours after the transfusion. Antipyretics, however, may be administered to shorten the duration of the fever and make the patient more comfortable. Acetaminophen (325 to 650 mg orally for adults or 10 to 15 mg/kg/dose orally for children) is effective for this purpose. Antihistamines are not indicated, because FNHTRs are not mediated by histamine release. For adult patients who experience shaking chills or rigors, meperidine (25 to 50 mg, IV) may be used if not otherwise contraindicated (Davenport 1999). Rigors are less common among children receiving blood components, and this intervention is rarely used in pediatrics to treat FNHTRs.

Whether the remainder of an implicated unit may be safely transfused to the patient after ruling out an acute hemolytic reaction has been debated (Widmann 1994; Oberman 1994). The primary argument against resuming the transfusion is the lingering possibility that the fever is the first manifestation of a septic reaction to a bacterially contaminated unit as well as the probability that, whatever the cause, continued administration of

the offending unit will aggravate the patient's condition. An alternative opinion is that the decision to restart a transfusion should be guided by the patient's status and the results of the transfusion reaction investigation.

Prevention

Premedication

Administration of antipyretics (for example, acetaminophen) 30 to 60 minutes before starting a transfusion may prevent FNHTRs and is not likely to mask a more serious septic or hemolytic transfusion reaction (see Table 26.3). Premedication with acetaminophen has not been systematically studied for prevention of FNHTRs but is generally indicated for patients who have had two or more febrile reactions to transfusion. The routine use of premedications for all intended transfusion recipients is excessive and unnecessary. Antihistamines are not indicated for prophylaxis of FNHTRs.

Leukocyte-Reduced Blood Components

FNHTRs may be prevented by transfusing leukocyte-reduced blood components. High-efficiency leukoreduction filters are capable of reducing the leukocyte content of blood components to less than 1×10^6; alternatively, the leukocyte content may be minimized with special apheresis collection methods (Brecher 2002). Poststorage leukocyte reduction is effective for reducing FNHTRs due to leukocyte alloimmunization but is ineffective in preventing FNHTRs due to cytokine accumulation during storage. The bulk of the available evidence, although not every clinical study, supports the advantage of prestorage leukoreduction in reducing the incidence of FNHTRs following red cell and platelet transfusion (Uhlmann et al. 2001; Vamvakas and Blajchman 2001).

For patients who continue to experience distressing FNHTRs despite premedication and provision of prestorage leukoreduced units, a trial of washed blood components may be indicated. Most of the residual plasma, and some of the leukocytes, can be removed by washing a unit of red cells with 1 to 2 L of sterile normal saline (Brecher 2002). Plasma can be removed from platelet units by volume reduction and saline suspension, or the unit can be washed by automated or manual methods. The major disadvantage to washing in general is the physical loss of 20% to 30% of red cells or platelets in the unit during preparation and the possible requirement for additional units to achieve the clinical transfusion goals. Platelets are more affected by the centrifugation and manipulation that occurs with washing, as evidenced by the impairment of platelet function and decreased recovery following transfusion (Pineda et al. 1989). Because of these limitations inherent in washed blood components, the effectiveness of this approach in preventing febrile reactions should be used for a trial period during which each transfusion is assessed. Washing should be discontinued if a patient continues to experience FNHTRs.

Patient-related factors likely account for coincidental fevers or predispose to febrile reactions following transfusion and may underlie the relatively common observation that some patients continue to experience FNHTRs despite the recent change in practice to prestorage leukoreduced units and other preventive measures.

ALLERGIC TRANSFUSION REACTIONS

Allergic reactions to blood components span a wide continuum, from the common occurrence of mild localized cutaneous manifestations to more severe systemic reactions to rarely encountered but life-threatening anaphylaxis. "Anaphylactoid" is often used in distinction to anaphylactic to describe systemic but clinically less severe allergic reactions that are caused by a different immune mechanism (Sandler et al. 1994). Regardless of the etiology of the allergic reaction, the treatment approach is guided by the severity of respiratory and cardiovascular symptoms.

Incidence

Allergic reactions complicate up to 1% to 3% of transfusions overall and are more likely to occur with platelets or plasma-derivatives than with red cells (Brecher 2002). In a study of pediatric transfusion recipients, allergic reactions occurred in 5% of platelet transfusions (Couban et al. 2002). Allergic reactions are exceedingly rare in infants, and anaphylaxis is rare in both children and adults, occurring in one in 20,000 to 50,000 transfusions. Only eight deaths in a 9-year period were attributed to transfusion-induced anaphylaxis, although severe allergic reactions to blood components may have played a role in the death of other critically ill patients (Sazama 1990; Williamson et al. 1999).

Pathophysiology

Anaphylaxis is mediated by preformed IgE antibodies on mast cells that are activated to immediately release histamine and synthesize additional mediators upon exposure to corresponding antigens in donor plasma. Mast cell activation in allergic reactions can also occur by IgE-independent mechanisms, such as by IgG-, immune complex-, and complement (C3a, C5a)-mediated pathways. Histamine release from mast cells causes urticaria, pruritis, bronchospasm, abdominal discomfort, and hypotension. Generation and release of leukotrienes, such as leukotriene D4, and prostaglandins contribute to the immediate symptoms, provoking acute bronchospasm, urticaria, and increased vascular permeability.

Implicated antigens in immediate allergic reactions include plasma proteins such as IgA, C4 (Chido/Rogers blood group antigens), or haptoglobin; chemicals that leech from administration tubing, such as ethylene oxide used in sterilization procedures; or drugs, such as penicillin, if the predonation screening process failed to eliminate donors who recently ingested the drug (Vyas et al. 1968; Westhoff et al. 1992; Koda et al. 2000). The best characterized examples of anaphylaxis occur in IgA-deficient (<0.05 mg/dL) patients with detectable IgG class-specific anti-IgA following administration of plasma or intravenous immunoglobulin (Vyas et al. 1968; Sandler et al. 1995). Paradoxically, IgA deficiency is relatively common, affecting approximately one in 900 healthy blood donors, and anti-IgA is detectable in 20% to 40% of these individuals, but anaphylaxis is a rare event. This variability in the response to transfusion of IgA-deficient patients with anti-IgA antibodies may reflect the degree of IgA deficiency or the sensitivity of the assay used to measure IgA, the characteristics of the patient's anti-IgA, and the amount of administered IgA. Less severe allergic reactions have been attributed to subclass- or allotype-specific IgA antibodies in persons with a normal concentration of total IgA (Sandler et al. 1994). In sensitive patients, anaphylaxis to plasma may occur after administration of as little as 5 to 10 mL.

Diagnosis

Clinical Evaluation

If a patient experiences an allergic reaction, the transfusion should be stopped and the patient should be evaluated to determine the most appropriate course of action, as described in a later section on management. Allergic reactions to blood components have variable severity and onset, but the vast majority are mild and usually begin within an hour of starting the transfusion. In general, anaphylaxis is immediate and severe reactions occur within minutes; whereas, mild allergic reactions may not become apparent until several hours after the transfusion. Cutaneous manifestations of an allergic reaction may be limited to localized urticaria (hives), generalized flushing or rash, itching, localized swelling, or angioedema in severe cases. Systemic symptoms predominate in more serious reactions, which may or may not be preceded by urticarial lesions. Respiratory manifestations may include dyspnea, cough, hoarseness, stridor, wheezing, chest tightness, or pain. Gastrointestinal disturbance may be evident as nausea, cramps, vomiting, diarrhea; cardiac involvement manifests as tachycardia, arrhythmia, and rarely cardiac arrest. The absence of fever and the specificity of urticaria help in distinguishing allergic reactions from other types of transfusion reactions. Allergic transfusion reactions are often isolated incidents, but some patients experience recurrent reactions. Patients with a history of atopy, asthma, or allergic reactions are more likely to develop urticaria following plasma transfusion than are nonatopic recipients. Allergic reactions usually do not increase in severity if they recur with subsequent transfusions.

Anaphylaxis causes immediate symptoms, with marked hypotension and respiratory distress due to laryngeal edema or bronchospasm, and could lead to death if airway obstruction and shock are not effectively managed. Anaphylaxis will recur with subsequent transfusion, unless precautionary measures are taken, and blood components should be avoided if possible.

Laboratory Evaluation

Systemic allergic reactions should be investigated as described for acute transfusion reactions (Table 26.1). Reactions strictly limited to urticaria need not be investigated in this manner but should be recorded in the patient's medical or blood bank record to monitor the frequency and severity of their occurrence and determine if preventive measures are necessary. Further laboratory investigation of allergic reactions to blood components is not necessary or useful, as the offending plasma constituent is rarely identified and the reaction may not recur.

Screening transfusion recipients for IgA deficiency and/or anti-IgA antibodies is not practical, cost-effective, or clinically useful because the presence of anti-IgA in an individual who has not been transfused does not predict anaphylaxis. In contrast, a patient who experiences an anaphylactic reaction to blood component transfusion should be tested for IgA deficiency and the presence of anti-IgA antibodies, because blood components lacking IgA will need to be obtained for future transfusion.

Differential Diagnosis

Allergic reactions to food, drugs, or latex may mimic plasma sensitivity if exposure to the incriminated allergens occurs in proximity to blood component transfusion. The use of latex gloves by health care workers may cause localized erythema or urticaria at the IV site, as well as acute respiratory distress in a latex-sensitive individual. A history of allergies should be elicited from the patient, and the possibility of exposure to known allergens within four hours before the transfusion should be investigated.

Hypotensive reactions accompanied by bronchospasm have been reported in patients undergoing concurrent therapy with angiotensin converting enzyme (ACE) inhibitors and therapeutic apheresis or dialysis (Brecher et al. 1993). These reactions are not due to transfused blood components but to the activation of the kinin cascade by surface contact in the extracorporeal circuits and impaired degradation of bradykinin in the presence of the ACE inhibitor that prolongs its vasoactive effects. Treatment is supportive, and hypotension resolves after termination of the procedure and does not recur if ACE inhibitors are held before future procedures for these patients. Hypotensive reactions during transfusion also occur with other acute transfusion reactions, vasovagal reactions, underlying disease, or drug reactions.

Management

Allergic Reactions

If an allergic reaction to blood component transfusion is limited to the development of hives or urticaria, the transfusion may be stopped temporarily for administration of an antihistamine to allow for resolution of the cutaneous symptoms. This approach is only acceptable for mild, cutaneous reactions during transfusion that resolve with antihistamine treatment, allowing completion of the transfusion within the requisite four-hour time limit.

If the reaction involves any symptoms other than limited urticaria, in particular, if associated with hypotension, generalized rash or swelling, respiratory compromise or gastrointestinal distress, the transfusion should be stopped and investigated as described for an acute transfusion reaction (see Table 26.1). In such cases, resuming a transfusion is not advisable and risks aggravating the patient's condition or missing an AHTR.

Treatment of allergic reactions is supportive, and the intensity of medical intervention should be guided by severity of symptoms (see Table 26.3). Antihistamines in age-appropriate doses (for example, diphenhydramine 1 to 1.5 mg/kg/dose orally or IV) may be sufficient for mild allergic reactions. Respiratory symptoms such as wheezing, cough, or chest tightness may be treated with nebulized albuterol or other beta-2 agonists. Supplemental inspired oxygen may alleviate hypoxia, as measured by pulse oximetry, or dyspnea. Hypotension may require rapid volume expansion with colloid or crystalloid (for example, 10 to 20 mL/kg IV normal saline). Severe allergic reactions may require epinephrine, steroids, and fluid resuscitation, as described for anaphylaxis in the next section.

Anaphylaxis

Anaphylaxis is a medical emergency and must be treated aggressively. The transfusion should be immediately discontinued, and fluid resuscitation should be initiated (see Table 26.3). Epinephrine in appropriate doses should be administered for severe hypotension, laryngeal edema, and respiratory failure. Bronchospasm may be treated with aerosolized or intravenous beta-2 agonists and supplemental inspired oxygen. Intubation and mechanical ventilation may be required for respiratory failure. Steroids are not helpful for the acute crisis but may be indicated if symptoms persist.

Prevention

Patients who experience recurrent allergic reactions to blood components may benefit from premedication with antihistamines (for example, diphenhydramine 1 mg/kg/dose orally or IV for children; 25 to 50 mg orally or IV for adults) at least 30 minutes before transfusion (see Table 26.3). If antihistamines fail to control allergic symptoms, corticosteroids may be added to the premedication regimen and are most effective if administered at least several hours before the transfusion. If severe reactions occur despite adequate premedication, the red cell units may be washed with saline to remove residual plasma or, less commonly, red cells units stored frozen may be thawed and deglycerolized for transfusion. Platelet units may be volume-reduced (saline-suspended) or saline-washed to remove plasma, as well, but these manipulations may be associated with a decreased efficacy of transfusion, as previously discussed for prevention of FNHTRs.

Patients who have experienced anaphylactic reactions with blood component transfusion and have documented IgA deficiency and/or anti-IgA antibodies should be transfused only if unavoidable and under close medical supervision. Blood components that lack IgA may be collected from IgA-deficient donors or prepared by extensive washing procedures (Davenport et al. 1992). Aggressive premedication regimens, such as those designed to prevent anaphylactic reactions to radiocontrast media, may be applied in this setting and entail administering corticosteroids at multiple intervals

in the 12 hours before transfusion, as well as giving diphenhydramine and ephedrine one hour before the transfusion (Greenberger and Patterson 1991). The clinical team should be prepared to provide emergent resuscitation and intensive supportive care, if these precautions fail to prevent anaphylaxis. The safest blood for these patients is likely their own blood, and arrangements should be made for autologous donation before elective procedures, if the clinical situation permits.

DELAYED HEMOLYTIC TRANSFUSION REACTIONS

Delayed reactions to blood component transfusion are immunologic consequences that occur more than 24 hours after transfusion of the offending unit. Delayed hemolytic transfusion reactions (DHTRs) result from red cell alloimmunization and subsequent antibody-mediated extravascular hemolysis of transfused red cells leading to unexpected anemia following transfusion. The platelet counterpart of DHTR, platelet refractoriness, is discussed in Chapter 22; transfusion-related graft-versus-host disease and immunomodulatory effects, in Chapter 23.

Incidence

Red cell alloimmunization following transfusion occurs in 0.2% to 2.6% of the general population, at higher rates among chronically transfused children, and rarely among infants younger than four months (Brecher 2002). The risk of red cell alloimmunization depends on the patient's age at first transfusion, the number of donor exposures, and the antigen disparity between donor and recipient populations. Children with sickle cell disease who receive transfusions before the age of 10 are less likely to become alloimmunized than older children and adults despite exposure to more red cell units in their lifetime, possibly due to the induction of immune tolerance in young children (Rosse et al. 1990). The frequency of alloimmunization among chronically transfused children with sickle cell disease is higher than among chronically transfused children with thalassemia, with population estimates ranging from 18% to 47% compared to 5% to 11%, respectively (Smith-Whitley 2002). This discrepancy is largely accounted for by the greater probability of sensitization to foreign red cell antigens among transfused sickle cell patients due to the genetic differences in red cell antigen expression between the predominately African-American patient population and the mostly Caucasian blood donor population.

Reflecting their higher rates of red cell alloimmunization, chronically or episodically transfused patients are more likely to experience DHTRs than the general hospitalized population. The incidence of detected DHTR is estimated as one in 1500 red cell units at a tertiary-care medical center (Vamvakas et al. 1995). Delayed hemolytic transfusion reactions complicate 4% to 22% of transfusions among transfused sickle cell patients (Smith-Whitley 2002).

Pathophysiology

Delayed Serologic Transfusion Reactions

In contrast to "naturally occurring" IgM antibodies against carbohydrate blood group antigens, like anti-A and anti-B, antibodies against Rhesus, Kell, Duffy, Kidd, and other protein blood group antigens only rarely arise without prior exposure to red cells, either during pregnancy or with transfusion. Exposure to these foreign red cell antigens may stimulate a primary immune response in the recipient, during which IgM antibodies are produced, followed several weeks later by an allotype switch to IgG antibodies. Consequently, primary alloimmunization becomes apparent weeks to months after transfusion and is usually an asymptomatic serologic finding (Vamvakas et al. 1995). Once alloimmunization has occurred, the red cell alloantibodies diminish to undetectable levels, until subsequent reexposure to incompatible red cells elicits a secondary or anamnestic immune response, characterized by rapid production of IgG antibodies with enhanced affinity for the red cell antigen. The antibody typically appears within two days to two weeks after the implicated transfusion. If further investigation reveals the presence of the antibody is not associated with evidence of accelerated red cell destruction, the reaction is an isolated serologic finding, classified as a delayed serologic transfusion reaction.

Delayed Hemolytic Transfusion Reactions

The diagnosis of DHTR requires clinical correlation to determine that the newly discovered red cell alloantibody is causing extravascular hemolysis. The pathogenicity of red cell alloantibodies is influenced by several factors, including the thermal range, antibody specificity, and subclass. Antibodies that are reactive at 37°C in vitro are more likely to be clinically significant, and Duffy (Fy^a) and Kidd (Jk^a) antibodies are more likely to cause DHTRs than other antibody specificities (Vamvakas et al. 1995). The immunoglobulin subclasses IgG1 and IgG3 have greater affinity for Fc receptors on phagocytic cells than IgG2 and IgG4, which may account for their association with more severe hemolysis.

Extravascular hemolysis results from the binding of IgG red cell antibodies and, in some cases, incomplete complement activation with deposition of C3b on the surface of red cell. Macrophages and other cytotoxic effector cells in the reticuloendothelial system, primarily in the spleen, recognize these targeted red cells via their corresponding Fc and complement receptors resulting in red cell destruction. The progressive removal of portions of the red cell membrane by phagocytic cells in the spleen results in the appearance of spherocytes in the peripheral circulation. Accelerated destruction of transfused red cells is also evident by reticulocytosis, unconjugated hyperbilirubinemia, and increased serum LDH.

Diagnosis

Clinical Evaluation

Delayed hemolytic transfusion reactions are characterized by an unexpected anemia or a less than expected posttransfusion increment in hematocrit following transfusion. The onset is variable but usually occurs within the first two weeks following transfusion. Because the red cell alloantibodies cause varying degrees of red cell destruction and may persist in the peripheral circulation for several weeks, signs or symptoms may not be recognized for four to eight weeks following transfusion. Symptoms of extravascular hemolysis include fever or chills, jaundice, malaise, and back pain; renal failure rarely occurs.

Delayed hemolytic transfusion reactions are often mild in patients without underlying hematologic illness but may be severe in chronically transfused, sickle cell patients. Sickle cell patients who develop red cell alloantibodies with transfusion are more likely to develop red cell autoantibodies and are susceptible to sickle cell crises or hyperhemolysis following red cell transfusion, which is also referred to as the sickle cell hemolytic transfusion reaction syndrome. In this clinical setting, transfused sickle cell patients develop severe anemia following transfusion, with evidence of bystander hemolysis of autologous red cells as well as transfused cells, often in the absence of identifiable red cell alloantibodies, that ultimately results in a lower hemoglobin concentration after the transfusion than before it was begun (Win et al. 2001). Suppressed erythropoiesis and autoantibody formation may contribute to the profound posttransfusion anemia in these patients (Smith-Whitley 2002).

Laboratory Evaluation

The cardinal sign of DHTR is a lower than expected hemoglobin/hematocrit following red cell transfusion, and laboratory evaluation may provide additional evidence of extravascular hemolysis. The appearance of spherocytes on the peripheral blood smear and reticulocytosis are consistent with destruction of red cells in the reticuloendothelial system, as are alterations in serum chemistry such as increased total and unconjugated bilirubin, increased LDH and decreased or absent haptoglobin. Urinalysis may reveal urobilinogen, which reflects increased serum unconjugated bilirubin. Supplemental evidence of the shortened lifespan of transfused red cells in sickle cell patients include an unexpected increase in the relative proportion of HbS and decrease in the relative proportion of HbA following transfusion.

If a DHTR is suspected on the basis of clinical presentation or laboratory findings, the DAT and antibody screen should be repeated on a newly drawn specimen from the patient. With DHTRs, the DAT is often positive in a mixed field pattern, indicating the presence of alloantibody or complement on the transfused red cells but not on the patient's own (autologous) red cells. The DAT, however, may be negative if all of the transfused cells have been eliminated from the patient's peripheral circulation. Serial testing over several weeks may be required to detect the appearance of the offending red cell alloantibody in the serum. The diagnosis of DHTR is supported by demonstration of the corresponding antigen on the red cells from a retained segment from one or more of the recently transfused units (Brecher 2002).

Differential Diagnosis

The development of autoantibodies and possible autoimmune hemolytic anemia complicates the evaluation of DHTRs, for the dual reason of confounding detection of underlying alloantibodies and contributing directly to ongoing hemolysis. Bleeding or infection may also cause unexpected anemia. Cephalohematomas may also be associated with laboratory findings that overlap those characteristic of DHTRs, including increased unconjugated bilirubinemia, increased LDH, and decreased haptoglobin. In distinction from DHTRs, bleeding or other causes of nonspecific red cell destruction will be associated with the persistence of transfused cells with mixed field reactivity in red cell phenotyping reactions. In chronically transfused sickle cell patients, DHTRs often must be distinguished from other causes of increasing transfusion requirements, such as underlying disease and hypersplenism, including the possibility of accessory spleens or splenic regeneration in previously splenectomized patients.

Management

Because DHTRs are usually mild, specific treatment is rarely necessary. The development of anemia sooner after transfusion than expected may require additional red cell transfusion, and red cell units should be selected that lack the antigen corresponding to the newly discovered antibody. If transfusion is ordered by the clinical service before identification of the specificity of the underlying alloantibody, this need must be balanced against the risk of additional transfusion of potentially antigen-positive blood with aggravation of ongoing extravascular hemolysis. Clinical judgment and close communication between the blood bank and clinical service taking care of the patient are essential in such cases. Adequate hydration in light of ongoing hemolysis and monitoring of renal function is prudent, but renal failure is extremely rare in the setting of DHTRs.

In contrast, any further transfusions must be carefully considered in chronically transfused sickle cell patients with multiple red cell alloantibodies and/or autoantibodies. For these complicated patients, treatment must be carefully tailored to the individual, balancing the risk of severe DHTRs and possibly fatal hyperhemolysis against the possible benefits of red cell transfusion to alleviate acute complications or prevent chronic morbidity of sickle cell disease. If transfusion is medically necessary, not only should the red cell units provided for these patients lack the antigens corresponding to any current or historical red cell alloantibodies, but extended phenotype matching for all clinically significant antigens (for example, Rh, Kell, Kidd, Duffy, MNSs) should be performed in an effort to prevent further alloantibody production. In general, red cell transfusion should be avoided in sickle cell patients who have experienced severe hemolytic transfusion reactions, and these patients may instead benefit pharmacological therapy with steroids, IVIg, erythropoietin, or hydroxyurea (Smith-Whitley 2002).

Prevention

If a patient develops a clinically significant alloantibody, red cell units lacking the corresponding antigen should be provided for all future red cell transfusions. The prompt identification and accurate record-keeping of clinically significant red cell alloantibodies are requisite blood bank practices. Pretransfusion testing requirements for all patients, except newborn infants, specify antibody detection and compatibility tests before each transfusion, with repeat testing within three days of the next scheduled transfusion for patients transfused or pregnant within the preceding three months. The three-day interval requirement for repeat testing of transfused patients is based on the finding that new antibodies may be detected within one to two days after transfusion due to an anamnestic response (Schulman 1989). AABB Standards require documentation of clinically significant alloantibodies in a permanent record, as well as review of previous transfusion history before transfusion to safeguard against anamnestic antibody responses in individuals whose alloantibody becomes undetectable in the intervals between transfusions (Gorlin 2002).

Additional pretransfusion testing and other special measures may be undertaken for patients with sickle cell disease or warm autoimmune hemolytic anemia. The primary cause of high red alloimmunization rates in children with sickle cell disease is genetic differences between the predominantly African-American patient population and the predominantly Caucasian general blood donor population. Many red cell antigens, in particular C and E in the Rhesus blood group and K1 in the Kell blood group, are more frequently expressed in Caucasians than African-Americans, and about two-thirds of the clinically significant red cell alloantibodies found in chronically-transfused sickle cell patients are directed against these antigens (Smith-Whitley 2002).

By prospectively avoiding incompatibility to C, E, and K1 blood group antigens, the alloimmunization rate among chronically transfused sickle cell patients was reduced from 3% to 0.5% per unit, and hemolytic transfusion reactions were reduced by 90% in one study (Vichinsky et al. 2001). Prophylactic red cell phenotype matching for C, E, and K1 has been recommended for all chronically transfused patients with sickle cell disease (Smith-Whitley 2002). This practice, however, has not been universally accepted, because of the expense, logistic difficulty, and potential for diverting relatively rare antigen-negative red cell units to patients who are not yet demonstrating red cell alloantibodies. An alternate approach restricts the use of phenotypically matched red cells to patients who have already become immunized to one red cell antigen and are at risk of developing additional alloantibodies and autoantibodies. Programs designed to specifically recruit African-American donors and direct this blood to sickle cell patients facilitate the identification of phenotypically matched units and may further reduce the risk of alloimmunization among chronically transfused children with sickle cell disease (Smith-Whitley 2002).

Patients with warm autoimmune hemolytic anemia (WAIHA) do not require prophylactic antigen matching as recommended for sickle cell patients but may benefit from extended red cell phenotyping before transfusion, if prolonged transfusion support for chronic anemia is likely. The presence of red cell autoantibodies that react with all reagent red cells may interfere with the detection of underlying red cell alloantibodies, which develop in 12% to 40% of transfused patients with WAIHA (Brecher 2002). Determination of the patient's red cell phenotype at Rh (Cc, Ee), Kell, Kidd, Duffy, and other clinically significant loci before transfusion will facilitate future alloantibody identification, by identifying those red cell antigens absent from the patient's red cells, against which alloantibodies may be produced.

SPECIAL CONSIDERATIONS FOR NEWBORN INFANTS

Pretransfusion testing to determine a patient's ABO/Rh type and screen for red cell antibodies is performed for all transfusion recipients to ensure transfusion of compatible blood. AABB Standards, however, allow for abbreviated testing procedures for newborn infants younger than four months, because of their unique immunologic status and susceptibility to iatrogenic anemia with frequent blood sampling (Gorlin 2002). Although forward grouping must be performed to determine the infant's ABO/Rh type, the reverse group to detect maternal IgG anti-A or anti-B antibodies need not be performed if exclusively group-O red cells will be transfused. Type-specific red cells may be safely transfused in the absence of IgG anti-A or anti-B and permit more effective utilization of a limited natural resource, especially at pediatric facilities that transfuse many critically ill infants. Crossmatching before transfusion to newborn infants is not required if the initial sample, collected either from the mother or the infant, lacks detectable red cell antibodies. If a clinically significant red cell alloantibody is detected in the screen, crossmatch procedures may still be omitted if antigen-negative red cells are provided, otherwise units selected for transfusion should be fully crossmatch-compatible utilizing a method that detects IgG antibodies (for example, a "Coombs" crossmatch). The antibody screen does not need to be repeated during the hospitalization, even if the infant receives multiple transfusions, because of the extremely low probability of active alloimmunization during the neonatal period (Ludvigsen et al. 1987; Strauss et al. 2000).

Infants have lower plasma complement and ABO isohemagglutinin titers than children and adults, factors which may mitigate the severity of acute hemolytic transfusion reactions but do not eliminate the possibility of fatal outcomes in response to ABO-incompatible red cells. Transfusion of incompatible plasma is more likely to cause acute hemolysis in infants than adults, due to small blood volumes, and fatal hemolytic transfusion reactions have occurred when group O whole blood was administered to group A infants, with the mistaken belief that this was acceptable as "universal donor" blood. Allergic and febrile transfusion reactions are rare in infants; premedication is not indicated for neonatal transfusion.

References

Brecher ME, Owen HG, and Collins ML. 1993. Apheresis and ACE inhibitors. *Transfusion* 33:963–964.

Brecher ME. 2002. "Technical manual," 13th ed. Bethesda, MD: American Association of Blood Banks.

Couban S, Carruthers J, Andreou P, et al. 2002. Platelet transfusions in children: results of a randomized, prospective, crossover trial of plasma removal and a prospective audit of WBC reduction. *Transfusion* 42:753–758.

Davenport RD. 1999. Management of transfusion reactions. In *Transfusion therapy: clinical principles and practice*, Minz PD, ed. Bethesda MD: AABB Press.

Davenport RD, Burnie KL, and Barr RM. 1992. Transfusion management of patients with IgA deficiency and anti-IgA during liver transplantation. *Vox Sang* 63:247–250.

Dodd RY, Notari EP, and Stramer SL. 2002. Current prevalence and incidence of infectious disease markers and estimated window-period risk in the American Red Cross blood donor population. *Transfusion* 42:975–979.

Eder AF and Manno CS. 2001. Does red cell T activation matter? *Br J Haematol* 114:25–30.

Food and Drug Administration, HHS (21 Code of Federal Regulations Parts 201, 606 and 610). 2003. Bar Code Label for Human Drug Products and Blood; Proposed Rule. *Federal Register*. 68:12500–12534.

Gorlin JB, ed. 2002. Standards for blood banks and transfusion services, 21st ed. Bethesda MD: American Association of Blood Banks.

Greenberger PA and Patterson R. 1991. The prevention of immediate generalized reactions to radiocontrast media in high-risk patients. *J Allergy Clin Immunol* 87:867–872.

Heddle NM, Klama L, Singer J, et al. 1994. The role of the plasma from platelet concentrates in transfusion reactions. *N Engl J Med* 331:625–628.

Kaplan HS, Battles JB, Van der Schaaf TW, et al. 1998. Identification and classification of the causes of events in transfusion medicine. *Transfusion* 38:1071–1081.

Koda Y, Watanade Y, Soejima M, et al. 2000. Simple PCR detection of haptoglobin gene deletion in anhaptoglobinemic patients with antihaptoglobin antibody that causes anaphylactic transfusion reactions. *Blood* 95:1138–1143.

Linden JV, Wagner K, Voytovich AE, and Sheehan J. 2000. Transfusion errors in New York State: an analysis of 10 years' experience. *Transfusion* 40:1207–1213.

Ludvigsen C, Swanson JL, Thompson TR, and McCullough J. 1987. The failure of neonates to form red cell alloantibodies in response to multiple transfusions. *Am J Clin Pathol* 87:250–251.

Miller JP and AuBuchon JP. 1999. Leukocyte-reduced and cytomegalovirus-reduced-risk blood components. In *Transfusion therapy: clinical principles and practice*, Mintz PD, ed. Bethesda MD: AABB Press.

Oberman HA. 1994. Controversies in transfusion medicine: should a febrile transfusion response occasion the return of the blood component to the blood bank? *Con. Transfusion* 34:353–355.

Pawlowski N. 1998. Anaphylaxis. In *Pediatrics at a Glance*, Altschuler SM and Ludwig S, eds. Philadelphia: Appleton and Lange.

Phipps RP, Kaufman J, and Blumberg N. 2001. Platelet derived CD154 (CD40 ligand) and febrile responses to transfusion. *Lancet* 357:2023–2024.

Pineda AA, Brzica SM, and Taswell HF. 1978. Hemolytic transfusion reaction. *Mayo Clinic Proceedings* 53:378–390.

Pineda AA, Zylstra VW, Clare DE, et al. 1989. Viability and functional integrity of washed platelets. *Transfusion* 29:524–527.

Reid ME and Lomas-Francis C. 1997. *The blood group antigen facts book*. San Diego, CA: Academic Press.

Rosse WF, Gallagher D, Kinney TR, et al. 1990. Transfusion and alloimmunization in sickle cell disease. The Cooperative Study of Sickle Cell Disease. *Blood* 76:1431–1437.

Sandler SG, Eckrich R, Malamut D, and Mallory D. 1994. Hemagglutination assays for the diagnosis and prevention of IgA anaphylactic transfusion reactions. *Blood* 84:2031–2035.

Sandler SG, Mallory D, Malamut D, and Eckrich R. 1995. IgA anaphylactic transfusion reactions. *Transfus Med Rev* 9:1–8.

Sazama K. 1990. Reports of 355 transfusion-associated deaths: 1976 through 1985. *Transfusion* 30:583–590.

Schulman IA. 1989. Controversies in red blood cell compatibility testing. In *Immune destruction of red blood cells*, Nance SJ, ed. Arlington VA: American Association of Blood Banks.

Smith-Whitley K. 2002. Alloimmunization in patients with sickle cell disease. In *Pediatric transfusion therapy*, Herman JH, Manno CS, eds. Bethesda, MD: AABB Press.

Strauss RG, Johnson K, Cress G, and Cordle DG. 2000. Alloimmunization in preterm infants after repeated transfusions of WBC-reduced RBCs from the same donor. *Transfusion* 40:1463–1468.

Uhlmann EJ, Isgriggs E, Wallhermfechtel M, and Goodnough LT. 2001. Prestorage universal WBC reduction of RBC units does not affect the incidence of transfusion reactions. *Transfusion* 41:997–1000.

Vamvakas EC and Blajchman MA. 2001. Universal WBC reduction: the case for and against. *Transfusion* 41:691–712.

Vamvakas EC, Pineda AA, Reisner R, et al. 1995. The differentiation of delayed hemolytic and delayed serologic transfusion reactions: incidence and predictors of hemolysis. *Transfusion* 35:26–32.

Vichinsky EP, Luban NLC, Wright E, et al. 2001. Prospective RBC phenotype matching in a stroke-prevention trial in sickle cell anemia: a multicenter transfusion trial. *Transfusion* 41:1086–1092.

Vyas GN, Perkins HA, and Fudenberg HH. 1968. Anaphylactoid transfusion reactions associated with anti-IgA. *Lancet* 2:312–315.

Westhoff CM, Sipherd BD, Wylie DE, and Toalson LD. 1992. Severe anaphylactic reaction following transfusion of platelets to a patient with anti-Ch. *Transfusion* 32:576–579.

Widmann FK. 1994. Controversies in transfusion medicine: should a febrile transfusion reaction occasion the return of the blood component to the blood bank? *Pro. Transfusion* 34:356–358.

Williamson LM, Lowe S, Love EM, et al. 1999. Serious hazards of transfusion (SHOT) initiative: analysis of the first two annual reports. *Br Med J* 319:16–19.

Win N, Doughty H, Telfer P, et al. 2001. Hyperhemolytic transfusion reaction in sickle cell disease. *Transfusion* 41:323–328.

27

Noninfectious Complications of Pediatric Transfusion

JED B. GORLIN, MD

INTRODUCTION

This chapter addresses potential adverse noninfectious consequences of transfusion. It is imperative that the ordering practitioner considers both the risks and the intended benefits of any potential transfusion. These should be shared with the patient, if age appropriate, and family as part of the consent process for a pediatric recipient. Discussions of risks should include both common outcomes and risks as well as those that are less common but profound clinical consequence should they occur (Sazama 1997). While many families express concerns about potential infectious complications of transfusion, they are rarely aware of the more frequent noninfectious complications. This chapter will provide a broad overview of some of the more clinically relevant potential adverse noninfectious outcomes to allow the practitioner to have an honest discussion with parents and, when appropriate, pediatric transfusion recipients (Table 27.1). In addition, awareness of potential reactions can permit early recognition and intervention that may ameliorate some of the consequences should a reaction occur. For example, early recognition of a hemolytic transfusion reaction or transfusion-related acute lung injury (TRALI) may facilitate early termination of the transfusion and immediate resuscitative steps to prevent further harm.

TRANSFUSION REACTIONS

Transfusion Safety

Any discussion of transfusion reactions should include transfusion safety; 10 to 20 transfusion recipients a year in the United States die of hemolytic transfusion reactions, the majority of them from incompatible red cells (Sazama 1986). The events that result in mistransfusion, often include a mislabeled sample from the patient used for ABO typing or misidentification of the patient at the time of transfusion (Linden et al. 2000). As the FDA considers what steps to take to further reduce transfusion risk, it is imperative that focus be placed on the entire transfusion sequence as opposed to the safety of the transfused product alone (Dzik 2003). Hospitals are increasingly utilizing total quality approaches to further reduce errors, including barcode labeling of medications as well as blood units. Only when the potential for error throughout the entire transfusion sequence is systematically addressed can one hope to appreciably reduce the global risks of transfusion.

Patient identification is the first critical step of the transfusion process. Every teaching hospital has the experience of an increased number of mislabeled specimens upon the annual influx of a cohort of new trainees who collect samples. Mislabeled, incompletely labeled or discrepantly labeled specimens have a forty-fold higher rate of incorrect blood types in mislabeled tubes (Lumadue et al. 1997). While the majority of blood types in mislabeled tubes match the historical patient type, this data supports having a strict labeling policy for transfusion specimens. A recent international study demonstrated a wide variation of frequency of "wrong blood in tube"; approximately ~1/1000 specimens were documented to not contain the labeled patient's blood (Dzik et al. 2002). Practices that decrease incorrect samples include use of dedicated phlebotomy staff, transfusion safety officers, and rigid

TABLE 27.1   Relative Risks of Transfusion
Complications Per Unit

Risk	Risk Ratio	Reference
Hepatitis C	1:1,600,000	(Busch 2001)
HIV	1:1,900,000	(Busch 2001)
Wrong blood in tube	1:1000	(Dzik et al. 2002)
Wrong recipient of auto unit	1:16,000	(Linden et al. 2000)
TRALI	1:5000	(Papovsky 2001)
Metabolic reaction neonate	1:1000	(Stranss 2000)
Hemolysis from ABO-incompatible plasma in an apheresis platelet	1:10,000–46,000	(Mair and Benson 1998, Larsson et al. 2001)

TRALI = Transfusion-related acute lung injury.

adherence to specimen labeling guidelines (Linden et al. 2000; Lumadue et al. 1997; Cursio and Fountas 2002).

The most risk-free transfusion, of course, is the one that never happens. Hence, it is imperative to consider whether each and every transfusion is truly indicated. All transfusion service medical directors can recall an anecdote of a transfusion complication that followed a transfusion of dubious clinical necessity or value. Transfusing multiple units of fresh frozen plasma (FFP) to correct minimally prolonged coagulation parameters or platelets before a minimally invasive procedure are examples (Dzik 1999).

Special transfusion requirements are generally far more apparent to the ordering physician than the blood bank personnel. Ordering staff often know far more about the specific patient needs (for example, requirements for cytomegalovirus [CMV] reduced risk or irradiated components) than the blood bank, which may have no prior transfusion history on a patient at increased risk for some transfusion complication. While blood bank personnel are available for consultations, it is imperative to minimize risks by ensuring that the ordering physician selects the optimal type and quantity of transfusible component. Furthermore, the mechanisms in place to issue those components to the patient must ensure proper storage and handling and proper identification of the patient to be transfused. Finally, the system used to infuse the actual components must be fully validated and the personnel administering the transfusion appropriately trained. Mechanisms to ensure close and careful observation during and following a transfusion must be in place and followed religiously. Should something untoward happen or vital signs not meet expected measures, there needs to be clear and explicit instructions to staff on how to manage and evaluate the reaction.

Hemolytic-Acute

Acute hemolytic transfusion reactions in pediatric transfusion are less common than in adult transfusions. When comparing absolute numbers of serious reactions, far fewer transfusions are given to pediatric than adult recipients. In addition, far fewer pediatric transfusions occur in uncontrolled circumstances, such as emergency room, obstetric emergencies, and emergent surgeries, all of which lend themselves to higher risks of adverse outcomes. From a physiological standpoint, newborns do not have measurable circulating isohemagglutinins; by age three months, infants have developed about 25% of adult levels. In one study, anti-A titer was four at three to six months and 64 by 12 to 18 months (Fong et al. 1974). Adult levels are reached by age five to 10 years and fall slowly with age (Grundbacher 1967). Hence, one could theoretically administer a full volume AB red cell transfusion to a group O neonate without any clinical consequence.

Despite anecdotal observations of fewer acute hemolytic episodes following red cell administration, there are more frequent episodes of passive hemolysis from incompatible plasma components (Lozano and Cid 2003). Specifically, since pediatric transfusion recipients commonly receive large volumes of plasma-containing components (commonly 15 to 20 mL/kg), the risk is greater should they receive plasma with a high-titer ABO antibody (Duguid et al. 1999). This may occur, for example, in a pediatric oncology patient receiving plasma-incompatible apheresis platelets or a neonate receiving a single platelet from a whole blood donation (Larsson et al. 2001). While it is possible to concentrate platelets by centrifugation, it is generally not warranted since one can typically achieve a 50 to 100,000 count increment following transfusion of unconcentrated platelet-rich plasma alone (Strauss 2000). However, when faced with the potential for passive immune hemolysis from incompatible plasma, many large pediatric hospitals have validated protocols to volume reduce plasma incompatible platelets, if plasma compatible ones are unavailable (Moroff et al. 1984).

One type of passive and acute red cell hemolytic transfusion reaction unique to newborns is worthy of specific mention. If a nongroup-O neonate is to receive nongroup-O red blood cells (RBCs) that are not compatible with the maternal ABO group, the neonate's serum or plasma must be tested for anti-A or anti-B (AABB Standard #5.15.2) (Gorlin 2002). This is because a negative antibody screen (always performed against group O cells) does NOT rule out the presence of a high titer anti-ABO group-specific IgG antibody crossing the placenta to the infant and causing profound hemolysis should that blood type be transfused to the infant. Since newborn cells express less A and B substance per cell than adult cells, it is possible for the infant not to manifest stigmata of ABO-hemolytic disease of the newborn (HDN) but still manifest vigorous hemolysis following transfusion of red cells of the same ABO group as the infant.

Hemolytic-Delayed

Alloimmunization against other red cell antigens is also less frequent in pediatric recipients, especially in the neonatal period. Indeed AABB Standards have always excluded neonates (that is, younger than four months) from the requirement for crossmatching, as development of alloantibodies is rare at that age (Ludvigsen et al. 1987). Still, anecdotal reports of development of antibodies in infants exist.

For older children, red cell alloimmunization certainly occurs in transfusion settings analogous to adult recipients of transfusion. One interesting observation is the relatively low rate of alloimmunization among patients receiving lifelong and frequent chronic transfusions for thalassemia in contrast to those with sickle cell disease who receive far fewer and typically intermittent transfusions starting at an older age. Whether frequent transfusion from infancy results in some degree of immune tolerance or the greater antigenic disparity between donors and recipients yields a greater opportunity for antibody development in the sickle cell recipients remains controversial (Rosse et al. 1998; Vichinsky et al. 1990). The recommended methods to ameliorate this tendency towards alloimmunization among the chronically or frequently transfused patients with sickle cell anemia is dealt with in Chapter 19.

Febrile

Febrile transfusion reactions also occur less frequently in pediatric patients than their adult counterparts, but are second only to allergic-type transfusion reactions in frequency. White cells and the inflammatory cytokines (Il-1β, Il-6, Il-8, TNF) are released during storage and appear to be the prime mediators of this type of reaction; prestorage leukoreduction of red cells and platelets appears to abrogate a majority of this type of reaction. Leukoreduction filtration of pooled, whole blood-derived platelets at time of transfusion may slightly reduce but does not prevent febrile reactions and the myriad cytokine mediators of fever have already been released into the plasma. Hence, volume reduction and even washing may be used to further reduce these reactions when prestorage leukoreduced components are not readily available and such reduc-

tion is considered warranted for a particularly vulnerable recipient population (Couban et al. 2002; Heddle et al. 2002).

Staff should not restart transfusions following a febrile reaction as fever may also be a symptom of either a hemolytic reaction or bacterial contamination; in those cases, administration of additional product may worsen the consequences. Practically stated, most nursing protocols for dealing with a transfusion reaction, only permit continuation of a transfusion after a significant reaction if the product is vital (for example, bone marrow or peripheral stem cell infusion) or following a mild to moderate allergic type reaction when the patient has been appropriately medicated (Heddle et al. 2001).

Allergic

Allergic reactions vary from mild to severe (anaphylactic) reactions and range from common to rare, respectively. Mild cutaneous hypersensitivity reactions (itching, rash, redness) are common and generally respond to appropriate doses of antihistamines. The antihistamines generally serve to prevent propagation of the symptoms; despite adequate treatment, the symptoms may persist for many minutes after administration. Severe allergic type reactions are often, though not always, associated with development of anti-IgA antibodies in an IgA-deficient patient. Newer, far more sensitive IgA antigen assays have documented that many individuals have low levels; this explains why severe reactions are far rarer than the observed prevalence of IgA "deficiency," which may be relative and not absolute in most cases (Sandler et al. 1995). While anaphylactic reactions have been documented in all age groups, they are exceedingly rare in younger pediatric patients and hence the requirement to be able to rapidly obtain IgA-deficient plasma products for pediatric patients is rarely necessary.

Transfusion-Associated Lung Injury (TRALI)

TRALI is the third most frequent cause of a fatal transfusion reaction in adults, but rare in pediatric patients (Zoon 2001).

TRALI has historically been characterized as the constellation of symptoms including: dyspnea, hypoxemia, hypotension, and fever (Popovsky 2001). In a retrospective series, 76% of patients had respiratory distress, 15% had hypotension, and 15% had frank hypertension (Popovsky 2000). Chest radiographs typically reveal bilateral "white out," that is, pulmonary infiltrates in the absence of cardiac compromise or overload. Symptoms typically manifest within one to two hours after initiation of transfusion and become most severe within six hours. All blood components have been implicated in TRALI episodes, but they are most frequently associated with larger infusions of plasma from a single plasma donor, that is, whole blood-derived platelets, frozen plasma units, or apheresis platelets (Popovsky et al. 1992; Silliman et al. 2003). Even intravenous gamma globulin (IVIG) has been associated with inducing TRALI although reports are rare (Rizk et al. 2001).

The syndrome of TRALI needs to be distinguished from the clinically more common scenario of cardiac overload. Both small infants and the elderly are at greater risk of cardiac overload either because of the sheer volume we infuse into small infants, or the endogenous cardiac compromise of the elderly, whereas TRALI is not more frequent in any particular risk group. The pathophysiology of cardiac overload implies that the heart cannot adequately pump fluid through the lungs either because of increased pulmonary resistance or cardiac pump failure, which results in pulmonary edema. Cardiac overload can be confirmed by documenting an elevated pulmonary capillary wedge pressure, whereas it is normal in TRALI.

TRALI is clearly underdiagnosed, as was elegantly demonstrated by a retrospective study performed following a well-documented TRALI fatality. The implicated transfused product had come from a frequent jumbo plasma donor who had donated at least 50 products. Of 36 charts reviewed, seven recipients had mild/moderate reactions, eight had severe reactions, two had had multiple reactions, but only two of the eight severe reactions had been reported to the regional blood collection facility (Kopko et al. 2002). TRALI is very rarely reported in pediatric recipients. Whether this is a result of a negative diagnostic bias (few pediatric residents are knowledgeable about this potential outcome of transfusion), greater pediatric pulmonary reserve (hence the disease when present may be less clinically apparent), the rarity of pediatric patients receiving large volumes of plasma, or whether there really is a physiological basis for fewer reactions is unknown.

The first proposed explanation for the pathophysiology of TRALI was that antibodies in the donor plasma bind to and activate the host white cells, which adhere to pulmonary endothelium resulting in functional blockade of the capillaries and fluid leak into the alveoli. Evidence in support of this theory includes the observations that implicated donors are typically multiparous females and have clearly demonstrable anti-HLA or anti-neutrophil antibodies that bind to the host white cells in almost 90% of cases (Silliman 1999). A prospective randomized trial of plasma transfusions in

intensive care unit (ICU) patients who received at least two units of FFP compared the pulmonary effects of plasma transfused from multiparous donors (≥ three live births) versus control donors. Temperature, blood pressure, and heart rate were monitored and blood samples obtained. Transfusion from the multiparous donors resulted in significantly lower oxygen saturations and higher TNFα concentrations (Palfi et al. 2001). This study and others suggest that there are multiple physiological mediators of this disorder including but not limited to passive transfer of anti-white blood cell (WBC) antibodies against HLA class I, class II, and neutrophil-specific antigens (Kao et al. 2003; Kopko et al. 2001). Plasma cytokines capable of priming host leukocytes may mediate TRALI even in the absence of WBC antibodies; other mediators like interleukin 6 (IL-6) and lipid priming activity (neutral lipids and lysophosphatidylcholines) may also be involved (Silliman et al. 2003). A very unique report documented HLA-specific antibodies against a transplanted lung mediating TRALI in the transplanted lung but not the transplant recipient; this study suggests that the antibodies can interact with both the host endothelium and the host WBCs (Dykes et al. 2000).

Some investigators have observed a higher anti-HLA antibody prevalence among multiparous donors and thus suggest banning them from donation, or at least the donation of large plasma volume products. Densmore et al. (1999) studied the frequency of HLA sensitization among female plateletpheresis donors and observed a frequency of sensitization of 7.8%, 14.6%, and 26% among women with 0, one to two, and ≥ three pregnancies, respectively. Because three-fourths of the multiparous donors were not sensitized, eliminating multiparous donors would decrease donor availability and would not prevent the transfusion of plasma from the far larger number of female donors who had two or less pregnancies (Densmore et al. 1999). Popovsky and Davenport editorialized that deferring all multiparous donors is not warranted in that blood and blood products from none of the 324 women making 9000 donations had resulted in a single report of TRALI (2001). In a subsequent retrospective study, measuring the presence of anti-leukocyte antibodies was felt to be warranted in frequent donors who have a greater likelihood of sensitization (Popovsky 2002).

METABOLIC COMPLICATIONS

Infants are at particular risk of metabolic complications including hyperkalemia, hyper- and hypoglycemia, hypocalcemia, hypothermia, and hemolysis. For many of these complications, the larger relative volume transfused into pediatric compared to adult patients (for example, 15 to 20 mL/kg of red cells, platelets, plasma) and the practical ease of very rapid transfusions appear to contribute the elevated risk.

Hyperkalemia

Potassium (K+) is normally sequestered within cells. Hence, the concentration within cells may be thirty to forty fold higher than the extracellular concentration. As cells age and undergo storage damage, the Sodium (Na+) K+-dependent proton pumps become less efficient, potassium leaks out of the cells, and the extracellular K+ concentration rises. There is no net change in the amount of K+ between a fresh and older unit, rather there simply is a leakage of K+ out of the cell. The K+ concentration of a unit of blood is dependent on the age, irradiation status and the type and volume of anticoagulant. For example, red cells collected in CPDA-1 have a potassium concentration of about 78 mmol/L at day 35 versus 45 to 50 mmol/L at day 42 in Adsol units (Harmening 1994). The extra volume of anticoagulant preservative solution in an additive cell results in both a lower hematocrit and a larger supernatant volume into which K+ is distributed. Other factors that may increase K+ leakage include exposure to storage conditions either too warm (resulting in more rapid loss of glucose to fuel the pumps) or too cold. Particularly relevant to pediatrics is hemolysis and subsequent K+ leakage that may result from exposure to excess heat from malfunctioning blood warmers or inappropriate warming devices. Warming blood in commercial food microwaves (as opposed to Food and Drug Administration [FDA]-cleared devices for blood warming) and exposure of blood to external warming lamps has caused morbidity either from hemolysis or hyperkalemia (McCullough et al. 1972). Rapid transfusions put the infant at greatest risk as they have little chance to metabolize the free potassium. In contrast, in slow or low volume transfusions, the transfused cells regenerate their energy stores and are able to subsequently reduce extracellular K+ concentrations by pumping potassium back into the cell. Hence, high capacity fluid warmers put infants at particular risk; such warmers may be used in bypass, exchange transfusion, or extra corporeal membrane oxygenation (ECMO) (Jameson et al. 1990). Gamma irradiation of RBCs followed by refrigerated storage dramatically affects the rate of K+ leak (Button et al. 1981; Pisciotto and Luban 2001). At 14 days post-collection and irradiation, irradiated red cells had a mean K+ level of 68 mmol/L versus 31 mmol/L for unirradiated blood (Ramirez et al. 1987).

The clinical significance of elevated "plasma" potassium in stored red cells is a function of the volume and rate of transfusion. In one study, no significant changes in plasma K+ levels were observed in neonates transfused small volumes (10 to 20 mL/kg over two to three hours) even when using blood close to 42 days old (Strauss et al. 1994). When small volumes of irradiated blood are transfused to neonates, washing to remove supernatant K+ is not routinely warranted (Strauss 1990). Of note, the studies supporting no further manipulation to reduce K+ levels used inverted spin (that is, plasma depleted) aliquots that were irradiated at the time of issue and not refrigerator stored following irradiation.

Hypoglycemia

Hypoglycemia is another paradoxical potential complication of transfusion, especially for neonates. Hypoglycemia may occur following the initiation of red cell transfusion that contains relatively less glucose than an infusion of dextrose containing crystalloid. Alternatively, hypoglycemia may follow cessation of the red cell transfusion if the neonate has not previously received an intravenous infusion. All anticoagulant preservative solutions contain supraphysiological concentrations of glucose. An infusion of a dextrose-containing solution (D5) provides about 4 to 8 mg/kg/minute of dextrose. If that infusion is stopped to switch to a concentrated red cell solution, between 0.2 to 0.5 mg/kg/minute of dextrose is provided, depending upon whether CPDA-1 or additive solution red cells are transfused (Rock et al. 1999). Therefore, transfusion may result in a transient rise in serum glucose concentrations to 200 mg/dL or greater and spilling of glucose in urine. Transfusion of significant volumes of blood may result in hyperglycemia, which increases insulin release. When transfusion stops, the source of glucose is gone, but insulin remains temporarily elevated resulting in rebound hypoglycemia (Goodstein et al. 1999).

Hypothermia

During cardiac bypass surgery, cardiothoracic surgeons typically stop the heart by infusing a solution cooled to (2° to 6°C) containing a high K+ concentration. The rapid infusion of cold blood units of older age with high K+ provides the potential for arrythmias. To prevent hypothermic arrest, rapid large volume transfusions require the use of a blood warmer. Use only FDA-cleared warmers with alarms as mandated by AABB Standards.

Hypocalcemia

Blood collection occurs into a solution containing citrate, which binds (chelates) free calcium ions and prevents blood clotting. Since calcium is a required constituent of the clotting cascade, removal of all free calcium prevents clotting. It is not possible to anticoagulate a patient with citrate, however, as the loss of free calcium ions would result in asystole long before anticoagulation is achieved. Hence, the risk of excessive infusion rates of citrate-containing blood components is not anticoagulation but heart rate irregularities (Dzik and Kirkley 1988; Uhl and Kruskall 2001). On electrocardiogram, a prolonged QT interval may be observed due to hypocalcemia. Patients undergoing apheresis with albumin replacement are especially at risk for hypocalcemia as the albumin contains no calcium and acts as a metabolic sink (Weinstein 1996; Strauss and McLeod 2001).

Alkalosis is another metabolic complication of large volume infusion of citrate. Since citrate is added in its base form (citrate, not citric acid), it becomes bicarbonate when metabolized. If the rate of citrate administration exceeds the body's capacity to get rid of bicarbonate (as carbon dioxide), profound metabolic alkalosis may result. This phenomenon has been observed following liver transplantation (Driscoll et al. 1987).

Hemolysis

Hemolysis of red cells may result from improper storage, exposure to hypo-osmotic conditions, or other mechanisms of red cell membrane damage including mechanical, bacterial contamination, and irradiation. AABB standards require storage between 2° to 6°C and between 2° to 10°C during transport. Heating of blood using unapproved warmers or malfunctioning warmers has resulted in fatal hemolysis. In the neonatal intensive care unit (NICU) or pediatric ICU (PICU), where a tiny infant may be kept on a warming bed with external heat lamps, there is the unique hazard of hemolysis of the blood product if the container or tubing is in proximity to a heat lamp (Strauss et al. 1986).

It is imperative that only compatible solutions such as normal saline (NS) be used in conjunction with blood transfusions. While initially Isosmotic solutions like Ringer's lactate, D5 50% or 25% NS rapidly become hypo-osmotic as the glucose is metabolized by the red cells. Since red cells have no organelles, they use the glycolytic cycle and have a considerable rate of glucose metabolism.

Since hemolysis can be a sign of bacterial contamination, it is important to examine blood units before transfusion for any signs of contamination, such as unusual discoloration or hemolyzed segments. Because blood drawn up into a syringe is difficult to examine and the aliquoting process may result in contamina-

tion, it is important that blood once aliquoted be transfused promptly and not left in nonmonitored, nontemperature-controlled environments awaiting a convenient opportunity to be transfused (Goldman and Blajchman 2001).

CARDIAC OVERLOAD

Volume shifts from blood drawing and infusion may be profound and result in either hypotension or cardiac overload. While a premature newborn has a blood volume in excess of 100 mL/kg, the miniature size of the most premature infants puts them at special risk of volume shifts. For example, an infant weighing 500 g based on the formula above has a total blood volume of 50 to 60 mL; hence a red cell transfusion of 20 mL/kg represents 10 mL. Recognizing that it may take far more volume than that simply to prime the tubing and 170-micron blood filter, the blood bank will likely dispense a 20 to 50 mL bag or syringe. Nursing staff must ensure that only the 10 mL transfusion volume and not the entire issued volume is transfused. Signs of overload may also be more subtle (increased FiO_2 requirement) and sudden in infants than for adult or older patients (Kevy 1998).

Transfusion-associated graft-versus-host disease (TA-GVHD) is addressed in Chapter 9 and will not be further discussed here.

ALLOIMMUNIZATION

Red cell alloimmunization, may cause delayed hemolytic transfusion reactions (see Chapter 26); transfusion practices for thalassemia and sickle cell patients will be reviewed in Chapters 19 and 21 as well as red cell alloimmunization risks.

Platelet alloimmunization may be as problematic for pediatric cancer patients as for adults, although the prevalence of antibody development may be less. Many multiparous females have some degree of platelet alloimmunization even before receiving any transfusions, which is clearly not an issue for most pediatric transfusion recipients.

Approaches to preventing and optimizing response to transfusion in alloimmunized pediatric patients is reviewed in Chapter 22.

T ANTIGEN ACTIVATION

Necrotizing enterocolitis (NEC) is a severe complication of prematurity that results in ischemic damage to the colonic mucosa and is a potential source of bacterial sepsis. An unusual hematological manifestation of severe cases of NEC is hemolysis that results from bacterial release of sialidases that cleave sialic acid residues creating neoantigens. Naturally occurring complement-dependant antibodies in transfused products may result in lysis of the neoantigen exposed red cells (Pisclotto and Luban 2001). While NEC is reasonably common among extremely premature infants (<750 grams birthweight), T antigen-mediated hemolysis is not. The "attending" antibodies are present in plasma and plasma-containing blood products. Since the hemolysis is complement mediated, one of the factors limiting the rate of hemolysis may be the supply of endogenous complement. Providing washed red cell and platelet components and awaiting plasma and cryoprecipitation would limit the hemolysis, since no additional complement or naturally occurring antibodies would be transfused. Looking for T antigen expression is not practical, in that many healthy neonates have T expression. One recent study documented that about 13% of infants in a NICU had T or T-variant activated red cells when assessed using a lectin panel. T activation was not particularly well correlated with episodes of documented sepsis; many infants who had T activation were "well" infants in convalescence. Despite significant transfusion requirements among infants demonstrating T activation in this study, no episodes of hemolysis were observed. These authors suggest that routine provision of washed components to infants with NEC or routinely measuring T activation in infants with presumed NEC was not warranted (Boralessa et al. 2002). Routine avoidance of all plasma-containing transfusions may cause more harm than good if a subsequent dilutional coagulopathy occurs in extremely premature infants at risk of intracranial hemorrhage (Crookston et al. 2000). Provision of washed components and avoidance of plasma-containing products for NEC patients is only warranted in the setting of documented hemolysis clearly identified to have resulted from T activation (Eder and Manno 2001).

DILUTIONAL COAGULOPATHY

Massive transfusion of any patient may result in a dilutional coagulopathy. Typically, the fall in platelet count contributes to the coagulopathy before loss of coagulation factors become limiting, but neonates may be an exception. This is because neonatal levels of vitamin K-dependent factors are normally low at birth and any further decrease can exacerbate this hemorrhagic disease of the newborn. For example vitamin-dependent factors II, X, and VII typically are at 44%,

40%, and 52% of adult levels, respectively in a full-term newborn, before intramuscular vitamin K is administered (Andrew et al. 2000). The typical setting in which massive dilution occurs is when large volume extracorporeal circuits for cardiopulmonary bypass ECMO or neonatal exchange are required. Transfusion support in pediatric surgery, trauma, and the ICUs requires special attention to the potential for dilutional coagulopathies (Festa et al. 2002). Some programs have developed protocols whenever specific ratios of plasma or platelet-containing products are added to prime circuits to prevent excessive dilutional coagulopathy. The special requirements and consumptive coagulopathies engendered by use of ECMO circuits is discussed in Chapter 17.

COMPLICATIONS OF IRON OVERLOAD FROM CHRONIC TRANSFUSION

While we have the capacity to regulate iron uptake to some degree, there is no natural mechanism for excreting excess iron to maintain iron balance. To compound the dilemma of maintaining iron balance, iron is an essential nutrient for hemoglobin synthesis and for other iron-requiring physiological functions, including myoglobin synthesis and other iron-containing proteins. While too little iron leads to iron deficiency anemia, excess iron can cause myriad problems discussed later; iron stores must be maintained within a narrow safe range. Physiologically, most regulation of iron stores occurs in the gastrointestinal enterocytes that normally only allow uptake of about 10% of dietary iron (Finch 1994). Iron is absorbed predominantly in the duodenum and upper jejunum. Dietary iron is oxidized to the ferric form, which is made more soluble by stomach acids (Andrews and Bridges 1998). Heme iron (as found in meat) is much better absorbed than ferrous ($Fe2+$) iron salts, which in turn are better absorbed than ferric iron ($Fe3+$). Iron salt absorption is facilitated by co-administration with vitamin C, citrate, and amino acids (Andrews 2000). Hence, having a glass of orange juice with a high iron cereal in the morning may substantially increase iron absorption. If body iron stores are sufficient, gastrointestinal absorption is decreased whereas iron deficiency stimulates uptake (Andrews 2002). If iron stores are excessive, intestinal absorption is decreased but not eliminated, making it imperative not to administer multivitamins with iron to patients with transfusional iron overload.

The amount of iron uptake or loss is trivial in comparison to total body stores. Dietary iron uptake and loss (from sloughed skin and mucosa and blood loss) are usually balanced at about one mg/daily. Total body stores are about three to five grams of which about one gram is in the liver and two grams in blood. Each mL of packed red blood cells (PRBCs) contains about one mg of iron, hence significant loss or receipt of blood can significantly disturb the body's iron balance. Because so much of the body stores of iron are relegated to red cell production, there is a significant linkage of erythropoiesis to iron absorption. Hence, congenital disorders of dyserythropoiesis, for example thalassemia major, have increased gastrointestinal iron uptake, despite total body iron overload unless transfusion therapy restores the hematocrit to a level that would suppress the erythropoietic drive. Once absorbed, free iron is toxic, hence virtually all iron is bound to a carrier molecule. In the blood, circulating transferrin acts as the carrier molecule, and in the liver, ferritin acts as the storage molecule. Transferrin allows iron to remain soluble, prevents iron-mediated free radical formation, and enhances transport. Measurement of transferrin is useful to determine saturation, but the absolute concentration is not reflective of iron overload. Similarly elevated serum ferritin levels may be correlated with total body iron overload, but are not linearly correlated with total body iron stores. Only quantification of iron through liver biopsy or measurement using total body iron scans (SQUID) is reflective of total body iron stores.

Iron overload syndromes in pediatrics are largely all iatrogenic. There are rare congenital hemochromatosis syndromes that provide significant insight into the genetics of iron balance, but they are quite rare. In contrast, the gene for hereditary hemochromatosis is quite common, especially among Caucasians with a heterozygous gene frequency between 5% to 10% and homozygous gene frequency of about 1 : 200–400. The heterozygous state rarely has clinical significance, the homozygous state is of significant clinical consequence, but symptoms do not appear until the fourth or fifth decade. The gene for the most frequent form of hereditary hemachromatosis (HLA-H) has been cloned and appears to be responsible for the vast majority of cases in the Caucasian population, thereby raising the intriguing prospect of using widespread genetic screening (McLaren et al. 2003). Genetic screening could be used both to identify homozygous patients at risk of manifesting significant disease and to identify potential highly motivated blood donors as their donation would also provide a health benefit. In fact, some blood centers offer hemachromatosis gene detection, as a tool to increase donations from eligible donors who carry the trait.

Consequences of iron overload are profound. It may take several years of regular transfusion to achieve sig-

nificant oversaturation of transferrin and clinical effects. The organ systems most profoundly affected by chronic iron overload are the liver, heart, and endocrine systems. Other stigmata of thalassemia such as facial or bony deformaties or enlarged liver and spleen result from ineffective erythropoiesis and should be distinguished from consequences of iron overload. The liver is the major site of physiological iron storage and shows the first sequelae of overload; hepatomegaly, which may progress to a small liver when cirrhosis and fibrosis predominate. Because there is little inflammation, profound damage may occur in the absence of significant elevations of serum transaminase levels.

Cardiac toxicity due to iron overload may be manifested as congestive cardiomyopathy, but also restrictive cardiomyopathy and angina in the absence of coronary occlusion. Unfortunately, cardiac symptoms may develop suddenly and harbor the onset of profound and irreversible damage. Routine echocardiography is usually recommended to follow cardiac function of children receiving chronic transfusions.

Endocrine consequences of transfusional iron overload include pancreatic and pituitary dysfunction. Glucose intolerance and insulin dependent diabetes may result from inadequately controlled iron overload. Delayed growth, shortened stature, and delayed sexual maturation are other endocrine disorders that may occur. Cutaneous deposition of iron results in increased melanin production and contributes to the classical tanned appearance of patients with transfusional overload.

Approaches to iron overload include phlebotomy and iron chelation. Phlebotomy is the treatment of choice for adults with genetic homozygous hemochromatosis mutations. By utilizing blood from these donors, the donation contributes to the blood supply and is a lifesaving prophylactic treatment at minimal cost to the health care system. Another approach to phlebotomy is the removal of older cells at the same time that additional transfusions with banked young RBCs are administered; net iron burden can be dramatically reduced. This is most easily achieved by performing automated erythrocytopheresis. This mode of chronic red cell transfusion has been used for sickle patients requiring chronic transfusion who are poorly compliant with iron chelation therapy. In the study of this technique, red cell exchange resulted in a 77% greater donor exposure, but 83% lower total iron burden over time (Kim et al. 1994). In a pilot trial, ferritin levels dropped by 26% over six months of erythrocytopheresis in the absence of chelation, during which time one would otherwise have expected a net increase (Singer et al. 1999).

Chelation of iron in overloaded patients relies on the use of Desferal (deferoxamine) therapy, by constant SQ infusion using a portable pump. Since only a small portion of iron is readily accessible for chelation at any given time, the continuous infusion pump is required. Chelation therapy is best started before signs and symptoms of iron overload have begun, as not all organ damage is reversible even with chelation begun early in life. Hepatocyte iron is the chief target of chelation therapy as the liver is the major site of body iron storage. When chelated with deferoxamine, iron is excreted in stool. The other pool of chelatable iron is the ~20 mg of iron being recycled daily by macrophage turnover of senescent red cells. Iron released from macrophages may be accessible to deferoxamine and is likely the major source of urinary iron excretion on therapy (Parter and Olivieri 2001).

Desferal therapy has dual roles. In addition to iron chelation and removal, it also detoxifies free iron in the circulation. Hence, even in a state of significant overload, continuous infusion deferoxamine therapy binds free iron that could do further damage and prevents additional free radical formation. This may well be why clinical improvements have been noted in patients with end-stage iron overload complications when aggressive chelation therapy is started. Deferoxamine is too large a molecule to be well absorbed orally. It was first administered by bolus intramuscular (IM) injections, which yielded significant urinary iron excretion; however, the IM route was not able to achieve significant iron burden reduction when administered at a frequency that patients can comply with. Subcutaneous infusion of about 40 mg/kg over 8 to 12 hours is now the most common dosing regimen (Propper et al. 1977). Although it has a wide therapeutic index, significant toxicities are associated with chronic use. These include retinal toxicity, auditory toxicity, and effects on growth. The former can generally be avoided by limiting doses to 40 mg/kg/day and titrating the dose to the serum ferritin level (Porter and Olivieri 2001). Because compliance with deferoxamine therapy is difficult (both due to side effects and the stigma of having to wear a continuous infusion pump), there has long been interest in alternative regimens. Short term IV therapy has been demonstrated to achieve significant iron unloading. Alternatively twice daily subcutaneous injections have been shown to approach the same effect as continuous infusion administration. Even two to three times per week administration of deferoxamine achieves better iron excretion than the oral chelator deferiprone, hence deferoxamine remains the treatment of choice (Weatherall and Clegg 2001). The most widely studied oral chelating agent is deferiprone, which has had mixed reviews. Its advantage is that it may be administered orally. Problems include side effects such as neutropenia and even agranulocytosis (<5%) and arthropathy.

Other adverse reported symptoms include nausea (8%), zinc deficiency (14%), and abnormal liver function tests (44%). Perhaps most problematic is limited capacity to unload iron in a significant proportion of patients (Olivieri and Brittenham 1997).

Iron overload in chronically transfused sickle cell patients is similar to that observed in thalassemia patients, although most sickle cell patients do not begin chronic transfusion until late childhood and hence are likely to develop symptoms later. This raises the challenge that it may be harder to achieve compliance in a teenager required to start a new intensive medical regimen than in a young child who has grown up with this requirement. This has lead to red cell exchange transfusion programs previously described.

References

Andrew M, Monagle PT, and Brooker L. 2000. Developmental hemostasis: relevance to thrombotic complications in pediatric patients. In Andrew M, Monagle PT, and Brooker L, eds. *Thromboembolic complications during infancy and childhood.* Hamilton: BC Decker Inc.

Andrews NC. 2002. A genetic view of iron homeostasis. *Semin Hematol* 39:227–234.

Andrews NC. 2000. Iron metabolism:iron deficiency and iron overload. *Annu Rev Genomics Hum Genet* 1:75–98.

Andrews NC and Bridges KR. 1998. Disorders of iron metabolism and sideroblastic anemia. In Nathan DG and Orkin S, eds. *Nathan and Oski's hematology of infancy and childhood.* Philadelphia: WB Saunders.

Boralessa H, Modi N, Cockburn H, et al. 2002. RBC T activation and hemolysis in a neonatal intensive care population: implications for transfusion practice. *Transfusion* 42:1428–1434.

Busch MP. 2001. Closing the windows on viral transmission by blood transfusion. In Stramer S, ed. *Blood safety in the new millennium.* Bethesda, MD: American Association of Blood Banks.

Button LN, DeWolf WC, Newburger PE, Jacobson MS, and Kevy SV. 1981. The effects of irradiation on blood components. *Transfusion* 21:419–426.

Couban S, Carruthers J, Andreou P, et al. 2002. Platelet transfusions in children: results of a randomized, prospective, crossover trial of plasma removal and a prospective audit of WBC reduction. *Transfusion* 42:753–758.

Crookston KP, Reiner AP, Cooper LJN, Sacher RA, Blajchman MA, and Heddle NM. 2000. RBC T activation and hemolysis: implications for pediatric transfusion management. *Transfusion* 40: 801–812.

Cursio C and Fountas M. 2002. Building relationships to improve quality (abstract). *Clin Leadersh Manag Rev* 16:158–161.

Densmore TL, Goodnough LT, Ali S, Dynis M, and Chaplin H. 1999. Prevalence of HLA sensitization in female apheresis donors. *Transfusion* 39:103–106.

Driscoll DF, Bistrian BR, Jenkins RL, et al. 1987. Development of metabolic alkalosis after massive transfusion during orthotopic liver transplantation. *Critical Care Medicine* 15:905–908.

Duguid JKM, Minards J, and Bolton-Maggs HB. 1999. Lesson of the week: incompatible plasma transfusions and haemolysis in children blood group compatibility must be observed when transfusing plasma or platelets to children to avoid haemolysis. *BMJ* 318:176–177.

Dykes A, Smallwood D, and Kotsimbos T. 2000. Transfusion-related acute lung injury (TRALI) in a patient with a single lung transplant. *Br J Haematol* 109:674–676.

Dzik S. 1999. The use of blood components prior to invasive bedside procedures: a critical appraisal. In Mintz PD, ed. *Transfusion therapy: clinical principles and practice.* AABB Press Bethesda, MD.

Dzik WH. 2003. Emily Coolss lecture 2002: transfusion safety in the hospital. *Transfusion* 43:1190–1199.

Dzik WH and Kirkley SA. 1988. Citrate toxicity during massive blood transfusion. *Transfusion Medicine Reviews* 2:76–94.

Dzik WH, Murphy MF, BEST working party. 2002. An international study of the performance of patient sample collection [abstract]. *Transfusion* 42:26S.

Eder AF and Manno CS. 2001. Does red cell T activation matter? *Br J Haematol* 114:25–30.

Festa CJ, Feng AK, and Bigos D. 2002. Transfusion support in pediatric surgery, trauma and the intensive care unit. In Herman JH and Manno CS, eds. *Pediatric transfusion therapy.* Bethesda, MD: AABB Press.

Finch C. 1994. Regulators of iron balance in humans. *Blood* 84:1697.

Fong SW, Qaqundah BY, and Taylor WF. 1974. Developmental patterns of ABO isohemagglutanins in normal children correlated with the effects of age, sex and maternal isohemagglutinins. *Transfusion* 14:551.

Goldman M and Blajchman MA. 2001. Bacterial contamination. In Popovsky M, ed. *Transfusion reactions.* Bethesda, MD: AABB Press.

Goodstein MH, Herman JH, Smith JF, and Rubenstein SD. 1999. *Pediatric Pathol Lab Med* 18:173–185.

Gorlin JB. 2002. AABB Standards for Blood Banks and Transfusion Services, 21st ed. Bethesda, MD: AABB Press.

Grundbacher FJ. 1967. Quantity of hemolytic anti-A and anti-B in individuals of a human population: correlations with isoagglutinins and effects of the individual's age and sex. *Z Immun Forsch* 134:317.

Harmening DM. 1994. Modern blood banking and transfusion practices. Philadelphia: F.A. Davis.

Heddle NM, Blajchman MA, Meyer RM, et al. 2002. A randomized controlled trial comparing the frequency of acute reactions to plasma-removed platelets and prestorage WBC-reduced platelets. *Transfusion* 42:556–566.

Heddle NM and Kelton JG. 2001. Febrile nonhemolytic transfusion reactions. In Popovsky MA, ed. *Transfusion reactions.* Bethesda, MD: AABB Press.

Jameson L, Popic PM, and Harms BA. 1990. Hyperkalemic death during use of a high-capacity fluid warmer for massive transfusion. *Anesthesiology* 73:1050–1052.

Kao GS, Wood IG, Dorfman DM, Milford EL, and Benjamin RJ. 2003. Investigations into the role of anti-HLA class II antibodies in TRALI. *Transfusion* 43:185–191.

Kevy SV. 1998. Blood products used in the newborn. In Cloherty JP and Stark AR, eds. *Manual of neonatal care.* Philadelphia: Lippincott-Raven.

Kim HC, Dugan NP, Silber JH, et al. 1994. Erythrocytapheresis therapy to reduce iron overload in chronically transfused patients with sickle cell disease. *Blood* 83:1136–1142.

Kopko PM, Marshall CS, MacKenzie MR, Holland PV, and Popovsky MA. 2002. Transfusion-related acute lung injury: report of a clinical look-back investigation. *JAMA* 287:1968–1971.

Kopko PM, Popovsky MA, MacKenzie MR, Paglieroni TG, Muto KN, and Holland PV. 2001. HLA class II antibodies in transfusion-related acute lung injury. *Transfusion* 41:1244–1248.

Larsson LG, Welsh VJ, and Ladd DJ. 2001. Acute intravascular hemolysis secondary to out-of-group platelet transfusion. *Transfusion* 40:902–906.

Linden JV, Wagner K, Voytovich AE, Sheehan J. 2000. Transfusion errors in New York State: an analysis of 10 years' experience. *Transfusion* 40:1207–1213.

Lozano M and Cid J. 2003. The clinical implications of platelet transfusions associated with ABO or Rh(D) incompatibility. *Transfusion Medicine Reviews* 17:57–68.

Ludvigsen CW, Swanson JL, Thompson TR, and McCullough J. 1987. The failure of neonates to form red blood cell alloantibodies in response to multiple transfusions. *Am J Clin Path* 87:250.

Lumadue JA, Boyd JS, Ness PM. 1997. Adherance to a strict specimen-labeling policy decreases the incidence of erroneous blood grouping of blood bank specimens. *Transfusion* 37:1169–1172.

Mair B and Benson K. 1998. Evaluation of changes in hemoglobin levels associated with ABO-incompatible plasma in apheresis platelets. *Transfusion* 38:51–54.

McCullough J, Polesky HF, Nelson C, and Hoff T. 1972. Iatrogenic hemolysis: a complication of blood warmed by a microwave device. *Anesthesia and Analgesia* 51:102–106.

McLaren CE, Barton JC, and Adams PC. 2003. Hemochromatosis and iron overload screening (HEIRS) study design for an evaluation of 100,000 primary care-based adults. *Am J Med Sci* 325:53–62.

Moroff G, Friedman A, Robkin-Kline L, Gautier G, and Luban NLC. 1984. Reduction of the volume of stored platelet concentrates for use in neonatal patients. *Transfusion* 2:144–146.

Olivieri NF and Brittenham GM. 1997. Final results of the randomized trial of deferiprone (L1) and deferoxamine (DFO). *Blood* 90:264a.

Palfi M, Berg S, Ernerudh J, and Berlin G. 2001. A randomized controlled trial of transfusion-related acute lung injury: is plasma from multiparous blood donors dangerous? *Transfusion* 41:317–322.

Pisciotto PT and Luban NLC. 2001. Complications of neonatal transfusion. In Popovsky M, ed. *Transfusion reactions.* Bethesda, MD: AABB Press.

Popovsky MA. 2002. Breathlessness and blood: a combustible combination. *Vox Sang* 83 Suppl 1:147–150.

Popovsky MA. 2001. Transfusion-related acute lung injury (TRALI). In Popovsky MA, ed. *Transfusion reactions.* Bethesda, MD: AABB Press.

Popovsky MA. 2000. Transfusion-related acute lung injury. *Curr Opin Hematol* 7:402–407.

Popovsky MA and Davenport RD. 2001. Transfusion-related acute lung injury: femme fatale? *Transfusion* 41:312–314.

Popovsky MA, Chaplin HC, and Moore SB. 1992. Transfusion-associated lung injury: a neglected serious complication of hemotherapy. *Transfusion* 32:589–592.

Porter JP and Olivieri NF. 2001. Secondary iron overload. In Anonymous. Hematology 2001: American Society of Hematology Education Program Book. Orlando: ASH.

Propper RL, Cooper B, and Rufo RR. 1977. Continuous subcutaneous administration of deferoximine in patients with iron overload. *N Engl J Med* 297:418–423.

Ramirez AM, Woodfield DG, Scott R, and McLachlan J. 1987. High potassium levels in stored irradiated blood. *Transfusion* 27:444–445.

Rizk A, Gorson KC, and Kenney L. 2001. Transfusion-related acute lung injury after the infusion of IVIG. *Transfusion* 41:264–268.

Rock G, Poon A, and Haddad S. 1999. Nutricel as an additive solution for neonatal transfusion. *Transfusion Science* 20:29–36.

Rosse WF, Telen M, and Ware RE. 1998. Transfusion support for patients with sickle cell disease. Bethesda, MD: AABB.

Sandler SG, Mallory D, Malamut D, and Eckrich R. 1995. IgA anaphylactic transfusion reactions. *Transfusion Medicine Reviews* 9:1–8.

Sazama K. 1997. Informed consent for transfusion: an overview. In Stowell C, ed. *Informed consent for blood transfusion.* Bethesda, MD: American Association of Blood Banks.

Sazama K. 1990. Reports of 355 transfusion-associated deaths: 1976 through 1985. *Transfusion* 30:583–590.

Silliman C. 1999. Transfusion-related lung injury. *Transfusion* 39:177–186.

Silliman C, Boshkov LK, Mehdizadehkashi Z, et al. 2003. Transfusion-related acute lung injury: epidemiology and a prospective analysis of etiologic factors. *Blood* 101:454–462.

Singer ST, Quirolo K, Nishi K, Hackney-Stephens E, Evans C, and Vichinsky EP. 1999. Erythrocytapheresis for chronically transfused children with sickle cell disease: an effective method for maintaining a low hemoglobin S level and reducing iron overload. *J Clin Apheresis* 14:122–125.

Strauss RG. 2000. Neonatal transfusion. In Anderson K, Ness PM, eds. Scientific basis of transfusion medicine. Philadelphia: W.B. Saunders.

Strauss RG. 1990. Routinely washing irradiated red cells before transfusion seems unwarranted. *Transfusion* 30:675–677.

Strauss RG and McLeod BC. 2001. Complications of therapeutic apheresis. In Popovsky M, ed. *Transfusion reactions.* Bethesda, MD: AABB Press.

Strauss RG, Bell EF, and Snyder EL. 1986. Effects of environmental warming on blood components dispensed in syringes for neonatal transfusions. *J Pediatrics* 109:109–113.

Strauss R, Burmeister LF, and Johnson K. 1996. AS-1 red cells for neonatal transfusions: a randomized trial assessing donor exposure and safety. *Transfusions* 36:873–878.

Uhl L and Kruskall MS. 2001. Complications of massive transfusion. In Popovsky M, ed. *Transfusion reactions.* Bethesda, MD: AABB Press.

Vichinsky EP, Earles A, Johnson RA, Hoag MS, Williams A, and Lubin B. 1990. Alloimmunization in sickle cell anemia and transfusion of racially unmatched blood. *New Engl J Med* 322:1617–1621.

Weatherall DJ and Clegg JB. 2001. *The thalassaemia syndromes.* Oxford: Blackwell Science.

Weinstein R. 1996. Prevention of citrate reactions during therapeutic plasma exchange by constant infusion of calcium gluconate with the return fluid. *J Clin Apheresis* 11:204–210.

Zoon KC. Transfusion associated lung injury. 8-13-2001. FDA guidance CBER. Bethesda, MD.

28

Infectious Complications

FARANAK JAMALI, MD, AND PAUL M. NESS, MD

INTRODUCTION

The three major transfusion-transmitted viral agents that cause significant morbidity and mortality are: human immunodeficiency virus (HIV), hepatitis B, and hepatitis C. The risk of transfusion-transmitted hepatitis has diminished steadily since the early 1980s. Transfusion-transmitted HIV infection peaked in the early to mid 1980s and diminished significantly with the implementation of HIV antibody testing in 1985; the risk decreased further with the addition of molecular testing technology. The emergence of acquired immunodeficiency syndrome (AIDS) highlighted transfusion as a high-risk therapy. Advances in testing have reduced the risk of disease transmission in the United States (Table 28.1). However, there are still persistent risks of transmission of infections through transfusions in some cases due to emerging agents.

HUMAN IMMUNODEFICIENCY VIRUS

HIV History

The majority of HIV transmissions in the United States occurred before discovery of the virus and institution of blood donor screening. This early phase of transmission was attributable primarily to homosexual and bisexual men, groups who were eligible blood donors at the time. Studies have shown a rapid increase in the risk of transfusion-associated HIV-1 infection from 1978 to 1982 (Fiebig et al. 2003). When the first cases of transfusion-associated AIDS were reported (in

hemophiliacs and recipients of blood components) late in 1982, blood banks educated blood donors that individuals at high risk for AIDS (gay or bisexual males, intravenous drug users, hemophiliacs, Haitian immigrants) must abstain from donating blood. Subsequent studies of 18 cases of transfusion-associated AIDS collected by the Centers for Disease Control (CDC) confirmed the wisdom of this approach (Curran et al. 1984). The etiological agent for AIDS was identified later in 1984 (Gallo et al. 1984). The virus was originally called LAV by Montagnier and HTLV-3 by Gallo; later the nomenclature for the retrovirus universally accepted as the cause of AIDS was changed to human immunodeficiency virus (HIV).

These discoveries facilitated the development of a serological assay for the HIV virus, which was incorporated by blood banks as a part of routine donor screening in early 1985. Screening utilizes an ELISA method, which is sensitive but not adequately specific for definitive diagnosis, particularly in low-risk individuals such as blood donors. When ELISA testing is coupled with the more specific Western blot procedure for samples, which are repeatedly reactive by the ELISA test, HIV testing is highly sensitive and specific.

Although these interventions clearly reduced the risk of AIDS from transfusion, it became apparent that some transfusion-related infections continued to occur (Ward et al. 1988). Epidemiological studies of seropositive donors suggested that it was necessary to strengthen and redefine donor exclusion criteria. In September 1985, the Food and Drug Administration (FDA) changed the wording of deferral criteria so that "any male who has had sex with another male since 1977, even one time," would be excluded. In November

TABLE 28.1   Current Risk Estimates of Viral Transmission
by Transfusion

Virus/Infecton	Estimated Risk in 2002
Human immunodeficiency virus (HIV)	1:2,135,000[75]
Hepatitis C (HCV)	1:1,935,000[75]
Hepatitis B (HBV)	1:1,205,000[75]
Human T-cell lymphotropic virus (HTLV-I/II)	1:641,000[24]

1986, male and female prostitutes and their partners were included among those to be deferred. In December 1986, the confidential unit exclusion (CUE) form became a required procedure at most blood centers (Pindyck et al. 1985). This form provides an opportunity for high-risk persons who donate to indicate, in a confidential manner, that their blood should not be used for transfusion. Later, modifications of donor screening procedures evolved so that donor room personnel directly confront each donor with questions about his/her high-risk activity for AIDS.

HIV-2 is a second retrovirus that can cause immune deficiency in patients (O'Brien et al. 1992). Although it has not been identified as a cause of transfusion-transmitted AIDS as yet in the United States, the virus is sufficiently similar to HIV-1 to warrant its specific detection in blood donors and elimination from the blood supply. In 1992, an ELISA method, the HIV-1/HIV-2 combination test was introduced for blood donor screening; this test shortens the window period of undetectable infectivity for HIV infection since the test is reactive with both IgM and IgG antibodies to HIV viruses. It is currently believed that the window period from infection of a potential blood donor to antibody detection is approximately 25 days (Busch et al. 1995).

Additional laboratory steps to reduce HIV infection were implemented by blood banks in 1995 with testing for HIV p24 antigen (Table 28.2), which was proposed to shorten the seronegative window to approximately 20 days (Alter et al. 1990); the detection rate of donors who are positive for p24 antigen but negative for antibody is several cases per year among approximately 12 million annual blood donations. In 1999, additional screening with nucleic acid testing (NAT) for HIV (see Table 28.2) was implemented to potentially reduce the window by several additional days (Busch et al. 1997); although the detection rate is expected to be very low, several donors have been identified by NAT who were negative by p24 and antibody screening. These interventions have been very successful in making transfusion-associated AIDS a rare event. A study in multiply transfused cardiac surgery patients in

Baltimore and Houston from 1985–1991 provided prospective evaluation of 11,535 cardiac surgery patients who were transfused with 120,290 units of tested blood; only two HIV-1 seroconversions occurred, corresponding to an HIV-1 infection rate of 0.0017% or a residual risk per unit of about 1 in 60,000 units (Cohen et al. 1989). This study was supported by another investigation of 61,000 units of blood tested for HIV-1 that were studied with HIV-1 viral culture and polymerase chain reaction; only one positive unit was identified (Busch et al. 1991). Both of these studies overestimate the risk of posttransfusion AIDS since many patients die from their primary disease or natural causes in the latency period before clinical AIDS appears.

Transmission of HIV Infection through Transfusion

Refrigeration and freezing of blood products cannot inactivate HIV. The type of blood component transfused and its duration of storage are variables that have been identified to correlate with likelihood of HIV-1 transmission (Donegan et al. 1990). Early studies showed that washed red cell components and red blood cell (RBC) units stored longer than 26 days had lower rates of transmission than other components. Thus, component manipulations may reduce, but do not eliminate, infection by reducing free virus, viable white cells, or both in plasma.

Effect of Allogeneic Transfusion in Progression of HIV Disease

Interestingly, transfusion not only provides a mode of transmission for HIV but, according to some evidence, may also affect the course of disease in patients with AIDS. Retrospective studies and some small prospective studies suggest that allogeneic transfusions may accelerate progression of infection and cause shortened survival in HIV patients. In vitro studies of coculture of allogeneic leukocyte subsets, presumably through immunological activation of lymphocytes and macrophages, cause more active replication of HIV (Busch et al. 1992). However, a subsequent clinical trial did not confirm the in vitro observations (Collier et al. 2001).

HIV Subtypes

A characteristic feature of HIV and other lentiviruses is extensive genetic diversity. HIV-2 was initially discovered in 1985 in West Africa and was called HTLV-IV and lymphadenopathy-associated virus type-2. Viral penetrance into Western European countries has now been documented, with a significant number of infected blood donors, and some reported cases of transfusion-associated transmission (O'Brien et al. 1992). The prevalence of HIV-2 is <1:15,000,000 screened donations in the United States (Fiebig et al. 2003). There have been no cases of transfusion-transmitted HIV-2 reported in the literature up to 2001 (Fiebig et al. 2003). Routes of transmission of HIV-2 and HIV-1 are the same, including sexual contact, intravenous drug use, and transmission during fetal life. The rate of disease transmission appears to be lower, however, in persons infected with HIV-2. The resultant immunodeficiency state from HIV-2 infection is similar to that seen with HIV-1.

Diagnostic Tests

Nucleic acid similarity between HIV-1 and -2 causes crossreaction in most ELISA assays. Most HIV-2 infected persons react with anti-HIV 1 antisera; however, they show variable reactivity with HIV-1 Western blot. This crossreactivity was helpful in screening for HIV-2 when the combination HIV-1/HIV-2 screening tests were not available. Combination HIV-1/HIV-2 assays were developed in the late 1980s. Since 1992, FDA regulations require donor blood to be tested using either a combination HIV-1/HIV-2 test or a separate anti-HIV-2 test. Though NAT has been in use in Europe since 1998 and in the United States since 1999 to screen blood donors for HIV, donor screening for HIV-1 and HIV-2 antibodies is still required. Testing for HIV-1 Ag, (p24) Ag, which became mandatory in the United States in March 1996 (see Table 28.2), may be discontinued if NAT is being performed (LaGier and Murphy 2003).

Nucleic Acid Testing

The United States blood supply has been screened for HIV and HCV with the new technology, NAT, since 1999. The window period for HCV is considerably shortened, from 70 days using the most sensitive antibody assays to an estimated 10 to 30 days, with NAT. For HIV, NAT (in addition to p24 antigen testing) has shortened the 22-day window period to an estimated 10 days. It has been also used as an alternative to licensed HIV-1 p24 antigen tests. The estimated risks per unit of blood transfused in the United States after NAT testing are: HCV 1:1,935,000 units; HIV 1:2,135,000; HBV 1:205,000; and HTLV-1 and -2 1:641,000 units (Dodd et al. 2002). (See Table 28.1) Rare, unfortunate cases of HIV transmission from NAT screened units have occurred, demonstrating the potential infectiousness of such units.

TABLE 28.2 Chronology of Blood Donor Screening Tests in the United States

Year	Disease Marker	Method Presently in Use
1940s	Syphilis	Reagin-based tests (RPR)
1974	HBs Ag	RIA*
1979–80	HBs Ag	EIA*
1985	Anti-HIV	EIA
1986–87	ALT	Enzymatic
1986–87	Anti-HBc	EIA
1988	Anti-HTLV-1	EIA
1992	Anti-HIV-I/II	EIA
1995	Syphilis	PK-TP*
1996	HIV-p24	EIA
1996	Anti-HCV (3rd generation)	EIA
1998	Anti-HTLV-I/II	EIA
1999	HCV RNA	PCR/NAT*
2000	HIV RNA	PCR/NAT
2003	West nile virus	PCR/NAT
2003	Bacteria in platelet products	Pending

RIA = Radioimmunology, EIA = Enzyme-linked immunosorbent assay, PK-TP = An automated test for treponemal antibodies, PCR = Polymerase chain reaction, NAT = Nucleic acid amplification test.

Supplemental Testing

HIV-EIA is a very sensitive screening test most commonly used for viral antibodies. Samples that show no reactivity in the initial screening test should be considered negative and need not be tested further. Despite the excellent specificity, the majority of positive screening results are false-positive. Supplemental assays are therefore essential to confirm positive screening results. Currently, United States blood banks employ either HIV-1 Western blot test or immunofluorescence (IFA) assays for confirmation of a repeatedly reactive anti-HIV-EIA, in combination with a licensed anti-HIV-2 EIA and an unlicensed HIV-2 supplemental assay. For currently licensed assays, a positive interpretation requires antibody reactivity to two of the following three HIV antigens: p24 (the major gag protein), gp41 (transmembrane env protein), and gp 120/160 (external env protein/env precursor protein). All donors who test repeat reactive with persistently indeterminate Western blot patterns are deferred from further blood donations and are notified of their screening and supplemental test result.

HEPATITIS

Etiologies and Definitions

Although AIDS has received most of the attention, posttransfusion hepatitis (PTH) has been a more common problem and an under-recognized cause of morbidity in transfusion recipients. Transfusion-associated hepatitis (TAH) is almost exclusively caused by viruses. These viruses include hepatitis viruses A through E (HAV, HBV, HCV, HDV, HEV), cytomegalovirus (CMV), Epstein-Barr virus (EBV), and possibly newly described hepatitis viruses (such as GBV-C, TTV, and SEN-V). Blood donors with HBV and HCV can have a prolonged asymptomatic carrier state, whereas HAV and HEV are enterically transmitted viruses, which circulate only transiently during the acute phase of infection with no carrier state, and the viremic individual is usually clinically ill and symptomatic. Therefore HAV and HEV are not considered as important etiologies for TAH. Hepatitis D virus is a defective virus, which is found only in the presence of HBV infection; therefore screening donors for HBV simultaneously eliminates the risk of HDV.

Non-A, Non-B Hepatitis

A number of studies have shown that <1% of cases of community-acquired hepatitis in the United States are caused by transfusion (Alter 2002). A relatively large number of TAH patients showed transaminase elevations after transfusion but were not attributable to any underlying disease, medication, or other forms of viral hepatitis (Dienstag et al. 1977). In the early 1970s the virus that caused hepatitis B was identified as the Australia antigen, and tests using sensitive radioimmunoassay systems were instituted (Hollinger et al. 1974). These measures took place at about the same time that most hospitals switched to an "all volunteer blood supply," a measure that was expected to lower the incidence of PTH as well. Despite these actions, cases of PTH still occurred. Because these cases had no serological markers of infection with hepatitis A or B, the term non-A, non-B hepatitis (NANB) was coined for this disease with a presumed viral etiology. A major study of 1513 transfusion recipients and control patients (1974–1979) in four United States cities (Aach et al. 1981) was performed. Patients with NANB PTH were identified by fluctuating levels of alanine aminotransferase (ALT), the only means of identifying this disease since about 50% of patients do not develop overt jaundice or acute clinical illness. Approximately 10% of patients developed NANB PTH. The study also showed that 40% of cases could have been prevented by screening blood donors with the ALT test. When cohorts with NANB PTH were followed for longer periods of time, results of liver biopsies demonstrated that 50% of patients developed serious clinical problems such as chronic active hepatitis, cirrhosis, and hepatocellular carcinoma (Dienstag and Alter 1986). Based upon the evolving picture of NANB PTH, ALT screening was implemented in 1988 since a specific screening test remained elusive. When further studies demonstrated that the antibody to hepatitis B core antigen (anti-HBc) also identified a separate population of blood donors more likely to transmit disease, both of the assays were introduced as surrogate tests for NANB PTH. Together with the more rigid screening of donor histories for high-risk activities associated with AIDS and HIV screening, these surrogate tests probably reduced the rate of NANB PTH by 50%.

10% to 30% of chronic NANB hepatitis cases are caused by agents other than HCV, some of which include: GB virus (GBV), the hepatitis G virus (HGV), the TT virus, and the SEN virus; these other agents have solved part of the NANB hepatitis enigma but not all of it.

Hepatitis C

Like HBV and HIV, HCV has a sexual as well as perinatal and parenteral route of transmission. However, HCV is much less efficiently transmitted through the nonparenteral routes. Only 25% of patients with HCV present with clinical symptoms; the rest of them show only laboratory abnormalities, primarily ALT elevation. However, even silent TAH can cause significant hepatic consequences. Patients with HCV rarely present with sufficient symptoms to seek medical care. HCV is the cause of at least 90% of cases of TAH. Since the introduction of specific anti-HCV antibody screening test in 1990, there has been a marked reduction in the incidence of TAH. The most important characteristic of HCV is its potential to evolve into a chronic hepatitis, cirrhosis, or rarely hepatocellular carcinoma. Recent data show that about 70% of HCV-infected patients eventually progress to chronic hepatitis.

Plasma-derived clotting factor concentrates are made virologically free by the combination of multiple viral screening steps and inactivation by filtration, heat, and solvent-detergent treatment. At least 50% of adult hemophiliacs **>18 years of age** have ALT elevations and serological evidence of chronic hepatitis C (Seeff 1981). This is mostly because of extensive use of pooled plasma components before the application of viral inactivation.

Studies done by Kiyosawa et al. (1982) showed that there is a long interval between transfusion and development of chronic liver disease (11.3 years). Long-term follow-up of children who underwent cardiac surgery demonstrated that children affected by posttransfusion HCV infection might have a higher rate of recovery and a milder course of disease than adults (Vogt et al. 1999). However, most adults who become infected with HCV will become chronic carriers. They have 75% to 85%

chance of showing persistent HCV RNA in the serum and liver. Despite this prolonged carrier state, most HCV-infected patients remain free of symptoms. However, 50% of chronic carriers will have either biochemical or histological evidence of inflammation in the liver.

Markers of HCV Infection in Blood Donors

Anti-HCV antibody is detected by EIA. This antibody is detectable about 10 weeks after infection. About 0.21% of blood donors in the United States show a repeatedly reactive EIA test (Brecher 2002). Positive EIA tests are confirmed by RIBA (recombinant immunoblot assays). A positive RIBA result is interpreted as a true HCV infection. However, regardless of RIBA result, a donor with repeatedly reactive EIA tests is excluded from future donation.

The most recent test approved by FDA for donor screening is HCV NAT, which reduces the window period for HCV detection to 10 to 30 days (Busch 2000). With initiation of NAT testing the estimated risk of HCV transmission per unit transfused in the United States is currently 1:1,935,000 (see Table 28.1).

Hepatitis B Virus

HBV belongs to the hepadnavirus family. It is an enveloped double-stranded DNA virus, which can be transmitted through sexual, percutaneous, and maternoneonatal routes. In adults, after the initial infection with HBV, about 95% of patients fully recover and develop protective anti-HBs antibody; whereas about 5% of them become chronic carriers. These patients produce HBsAg detectable in the plasma. In contrast, approximately 90% of infants with perinatal HBV infection become carriers and are at risk of progressing to cirrhosis and hepatocellular carcinoma (Brecher 2002).

Transfusion-Transmitted Hepatitis B

Studies from the 1970s show that the incubation period for transfusion-transmitted HBV infection is about 11 to 12 weeks (Dienstag 1983). HBV infections produce more symptomatology than HCV infection. ALT levels are also higher and jaundice occurs more frequently in HBV than with HCV infection. Currently the risk of HBV transmission through blood transfusion in the United States is 1:205,000 (see Table 28.1). Demographic studies showed that HBsAg was more common in males than in females, in blacks and Asian-Americans than in whites, and in paid donors than in volunteer donors (Tabor et al. 1980). Furthermore, paid

donors have a higher incidence of anti-HBc positivity. Each donation intended for allogeneic use must be tested for HBsAg and anti-HBc antibody.

Hepatitis A Transmission through Transfusion

Hepatitis A virus is a nonenveloped picornavirus. It has a single-stranded RNA genome. Its incubation period is about 4 to 6 weeks. The classic route of infection is fecal/oral. HAV viremia during the incubation period is asymptomatic, so the transmission of the virus could happen during this period. Isolated cases have been reported in association with transfusion of cellular components and plasma derived factor VIII concentrates (Dodd 1996). However, transmission of virus during the acute phase of the illness almost never happens because the donor is usually acutely ill and not a candidate to donate blood. Donors are not routinely tested for HAV.

HUMAN T-CELL LYMPHOTROPIC VIRUS (HTLV)

Both HTLV-I and HTLV-II can be transmitted by blood transfusion. They can produce chronic retroviral infection in humans, which can have serious outcomes. In the United States screening of blood donors for HTLV-I and II has been in place since 1988.

HTLV-I

Human T-cell lymphotropic virus, type 1 (HTLV-1) was the first recognized human retroviral agent. It was associated with a CD4+ T-lymphocytic lymphoma known as adult T-cell leukemia/lymphoma (ATL) (Hinuma et al. 1981; Hinuma et al. 1982). It is also associated with HTLV-associated myelopathy (HAM), also called tropical spastic paraparesis (TSP).

Areas endemic for HTLV-I are: subSaharan Africa; the Caribbean region, Peru, Columbia, and Brazil; South West Japan; and parts of Melanesia and Australia. Countries like Iran, India, and Taiwan with a large number of immigrants from the endemic regions of the world have reported some cases of HTLV-I. Transmission of HTLV-I occurs via breast milk, sexual contact (predominately male to female), and blood.

HTLV-II

Human T-cell lymphotropic virus, type 2 (HTLV-II) was discovered many years after the isolation of HTLV-I. HTLV-II infection is also prevalent among intravenous drug users and their sexual partners in the United States, Brazil, and Europe. It also has a high prevalence among some Native Americans. Its genome has >60% homology to that of HTLV-I. HTLV-II can cause HAM less frequently than HTLV-I. Both HTLV-I and HTLV-II, particularly the latter, can be associated with infectious processes, such as pneumonia, bronchitis, and urinary tract infections.

Transfusion-Related Transmission

Both viruses are markedly cell associated. Therefore, transfusion of infected viable lymphocytes can transmit infection. Plasma, which is almost acellular, does not transmit the infection. The risk of HTLV transmission is inversely related to the duration of refrigeration and storage, presumably related to the viability of lymphocytes. After about 10 days of refrigerated storage, there is much less risk of transmission of infection (Brecher 2002).

Soon after its discovery in Japan, transmission efficiency was shown to be at least 50%, whereas, in the United States, it appears to be much less efficient. Transmission of HTLV-I through transfusion has been associated with rapid onset HAM and at least one case of ATL (Brecher 2002).

A study on recipients of large volume transfusions in the United States showed that after 1988 and the institution of universal donor screening for HTLV antibodies, the estimated risks of HTLV transmission was 1.4 per 100,000 units (Nelson et al. 1992).

Donor Screening

In the United States, blood donor screening for anti-HTLV-I started in 1988. The initial test might have missed up to 50% of HTLV-II infected donors. In 1998, implementation of combination HTLV-I/II EIA tests for blood donor screening reduced further the risk of transmission of both viruses through transfusion. A donor with repeatedly reactive EIA screening tests should be deferred indefinitely. Because of significant homology of their genomes, there is significant crossreaction of HTLV-I with HTLV-II EIAs.

CYTOMEGALOVIRUS

CMV is a member of Betaherpesvirinae subfamily of the Herpesviridae family. Its genome consists of a linear double-stranded DNA molecule. It produces a self-limited infection in immunocompetent hosts; however, in immunosuppressed patients, it can cause significant morbidity and mortality. It can be isolated from solid organs and body secretions, including blood, urine, saliva, breast milk, semen, and cervical secretions, as well as from both fresh and anticoagulated stored blood and blood components. CMV, like other members of the Herpesviridae family, has an ability to remain latent in tissues after the acute infection.

Demographics

The prevalence of CMV infection varies significantly based on socioeconomic conditions (higher prevalence in lower socioeconomic groups), geographical region, and age (the older, the higher risk of CMV seropositivity). Approximately 1% of newborns become infected, transplacentally or through vaginal delivery or by breast milk. Early childhood CMV infection is often acquired through close contact, particularly in day-care settings.

Clinical Manifestations of CMV from Transfusion

CMV can be transmitted through transfusion. Depending on geographical locale, greater than 50% of blood donors may be CMV-seropositive (Brecher 2002). Posttransfusion CMV has a wide spectrum of manifestations ranging from a heterophile-negative mononucleosis syndrome to disseminated disease. The mononucleosis-like syndrome can present with fever, lymphadenopathy, lymphocytosis, pharyngitis, and hepatitis. Interstitial pneumonitis, meningoencephalitis, thrombocytopenia, hemolytic anemia, and polyneuropathy can also be present. There is a wide spectrum of manifestations in immunosuppressed patients including retinitis, nephritis, colitis, gastritis, and rash. Rejection of kidney allografts (Rubin et al. 1985) has been related to primary CMV infection.

Primary Infection

Primary infection is defined when a seronegative recipient receives blood or an organ transplant from a donor who has active or latent CMV infection. Community-acquired CMV infection is a common form of infection in children or adolescents, which is usually asymptomatic. Viruria and viremia do not clear until several months after seroconversion.

Secondary CMV Infection

Secondary infection is defined as either reactivation or coinfection. Reactivation of latent infection occurs when a previously CMV-seropositive individual receives either a seropositive or seronegative blood or organ. Reactivation of latent CMV remains subclinical for patients not severely immunosuppressed. In reactivation, there is a four-fold rise in CMV-IgG titer. Reinfection, or better called "coinfection," happens when a seropositive individual is exposed to a different stain of CMV from that which caused the original infection.

Donor and Recipient Factors Predisposing to Posttransfusion CMV Infection

Not all seropositive donor units can transmit infection. Predisposing factors depend on both recipient and donor. Studies have shown that posttransfusion CMV infection has occurred with all cellular blood components, but not with plasma products (Adler 1988; Bowden and Sayers 1990). CMV hyperimmune globulin (Preiksaitis et al. 1983) and IVIG (Winston et al. 1987) have been shown to protect against infection in transplant recipients, but they have had variable outcomes. Leukoreduction with washing (Luban et al. 1987) reduces and deglycerolization (Brady et al. 1984; Taylor et al. 1986; Simon et al. 1987) eliminates the risk of CMV transmission, when CMV-seropositive RBCs are transfused. Third generation leukoreduction filters also appear effective (Gilbert et al. 1989; De Graan-Hentzen et al. 1989; De Witte et al. 1990; Preiksaitis 1991).

Blood Product Variables Influencing Infectivity

Important variables in the transmissibility of CMV include the number of viable leukocytes containing CMV virions and the presence of neutralizing antibody in the recipient. Increase in storage time causes isolation of fewer lymphocytes from either RBC concentrates (McCullough et al. 1969) or platelet concentrates (Sherman and Dzik 1988). If CMV resides in polymorphonuclear cells or mononuclear cells, then the infectivity of older products might be less than with fresh products.

Preventive Methods for Transfusion-Transmitted CMV Infection

Leukofiltration with filters capable of reducing white blood cells (WBCs) greater than three logs has become a preventive method. Other factors effective in lowering the risk of CMV transmission include lower transfusion doses, use of CMV-seronegative blood, low seroprevalence of CMV-antibody in donor population, and the use of older blood.

Bowden et al. (1995) undertook a prospective randomized trial with CMV-seronegative marrow recipients to determine whether filtered, three-log, leukoreduced RBCs and platelets were as effective as CMV-seronegative blood products for the prevention of transfusion-transmitted CMV infection after bone marrow transplantation. The investigators demonstrated that the rates of infection from leukoreduced and seronegative products were comparable (2.4% versus 1.4%), but that the probability of development of CMV disease was higher among persons receiving the leukoreduced products (2.4% versus 0%, P = 0.03). These authors concluded that filtration was an effective alternative to use of seronegative blood products. However, a recent paper by Nichols et al. suggests that leukodepletion may not be equivalent to CMV-seronegative blood, suggesting that patients at very high risk of CMV infection may require seronegative blood (Nichols et al. 2003).

Recommendations

The use of CMV-reduced products is not indicated for general hospital patients, persons receiving nonneutropenia-promoting chemotherapy, persons receiving glucocorticoids, and term infants, regardless of CMV status. Persons who are CMV-seropositive and undergoing chemotherapy that may promote neutropenia, who are pregnant, or who have HIV infection do not need CMV-seronegative products (Collier et al. 2001). However, CMV-seronegative individuals should receive CMV-reduced-risk products for these indications.

Cytomegalovirus-seronegative recipients of CMV-seronegative solid-organ allografts should receive CMV-reduced products. For persons undergoing bone marrow or hematopoietic stem cell transplantation, the CMV status of the donor is not considered, and the use of CMV-reduced-risk blood is indicated except when the allograft recipient is CMV-seropositive. All low birth weight or premature infants (<1200 g), seronegative pregnant women, and fetuses receiving intrauterine transfusion need CMV-reduced-risk blood products.

PARVOVIRUS B19

Parvovirus B19 is a nonenveloped, single-stranded DNA virus. About 30% to 60% of blood donors show antibodies to parvovirus B19, which indicates immunity, not chronic, persistent infection (Brecher 2002). The clinical manifestations of parvovirus B19 infection include erythema infectiosum, polyarthropathy, chronic bone marrow failure, and transient aplastic crisis. It can rarely cause vasculitis, myocardiopathy, and neurological disease. Immunodeficient patients might not have any systemic manifestations. During pregnancy, it may cause severe hydrops fetalis, fetal death, myocardiopathy, and congenital malformations. Intrauterine fetal transfusions have been shown to be effective in saving fetal lives and completing the pregnancy. A chronic carrier state is unusual. Viremia is found in early phases of infection. Normal blood donors who are in the viremic phase are estimated to range from $1:20{,}000$ to $1:50{,}000$ (Brown et al. 2001). Any plasma derivative, especially pooled plasma products, can transmit the virus. Since parvovirus lacks a lipid envelope, it is not inactivated by solvent/detergent treatment or heat inactivation. Its presence in clotting factor concentrates has caused transmission to hemophiliacs, but with limited clinical adverse effect.

Transmission through cellular blood components (red cells, and platelets) (Jordan et al. 1998; Cohen et al. 1997; Zanella et al. 1995) has been rarely reported. Nucleic acid testing of pooled plasma products reduces the risk of parvovirus transmission and has been implemented by many manufacturers of those products. Universal donor testing for this virus, especially for cellular products has not been instituted, until more cases are reported.

PARASITIC INFECTIONS

Malaria

The most commonly reported transfusion-transmitted parasitic disease in the United States is malaria. According to the CDC, from 1958 to 1998, 103 cases of transfusion-related malaria were reported in the United States (1999). Any blood product containing red cells can transmit infection by the intraerythrocytic parasite. Transfusion-transmitted malaria often manifests with nonspecific clinical signs and symptoms. Fever, chills, headache, and hemolysis occur a week to several months after the transfusion. Asymptomatic parasitemia in blood donors can be missed at the time of donor screening, particularly among residents of endemic areas or donors traveling from an endemic area to the United States Each malaria species has a different period of asymptomatic infection: *Plasmodium falciparum* and *P. vivax*, up to five years; *P. ovale* for seven years; and *P. malariae* can remain transmissible in blood lifelong. Malaria parasites can survive at least one week in blood stored at 4°C (Brecher 2002). There have been reported cases of *P. falciparum* that have been transmitted by blood stored more than 10 days (Turc 1990).

Since there is no blood donor screening test approved for malaria in the United States, careful donor questioning is the only way to exclude asymptomatic donors at risk of transmitting malaria. There are established recommendations to defer persons at risk from donating blood based on donors' travel in endemic areas. These areas are identified in the CDC Health International Travelers, also known as the CDC yellow book. It is revised periodically and can be found on the CDC website *http://www.cdc.gov/*.

Babesiosis

Babesiosis is the most common transfusion-transmissible tickborne disease reported in the United States (McQuiston et al. 2000). Babesiosis occurs in areas where deer are commonly infested with ticks carrying the organism. In the United States most cases are reported in the northeastern states (Martha's Vineyard, Long Island, and Cape Cod), and the Midwest. The increase in its incidence is mainly because of expanding population of deer in the northeast United States.

Usually after one to four weeks of transfusion of the infected blood product, fever accompanied by chills and headache occurs; occasionally, hemolysis and hemoglobinuria result. Rarely severe complications, such as renal failure, coagulopathy, and life threatening hemolytic anemia, can occur. Immunosuppressed patients, or patients without a spleen, are most likely to develop a severe form of the disease.

Babesia are parasites living within the red cells; therefore RBCs and platelets (which contain some RBCs), are the components most likely to be responsible for transmission of infection. Babesia can survive in stored blood for up to at least 35 days (Brecher 2002). Even frozen-thawed blood has been reported to transmit the organism (Grabowski et al. 1982).

Because of the chance of asymptomatic parasitemia, any history of babesiosis requires permanent deferral of the donors (Gorlin 2000; Popovsky 1991; Krause et al. 1998). A recent tick bite or residence in the endemic area does not defer prospective donors. Currently, there is no approved test to screen donors for asymptomatic parasitemia.

Trypanosoma Cruzi

Trypanosoma cruzi is the etiological agent causing Chagas' disease, which is endemic in parts of South and Central America and Mexico, but is extremely rare in the United States. It is a hemoflagellated protozoan, and the usual route of transmission of infection is through contact with Triatoma (Reduviidae) bugs that have *T. cruzi* infection of their digestive tract. The parasite can be transmitted to human or other mammalian hosts after a bug bite and blood meal, during which the bug excretes feces containing *T. cruzi*. The transmission of infection via the infected feces occurs either through the skin or mucous membranes. In Latin America, transfusion transmission is the second most common mechanism of spread of *T. cruzi* (Schmunis 1991; Schmunis 1999). Blood components other than plasma derivatives can carry the parasite and transmit the infection to the recipients (Gudino and Linares 1990). In the United States, emigrants from endemic countries may be an important source of transmission of infection to blood recipients (Wendel 1994).

It is possible that volunteer blood donors may be screened for *T. cruzi* when a licensed test becomes available in the United States.

CREUTZFELDT-JAKOB DISEASE

Creutzfeldt-Jakob disease (CJD) is a neurodegenerative disorder caused by prions, typically affecting individuals between 50 and 75 years of age. There is about one case of CJD per million people per year reported worldwide. Most cases occur randomly (sporadic); 10% of cases are familial. The rest of the cases are acquired iatrogenically, from neurosurgical crosscontamination, cadveric pools of dura mater grafts or pituitary extracts, by administration of growth hormone and gonadotropic hormone derived from pooled human pituitary tissue, or through reuse of intercerebral electroencephalographic electrodes. Another route of transmission recently described is oral exposure to animal products contaminated with bovine spongiform encephalopathy (BSE, "mad cow disease"), causing a variant form of CJD (vCJD).

Transmission of CJD through Blood Components

It is still uncertain whether blood components can transmit the infection to humans; however, in experimental animals, the disease has been transmitted through blood (Brown 2002). Early studies implied that B lymphocytes may have a role in transmission of variant CJD, but later studies implied that plasma and platelets are the major source of prion protein (Williamson 2000). Whole blood and plasma have been shown to have extremely low levels of infectivity. Experiments have shown that buffy coats have a higher infectivity than plasma or whole blood (Brown 2002). So far, CJD transmission through blood components has not been shown in humans. Surveillance of recipients of blood products from patients who died of CJD and of hemophiliacs receiving either crude or purified plasma-derived factor VIII, or in patients with congenital immune deficiency who were treated with immune globulin have shown no evidence of neurological disease after an average survival of 12 years (Brown 2002).

Inability to demonstrate transmission through blood components in humans might be related to the very low concentration in the blood.

Prevention

Donors who are at risk for CJD are deferred. These include recipients of tissues or tissue derivatives known to be a source of CJD agent: dura mater allograft, pituitary (not recombinant) growth hormone and persons with a family history of CJD. Positive family history includes a blood relative, who has had this disease.

An unusual outbreak or cluster of vCJD was described in the United Kingdom in 1996. The most reasonable route of spread of vCJD in the United Kingdom is believed to be the consumption of beef products contaminated by the central nervous system tissue from cattle affected with "bovine spongiform encephalopathy" (BSE). Most infections in England occurred during the period of time from 1980 to 1996. The blood from patients with vCJD might pose a higher risk than blood from donors with other forms of CJD. This has been suggested by the fact that in vCJD, the lymphoreticular tissues contain the proteinase-resistant isoform of PrP, which is the most important factor for infectivity (Hill et al. 1999). Risk of transmission of vCJD through blood transfusion is unknown in humans. There have been no reported cases of vCJD or BSE in the United States.

Preventive methods used by the British government include leukoreduction of blood components, such as whole blood or labile components, and use of imported plasma products and plasma derivatives. In the United States, donors who lived at least three months in the United Kingdom from 1980 to 1996, cumulatively travelled or resided in Europe for six months, or who have received any blood component in the United Kingdom from 1980 to the present are deferred.

BACTERIAL INFECTIONS

Bacterial infection is the most common form of post-transfusion infection. It is also the most common cause of posttransfusion morbidity and mortality. No matter how carefully blood collection, processing, and storage have been performed, it is extremely difficult to completely eliminate bacterial agents. Donors are considered the main source of bacterial contamination, either from the venipuncture site or from asymptomatic (unrecognizable) bacteremia in the donor. Platelets, which are stored at room temperature, are more likely to cause infection than RBCs (refrigerated) or fresh frozen plasma (FFP) (frozen). Organisms that multiply in the refrigerated blood components (RBCs) are often psychrophilic gram-negative organisms. Bacterially contaminated platelets have a risk of about 50 to 250 fold higher than the combined risk of HIV-I, HCV, HBV, HTLV-I, and HTLV-II to cause posttransfusion infection in recipients (AABB 1996).

Red Blood Cells

The contamination rate estimated by the CDC is approximately 22 cases per 28 million units (Brecher 2002). The most common organism responsible for bacterial contamination of the RBCs is *Yersinia enterocolitica*, usually serotype 03, followed by Serratia genus (*S. liquefaciens* or *S. marcescens*) and Pseudomonas genus. These organisms have the ability to grow at 1°C to 6°C. Donors with asymptomatic transient bacteremia are the most probable source of posttransfusion *Yersinia enterocolitica* septicemia. The storage time is an important factor in the rate of occurrence of posttransfusion *Yersinia* infection. Most cases occur in blood units older than 25 days (Arduino et al. 1989).

Clinical signs and symptoms of infection due to transfusion of contaminated RBCs are usually more severe and more rapid than those due to contaminated platelets. Factors that affect the severity and outcome of the posttransfusion infection include bacterial virulence, immune status of the recipient, the concentration of the transfused bacteria, how rapidly antibiotic therapy was started, patient monitoring, and preexisting antibiotic therapy. Darkening of the RBC units and presence of either hemolysis or clots in the RBC bags may suggest bacterial contamination (Brecher 2002).

Platelets

Contaminated platelets are the most common cause of transfusion-transmitted disease. The most important reason is that platelets are kept at room temperature (20°C to 24°C) for up to five days. The organisms most commonly causing infection in platelet transfusions are skin flora, such as *Staphylococcus epidermidis* and *Bacillus cereus*. About one-third of cases arise from donors with asymptomatic bacteremia (Ness et al. 2001). These organisms multiply easily at 20°C to 24°C, which is the platelet storage temperature. A major factor responsible for the contamination of platelets is thought to be the inadequate disinfection of the phlebotomy site. Since most recipients of platelet transfusions are immunocompromised from their underlying diseases and therapies, bacterial sepsis due to platelets is unrecognized and is attributed to the underlying condition. The risk of sepsis is about six to ten times greater with

pooled platelet concentrates than with single donor apheresis units, because of higher donor exposure.

Diagnosis and Management

As soon as the bacterial contamination due to the transfused blood product is suspected, the transfusion must be stopped. Gram stain and culture both from the unit and the recipient should be sent immediately to the microbiology laboratory. Treatment should not be delayed for the results of these tests. The recipient should be treated promptly with broad spectrum intravenous antibiotics and proper therapy for shock, disseminated intravascular coagulation (DIC), and renal failure, if indicated.

Prevention

The first step and the most important preventive measure is the proper selection of the donors. This requires obtaining a detailed medical history and performing a physical examination with attention to phlebotomy site and skin. If the donor has had any history of procedures likely to produce bacteremia, he or she should be deferred from blood donation on that visit; these procedures include recent dental work, gastrointestinal or genitourinary manipulation, and breastfeeding (because of possible skin cracks). All of these can be associated with bacteremia in the donor. Upon retrospective questioning, almost 50% of the donors implicated in *Yersinia* sepsis due to contamination of RBC units recalled having had some gastrointestinal symptoms, including diarrhea, in the 30-day period before donation (Tipple et al. 1990; CDC 1997).

The next step is proper selection and cleansing of the phlebotomy site. The presence of any scar or lesion at the phlebotomy site should be considered a risk for contamination. These areas are often difficult to disinfect because of skin dimpling.

A number of other steps are being considered to reduce the risk of bacterial contamination including diversion of the first donor aliquot, reducing the length of storage, reducing donor exposure by using single-donor platelets rather than pools, and using refrigerated storage of platelets (Hoffmeister 2003). In the near future, platelets will require additional testing to identify possible bacterial contamination to reduce the risk of this important problem.

Syphilis

Syphilis is an infectious disease caused by *Treponema pallidum*, which can be transmitted sexually, transplacentally, or through other mucous membranes. During primary syphilis, about 25% to 30% of patients show a positive VDRL test, and 50% of them have a positive fluorescent treponemal antibody-absorption(FTA-ABS) test. These tests usually become positive about three weeks after the onset of symptoms. Therefore, many infectious donors are asymptomatic and have negative serological tests for syphilis.

There is an extremely low rate of transmission through transfusion for many reasons. These include donor questions, which focus on high-risk behavior, screening serological tests, and loss of viability of *T. pallidum* during the refrigerated storage. Spirochetes can survive up to 96 to 120 hours in the refrigerator. Platelets, that are stored at 20°C to 24°C can provide a suitable environment for *T. pallidum* to grow. Therefore, a fresh unit of RBCs or platelets could be responsible for those rare cases of transfusion-associated syphilis infection. However, since 1969, there have been only three cases reported in literature (Menitove and Tegtmeier 2003).

Presently, donors with reactive syphilis serology are deferred for one year, if their confirmatory testing is positive. After this period, they are permitted to donate, if they have been treated and have tested negative.

References

AABB. 1996. Bacterial contamination of blood components. Bulletin 96-6. AABB Faxnet 294. Bethesda, MD: American Association of Blood Banks.

Aach RD, Szmuness W, Mosley JW, Hollinger FB, Hahn RA, Stevens CE, Edwards VM, and Werch J. 1981. Serum alanine aminotransferase of donors in relation to the risk of non-A, non- B hepatitis in recipients. The transfusion-transmitted viruses study. *N Engl J Med* 304:989–994.

Adler SP. 1988. Data that suggests that FFP does not transmit CMV [letter]. *Transfusion* 28:604.

Alter HJ. 2002. Transfusion-transmitted hepatitis C and non-A, non-B, non-C virus infections. In Simon TL, Dzik WH, Snyder EL, et al., eds. *Rossi's principles of transfusion medicine*, 3rd ed. Philadelphia: Lippincott Williams and Wilkins.

Alter HJ, Epstein JS, Swanson SG, et al. 1990. Prevalence of immunodeficiency virus type I p24 antigen in U.S. blood donors—an assessment of the efficacy of testing in donor screening. *N Eng J Med* 323:1312–1318.

Arduino MJ, Bland LA, Tipple MA, et al. 1989. Growth and endotoxin production of *Yersinia enterocolitica* and *Enterobacter agglomerans* in packed erythrocytes. *J Clin Microbiol* 27:1483–1485.

Bowden RA and Sayers M. 1990. The risk of transmitting cytomegalovirus by fresh frozen plasma. *Transfusion* 30:762–763.

Bowden RA, Slichter SJ, Sayers MH, et al. 1995. A comparison of filtered leukocyte reduced and cytomegalovirus (CMV) seronegative blood products for the prevention of transfusion-associated CMV after marrow transplant. *Blood* 86:3598–3603.

Brady MT, Milam JD, Anderson DC, et al. 1984. Use of deglycerolized red blood cells to prevent post-transfusion infection with cytomegalovirus in neonates. *J Infec Dis* 150:334–339.

Brecher ME, ed. 2002. Technical manual, 14th ed. Bethesda, MD: American Association of Blood Banks.

Brown KE, Young NS, Alving BM, et al. 2001. Parvovirus B19: implications for transfusion medicine. Summary of a workshop. *Transfusion* 41:130–135.

Brown P. 2002. Transmission Creutzfeldt-Jakob disease by transfusion. In Simon TL, Dzik WH, Snyder EL, et al., eds. *Rossi's principles of transfusion medicine*, 3rd ed. Philadelphia: Lippincott Williams and Wilkins.

Busch MP. 2000. Closing the windows on viral transmission by blood transfusion. In Stramer S, ed. *Blood safety in the new millennium*. Bethesda, MD: American Association of Blood Banks 33–54.

Busch MP, Eble BE, Khayam-Bashi H, et al. 1991. Evaluation of screen blood donations for human immunodeficiency virus type I infection by culture and DNA amplification of pooled cells. *N Eng J Med* 325:1–5.

Busch MP, Lee LL, Satten GA, et al. 1995. Time course of detection of viral and serologic markers preceding human immunodeficiency virus type 1 seroconversion: implications for screening of blood and tissue donors. *Transfusion* 35:91.

Busch MP, Lee TH, and Heitman J. 1992. Allogeneic leukocytes but not therapeutic blood elements induce reactivation and dissemination of latent human immunodeficiency virus type I infection: implications for transfusion support of infected patients. *Blood* 80:2128.

Busch MP, Stramer SL, and Kleinman SH. 1997. Evolving applications of nucleic acid amplification assays for prevention of virus transmission by blood components and derivatives, In Garratty G, ed. *Applications of molecular biology to blood transfusion medicine*. Bethesda, MD: American Association of Blood Banks.

Centers for Disease Control and Prevention. 1999. Transfusion-transmitted malaria—Missouri and Pennsylvania, 1996–1998. *MMWR Morb Mortal Wkly Rep* 48:253.

Centers for Disease Control and Prevention 1997. Red blood cell transfusions contaminated with *Yersinia enterocolitica*—United States 1991–1996—and initiation of a national study to detect bacteria-associated transfusion reactions. *MMWR* 20;46:553–555.

Cohen BJ, Beard S, Knowles WA, et al. 1997. Chronic anemia due to parvovirus B19 infection in a bone marrow transplant patient after platelet transfusion. *Transfusion* 37:947–952.

Cohen ND, Munoz A, Reitz BA, Ness PM, et al. 1989. Transmission of retroviruses by transfusion of screened blood in patients undergoing cardiac surgery. *N Eng J Med* 320:1172–1176.

Collier AC, Kalish LA, Bush MP, et al. 2001a. Leukocyte-reduced red blood cell transfusions in patients with anemia and human immunodeficiency virus infection. *JAMA* 284:1592–1601.

Collier AC, Kalish LA, Busch MP, et al. 2001b. Leukocyte-reduced red blood cell transfusions in patients with anemia and human immunodeficiency virus infection. The Viral Activation Transfusion Study: a randomized controlled trial. *JAMA* 285:1592–1601.

Curran JW, Lawrence DN, Jaffe H, et al. 1984. Acquired immunodeficiency syndrome (AIDS) associated with transfusions. *N Eng J Med* 310:69–75.

De Graan-Hentzen YCE, Gratama HW, et al. 1989. Prevention of primary cytomegalovirus infection in patients with hematologic malignancy by intensive white cell depletion of blood products. *Transfusion* 29:757–760.

De Witte T, Schattenberg A, Van Dijk BA, et al. 1990. Prevention of primary cytomegalovirus infection after allogeneic bone marrow transplantation by using leukocyte-poor random blood products from cytomegalovirus unscreened blood bank donors. *Transplantation* 50:964–968.

Dienstag JL. 1983. Non-A, non-B hepatitis. I. Recognition, epidemiology, and clinical features. *Gastroenterology* 85:439–462.

Dienstag JL and Alter HJ. 1986. Non A, non-B hepatitis: evolving epidemiologic and clinical perspective. *Semin Liver Dis* 6:67.

Dienstag JL, Feinstone SM, Purcell RH, et al. 1977. Non-A, non-B post-transfusion hepatitis. *Lancet* 1:56–562.

Dodd R. Hepatitis. In Petz LD, Swisher SN, Kleinman S, et al., eds. 1996. *Clinical practice of transfusion medicine*, 3rd ed. New York: Churchill Livingstone 847–873.

Dodd RY, Notari EP, and Stramer SL. 2002. Current prevalence and incidence of infectious disease markers and estimated window—period risk in the American Red Cross blood donor population. *Transfusion* 42:975–979.

Donegan E, Lenes BA, Tomasulo RA, Mosley JW, and the Transfusion Safety Study Group. 1990. Transmission of HIV-I by component types and duration of shelf storage before transfusion. *Transfusion* 30:851.

Fiebig EW, Murphy EL, and Busch MP. 2003. Human immunodeficiency virus, human T-cell lymphoptrophic viruses, and other retroviruses. In Hillyer CD, Silberstein LE, Ness PM, et al., eds. *Blood banking and transfusion medicine. Basic principles and practice*, 1st ed. Philadelphia: Churchill Livingstone.

Gallo RC, Salahuddin SZ, Popovic M, et al. 1984. Frequent detection and isolation of cytopathic retroviruses (HTLV-III) from patients with AIDS and at risk for AIDS. *Science* 244:500–503.

Gilbert GL, Hayes K, Hudson IL, et al. 1989. Prevention of transfusion-acquired cytomegalovirus infection in infants by blood filtration to remove leukoctyes. Neonatal Cytomegalovirus Infection Study Group. *Lancet* 1:1228–1231.

Gorlin JB, ed. 2000. Standards for blood banks and transfusion services, 20th ed. Bethesda, MD: American Association of Blood Banks.

Grabowski EF, Giardina PJV, Goldbery D, et al. 1982. Babesiosis transmitted by a transfusion of frozen-thawed blood. *Ann Intern Med* 96:466–467.

Gudino MD and Linares J. 1990. Chagas' disease and blood transfusion. In Westphal RG, Carlson KB, Turc JM, eds. *Emerging global patterns in transfusion-transmitted infections*. Arlington, VA: American Association of Blood Banks.

Hill AF, Butterworth RJ, Joiner S, et al. 1999. Investigation of variant Creutzfeldt-Jakob disease and other human prion diseases with tonsil biopsy samples. *Lancet* 353:183–189.

Hinuma YK, Komoda H, Chosa T, et al. 1982. Antibodies to adult T-cell leukemia: virus-associated antigens (ATLA) in sera from patients with ATL and controls in Japan: a nationwide seroepidemiologic study. *Int J Cancr* 29:631.

Hinuma YK, Nagata K, Hanoaka M, et al. 1981. Adult T-cell leukemia: antigen in an ATL cell line and detection of antibodies to the antigen in human sera. *Proc Natl Acad Sci U S A* 78:6476.

Hoffmeister KM, Felbinger TW, Falet H, et al. 2003. The clearance mechanism of chilled blood platelets. *Cell* 112:1–20.

Hollinger FB, Werch J, and Melnick JL. 1974. A prospective study indicating that double-antibody radioimmunoassay reduces the incidence of post-transfusion hepatitis B. *N Engl J Med* 290:1104–1109.

Jordan J, Tiangco B, Kiss J, et al. 1998. Human parvovirus B19: prevalence of viral DNA in volunteer blood donors and clinical outcomes of transfusion recipients. *Vox Sang* 75:97–102.

Kiyosawa K, Akahane Y, Nagata A, et al. 1982. Significance of blood transfusion in non-A, non-B chronic liver disease in Japan. *Vox Sang* 43:45–52.

Krause PJ, Spielman A, Telford SR, III, et al. 1998. Persistent parasitemia after acute babesiosis. *N Engl J Med* 339:160–165.

LaGier BO and Murphy S. 2003. Testing changes for blood products. American Red Cross Blood Services, Penn Jersey Region, Philadelphia PA. Memo to Blood Bank Directors, Managers and Supervisors.

Luban NLC, Williams AE, MacDonald MG, et al. 1987. Low incidence of acquired cytomegalovirus infections transfused with washed red blood cells. *Am J Dis Child* 141:146–149.

McCullough J, Yunis EJ, Benson SJ, et al. 1969. Effect of blood bank storage on leukocyte function. *Lancet* 1:1333.

McQuiston JH, Childs JE, Chamberland ME, et al. 2000. Transmission of tickborne agents of disease by blood transfusion: a review of known and potential risks in the United States. *Transfusion* 40:274–284.

Menitove JE and Tegtmeier GE. 2003. Additional infectious complications. In Hillyer CD, Silberstein LE, Ness PM, et al., eds. *Blood banking and transfusion medicine, basic principles and practice*, 1st ed. Philadelphia: Churchill Livingstone.

Nelson KE, Donahue JG, Munoz A, et al. 1992. Transmission of retroviruses from seronegative donors by transfusion during cardiac surgery: a multicenter study of HIV-I and HTLV-I/II infections. *Ann Intern Amed* 117:554.

Ness PM, Hayden B, King K, et al. 2001. Single-donor platelets reduce the risk of septic platelet transfusion reactions. *Transfusion* 41:857–861.

Nichols WG, Price TH, Gooley T, et al. 2003. Transfusion-transmitted cytomegalovirus infection after receipt of leukoreduced blood products. *Blood* 101(10):4195–4200.

O'Brien TR, George JR, Holmberg SD, et al. 1992. Human immunodeficiency virus type II infection in the United States: epidemiology, diagnosis, and public health implication. *JAMA* 2775:9.

Pindyck J, Waldman A, Zang E, et al. 1985. Measures to decrease the risk of acquired immunodeficiency syndrome transmission by blood transfusion: evidence of volunteer blood donor cooperation. *Transfusion* 25:3–9.

Popovsky MA. 1991. Transfusion-transmitted babesiosis. *Transfusion* 31:296–297.

Preiksaitis JK. 1991. Indication for the use of cytomegalovirus-seronegative blood products. *Transfus Med Rev* 5:1–17.

Preiksaitis JK, Rosno S, Grumet C, et al. 1983. Infections due to herpes viruses in cardiac transplant recipients: role of the donor heart and immunosuppressive therapy. *J Infect Dis* 147:974–981.

Rubin RH, Tolkoff-Rubin NE, Oliver D, et al. 1985. Multicenter seroepidemiologic study of the impact of cytomegalovirus infection on renal transplantation. *Transplantation* 40:243–249.

Schmunis GA. 1991. *Trypanosoma cruzi*, the etiologic agent of Chagas' disease: status in the blood supply in endemic and nonendemic countries. *Transfusion* 31:547–557.

Schmunis GA. 1999. Prevention of transfusion *Trypanosoma cruzi* infection in Latin America. *Mem Inst Oswaldo Cruz* 94:[Suppl]:93–101.

Seeff LB. 1981. Hepatitis in hemophilia: a brief review. In Burk BP, Chalmers C, eds. *Frontiers in liver disease*. New York: Thieme-Stratton.

Sherman ME and Dzik WH. 1988. Stability of antigens in leukocytes in banked platelet concentrates: decline in HLA-DR antigen expression and mixed lymphocyte culture stimulating capacity following storage. *Blood* 72:867–872.

Simon T, Johnson J, Koffler H, et al. 1987. Impact of previously frozen deglycerolized red blood cells in cytomegalovirus transmission to newborn infants. *Plasma Ther Transfu Technol* 8:51–56.

Tabor E, Goldfield M, Black HC, et al. 1980. Hepatitis B antigen in volunteer and paid donors. *Transfusion* 20:192–198.

Taylor BJ, Jacobs RF, Baker RL, et al. 1986. Frozen deglycerolized red cells prevents transfusion acquired cytomegalovirus transmission in neonates. *Pediatr Infect Dis J* 5:188–191.

Tipple MA, Bland LA, Murphy JJ, et al. 1990. Sepsis associated with transfusion of red cells contaminated with *Yersina enterocolitica*. *Transfusion* 30:207–213.

Turc JM. 1990. Malaria and blood transfusion. In Westphal RG, Carlson KB, Turc JM, eds. *Emerging global patterns in transfusion-transmitted infections*. Arlington, VA: American Association of Blood Banks.

Vogt M, Lang T, Frosner G, et al. 1999. Prevalence and clinical outcome of hepatitis-C infection in children who underwent cardiac surgery before the implementation of donor screening. *N Engl J Med* 341:866–870.

Ward JW, Holmberg SD, Allen JR, et al. 1988. Transmission of human immunodeficiency virus (HIV) by blood transfusion screened as negative for HIV antibody. *N Eng J Med* 318:473–478.

Wendel S. 1994. Current concepts on transmission of bacteria and parasites by blood components. *Vox Sang* 67[Suppl]:161–174.

Williamson LM. 2000. *British Journal of Haematology* 110(2):256–272.

Winston DJ, Ho WG, Lin C, et al. 1987. Intravenous immune globulin for prevention of cytomegalovirus infection and interstitial pneumonia after bone marrow transplantation. *Ann Intern Med* 106:12–18.

Zanella A, Rossi F, Cesana C, et al. 1995. Transfusion-transmitted human parvovirus B19 infection in a thalassemic patient. *Transfusion* 35:769–772.

CHAPTER

29

Therapeutic Apheresis

BRUCE C. McLEOD, MD, AND HAEWON C. KIM, MD

INTRODUCTION

The term "apheresis" refers to a group of technologies for separating peripheral blood components from each other, continuously and on-line, on the basis of density and/or size. The first apheresis instruments were marketed in the 1970s; they required frequent operator intervention for optimal outcomes. Later designs have been more and more automated. The principal use of these instruments has always been blood donations, platelet and plasma donations being the most common, but they have found numerous therapeutic applications as well. All apheresis instruments are designed on an adult scale, but all have been adapted for therapeutic use in children.

GOALS OF THERAPEUTIC APHERESIS

Therapeutic apheresis is usually intended to be therapeutic depletion of some noxious substance believed to be circulating in peripheral blood. In this sense, therapeutic apheresis is conceptually related to the medieval practice of bloodletting to release evil humors. The present authors try to observe three principles whenever possible to keep applications of therapeutic apheresis on a firmer evidential footing than bloodletting. These are listed in Box 29.1. First, there should be firm evidence that a cell or substance circulating in the blood is important in the pathogenesis of the disease being treated. Second, there should be convincing evidence that meaningful and lasting depletion of the pathogenic constituent can be achieved with therapeutic apheresis. Among plasma constituents, such evidence

has only been adduced for large macromolecules that are substantially intravascular and have slow turnover, such as IgG antibodies and low density lipoproteins. Third, there should be evidence, ideally from controlled clinical trials, that lasting depletion of the pathogenic cell or substance brings meaningful clinical benefit. In some instances, such as red cell exchange, part of the benefit of therapeutic apheresis also accrues from infusing a large amount of a normal blood component without producing volume overload.

There are four classic modalities of therapeutic apheresis, which are listed in Table 29.1. All of them can be carried out in children.

Cell Removal Procedures

Cell removal procedures may be helpful for patients who have elevated white cell or platelet counts as well as for "autologous apheresis donation" of peripheral blood progenitor cells. Because a "minor" blood component is being removed, it is usually not necessary to provide any replacement for lost blood volume other than the crystalloid solution being added as an anticoagulant (see Anticoagulation section). Cell removal procedures often have a desired endpoint, expressed in terms of a postprocedure white count or platelet count. It is difficult to predict how much blood should be processed or even how many cells should be removed to reach the desired goal. Blood volume may be greater than estimated, efficiency of cell removal by the instrument may not match expectations, and cells may be mobilized from extravascular sites during the procedure. STAT blood counts during the procedure, if available, provide a better guide to when it is permissible to

TABLE 29.1   Classic Therapeutic Apheresis Procedures

Cell Depletions

Plateletpheresis
Leukapheresis

Blood Component Exchanges

Red cell exchange
Plasma exchange

TABLE 29.2   Metabolic Characteristics of Selected
Plasma Proteins

Protein	% Intravascular	FCR	TER
IgG	45	6.7	3
IgM	76	18	1–2
Albumin	40	10	5–6
Fibrinogen	80	25	2–3
C3	53	56	—

FCR = Fractional catabolic rate as % intravascular mass per day;
TER = Transcapillary escape rate: transfer from intra- to extravascu-
lar compartment as % intravascular mass per hour.
(Adapted with permission from Chopek and McCullough [1980].)

stop the procedure than do predictions based on preprocedure cell counts and blood volume. Cell removal
procedures are discussed in detail in Chapter 30.

Blood Component Exchanges

Blood component exchange is the other major therapeutic apheresis modality. Because these procedures
entail removal of components that make up a large fraction of total blood volume, some replacement must be
provided for the material removed to avoid profound
hypovolemia. Removal of a pathogenic substance
becomes less efficient as a procedure progresses because
an increasing proportion of the material removed is
replacement medium infused earlier. This phenomenon
is accurately described by the formula $y_x = y_o e^{-x}$, in which
y_o is the starting concentration of the material, e is the
base natural logarithm, and y_x is the concentration of the
substance after x patient component volumes have been
exchanged. As shown later, a patient's red cell and/or
plasma volume can be calculated based on (1) an estimate of total blood volume derived from body size
and (2) hematocrit (Hct). This formula predicts that
exchange of one patient component volume (that is, one
red cell volume or one plasma volume) will lower the
concentration of a pathogenic constituent by about two-
thirds, and most practitioners choose to stop exchanges
at this point or, at most, at 1.5 component volumes.

Red Cell Exchange

The duration of the effect of a component exchange
depends on the half-life of the pathogenic substance and
the extent to which it is confined to the intravascular

compartment. Since red cells are entirely intravascular
and have a half-life measured in weeks or months, a
single red cell exchange is expected to produce a
marked therapeutic effect that will persist for at least
several weeks. Specific applications for red cell
exchange are discussed in Chapter 15.

Plasma Exchange

Plasma exchange (also called plasmapheresis; the
two terms are interchangeable) is the most frequently
performed therapeutic apheresis procedure. It is often
carried out to deplete a pathogenic IgG antibody. IgG
has a half-life of about four weeks and is thus a plausible candidate for depletion by plasma exchange.
However, since IgG is only about 45% intravascular
(Table 29.2) and does not reequilibrate to any meaningful extent during an exchange, fully 55% of the total
body burden of a pathogenic IgG is not accessible for
removal in any given exchange. The accepted strategy is
to perform a series of exchanges separated by intervals
of one to three days. Intravascular and extravascular
IgG reequilibrate during the interval between treatments, restoring the 45% intravascular distribution. This
allows more efficient removal at the beginning of the
next treatment than at the end of the preceding one. An
interval of one to two days between treatments also
allows time for resynthesis of normal plasma proteins if
these have been depleted by the exchange. A series of
six exchanges, three times per week, will usually lower
IgG levels to one-fifth to one-sixth of baseline. Further
exchanges are relatively inefficient at these low IgG
levels; the rate of IgG removal approximates the rate of
resynthesis so that no further net depletion occurs.

Replacement Cells/Fluids

As mentioned, a replacement must be provided for
the component removed in an exchange procedure

TABLE 29.3 Replacement Fluids For Plasma Exchange

Fluid	Advantages	Disadvantages
5% Albumin	Viral inactivation	High cost
	Ease of use	Most proteins not replaced
	Reactions rare	
Single-donor plasma*	All proteins replaced	High cost
		Inconvenient[†]
		Citrate reactions
		Urticaria
		Viral infection risk

*Fresh-frozen plasma or cryoprecipitate-poor plasma.
[†]Must be thawed before use; must match patient ABO type.

TABLE 29.4 ASFA and AABB Indication Categories for Therapeutic Apheresis

Category	Brief Description
I	Standard primary therapy
II	Standard adjunctive therapy
III	Controversial
IV	No benefit in controlled trials

(Table 29.3). In a red cell exchange this will be packed red blood cells (PRBCs) from normal donors who may, in the context of treating sickle cell disease, be documented to be sickle trait-negative. In plasma exchange there are two major choices: fresh frozen plasma (FFP) and 5% human albumin in saline. Their respective advantages and disadvantages are summarized in Table 29.3.

FFP

FFP from normal donors is the most physiological replacement. It replenishes coagulation factors, complement components, transport proteins, and other minor plasma constituents, as well as albumin and immunoglobulins directed against a normal spectrum of prevalent microbes. FFP has significant drawbacks, however. It must be type-specific and must be thawed just before use. It may cause allergic reactions in some patients, and it has a high citrate content, which may lead to hypocalcemic reactions as described later. Finally, it carries a small but finite risk of transmitting bloodborne pathogens.

5% Albumin

An alternative to donor plasma is 5% albumin in saline. This plasma substitute is commercially available from several fractionators. It can be stored in the liquid state at room temperature and can be administered without regard to blood type. Allergic reactions to albumin are extremely rare, and it is pasteurized to inactivate pathogens. Obviously most of the plasma proteins that are removed are not replaced in an albumin exchange; however, all except immunoglobulins are rapidly resynthesized in the normal course of protein metabolism, and functional deficiencies do not generally develop in well-nourished patients if exchange frequency is limited to every other day or three times per week. Thus a series of plasma exchanges with 5%

albumin replacement results in fairly selective depletion of the patient's immunoglobulins, including pathogenic antibodies.

Other Exchanges

Not listed in Table 29.1 are two other complementary procedures that can be performed with apheresis instruments but are very seldom requested. One is rapid RBC transfusion, in which patient plasma is removed while RBCs are simultaneously infused. This can allow rapid RBC transfusion without any risk of volume overload. The other is RBC depletion, in which RBCs are removed from the patient while 5% albumin (or FFP) is reinfused. This can allow rapid reduction of the hematocrit (Hct) in polycythemic patients, such as children with cyanotic heart disease and secondary erythrocytosis, without any risk of hypovolemia.

INDICATION CATEGORIES

Both the American Society for Apheresis (ASFA) and the American Association of Blood Banks (AABB) have issued guidelines concerning the use of therapeutic apheresis in specific disorders. These guidelines represent consensus assessments prepared by committees of members with recognized expertise and approved by the boards of directors of the respective organizations. The recommendations are based almost entirely on the results of clinical studies. They are periodically updated based on accumulated new data, and physicians using them should also take care to be aware of the latest published data. The guidelines take the form of indication category assignments, which are summarized in Table 29.4 and explained in more detail below.

Category I

This category comprises diseases for which therapeutic apheresis is considered a standard, if not a mandatory measure, either as a primary therapy or an early adjunctive therapy. This designation is usually based on positive results in randomized controlled trials or on a broad, noncontroversial base of favorable published experience.

Category II

This category includes disorders in which therapeutic apheresis is generally deemed to be beneficial but is considered to be supportive and/or adjunctive to other more definitive treatments, rather than a first-line therapy. Some category II indications are supported by randomized, controlled data, while others are supported only by small series or very informative and compelling case studies.

Category III

This category includes illnesses in which there is evidence of benefit that is not conclusive. In some instances controlled trials have produced conflicting results. In others the evidence consists entirely of anecdotal reports. In still others the risk:benefit ratio is not clear. Therapeutic apheresis is considered a reasonable option in these illnesses when conventional treatments have failed and is also considered an appropriate subject for controlled trials.

Category IV

This designation indicates that controlled trials have been done and were negative, or that published experience is uniformly discouraging. Therapeutic apheresis should only be done as part of an internal review board (IRB)-approved research protocol, if at all.

PHYSIOLOGY AND ADVERSE EFFECTS

Undergoing an apheresis procedure entails several departures from normal homeostasis. Instrument design and procedure planning are intended to minimize the magnitude of these departures; however, the procedures can sometimes result in clinically significant adverse effects despite these measures.

Anticoagulation

An anticoagulant must be added to blood withdrawn from the body in an apheresis procedure to prevent clotting in the instrument and/or associated tubing. Citrate, which prevents clotting by chelating ionized calcium, thereby blocking numerous steps in the coag-ulation cascade and inhibiting platelet aggregation, is the standard anticoagulant in adults. Citrate is present in blood returned to the patient during an apheresis procedure. Redistribution, excretion, and metabolism in the Krebs cycle moderate citrate levels in the patient's bloodstream; nevertheless, if citrate is infused too rapidly, it will produce ionized hypocalcemia sufficient to cause acral and circumoral paresthesias or even tetany. Most adults will notice and report paresthesias, allowing apheresis operators to reduce citrate infusion rates while symptoms are mild. Ionized calcium levels low enough to interfere with clotting are never encountered, so that citrate anticoagulates the instrument but not the patient.

Monitoring the ionized hypocalcemia induced by citrate infusion by following symptoms is more difficult in children than adults, not only because of communication issues but also because the most prominent clinical manifestations in pediatric patients may be abdominal pain and/or hypotension. Concern that profound hypocalcemia might go undetected may underlie more frequent use of heparin, or combinations of heparin and reduced quantities of citrate, in children, even though this will entail anticoagulation of the patient as well as the apheresis instrument.

Volume Shifts

An apheresis procedure requires that a certain amount of patient blood be in the instrument for processing, and therefore outside the body. Thus an intravascular volume deficit develops early in a procedure, persists during the procedure, and is corrected at the end of the procedure when blood in the instrument is returned to the patient. The absolute magnitude of obligate extracorporeal volume depends on the instrument, the procedure, and, in some cases, the patient's Hct. Also, in certain instances the impact on red cell volume (RCV) may be greater than on total blood volume (TBV). As a rule, the likelihood of hypotension is low if extracorporeal volume (ECV) is <10% of patient blood volume and rises steeply if it is >15% of patient blood volume. Rules of thumb for estimating blood volume in pediatric patients are given in Table 29.5. ECV is in the range of 200 to 400 mL for most modern continuous-flow apheresis instruments. This is well below 10% of total blood volume for most adults but may exceed even the 15% level for small children. Also, accumulation of red cells in the instrument may lower a pediatric patient's Hct substantially during the procedure. Such instruments can nevertheless be used for small children if the fluid in the tubing and blood

TABLE 29.5   Total Blood Volume Estimate According to Age of Child

Age Group	Total blood volume (mL/kg)
Premature infants	89–105
Term newborn infants	82–86
Infants >3 months old to preschool children	73–82
School-age children until puberty	70–75
After puberty	Same as adult

separation chamber is infused to the patient as blood enters the instrument. For very small children or those in whom the RCV deficit would be problematic, the instrument can be filled ("primed") with donor PRBCs or reconstituted whole blood (RBCs + FFP) before being connected to the patient. This can prevent any volume deficit from occurring during the procedure but requires that patient blood be left in the instrument at the conclusion of the procedure. See the following for a more detailed discussion of blood priming.

Immediate Adverse Effects

Hypocalcemia

Apheresis is generally well tolerated. Indeed, it is so intrinsically safe that it is widely used for blood donation in adults. Adverse effects are nevertheless noted in some therapeutic procedures. Paresthesias due to citrate infusion are quite common in adults, though they are often so trivial and so easily managed by lowering the citrate infusion rate that it is debatable whether they should be called an adverse effect or merely a physiological effect. Troublesome manifestations of ionized hypocalcemia, such as tetany, seizures, or cardiac arrhythmias, are rare in adults. As mentioned, troublesome hypocalcemia is felt to be more frequent in children, but the incidence has not been studied systematically in any pediatric age group. Severe symptoms or very low ionized calcium levels (for example, <0.7 mmol/liter) should be treated with intravenous calcium infusion at a dose and rate appropriate for the patient's body size.

Hypotension

Significant hypotension with a systolic blood pressure <80 mm Hg occurs in 2% to 4% of adult therapeutic apheresis procedures. This can be prevented in pediatric patients in some instances by priming the instrument with blood. If it occurs it should be treated by lowering the patient's head below trunk and leg level and infusing extra saline or colloid.

Transfusion Reactions

Other adverse effects, such as fever, urticaria, anaphylaxis, or hemolytic transfusion reactions, may be seen in procedures that require FFP or RBC replacement. Such reactions should be managed in the same manner as transfusion reactions occurring in any other context, except that it is generally permissible to continue a clinically indicated plasma exchange in the face of an urticarial reaction once symptoms have been controlled with an antihistamine.

GENERAL APPROACH TO PEDIATRIC PATIENTS NEEDING THERAPEUTIC APHERESIS

There is a lack of universally accepted indications for application of therapeutic apheresis in pediatric patients. However, apheresis procedures have been shown to be an effective first-line treatment or valuable adjunctive therapy in children with selected diseases. The ASFA and AABB guidelines on the use of therapeutic apheresis shown in Table 29.4 are based primarily on data and experience derived from adults. Nevertheless, these guidelines have provided direction for the application of therapeutic apheresis in children for many of the same diseases that also affect adults. Caution should be exercised, however, because the clinical course of a disease and response to treatment may differ in children compared to adults. When a child has a rare disease or life-threatening condition, especially one unresponsive to conventional therapy, it may not be found under category I (therapeutic apheresis is standard and acceptable) or category II (therapeutic apheresis is adjunctive therapy). If there is strong evidence in such cases that a circulating substance is pathogenic and that this pathogenic substance can be significantly depleted by apheresis, then the potential clinical benefit from removal of the pathogenic substance should be weighed against the potential for adverse effects. Potential complications related to catheter placement, which many children need, should be included. Finally, if it is determined that apheresis is medically indicated, an overall treatment plan should be developed, including a clinical endpoint. Children should not be denied therapeutic apheresis solely because it is difficult to perform technically or because they are very small, severely anemic, or hemodynamically unstable.

PLANNING PEDIATRIC PROCEDURES

Although therapeutic apheresis can be performed safely in children, many difficulties can be encountered. In close consultation with the patient's physician, each procedure should be planned to maximize the therapeutic effects while minimizing the chances of adverse effects. To ensure safety and efficacy, both technical and psychological factors unique to pediatric patients must be considered.

Technical Considerations

The principle of the apheresis procedure in children is the same as in adults. However, since the instruments are designed on an adult scale, procedures may need to be modified for pediatric application. As mentioned, the key factors for safety are maintenance of a satisfactory TBV and RCV and prevention or prompt recognition of adverse reactions related to volume shifts, anticoagulants, and replacement fluids.

Apheresis instrument disposables have a fixed ECV and extracorporeal red cell volume (ECRCV). These are, respectively, the volumes of whole blood and red cells that fill the instrument's centrifuge module and associated tubing during a procedure. Intravascular volume depletion during the procedure is primarily dependent on ECV. The fraction of TBV lost in the extracorporeal circuit is greater in a child than in an adult. Without modifications, a small patient's intravascular volume status would become quite negative during the procedure. Before the procedure, one must establish maximum safe limits for ECV and ECRCV for the patient. To determine these safe limits, the patient's TBV, RCV, and plasma volume (PV) must be estimated. This also serves as a basis for calculating the extent of exchange or collection needed to achieve the desired goal of the procedure.

TBV varies with age and body composition as well as size. In general, TBV in healthy individuals is related to lean body mass; thus, TBV in relation to body weight and height is greater in males than in females. There is also a difference between newborns and adults, with newborns having a higher TBV and RCV than adults. TBV can be estimated using several formulas. For simplicity, TBV can be calculated by multiplying the blood volume in mL per kg of body weight (see Table 29.5) for the appropriate age group by the patient body weight in kg. Some automated apheresis instruments utilize more complex algorithms to calculate TBV based on gender, height, and weight; however, the calculation performed by the COBE Spectra is inaccurate in male children. For example, prepubescent boys weighing less than 30 kg have smaller TBVs than those predicted by the algorithm. Consequently, the TBV estimated by the instrument must be verified against a manually calculated weight-based estimate for pediatric patients, especially for boys <10 years old or <30 kg. Simple formulas for estimation of PV and RCV are shown below, with Hct expressed as a decimal fraction:

$$PV = TBV \times (1 - Hct)$$

$$RCV = TBV \times Hct \text{ or } RCV = TBV - PV$$

Psychological Considerations

A child's anxiety or fear often results in a lack of cooperation with the apheresis team. An understanding of age-appropriate behavior, pediatric psychology, and social dynamics among children and their parents or caretakers is essential to alleviate fear and anxiety and can facilitate a successful procedure. Also extremely helpful is communication by the apheresis operator with the patient's staff nurse, play therapist, and family. Sedation of a child should be avoided whenever possible unless medically indicated (for example, a child who experiences frequent seizures or with emotional or neuromuscular instability), since sedation may interfere with early recognition of adverse effects of apheresis.

VASCULAR ACCESS

Adequate vascular access is a prerequisite for therapeutic apheresis. The antecubital veins of large children and adolescents may accommodate the necessary needles or catheters. Continuous-flow apheresis devices require dual vascular access with at least an 18-gauge needle for the draw line and at least a 22-gauge intravenous catheter for the return line. Unfortunately, peripheral access is often not an option for young children and infants. In this situation, a central venous catheter (CVC) may be required. However, these can cause local or systemic infection, thrombosis, or other serious complications that can result in increased morbidity and mortality, prolonged hospitalization, and increased medical costs. For each pediatric candidate, the risks associated with CVC placement and maintenance should be weighed against the potential benefits of therapeutic apheresis. Depending on the urgency and expected frequency and duration of treatment, a femoral line may be placed for either single or short-term use until a CVC suitable for longer-term use can be surgically implanted. Curved extensions are more comfortable than straight extensions, particularly for internal jugular vein catheters. All CVCs require ongoing maintenance care to prevent catheter-related complications. If a patient requires chronic apheresis

TABLE 29.6   Guidelines for Catheter Size According to Patient's Weight

Patient's Weight (kg)	Dual-Lumen Central Venous Catheter
<10	7 Fr. MedComp (MedComp, Inc., Harleysville, PA)
10–20	8 Fr. MedComp or 8 Fr. Mahurkar (Quinton Instrument Co., Seattle, WA)
20–50	9 Fr. MedComp or 10 Fr. Mahurkar
>50	9 Fr. or 11.5 Fr. MedComp, 10 Fr. , 11.5 Fr., or 13.5 Fr. Mahurkar, adult PermCath (Quinton Instrument Co., Seattle, WA) or adult Vas-Cath (Vas-Cath Inc., Mississauga, Ontario, Canada)

therapy, an arteriovenous shunt or fistula can provide a lower maintenance, long-term vascular access, especially for older children.

A CVC must be rigid enough to withstand the negative pressure generated when blood is withdrawn at a high-flow rate. In general, large-bore, dual-lumen hemodialysis catheters provide adequate access for apheresis procedures. MedComp (MedComp, Inc., Harleyville, PA), Vas Cath (Vas-Cath, Inc., Mississanga, Ontario, Canada), and Quinton (Mahurkar or PermCath catheters, Quinton Instrument Co., Seattle, WA) all offer a suitable CVC device. Hickman (C.R. Bard, Inc., Cranston, RI), Infuse-a-Port (Shiley Infusaid, Inc., Norwood, MA) or peripherally inserted central cathetis (PICC) lines are not suitable for drawing blood but can be used for returning blood from the instrument. Peripherally inserted central catheters (PICC lines) are not suitable for drawing blood but can be used for returning blood from the instrument. Guidelines for catheter size according to patient weight are given in Table 29.6.

HEMODYNAMIC CHALLENGES

Intravascular Fluid Volume

At the beginning of a standard procedure, as whole blood is drawn into the apheresis circuit displacing saline that is diverted to a waste bag, both TBV and RCV decrease. The resulting deficit persists throughout the procedure until the red cells and plasma in the extracorporeal circuit are returned to the patient ("rinsed back") at the end of the procedure. The TBV and RCV deficit during a procedure and the volume load associated with rinseback depend on the instrument and the procedure, but, for a given instrument and procedure, each is a fixed quantity that is known precisely in advance. ECV is significantly larger with intermittent-flow (IF) instruments, especially in single-needle procedures, than with continuous-flow (CF) cell separators, which require dual-needle procedures. In addition, with an IF instrument, the maximum ECV per pass is not reached until the centrifuge bowl is filled with red cells. It therefore depends on the patient's Hct; that is, the lower the Hct, the larger the ECV. A CF system is generally a better choice for small and/or hemodynamically unstable children.

The volume deficit that develops during a procedure may cause hemodynamic changes. Signs and symptoms of hypovolemia that may occur include tachycardia, orthostatic hypotension, and narrow pulse pressure, and, clinically, apprehension, weakness, nausea, light-headedness, pallor, thirst, cool skin, and even loss of consciousness. Conversely, the volume overload that may result from rinseback at the end of a standard procedure should be avoided in small children or patients with anemia or impaired cardiac or renal function.

Circulating Red Cell Volume

As stated previously, drawing patient blood into the apheresis instrument results in a reduction in circulating red cell mass, the extent of which depends on the instrument chosen and the procedure to be performed. Red cells remain in the extracorporeal circuit throughout the procedure until the blood in the apheresis instrument is rinsed back to the patient. In general, IF instruments deplete more red cells than CF instruments. Also, with the COBE Spectra, leukapheresis requires more red cells in the centrifuge than other procedures. Depending on the volume lost, the size of the patient, and the patient's starting Hct, this temporary red cell deficit may reduce RCV to unacceptable levels. With a lower RCV due to small size, anemia, or both, the fractional RCV depletion is greater for a fixed red cell loss into the apheresis instrument. One must also consider the patient's underlying medical status. Before the procedure, a safe minimum Hct during the procedure should be established based on patient-specific variables. If the expected RCV reduction would cause the Hct to fall below this limit, it is necessary to prime the circuit with blood before the procedure to maintain a safe Hct throughout the procedure.

PROCEDURAL MODIFICATIONS FOR PEDIATRIC PATIENTS

Intravascular Volume Shift

As mentioned previously, the ECV for apheresis instruments varies from 200 to 400 mL. Since ECV should be <15% of TBV (and preferably <10%) for a safe procedure, this may be too much for a small child. Although operating procedures differ according to the type of apheresis instrument, the principles can be applied to all types of instrument. In older children with a body weight >30 kg (TBV >[30 kg × 75 mL/kg = 2250 mL]), apheresis technique need not differ from that used in adult patients. However, in small children, significant modifications of technique are required to provide safe and effective therapy.

The goal of procedural modification in a small child should be to avoid any change in the patient's blood volume or red cell mass during the procedure. At the beginning of the procedure, diverting the saline in the instrument into a waste bag would result in a negative fluid balance with removal of whole blood into the centrifuge. One should calculate whether the volume deficit that would be induced in this way is >10% to 15% of TBV or greater than the safe limit for the individual patient. If the patient may not tolerate this degree of volume deficit, the saline in the instrument may be returned to the patient without diverting. Using the COBE Spectra, the volumes of saline diverted to the waste bag with plasmapheresis and erythrocytapheresis procedures are 150 mL and 100 mL, respectively. To illustrate, consider two examples. A plasmapheresis procedure in a 50-kg adolescent boy with a TBV of about 3750 mL, which requires losing 150 mL to the extracorporeal circuit, would be well tolerated as this comprises only 4% of his TBV. In contrast, in a 10-kg child with a TBV of about 700 mL, the acute loss of 150 mL of whole blood into the apheresis circuit would mean an unacceptable 21% drop in TBV.

One must also decide whether rinseback can be performed. If rinseback would result in a volume load >15% of TBV, or greater than the individual child could tolerate, rinseback should be omitted even though a significant red cell loss will occur. With the COBE Spectra, the default rinseback would result in net positive volume shifts of 195 mL with plasmapheresis and 245 mL with erythrocytapheresis. Again consider two examples. For a 50-kg boy with a TBV of about 3750 mL, the 195 mL fluid load from rinseback would comprise only 5% of TBV and would likely be well tolerated. In contrast, in a 10-kg child with a TBV of about 700 mL, this would correspond to 28% of TBV. To avoid a volume shift of this magnitude, the standard procedure must be modified by omitting rinseback.

Blood Prime

As stated previously, an apheresis procedure results in a temporary reduction of RCV. This red cell deficit persists throughout the procedure until the blood in the centrifuge is rinsed back to the patient. Maintenance of

a red cell mass adequate to provide oxygen delivery to tissues is also important for a safe procedure. The first step is to evaluate whether a patient could tolerate the temporary red cell loss during the procedure if isovolemia was maintained. If evaluation of the patient's RCV, and cardiovascular and pulmonary reserve suggests that the patient would not tolerate this degree of red cell loss, blood priming is indicated. In general there are two indications for red cell priming: (1) RBC depletion is >30% of the original circulating RBC volume in a hemodynamically stable patient with normal hemoglobin level and (2) when further reduction of the circulating RCV is not desirable in a hemodynamically unstable or anemic patient, or in a patient at a risk of organ ischemia. With the COBE Spectra, the ECRCV is 68 mL during plasmapheresis or erythrocytapheresis and 114 mL with leukapheresis. For a 10-kg child with Hct = 0.30, the Hct would drop to 0.20 with plasmapheresis or erythrocytapheresis and to 0.14 with leukapheresis. This would not be acceptable and therefore necessitates blood priming. To maintain Hct = 0.30 throughout a plasmapheresis, at least 70 mL red cells should be available to compensate the temporary red cell loss. Assuming the Hct of packed red cell units is 0.60, 117 mL (70 mL/0.6 = 117 mL) will be needed. If a red cell unit is divided using a sterile technique, the remainder can be dedicated to this patient for later transfusion to limit the number of donor exposures the patient receives.

Blood priming can be done in several different ways. One method is to fill the centrifuge chamber with donor PRBCs at a predetermined Hct while the prime saline is diverted into the waste bag. Another method is to prime only the return line with PRBCs rather than the entire instrument. The latter approach is recommended because it is technically simple and the volume of red cells infused to the patient can be accurately determined. In general, blood priming is indicated for children weighing <20 kg for plasmapheresis and <25 kg for leukapheresis. Blood priming may also be indicated for a larger patient that is anemic or hemodynamically unstable. Therefore, a half to a full unit of red cells must usually be ordered and available before an apheresis procedure on such a patient can be started.

The next step is to decide whether rinseback of red cells along with fluid from the centrifuge can be performed at the end of the procedure. Although rinseback will return most of the red cells from the extracorporeal circuit, this decision should include consideration of whether the patient can tolerate the additional volume that would be gained with rinseback. If the child cannot tolerate the expected volume load, rinseback should not be performed even though a significant volume of red cells will remain in the instrument. With the COBE

Spectra, the default rinseback would return 53 mL out of the 68 mL of red cells in the circuit with both plasmapheresis and erythrocytapheresis and 90 mL of the 114 mL of red cells with leukapheresis. In general, to avoid fluid overload, rinseback is not indicated for children who require blood priming. In addition, there is usually no need to increase RCV in such patients since Hct will remain at the preprocedure level throughout the procedure.

Leukapheresis

Leukapheresis is distinctly different from plasmapheresis and red cell exchange procedures. Leukapheresis is a *collection* procedure rather than an *exchange* procedure; thus, a replacement line is not provided. When additional fluids, such as FFP, red cells, or platelets are needed during the procedure, these products should be given through a separate intravenous (IV) line or, if no other IV line is available, through a Y-connector that is attached to the return needle. Citrate toxicity is a greater problem with leukapheresis than with other procedures because virtually all of the anticoagulant administered during the procedure is returned to the patient, with <5% remaining in the collection bag. Therefore, when citrate anticoagulation is used for leukapheresis, the ionized calcium level must be monitored frequently, and supplemental calcium infusions should be given as needed to prevent serious reactions secondary to hypocalcemia. As an alternative, a combination of citrate and heparin or heparin alone may be used for anticoagulation to reduce or prevent citrate toxicity. One method is to add 5000 units of heparin to a 500-mL bag of ACD-A and add this mixture to whole blood being withdrawn at a ratio of 1:30–1:50. Unless the patient has a pre-existing coagulopathy, this citrate and heparin combination may offer an advantage over the standard citrate anticoagulation, since hypocalcemia can be prevented and therefore supplemental calcium will not be needed. Readers are referred to a previous publication (Kim 2000) providing detailed guidelines on blood priming and technical modifications of apheresis procedures for pediatric patients.

SUMMARY

The goal of this chapter is to provide basic practical guidelines for pediatric apheresis. Despite its technical difficulties, apheresis can be performed safely in critically ill small children if special attention is paid to vascular access, control of intravascular fluid volume and circulating red cell mass, anticoagulation, and psychological factors unique to pediatric patients. Although apheresis is not considered standard therapy for many pediatric diseases that are actually treated with apheresis, its effectiveness as a first-line or adjunctive therapy has been shown in children with selected diseases. As approaches to treatment of children with various diseases are evolving, evidence of potential benefits of apheresis as adjunctive therapy should be sought. Multi-institutional prospective randomized clinical trials are warranted to identify patients who are most likely to benefit from therapeutic apheresis. For each disease for which it is applied, the proper role of apheresis should be defined and an optimal treatment schedule developed.

Suggested Reading

Chopek M and McCullough J. 1980. Protein and biochemical changes during plasma exchange. In *Therapeutic hemapheresis: a technical workshop.* EM Berkman and J Umlas, eds. Bethesda, MD: AABB Press.

Eder AF and Kim HC. 2002. Pediatric therapeutic apheresis. In *Pediatric transfusion therapy.* JH Herman and CS Manno, eds. Bethesda, MD: AABB Press.

Fosburg M, et al. 1983. Intensive plasma exchanges in small and critically ill pediatric patients: technique and clinical outcome. *J Clin Apheresis* 1:215–224.

Gorlin JB. 1998. Therapeutic plasma exchange and cytapheresis. In *Hematology of infancy and childhood,* 5th ed. DG Nathan and FA Oski, eds. Philadelphia: W.B. Saunders.

Gorlin JB, et al. 1996. Pediatric large volume peripheral blood progenitor cell collections from patients under 25 kg: a primer. *J Clin Apheresis* 11:195–203.

Grishaber JE, et al. 1992. Analysis of venous access for therapeutic plasma exchange in patients with neurological disease. *J Clin Apheresis* 7:119–123.

Hillman RS. 1995. Acute blood loss anemia. In *Williams hematology.* E Beutler, MA Lichtman, BS Coller, TJ Kipps, eds. 5th ed. New York: McGraw-Hill, Inc.

Kasprisin DO. 1989. Techniques, indications and toxicity of therapeutic hemapheresis in children. *J Clin Apheresis* 5:21–24.

Kim HC. 2000. Therapeutic pediatric apheresis. *J Clin Apheresis* 15:129–157.

Kevy SV and Fosburg M. 1990. Therapeutic apheresis in childhood. *J Clin Apheresis* 5:87–90.

McLeod BC, et al. 1983. Partial plasma protein replacement in therapeutic plasma exchange. *J Clin Apheresis* 1:115–118.

McLeod BC, et al. 1999. Frequency of immediate adverse effects associated with therapeutic apheresis. *Transfusion* 39:282–288.

McLeod BC. 2002. An approach to evidence-based therapeutic apheresis. *J Clin Apheresis* 17:124–132.

McLeod BC. 2000. Introduction to the third special issue: clinical applications of therapeutic apheresis. *J Clin Apheresis* 15:1–5.

Perkin RM and Levin DL. 1982. Shock in the pediatric patient, part I. *J of Pediatr* 101:163–169.

Richet H, et al. 1990. Prospective multicenter study of vascular catheter-related complications and risk factors for positive central-catheter cultures in intensive care unit patients. *J Clin Microbiol* 28:2520–2525.

Rogers R and Cooling LLW. 2003. Pediatric apheresis. In *Apheresis: principles and practice,* 2nd ed. BC McLeod, TH Price, R Weinstein, eds. Bethesda: AABB Press.

Spahn DR, et al. 1993. Acute isovolemic hemodilution and blood transfusion: effects on regional function and metabolism in myocardium with compromised coronary blood flow. *J Thorac Cardiovasc Surg* 105:694–704.

Smith JW, et al. 2003. Therapeutic apheresis—a summary of current indication categories endorsed by AABB and ASFA. *Transfusion*, in press.

Strauss RG, et al. 1993. An overview of current management. *J Clin. Apheresis* 8:189–194.

Thompson L. 1992. Central venous catheters for apheresis access. *J Clin Apheresis* 7:154–157.

Wayne AS and Fosburg MT. 1993. Therapeutic plasma exchange and cytapheresis. In *hematology of infancy and childhood* DG Nathan and FA Oski, eds. Philadelphia: W.B. Saunders.

Weinstein R. 2003. Principles of blood exchange. In *Apheresis: principles and practice*, 2nd ed. BC McLeod, TH Price, R Weinstein, eds. Bethesda: AABB Press.

Vamvakas EG. 2000. Evaluation of clinical studies of the efficacy of therapeutic apheresis. *J Clin Apheresis* 5:21–24.

30

Therapeutic Cytapheresis

ANNE F. EDER, MD, PhD, AND HAEWON C. KIM, MD

INTRODUCTION

Therapeutic cytapheresis procedures are performed to remove blood cells from patients with hematological, oncological, or infectious disorders (McLeod 2000; Grima 2000). Automated apheresis devices rely on principles of centrifugation to separate blood into its component parts, selectively remove fractions enriched in a specific blood cell type, and return autologous plasma and other cellular constituents to the patient. Erythrocytapheresis literally describes the removal of red cells from a patient's peripheral circulation, which may be the therapeutic goal, but more often a red cell exchange transfusion is performed to simultaneously or sequentially remove and replace the patient's abnormal red blood cells (RBCs) with red cells collected from healthy donors. Leukapheresis and plateletpheresis (thrombocytapheresis) are collection or depletion procedures for white cells and platelets, respectively. Pediatric patients undergoing these procedures receive replacement fluids such as albumin, plasma, or red cells as needed to maintain vascular volume, hemostasis, or red cell mass.

EVIDENCE-BASED PRACTICE GUIDELINES

Guidance on the use of therapeutic apheresis is available from two professional organizations, the American Association of Blood Banks (AABB) and the American Society for Apheresis (ASFA), as described in Chapter 29. Most therapeutic cytapheresis indications are considered "category I," accepted either as primary therapy or valuable first-line adjunct therapy, often not on the basis of well-designed randomized controlled studies but in light of a broad and noncontroversial base of published experience. Although several clear indications for the use of therapeutic cytapheresis have emerged, considerable variation still exists in practice because technical approaches and therapeutic goals are not uniformly defined. Transfusion therapy, in general, has been carefully scrutinized and widely debated in the management of sickle cell disease and is now clearly indicated for preventing first or recurrent stroke and treating certain acute complications (Cohen et al. 1996; National Institutes of Health 2002). Transfusion goals, however, can be met with either simple transfusion or exchange transfusion performed with manual or automated techniques. Few clinical studies of transfusion therapy systematically evaluate the advantages or cost-effectiveness of one method over the others, or even specify the preferred means to achieve specific hematological endpoints such as posttransfusion hematocrit, the target hemoglobin S percentage, or the desired extent of cellular removal (for example, the fraction of cells remaining (FCR) in the peripheral circulation). Moreover, therapeutic cytapheresis procedures are often used to treat children with critical illnesses with potentially catastrophic consequences, and few controlled trials have been performed to evaluate the efficacy of therapeutic cytapheresis in these settings. Because of increased demand and broadening transfusion recommendations, the number of transfusion centers performing automated therapeutic apheresis procedures has increased and more procedures are being performed in children each year.

GENERAL TECHNICAL CONSIDERATIONS FOR PEDIATRICS

For safe and effective treatment of infants and young children, practitioners must modify therapeutic apheresis procedures designed for adults (Eder et al. 2002). Adequate peripheral vascular access may be difficult or impossible to obtain in children, necessitating placement of a central venous line. Technical adjustments may include the need to prime the extracorporeal circuit with red cells to maintain red cell mass in small children and to prevent proportionately large intravascular volume shifts during the procedure. Red cell priming is generally recommended when (1) the red cell volume deficit at the beginning of the procedure is greater than 30% of the original circulating red cell volume or (2) when any degree of reduction in the circulating red cell volume is undesirable because of concomitant illness. For example, infants and children weighing less than 20 kg or younger than 5 or 6 years usually require blood priming before apheresis, as well as older children with anemia or underlying cardiopulmonary disease, hemodynamic instability, or tissue ischemia.

To prevent unacceptable intravascular volume shifts in small children, the saline used to prime the extracorporeal circuit may be infused to the patient rather than diverted into waste at the start of the procedure. Conversely, the rinseback phase may be eliminated at the end of the procedure to avoid fluid overload, if red cells were administered at the beginning of the procedure. As with adult practice, adverse reactions related to volume status, anticoagulants, transfused blood components, or anxiety must be promptly recognized and treated appropriately. This chapter addresses clinical applications of therapeutic cytapheresis procedures in pediatrics, with emphasis on those most commonly encountered in practice and the unique technical considerations in treating pediatric patients.

ERYTHROCYTAPHERESIS/RED CELL EXCHANGE TRANSFUSION

Erythrocytapheresis may be performed to remove red cells while providing fluid replacement to alleviate symptoms attributable to increased red cell mass (for example, secondary polycythemia resulting from cyanotic congenital heart disease) or to reduce iron stores in hemochromatosis (Grima 2000). While simple phlebotomy is more often performed for these indications, erythrocytapheresis may be preferable for hemodynamically unstable patients because red cells can be removed rapidly while simultaneously returning autol-

ogous plasma along with saline or colloid solution to maintain constant intravascular volume. Far more often, erythrocytapheresis is used for red cell exchange transfusion whereby abnormal red cells from patients with red cell disorders are removed and replaced with red cells collected from healthy donors. Red cell disorders amenable to treatment with this approach include intrinsic defects of red cells resulting from inherited disorders of hemoglobin, the erythrocyte membrane, or red cell enzymes, and extrinsic (acquired) disorders that destroy red cells or permanently impair their function (Table 30.1). The prototypical intrinsic red cell disorder, sickle cell disease, is the most common clinical indication for which red cell exchange is performed.

Intrinsic Red Cell Disorders

Sickle Cell Disease

Sickle cell disease is an inherited hemoglobinopathy due to genetic mutation of the major hemoglobin in adults, HbA, by a single amino acid substitution of valine for glutamate at position 6 in the β-chain. Predominance of HbS in patients with sickle syndromes (for example, homozygous HbS, heterozygous HbSC, or HbS-beta thalessemia) underlies the propensity of the red cells to become deformed and sickle under low oxygen tension and other conditions. Anemia results from ongoing hemolysis of red cells containing HbS, and the sickled red cells impede blood flow and increase blood viscosity, leading to complications of vascular occlusion and infarction, such as stroke, acute chest syndrome, painful crises, and splenic infarction. Stroke occurs in 6% to 10% of children with sickle cell disease, and two-thirds of patients who have had a stroke will suffer a recurrence without transfusion therapy, usually within two years of the initial stroke (Powars et al. 1978). Splenic sequestration crisis, the sudden accumulation of blood in the spleen, occurs often in young children with infection and causes life-threatening anemia, massive enlargement of the spleen, and severe pain. Individuals with sickle cell disease are also susceptible to bacterial infections and complications resulting from general anesthesia. Pregnant women with sickle cell disease are at risk of obstetrical complications.

Role of Red Cell Transfusion in Sickle Cell Disease

Red cell transfusions in general are indicated for sickle cell patients with symptomatic anemia or acute vaso-occlusive complications, such as acute chest syndrome, stroke, and acute multiorgan failure, or for chronic therapy, for example, to prevent first or recur-

TABLE 30.1 Indications for Erythrocytapheresis in Pediatrics

Disease	Clinical setting	Indication Category	
		AABB	**ASFA**
Sickle cell disease	Acute crises		
	Stroke	I	I
	Acute chest syndrome		
	Retinal infarction		
	Hepatopathy		
	Chronic complications		
	Stroke prevention		
	First stroke, abnormal TCD		
	Recurrent stroke		
	Debilitating pain		
	Prophylaxis in pregnancy	III	Not rated
Hyperparasitemia	Malaria	II	III
	Babesiosis	Not rated	III
Polycythemia/	Polycythemia vera	Not rated	II
erythrocytosis	Congenital heart disease, cyanotic		

TCD = Transcranial doppler. Category definitions: I, standard acceptable therapy; II, sufficient evidence to suggest efficacy usually as adjunctive therapy; III, inconclusive evidence or efficacy or uncertain risk/benefit ratio; IV, lack of efficacy in controlled trials.

rent stroke (see Table 30.1) (Grima 2000; Adams et al. 1998). For the management of sudden severe illness, red cell transfusion may improve tissue oxygenation and perfusion and limit the areas of vaso-occlusion by abrogating the sickling process. Although acute painful crises are usually managed without transfusion, aggressive transfusion regimens may improve the outcome in instances of multi-organ failure syndrome associated with severe painful crises (Hassell et al. 1994). Red cell transfusion is also used prophylactically in preparing patients for major surgery and general anesthesia, to prevent peri-operative complications (Vichinsky et al. 1995).

Chronic transfusion therapy to maintain the HbS level between 30% to 50% is warranted when the potential benefit of preventing serious medical complications from sickle cell disease outweighs the risks of transfusion-associated complications such as alloimmunization, infection, and iron overload (NIH 2002). Chronic transfusion therapy is recommended for primary prevention of stroke and prevention of stroke recurrence (Adams et al. 1998; Wang et al. 1991). Other accepted indications for chronic transfusion therapy include chronic, debilitating pain; pulmonary hypertension; and anemia associated with chronic renal failure or chronic heart failure. The use of prophylactic red cell transfusions in pregnancy is controversial; however, transfusion may be appropriate if complications arise in pregnancy, such as preeclampsia or septicemia, and possibly prior to general anesthesia and surgery (Koshy et al. 1991).

Simple Versus Exchange Transfusion

The choice of simple transfusion or red cell exchange transfusion for acute complications of sickle cell disease varies from center to center, and several factors influence the selection of a transfusion method for a given patient (Table 30.2). If a patient is severely anemic (Hb less than 6 to 7 g/dL) or if both the hemoglobin and the percentage of HbS is already low (less than 30%) in a recently transfused patient, simple or "top off" transfusions are generally preferred over red cell exchange transfusion. Simple transfusion not only will correct the anemia but also will retard the ongoing sickling process. If the hemoglobin S accounts for more than 25% of the patient's total hemoglobin, however, any increase in the hematocrit by simple transfusion disproportionately increases the blood viscosity and impairs oxygen delivery to tissues (Schmalzer et al. 1987). In a previously untransfused patient, the final hematocrit should not exceed 30% with simple transfusion. Simple transfusion will also be associated with volume expansion, which may not be well tolerated because of compromised cardiac and/or renal function.

The advantage offered by exchange transfusion over simple transfusion in previously untransfused patients is the simultaneous removal of sickled red cells as normal red cells are transfused; consequently, the HbS concentration can be rapidly reduced without raising hematocrit and causing volume overload (see Table 30.2). Exchange transfusion effectively avoids the co-existence of high HbS and high hematocrit, a condition

TABLE 30.2 Choice of Transfusion Method in Sickle Cell Disease

	Simple Transfusion	**Exchange Transfusion**
Advantages	• To rapidly increase Hb/HCT due to severe, acute anemia (e.g., acute splenic sequestration or aplastic crisis) • To prepare for general anesthesia to increase patient's hemoglobin to 10 g/dL	• To rapidly lower percentage HbS without increasing HCT or causing volume overload in patients with acute complications • To prevent iron overload in chronically transfused patients
Disadvantages	• Iron accumulation with chronic transfusion • Increased blood viscosity due to high hematocrit • Volume overload	• Technically more complex • Requires experienced apheresis team • Requires adequate peripheral vascular access • Increases the requirement for donor blood

that increases blood viscosity and the risk of further vascular occlusion. For chronically transfused patients, red cell exchange transfusion reduces or prevents iron accumulation and allows discontinuation of iron chelation therapy in some patients (Kim et al. 1994, Adams et al. 1996; Singer et al. 1999; Hilliard et al. 1998). The major disadvantage of red cell exchange transfusion is the need to use more donor blood. Annual blood requirements for long-term erythrocytapheresis compared to simple transfusion were increased by 73% and 23%, for conventional treatment (HbS < 30%) or less aggressive treatment (HbS < 50%), respectively (Kim et al. 1994). Disadvantages of red cell exchange may also include the lack of availability of an apheresis team experienced in treating children, as well as the increased technical complexity, and inadequate peripheral vascular access.

Increased donor exposure with exchange transfusion compared to simple transfusion is accompanied by increased risk of viral transmission and alloimmunization. Current estimates of the residual risk of viral transmission in blood donations from repeat donors are 1:1,935,000 for hepatitis C virus (HCV) and 1:2,135,000 for human immunodeficiency virus (HIV) (Dodd et al. 2002). The risk of red cell alloimmunization with chronic transfusion depends on the age of the patient, number of donor exposures, and probability of antigenic dissimilarity between blood donor and recipient. Red cell alloimmunization among children with sickle cell disease is estimated to occur at frequencies ranging from 18% to 47%, compared with 5% to 11% in chronically transfused thalassemia patients and 0.2% to 2.6% of the general population (Smith-Whitley 2002). Consequences of red cell alloimmunization include acute and delayed hemolytic transfusion reactions and decreased availability of compatible red cell units for subsequent transfusion. Delayed hemolytic transfusion reactions, usually mild in most clinical settings, may

result in severe life-threatening anemia in patients with sickle cell disease and occur at rates ranging from 4% to 22% (Smith-Whitley 2002). By prospectively avoiding incompatibility to C, E, and K1 blood group antigens, the alloimmunization rate among chronically-transfused sickle cell patients was reduced from 3% to 0.5% per unit, and hemolytic transfusion reactions were reduced by 90% in one study (Vichinsky et al. 2001).

Precise aggregate risk estimates and cost/benefit analyses for a child on chronic red cell exchange transfusion to prevent first or recurrent stroke are not available. Chronic transfusion therapy to prevent recurrent stroke is likely a lifetime commitment, because the risk of recurrent stroke even after five or more years of transfusion reverts to pretreatment levels after discontinuation of therapy (Wang et al. 1991). At many centers in the United States, exchange transfusion is countenanced because of the strong belief that the benefit of preventing stroke and delaying or preventing iron-induced organ damage outweighs the risks of viral transmission and alloimmunization and justifies the need for increased donor exposure as well as off-sets the increased cost of red cell exchange compared to simple transfusion with iron chelation therapy.

The choice of simple versus exchange transfusion has been investigated in the peri-operative management of patients with sickle cell disease requiring general anesthesia. A prospective multicenter study concluded that a conservative approach utilizing simple transfusion to increase hemoglobin to 10 g/dL was as effective in preventing peri-operative complications as an aggressive exchange transfusion regimen to decrease the percentage of hemoglobin S to less than 30% (Vichinsky et al. 1995). The aggressive regimen was associated with a higher rate of red cell alloimmunization, resulting from increased blood usage in this group. Consequently, a conservative approach with the use of simple transfusion to increase the patient's hemoglobin to 10 g/dL is recommended for prophylactic preoperative red cell transfusion in patients with sickle cell disease. Patients with HbSC or HbS-beta thalessemia who present with hemoglobin concentrations greater than 10 g/dL before surgery, however, are not candidates for simple transfusion but may benefit from exchange transfusion, to decrease the amount of sickle hemoglobin without increasing hematocrit and blood viscosity.

Technical Aspects

The goals of red cell transfusion in treating new onset of stroke, acute chest syndrome, and other acute complications of sickle cell disease are to reduce or prevent sickling by lowering the proportion of HbS-containing red cells and to increase oxygen delivery to tissues by increasing the hematocrit. As a general guideline, hemoglobin S should be reduced to less than 30% of the total hemoglobin, and the hematocrit at the end of the procedure should be at least 30%, but not more than 36%. High posttransfusion hematocrit levels in sickle cell patients have been linked to serious thrombotic complications, most likely due to increased viscosity and altered rheological properties of the blood (Schmalzer et al. 1987). If the patient has not been transfused within the previous three months, the hemoglobin S level should be assumed to be greater than 95%. Exchange of one blood volume should result in a HbA level of about 65% and lower HbS to about 35%; consequently, approximately 1.5 red cell volumes are exchanged. For example, a nine-year-old, 29-kg child with a blood volume of 2030 mL and a 25% hematocrit will require approximately four units of red cells (average unit volume 350 mL; hematocrit 57%). Alternatively, the volume of red cells needed to achieve the therapeutic goal, expressed as the FCR (fraction of the patient's cells remaining in circulation after the exchange transfusion), is calculated from the patient's weight and height, according to the preprogrammed algorithms in the available software of the apheresis equipment.

Following recovery from the acute event, patients with sickle cell disease may be started on chronic transfusion therapy. For patients who have had a stroke, the therapeutic goal in the first three years of chronic transfusion therapy is to maintain hemoglobin S below 30% of the total hemoglobin concentration. If the patient is stable on this regimen for three years, the threshold HbS may be increased to 50%, and the therapeutic interval may be lengthened to every four to five weeks instead of every three to four weeks (Cohen et al. 1992; Miller et al. 1992). Regardless of the target HbS level, the target hematocrit following red cell exchange is either 27% to 30% or the same as the patient's prehematocrit but less than 36%, to optimally prevent hyperviscosity

and iron accumulation. For most patients on chronic exchange transfusion, treatment goals are reached with regular partial red cell exchange of about 40% to 60% of their red cell volume, which requires two to four units of red cell replacement.

Technical difficulties with continuous-flow devices are rarely encountered in treating patients with sickle cell disease but have been reported with intermittent-flow devices. The discontinuous process and the high concentration of sickle cells in the early stages of the process may cause gel formation in the collection bowl (Grima 2000). If bedside leukocyte filters are used during an apheresis procedure, the amount of leukocyte reduction achieved may be inadequate if the in-flow rate exceeds the capacity of the filter (Grima 2000). This problem is obviated with the use of prestorage leukoreduced red cell units or poststorage leukoreduction of red cell units in the blood bank before the start of the procedure.

Component Selection and Modification

HbS-Negative Donor Blood and Storage Time Individuals with sickle trait (HbAS) are asymptomatic and qualify as blood donors, because their red cells function adequately under most physiological conditions. Blood intended for transfusion to sickle cell patients, however, is screened for HbS to avoid administration of red cells containing HbS, primarily for two reasons. In acute crises, hypoxia, acidosis, and dehydration may promote the sickling of transfused red cells from an individual with sickle trait, as well as the patient's own cells. Even in otherwise stable patients on chronic transfusion who are not in sickle crisis, blood from individuals with sickle trait is not used for transfusion because the therapeutic endpoint requires measuring the proportion of HbA and HbS, determined by hemoglobin electrophoresis. The transfusion of red cells from a donor with sickle cell trait confounds the calculation of the adequacy of the exchange, because the HbS from the donor cannot be distinguished from the HbS in the patient.

The storage time of the red cell units selected for transfusion to sickle cell disease deserves consideration because of two potential benefits of relatively fresh red cells. The concentration of 2,3-diphosphoglycerate (DPG) in donor red cells is depleted during storage. The use of relatively fresh red cells (< five to seven days old) with a less pronounced deficit in 2,3 DPG than older red cells may allow for improved oxygen delivery to hypoxic tissues. In addition, the more favorable posttransfusion survival of younger red cells compared to older red cells may translate into lower blood requirements and fewer transfusions for the patient over time. Although the theoretical benefits of using relatively fresh cells for exchange transfusion in acute crises or chronic therapy have not been substantiated in clinical studies, many transfusion services provide less than five to 14-day-old red cell units when possible.

Pretransfusion and Compatibility Testing Routine serological testing for transfusion recipients is limited to ABO/D typing and antibody screening; however, expanded testing to determine the antigenic phenotype at additional Rh (CcEe) and other loci (Kell, Duffy, Kidd, Lewis, Lutheran, P, and MNSs) is recommended for all patients with sickle cell disease older than 6 months (NIH 2002). Knowledge of the patient's extended red cell phenotype facilitates red cell alloantibody investigation by identifying those antigens foreign to the patient that may result in alloimmunization and permits optimal matching of red cell units for transfusion.

The primary cause of high red cell alloimmunization rates in children with sickle cell disease is the genetic disparity between the patient population and the general blood donor population. Many red cell antigens, such as C and E in the Rhesus blood group and K1 (Kell) are more commonly found in Caucasians than African-Americans (Smith-Whitley 2002). Because African-Americans are disproportionately affected in sickle cell disease but underrepresented in the general blood donor population, almost two-thirds of the clinically significant alloantibodies found in chronically-transfused sickle cell patients are directed against E, C, or K1 (Smith-Whitley 2002). Antibodies to Duffy, MNSs, and Kidd blood group antigens, primarily anti-Fy[a], –S, and –Jk[b], respectively, account for most of the remaining cases.

A transfusion strategy for sickle cell patients gaining experimental support and acceptance is prophylactic red cell matching at least for C, E, and Kell (K1) blood group antigens for all transfusions and more extensive antigen matching for patients who have developed red cell alloantibodies (Vichinsky et al. 2001). Prophylactic phenotype matching is still controversial, however, because of the expense, logistical difficulty, and potential for diverting relatively rare antigen-negative red cell units to patients not yet demonstrating red cell alloantibodies. An alternate approach restricts the use of phenotypically matched red cells to patients who have already become immunized to one red cell antigen and are at risk of developing additional alloantibodies and autoantibodies. Programs designed to specifically recruit African-American donors and direct this blood to sickle cell patients facilitate the identification of phenotypically matched units and may further reduce the risk of alloimmunization among chronically transfused children with sickle cell disease (Smith-Whitley 2002).

Modifications Leukocyte-reduction of cellular components is indicated in the setting of sickle cell disease, primarily for prevention of recurrent febrile nonhemolytic transfusion reactions and possibly for decreasing the risk of HLA alloimmunization. Formation of anti-human leukocyte antigen (HLA) antibodies is not a problem from the perspective of red cell transfusion, but is a concern if the patient is a candidate for bone marrow transplantation, which is an active area of investigation as a potential cure for sickle cell disease. Anti-HLA antibodies may interfere with the engraftment of transplanted marrow or decrease the response to transfused platelets during recovery from myeloablative treatment. Leukoreduction also decreases the risk of CMV transmission and may minimize immunomodulatory effects of transfusion, although the latter is controversial.

Gamma-irradiation of blood components is not required for patients with sickle cell disease, because most patients have normal immune responses and are not at risk of transfusion-associated graft versus host disease. Exceptions have medical indications for irradiation, which may occur in the setting of bone marrow transplantation for sickle cell disease. Cellular components intended for a transfusion recipient must be gamma-irradiated if the units are collected from related or HLA-matched donors or given to patients undergoing hematopoietic transplantation or candidates when transplantation is imminent.

Hemoglobin Variants and Thalassemia

Experience obtained with erythrocytapheresis for treating children with sickle cell disease has been applied to other hemoglobinopathies and intrinsic red cell disorders. Pre-operative red cell exchange has been utilized in patients with hemoglobin of abnormally high oxygen affinity, such as Hb Rainier (Francina et al. 1989) or Hb Bryn Mawr (Larson et al. 1997) to prevent potential tissue hypoxia, thromboembolic complications, and increased hemolysis during anesthesia.

Thalassemia major is associated with severe anemia resulting from absent or decreased synthesis of the beta globin chain of hemoglobin and ineffective production of red cells. Whereas complications in sickle cell disease result primarily from the presence of HbS rather than from anemia per se, the clinical picture in thalassemia major is dominated by the severe anemia and dependence on frequent red cell transfusion that inevitably leads to iron overload and associated morbidity. Red cell exchange transfusion as an alternative to simple transfusion is currently under clinical investigation as a means to reduce the rate of iron accumulation in thalassemia patients (Berdoukas et al. 1986; Cohen et al. 1998).

Extrinsic (Acquired) Red Cell Disorders

Extrinsic (acquired) red cell disorders, in contrast to intrinsic (inherited) disorders, usually affect a smaller number of the total circulating red cells but still may be associated with significant physiological sequelae. Red cell exchange has been used to treat patients with protozoal infection; poisoning; and incompatible transfusion; by removing infected, impaired, or foreign red cells and replacing them with compatible red cells from healthy donors (Phillips et al. 1990; Weir et al. 2000; Evenson et al. 1998). Erythrocytapheresis can be a lifesaving measure in patients with glucose-6-phosphate dehydrogenase deficiency following exposure to significant oxidant stress that causes refractory methemoglobinemia (Golden and Weinstein, 1998).

Malaria

Cerebral malaria is a life-threatening complication of *Plasmodium falciparum* infection. Although controlled comparative trials have not been performed, case studies suggest red cell exchange in conjunction with antimalarial regimens results in more rapid reduction in peripheral parasite load and clinical improvement than pharmacotherapy alone (Phillips et al. 1990). Because the degree of parasitemia correlates with mortality, red cell exchange transfusion has been recommended when parasites are detected in more than 10% to 20% of the patient's red cells (Grima 2000; Phillips et al. 1990; Weir et al. 2000). Red cell exchange has also been recommended for other patients, regardless of parasite burden, with severe complications of falciparum malaria, such as encephalopathy, disseminated intravascular coagulation, renal failure, and adult respiratory distress syndrome.

Due to the paucity of data, more specific recommendations regarding the extent of treatment are not possible. The minimum effective volume of red cell exchange in treating complications of falciparum malaria has not been determined; a little as one-half and as much as two blood volumes have been exchanged in published cases (Grima 2000; Phillips et al. 1990). Rather than aiming for a specified exchange volume, the goal of the procedure should be directed at attaining a low parasite load (for example, <5%) and achieving clinical improvement (Grima 2000; Weir et al. 2000).

Babesiosis

Babesiosis, a tickborne disease caused by a malaria-like parasite, *Babesia microti*, is endemic in the United States, notably New York, Massachusetts, Wisconsin, and Minnesota. Usually a mild, self-limiting illness in otherwise healthy individuals, babesiosis may cause serious disease in asplenic or immunocompromised patients with characteristic fevers, hemolytic anemia, hemoglobinuria, jaundice, disseminated intravascular coagulation, and renal failure. First-line therapy is appropriate antimicrobials; red cell exchange has been used as adjunctive therapy for severe infections (Evenson et al. 1998). Although published evidence is scant, red cell exchange may be indicated in cases of babesiosis if parasitemia exceeds 5% to 10%, or for lesser degrees of parasitemia in immunocompromised, asplenic, or critically ill patients.

Incompatible Blood Transfusion

Red cell exchange transfusion has also been used to treat life-threatening hemolytic transfusion reactions and to prevent alloimmunization following incompatible blood transfusion. This approach has been taken in anecdotal cases of medical error and in trauma settings when the supply of Rh-negative cellular components was exhausted and Rh-positive units had to be transfused to Rh-negative women (Werch and Todd 1993). Even after massive exposure, alloimmunization to the D antigen was prevented by performing red cell exchange transfusion to remove the vast majority of incompatible red cells before administering Rh immune globulin (RhIG). The potential benefit of averting Rh sensitization must be carefully weighed against the risks of high-dose intravenous RhIG and aggressive transfusion therapy. The value the woman or the guardian of a dependent-age girl places on preventing possible consequences of D alloimmunization in future pregnancies must be weighed against the current known risks of blood transfusion.

LEUKAPHERESIS

Therapeutic leukapheresis is performed to reduce the number of white blood cells (WBCs) that circulate in great excess in patients with hematological malignancies or myeloproliferative disorders (Table 30.3). Hyperleukocytosis with significant symptoms of leukostasis is associated with morbidity due to thrombosis and bleeding and early mortality. Cellular depletion with therapeutic apheresis is expected to provide temporary clinical improvement and symptomatic relief but may not be effective for this purpose and is not definitive treatment for the underlying disease. Consequently, therapeutic apheresis is used in conjunction with chemotherapy or as a temporizing measure if chemotherapy is contraindicated (see Table 30.3).

TABLE 30.3 Indications for Therapeutic Leukapheresis and Plateletpheresis

LEUKAPHERESIS

Disease	Clinical Setting	Indication Category	
		AABB	**ASFA**
LEUKEMIA			
AML	Hyperleukocytosis, with symptoms	I	I
CML, blast crisis, accelerated phase	Hyperleukocytosis, without symptoms	IV	Not rated
ALL (uncommon)			
CML, chronic phase with high percentages of immature myelocytes (uncommon)			
PLATELETPHERESIS			
Polycythemia vera	Symptomatic thrombocytosis	I	I
Essential thrombocythemia			
CML			
Myeloproliferative disorder			

ALL = Acute lymphocytic leukemia, AML = Acute myelogenous leukemia, CLL = Chronic lymphocytic leukemia, CML = Chronic myelogenous leukemia. Category definitions: I, standard acceptable therapy; II, sufficient evidence to suggest efficacy usually as adjunctive therapy; III, inconclusive evidence or efficacy or uncertain risk/benefit ratio; IV, lack of efficacy in controlled trials.

Symptomatic Leukocytosis

Extreme peripheral leukocytosis, often exceeding $100 \times 10^9/L$ and consisting almost entirely of blasts, occurs in various leukemias and may result in serious complications due to increased viscosity of whole blood and/or metabolic disturbances (Grima 2000; Lichtman and Rowe 1982). Leukostasis and hyperviscosity impair blood flow though vascular beds throughout the body, and symptoms reflect multi-organ involvement. Pulmonary and cerebrovascular symptoms predominate, with clinical presentations including tachypnea, dyspnea, pulmonary insufficiency, blurred vision, diplopia, dizziness, slurred speech, and coma. A dire complication of hyperleukocytosis is intracranial or pulmonary hemorrhage. Metabolic derangement resulting from increased cellular proliferation, metabolism, and cell death may result in hyperuricemia, hyperphosphatemia, hyperkalemia, hypocalcemia, and renal failure. Biochemical imbalances may be aggravated by chemotherapy-induced blast cell lysis.

Acute leukemias are more often associated with symptomatic hyperleukocytosis than stable, chronic leukemias, and myeloid malignancies are more prone to thrombotic or hemorrhagic complications than their lymphoid counterparts. This predilection is due to the larger size and unfavorable physical properties of myeloblasts compared to lymphoblasts. Myeloblasts range in size from 350 to 450 μm^3 and are relatively non-deformable; in contrast, lymphoblasts are 190 to 350 μm^3 and are less likely to sludge in capillaries and damage

vessels (Lichtman 1982; Bunin and Pui 1985). Consequently, complications of hyperleukocytosis are more common in acute myeloid leukemia (AML) or the accelerated or blast crisis of chronic myelogenous leukemia (CML) than acute lymphobastic leukemia (ALL), even though absolute blast counts may be extremely high in ALL. Symptomatic hyperleukocytosis is rare in chronic lymphocytic leukemia (CLL) and uncommon in chronic phase CML although leukostasis may occur due to the increased numbers of immature myelocytes. A higher incidence of hyperleukocytosis and leukostasis in CML in children than in adults has been reported (Rowe and Lichtman 1984).

Role of Leukapheresis

Leukapheresis to rapidly remove the offending WBCs is widely accepted as a means to alleviate symptoms and prevent complications due to hyperleukocytosis in children with leukemia (Bunin and Pui 1985; Grima 2000; Lane 1980). The threshold for treatment is extremely variable, as the degree of WBC elevation that poses risk to an individual patient depends on a number of factors including physical properties of the circulating leukocytes, the conditions in the microvasculature, the underlying diagnosis, and concomitant medical illness. As a general guideline, patients with peripheral WBC counts greater than $100 \times 10^9/L$, with a high percentage of blasts and promyelocytes, and neurological or pulmonary manifestations of leukostasis are candidates for leukocyte depletion. Different laboratory thresholds have been used to guide therapy, such as a fractional volume of leukocytes (leukocrit) above 10%, or circulating blasts above 50,000/μL (50 × 10⁹/L), and additional clinical criteria may have to be taken into account, such as the rate at which the WBC or blast count is rising, or the patient's coagulation status and general condition. Leukapheresis has also been used as an adjunct to chemotherapy to prevent metabolic complications associated with blast cell lysis in ALL and as a means to control leukocytosis in CML when cytotoxic therapy is contraindicated, as in early pregnancy (Caplan et al. 1978; Strobl et al. 1999).

Technical Aspects

Cytoreduction therapy with leukapheresis must be tailored to the individual. Leukapheresis may be repeated daily until the clinical manifestations of leukostasis and hyperviscosity resolve or the leukocyte count is substantially reduced. A therapeutic endpoint with respect to the postprocedure leukocyte count often cannot be specified in advance. Although some degree of cytoreduction is usually accomplished, the efficacy of the procedure is variable and reflects total-body tumor burden, cellular proliferative rate, physical properties of tumor cells, and response to concommitant chemotherapy. On average, a greater than 50% reduction in circulating WBCs is achieved with each procedure that processes at least two blood volumes, calculated as the patient's weight in kilograms multiplied by 75 mL/kg. Thus, two or more blood volumes must be processed, typically about 10 liters for a 70-kg adult or about five liters for a 35-kg child.

The total volume collected during a procedure depends on the patient's size and white cell count, but for a 30-kg child, the average collection removes about one liter of leukocyte-rich plasma. Calculation of expected volume shifts during the procedure and administration of appropriate replacement fluids is extremely important to avoid hypovolemia, dehydration, and acid-base imbalance, especially in small children. In addition, clinical and laboratory monitoring of citrate toxicity is important during leukapheresis procedures, and flow rates may need to be adjusted during the procedure, because large volumes of blood are processed and virtually all of the administered citrate is returned to the patient. Heparin may be used in conjunction with citrate anticoagulation, to decrease the amount of citrate administered to the patient, in the absence of specific contraindications such as coagulopathy or hemorrhage. Erythrocyte sedimenting agents, like 6% hydroxyethyl starch, may achieve more efficient extraction of mature and immature myeloid but are not commonly employed for therapeutic depletion of leukocytes and should not be used in patients with renal failure or cardiac disease, because of the associated risk of unacceptable volume expansion.

Most leukemic patients with hyperleukocytosis are also severely anemic and may be thrombocytopenic as well. Because red cell mass is proportional to blood viscosity, red cell transfusion should not be given before or during the acute hyperviscosity crisis, unless there is an acute need to increase oxygen-carrying capacity. Patients with hematocrits as low as 15% can tolerate the procedure with close monitoring and appropriate intravascular volume support, and red cell transfusion may be given after leukocyte depletion.

PLATELETPHERESIS (THROMBOCYTAPHERESIS)

Thrombocythemia occurs in myeloproliferative disorders, such as essential thrombocythemia, polycythemia vera, or chronic myelogenous leukemia (CML); thrombocytosis is frequently seen in conjunction with reactive or secondary megakaryocytic proliferation, as occurs following splenectomy or with underlying malignancy or inflammatory disease. Both malignant thrombocythemia and reactive thrombocytosis may result in platelet counts exceeding 1000 × 10⁹/L; however, reactive thrombocytosis in children is transient, asymptomatic, and rarely requires treatment. In contrast, thrombocythemia associated with myeloproliferative disorders is unremitting and may require treatment to prevent bleeding or thrombosis. Bleeding manifestions with thrombocytosis may be due to inherently abnormal platelet function of the malignant clone, administration of drugs, or vascular damage resulting from occlusion. Hemorrhage may be massive or minor and may occur locally or involve multiple locations, in particular mucocutaneous (epistaxis), gastrointestinal, genitourinary, or cerebrovascular sites. Thrombosis more commonly occurs in polycythemia vera and essential thrombocytosis; whereas, hemorrhage may be more frequent in CML and myelofibrosis, but both complications may occur in any of the diseases. The clinical course of these hemostatic complications is variable and unpredictable, and laboratory results rarely correlate with patterns of hemorrhage and thrombosis.

Role of Cytapheresis

Treatment of thrombocytosis is directed at controlling the underlying disease or targeted to symptomatic patients, with bleeding, thrombosis, or both complications (see Table 30.3). Symptomatic thrombocytosis can be controlled with single-agent chemotherapy, with hydroxyurea, interferon, busulfan, or anegrilide. Aspirin and/or dipyridamole and antiplatelet agents may also be given for thrombotic complications; however, their use, as well as systemic anticoagulation, are contraindicated in the presence of hemorrhagic complications.

Because pharmacological agents and chemotherapy take time to improve symptoms or suppress production by the bone marrow, plateletpheresis may be performed to immediately lower the platelet count. There is poor correlation between the platelet count and the risk of significant clinical problems. Plateletpheresis is generally recommended as an adjunct to chemotherapy for patients with platelet counts greater than $1000 \times 10^9/mL$ and for patients with markedly elevated platelet counts and manifestations of thrombosis or bleeding, irrespective of the platelet concentration (Grima 2000).

Technical Considerations

Plateletpheresis can achieve an immediate reduction in platelet counts, depending on the total blood volume processed, the preprocedure platelet count, and other technical factors specific to the apheresis device used for the procedure. In general more than one to two blood volumes must be processed to remove 30% to greater than 60% of circulating platelets, although the efficacy of the procedure may be highly variable. As with leukapheresis procedures, red cell mass and intravascular fluid balance must be monitored carefully, with appropriate blood component therapy or fluid replacement during or after the procedure.

ADVERSE REACTIONS

Apheresis personnel must recognize the early signs of adverse reactions and effectively manage these complications. Despite the critical nature of the indications for treatment, therapeutic cytapheresis procedures are relatively safe. A prospective, multicenter study reported significant adverse effects occur in approximately 5% of therapeutic apheresis procedures (McLeod 1999). Adverse reactions were more common in first-time procedures than in repeat procedures, and more common with blood component exchanges than for peripheral blood progenitor cell collections. Clinically troubling reactions, in descending order of occurrence, were transfusion reactions (1.6%); citrate-related nausea and/or vomiting (1.2%); vasovagal reactions such as hypotension (1%), nausea and/or vomiting (0.5%), pallor and/or diaphoresis (0.5%); and tachycardia (0.4%); respiratory distress (0.3%); citrate-related

tetany or seizure (0.2%); and chills or rigors (0.2%) (McLeod 1999). No deaths resulted from therapeutic apheresis; three deaths were attributed to the underlying primary disease.

The incidence of adverse reactions was not stratified according to the age of the patient population in this survey, however, the observations are consistent with the general experience in pediatric populations. Among children, citrate-related toxicity, transfusion reactions, and vasovagal reactions are the most frequently encountered adverse effects of therapeutic cytapheresis procedures. Citrate toxicity is a greater problem with leukapheresis than with exchange procedures because virtually all of the administered anticoagulant is returned to the patient. In addition, small infants or children with liver disease may demonstrate increased sensitivity to citrate. In adults, the first symptoms of citrate toxicity are peri-oral and peripheral paresthesias. Children may also report the sensation of mouth or finger tingling in response to citrate but more often manifest acute episodes of abdominal pain, nausea and/or vomiting, agitation, pallor, and sweating followed by tachycardia and hypotension. These toxic effects are usually avoided or minimized by careful monitoring of symptoms, blood pressure, and serum ionized calcium concentration and by reducing the flow rate and/or providing intravenous calcium supplementation as prophylaxis when citrate delivery exceeds 0.8 mL/min with the COBE Spectra.

If blood components, such as red cells or fresh frozen plasma, are used as replacement fluids, transfusion reactions may be difficult to distinguish clinically from other procedure-related reactions. Immediate adverse reactions to transfused components range from rare incidents of life-threatening, acute, intravascular hemolysis due to incompatible red cell units or mechanical hemolysis, to more commonly reported, and usually mild, allergic reactions to plasma constituents or febrile reactions mediated by the recipient's immune response to transfusion or cytokines in the donor units. The signs and symptoms of transfusion reactions may be suggestive of a certain type of response, such as hives and urticaria in allergic responses or back pain and disseminated intravascular coagulation in hemolytic reactions. More often, signs and symptoms are nonspecific with fever, chills, tachycardia, or hypotension being seen in hemolytic, allergic, febrile, and septic transfusion reactions. Consequently, an adverse reaction to blood product administration should not be attributed to a benign, febrile transfusion reaction, unless acute hemolysis and other, more serious reactions are eliminated as possibilities. The therapeutic apheresis procedure may need to be slowed, stopped temporarily, or discontinued until serological studies are completed by the blood bank and the patient's symptoms resolve. Pretreatment with acetaminophen and administration of leukocyte-reduced red cells and platelets may prevent recurrent febrile reactions. For patients with a history of allergic reactions, premedication with antihistamines or washing red cell units to remove residual plasma may be sufficient to prevent recurrent episodes.

Vasovagal reactions manifest as bradycardia, hypotension, diaphoresis, pallor, nausea, and apprehension. These reactions are managed by pausing the procedure and elevating the patient's legs. The changes in blood pressure usually resolve within a few minutes, allowing resumption and completion of the procedure. Vasovagal reactions may mimic reactions due to anxiety or hypovolemia with the exception that the former is usually associated with bradycardia, while the latter usually cause tachyardia. Distraction techniques, such as conversation, videos, or games, are used to keep the child's attention off the procedure and minimize anxiety.

References

Adams DM, Schultz WH, Ware RE, and Kinney TR. 1996. Erythrocytapheresis can reduce iron overload and prevent the need for chelation therapy in chronically transfused pediatric patients. *J Pediatr Hematol Oncol* 18:46–50.

Adams RJ, McKie VC, Hsu L, et al. 1998. Prevention of a first stroke by transfusions in children with sickle cell anemia and abnormal results on transcranial doppler ultrasonography. *New Eng J Med* 339:5–11.

Berdoukas VA, Kwan YL, and Sansotta ML. 1986. A study on the value of red cell exchange transfusion in transfusion dependent anaemias. *Clin Lab Haemat* 8:209–220.

Brecher ME, ed. 2002. *Technical manual*, 14th ed. Bethesda, MD: AABB.

Bunin NJ and Pui CH. 1985. Differing complications of hyperleukocytosis in children with acute lymphoblastic or acute nonlymphoblastic leukemia. *J Clin Oncol* 3:1590–1595.

Caplan SN, Coco FV, and Berkman EM. 1978. Management of chronic myelocytic leukemia in pregnancy by cell pheresis. *Transfusion* 18:120–124.

Cohen AR, Friedman DF, Larson PJ, Horiuchi K, Manno CS, and Kim HC. 1998. Erythrocytapheresis to reduce iron loading in thalassemia. 40th Annual Meeting, The American Society of Hematology, Miami, FL.

Cohen AR, Martin MB, Silber JH, et al. 1992. A modified transfusion program for prevention of stroke in sickle cell disease. *Blood* 79:1657–1661.

Cohen AR, Norris CF, and Smith-Whitley K. 1996. Transfusion therapy for sickle cell disease. In *New directions in pediatric hematology*. Capon SM and Chambers LA, eds. Bethesda, MD: American Association of Blood Banks.

Dodd RY, Notari EP, and Stramer SL. 2002. Current prevalence and incidence of infectious disease markers and estimated window-period risk in the American Red Cross blood donor population. *Transfusion* 42:975–979.

Eder AF and Kim HC. 2002. Pediatric therapeutic apheresis. In *Pediatric transfusion therapy*. Herman JH and Manno CS, eds. Bethesda, MD: American Association of Blood Banks.

Evenson DA, Perry E, Kloster B, et al. 1998. Therapeutic apheresis for babesiosis. *J Clin Apheresis* 13:32–36.

Francina A, Chassard D, Baklouti F, et al. 1989. Open heart surgery in a patient with a high oxygen affinity haemoglobin variant. *Anesthesia* 44:31–33.

Golden PJ and Weinstein R. 1998. Treatment of high-risk, refractory acquired methemoglobinemia with automated red blood cell exchange. *J Clin Apheresis* 13:28–31.

Grima KM. 2000. Therapeutic apheresis in hematological and oncological diseases. *J Clin Apheresis* 15:28–52.

Hassell KL, Eckman JR, and Lane PA. 1994. Acute multiorgan failure syndrome: a potentially catastrophic complication of severe sickle cell pain episodes. *Am J Med* 96:155–162.

Hilliard LM, Williams BF, Lounsbury AE, and Howard TH. 1998. Erythrocytapheresis limits iron accumulation in chronically transfused sickle cell patients. *Am J Hematol* 59:28–35.

Kim HC, Dugan NP, Silber JH, et al. 1994. Erythrocytapheresis therapy to reduce iron overload in chronically transfused patients with sickle cell disease. *Blood* 83:1136–1142.

Koshy M, Chisum D, Burd L, et al. 1991. Management of sickle cell anemia and pregnancy. *J Clin Apheresis* 6:230–233.

Lane TA. 1980. Continuous-flow leukapheresis for rapid cytoreduction in leukemia. *Transfusion* 20:455–457.

Larson PJ, Friedman D, Reilly MP, et al. 1997. The presurgical management of a patient with a high oxygen affinity, unstable hemoglobin variant (Hb Bryn Mawr) with erythrocytapheresis. *Transfusion* 37:703–707.

Lichtman MA and Rowe JM. 1982. Hyperleukocytic leukemias: rheological, clinical and therapeutic considerations. *Blood* 60:279–283.

McLeod BC, Sniecinski I, Ciavarella D, et al. 1999. Frequency of immediate adverse effects associated with therapeutic apheresis. *Transfusion* 39:282–288.

McLeod BC. 2000. Introduction to the third special issue: clinical applications of therapeutic apheresis. *J Clin Apheresis* 15:1–5.

Miller ST, Jensen D, and Rao SP. 1992. Less intensive long-term transfusion therapy for sickle cell anemia and cerebrovascular accident. *J Pediatr* 120:54–57.

National Institutes of Health; National Heart Lung and Blood Institute. 2002. *The management of sickle cell disease*, 4th ed. NIH Publication No. 02-2117.

Phillips P, Nantel S, and Benny WB. 1990. Exchange transfusion as an adjunct to the treatment of severe falciparum malaria: case report and review. *Rev Infect Dis* 12:1100–1108.

Powars D, Wilson B, Imbus C, et al. 1978. Thenatural history of stroke in sickle cell disease. *Am J Med* 65:461.

Rowe J and Lichtman M. 1984. Hyperleukocytosis and leukostasis: common features of childhood chronic myelogenous leukemia. *Blood* 63:1230–1234.

Schmalzer EA, Lee JO, Brown K, et al. 1987. Viscosity of mixtures of sickle and normal red cells at varying hematocrit levels. Implications for transfusion. *Transfusion* 27:228–233.

Singer ST, Quirolo K, Niski K, et al. 1999. Erythrocytapheresis for chronically transfused children with sickle cell disease: an effective method for maintaining a low hemoglobin S level and reducing iron overload. *J Clin Apheresis* 14:122–125.

Smith-Whitley K. 2002. Alloimmunization in patients with sickle cell disease. In *Pediatric transfusion therapy*. Herman JH and Manno CS, eds. Bethesda, MD: American Association of Blood Banks.

Strobl FJ, Voelkerding KV, and Smith EP. 1999. Management of chronic myeloid leukemia during pregnancy with leukapheresis. *J Clin Apheresis* 14:42–44.

Vichinsky EP, Haberkern CM, Neumayr L, et al. 1995. A comparison of conservative and aggressive transfusion regimens in the perioperative management of sickle cell disease. *N Engl J Med* 333:206–213.

Vichinsky EP, Luban NLC, Wright E, et al. 2001. Prospective RBC phenotype matching in a stroke-prevention trial in sickle cell anemia: a multicenter transfusion trial. *Transfusion* 41:1086–1092.

Wang WC, Kovnar EH, Tonkin IL, et al. 1991. High risk of recurrent stroke after discontinuance of five to twelve years of transfusion therapy in patients with sickle cell disease. *J Pediatr* 118:377–382.

Weir EG, King KE, Ness PM, and Eshleman SH. 2000. Automated RBC exchange transfusion: treatment for cerebral malaria. *Transfusion* 40:702–707.

Werch J and Todd C. 1993. Resolution by erythrocytapheresis of the exposure of an Rh-negative person to Rh-positive cells: an alternative treatment. *Transfusion* 33:530–532.

31

Therapeutic Plasma Exchange: Rationales and Indications

BRUCE C. McLEOD, MD

INTRODUCTION

Therapeutic plasma exchange (TPE) is an extracorporeal treatment modality in which a substantial fraction of a patient's plasma is removed and replaced with donor plasma or a plasma substitute such as 5% human serum albumin. The purpose of TPE is usually the removal of a toxic macromolecule that is wholly or partly confined to the intravascular space. The rationale for TPE in a specific disease is strongest when there is good evidence that a substance amenable to removable by this means contributes to pathogenesis and that it can be meaningfully depleted by TPE. As emphasized in Chapter 29, significant and lasting depletion by TPE can only be achieved for macromolecules that have a substantial intravascular distribution and a relatively long half-life. In many cases the target for removal is a harmful antibody, often an IgG antibody. IgG has a half-life of about four weeks. About half is intravascular, and the extravascular and intravascular IgG pools will re-equilibrate within 24 to 48 hours after intravascular levels are lowered by TPE. Thus IgG can be meaningfully depleted by a "standard" course of six TPEs of 1 to 1.5 patient plasma volumes over 10 to 12 days, as described in Chapter 29.

Evidence of clinical benefit will ideally be available to round out the rationale for TPE in a specific disease (McLeod 2002; Kim 2000). Much more knowledge of this nature has been gathered in adults than in the pediatric patients who are the focus of this volume. This may be cited as a potential barrier to reaching firm conclusions about the efficacy of TPE in pediatric illnesses. While this may be a valid concern in some cases, this author feels that if there is good evidence that the same

pathogenic mechanism is operative in children and adults with a specific disease and it involves a toxic macromolecule amenable to removal by TPE, then there is no reason not to extrapolate evidence of clinical benefit in adults to pediatric patients (Rogers et al. 2003).

Assessment of risk and therefore of risk:benefit relationships could be another matter. Still, as also emphasized in Chapter 29, TPE can be carried out quite safely in pediatric patients, even in very small children, with proper adjustments in technique (Eder and Kim 2002). Thus, even the risk:benefit argument may not be a compelling obstacle to extrapolating evidence of clinical benefit from adults.

The remainder of this chapter is devoted to discussion of the case of TPE in specific diseases. Information specific for the pediatric age group will be provided when possible, but knowledge gained from adult experience will also be liberally cited. Indication categories assigned by the American Society For Apheresis (ASFA) and the American Association of Blood Banks (AABB) and described in Chapter 29, which are based mainly on clinical data for adults, will also be quoted. These are summarized in Table 31.1.

GUILLAIN-BARRÉ SYNDROME

Clinical Features

Guillain-Barré syndrome (GBS) is an acute, progressive disease of the peripheral nervous system. Since the conquest of polio by vaccination it has been the most common cause of rapid-onset, areflexic paralysis

365

TABLE 31.1   Indication Categories for Therapeutic Plasma Exchange in Selected
Pediatric Disorders

Disorder	ASFA/AABB Indication Category*
Guillain-Barré syndrome	I
Chronic inflammatory demyelinating polyneuropathy	I
Sydenham's chorea	II
PANDAS**	II
Rasmussen's encephalitis	III
Thrombotic thrombocytopenic purpura	I
Hemolytic uremic syndrome	III
Familial hypercholesterolemia (LDL depletion)	I
Refsum's disease	I
Focal segmental glomerulosclerosis (recurrent in renal transplant)	III
Renal transplantation	
Rejection	IV
Presensitization	III
Heart transplant rejection	III
Liver failure	III
Myasthenia gravis	I
Goodpasture's syndrome	I
Systemic lupus erythematosus	
Nephritis	IV
Other	III
Autoimmune hemolytic anemia	III
Immune thrombocytopenic purpura (immunoadsorption)	II

*See Chapter 29 for definitions of indication categories.
**Pediatric autoimmune neuropsychiatric disorder associated with streptococcal infection.

in developed countries. In adults, a typical presentation is symmetrical leg weakness and distal parasthesias that worsen as similar symptoms appear in the arms and face. Leg and arm pain may be predominant symptoms in children, and resultant behavioral changes such as irritability and immobility may suggest encephalitis. Respiratory weakness and cranial nerve symptoms are found in more severe cases, and some patients have autonomic nerve dysfunction manifested as instability in pulse and blood pressure. Some patients remain ambulatory throughout the illness, and this mild course may be more common in children. At the other end of the severity spectrum are patients who require mechanical ventilation; the worst cases have quadriplegia, ophthalmoplegia, sensory deficits, and prolonged ventilator dependency. Taking all age groups together, mortality is about 5%, while another 5% are severely disabled and another 10% to 15% have residual weakness at one year. Outcomes have seemed generally more favorable in several pediatric series, but prolonged disability and death have been reported.

GBS patients are no longer regarded as a homogenous group from the standpoint of pathophysiology. While most have neurological deficits that are due to loss of peripheral nerve myelin (acute inflammatory demyelinating polyradiculoneuropathy [AIDP]), others

TABLE 31.2   Variants of Guillain-Barré syndrome

Acute inflammatory demyelinating polyradiculoneuropathy (AIDP)
Acute motor sensory axonal neuropathy (AMSAN)
Acute motor axonal neuropathy (AMAN)
Miller Fisher syndrome (MFS)
Acute panautonomic neuropathy

have axonal damage, sometimes limited to motor nerves (acute motor axonal neuropathy [AMAN]). In the Miller Fisher variant, which has been found in children, ophthalmoplegia and ataxia are the predominant findings and peripheral involvement may be limited to asymptomatic areflexia. Also, in a few patients, postural hypotension or other signs of autonomic neuropathy may predominate. Variant presentations of GBS are summarized in Table 31.2. Rapid progression followed eventually by spontaneous recovery is the rule in all types. Most patients reach a nadir within two weeks and virtually all do so within four weeks.

Clinical findings usually suggest GBS; however, two diagnostic tests may help to exclude other entities in the differential diagnosis, which in children may include tick paralysis and poliomyelitis. The cerebrospinal fluid (CSF) usually has a moderately elevated protein level and few cells. Electrophysiological studies usually show a conduction defect due to myelin loss, but inexcitable nerve fibers may also be demonstrated in the axonal forms. Chronic inflammatory demyelinating polyneuropathy can occur in children. It has similar electrodiagnostic findings but continues to progress well beyond the four-week outer limit usually set for GBS.

Pathogenesis

Inflammatory demyelination is the most prominent pathological finding in most patients with GBS although axonal changes may be found in those with inexcitable nerves. Old observations that GBS-like disease could be passively transferred to experimental animals by infusion of patient plasma suggested that an autoantibody was involved. Current evidence suggests that these antibodies arise from a misdirected humoral immune response to a preceding infectious illness. A number of infectious agents have been implicated, including cytomegalovirus (CMV), Epstein-Barr virus (EBV), and varicella-zoster virus, but the strongest links are to *Campylobacter jejuni*, with about 30% of patients having evidence of recent infection. An AMAN syndrome occurs in epidemics among rural Chinese children, and more than 90% of affected children have had recent *C. jejuni* infection.

Antibodies directed against microbial antigens probably crossreact with gangliosides on Schwann cell or axonal membranes (that is, so-called molecular mimicry). Several antiganglioside antibodies have been identified in GBS patients, some of which correlate with specific clinical variants. Most patients with Miller Fisher syndrome have antibody to GQ1b, a ganglioside enriched in ocular nerves. Anti-GQ1b may also be found in classic GBS patients who have ophthalmoplegia. Antibody to the ganglioside GM-1 is demonstrable in about 25% of classic GBS cases and, along with antibodies to gangliosides GD1a and GM1b, is even more common in patients with axonal variants. Antibody to one or another neuronal or myelin antigen can be found in almost all GBS patients, and fluctuations in antibody levels may correlate with changes in clinical state.

Therapy

Antibody levels in GBS patients probably decline as the immune response to the inciting microbe subsides. In any case, spontaneous recovery is the rule in GBS, probably even more so in children, and no specific treatment is recommended for children who remain ambulatory. Supportive care is a major challenge for more severely affected patients, especially those who require prolonged mechanical ventilation. A large trial conducted in adults indicated that corticosteroids are not helpful (McLeod 2003).

TPE can shorten the duration of symptoms in GBS (Jones 1995). Several large randomized trials have shown that TPE-treated patients fare better in terms of time to reach a number of clinical milestones, including weaning from ventilation and walking without assistance (Weinstein 2000). Randomized trials have not been done in children, and pediatric patients included in the large randomized trials were never analyzed separately. However, several open trials and comparisons with historical controls in children have been consistent with a similar disease-shortening effect. A typical course of TPE would include five to six treatments administered every other day or three times per week. GBS is a category I indication for TPE. Intravenous immunoglobulin (IVIG) has also been reported effected in GBS. A large randomized trial comparing TPE, IVIG, and TPE followed by IVIG in adults found no significant differences in efficacy. IVIG has also been reported to be beneficial in children.

CHRONIC INFLAMMATORY DEMYELINATING POLYNEUROPATHY

Clinical Features

Chronic inflammatory demyelinating polyneuropathy (CIDP) is an acquired disorder that may persist with either continuous or intermittent progression but in children may be monophasic and subside permanently after several months. Weakness is more evident than sensory loss, and both proximal and distal muscles are affected. A majority of children present with difficulty walking. Proximal weakness helps to differentiate CIDP from other chronic neuropathies, while progression for more than two months distinguishes it from GBS. Electrodiagnostic studies suggest demyelination, with slow conduction, conduction block, and prolonged distal latencies being found in multiple nerves. Examination of CSF usually reveals an elevated protein and a low white blood cell (WBC) count. Biopsies of superficial nerves typically show demyelination and patchy inflammatory cell infiltrates. MRI may reveal enhancement of the cauda equina. Detailed diagnostic criteria have been promulgated by the American Academy of Neurology for research but need not always be met for clinical diagnosis.

Pathogenesis

While the etiology of CIDP remains uncertain, the pathological findings and the similarity to GBS suggest a misguided humoral immune process. Antibodies to

GM-1 and other peripheral nerve myelin and protein antigens are found in some patients. In adults, CIDP may occur in the context of a monoclonal immunoglobulin that exhibits antinerve antibody activity. Many children with CIDP have a history of a recent infectious illness that might stimulate an antimicrobial response that could crossreact with peripheral nerve components.

Therapy

Recognition of CIDP is important because it responds to corticosteroid therapy in almost all cases. Several series suggest that a quarter to a third of children have a monophasic course, with complete recovery after several months of steroid therapy. The remaining children have a relapsing or slowly progressive course similar to the typical adult-onset picture, despite intermittent treatment with, and responses to, corticosteroids. Adults and children with progressive disease may be treated with other immunosuppressive agents, such as azathioprine, cyclophosphamide, or cyclosporine.

TPE has been studied in adults with sham-controlled trials that showed it to confer benefit in terms of motor function and electrodiagnostic findings. It has also been used in children with apparent good effect (Nevo 2000). CIDP is a category I indication for TPE. IVIG has also been shown to be effective in similar trials and is easier to administer than TPE in children; however, TPE should be offered to children with resistant disease. A typical TPE treatment schedule recommended in adults consists of three one-plasma volume TPEs per week for two weeks, followed by two TPEs per week for another four weeks.

SYDENHAM'S CHOREA AND PEDIATRIC AUTOIMMUNE NEUROPSYCHIATRIC DISORDERS ASSOCIATED WITH STREPTOCOCCAL INFECTIONS (PANDAS)

Clinical Features

Sydenham's chorea is a movement disorder that usually develops in childhood and has long been known to be related to an antecedent infection with Group A β-hemolytic streptoccocci. It is one criterion for the diagnosis of rheumatic fever, though it can occur in the absence of carditis. Some patients with Sydenham's chorea also have emotional lability, compulsions, and tics, and these observations fostered the hypothesis that childhood obsessive compulsive disorders (OCD)

and/or tic disorders might have a similar relationship to prior infection with Group A β hemolytic steptococci. Evidence of antecedent streptococcal infection can be found in many such children. In addition there is evidence that exacerbations of symptoms may be preceded by recurrence of streptococcal infection.

Swedo (2002) and her colleagues have identified a subgroup of children with neuropsychiatric disorders who meet the following five criteria: (1) OCD and/or a tic disorder; (2) onset of symptoms before puberty; (3) episodic course with exacerbations and spontaneous improvement or remission; (4) association with Group A β-hemolytic streptococcal infections, and (5) associated neurological abnormalities, particularly hyperactivity or adventitious movements during exacerbations. These children are said to have pediatric autoimmune neuropsychiatric disorders associated with streptococcal infection (PANDAS).

Pathogenesis

Both Syndenham's chorea and PANDAS are believed to be caused by antistreptococcal antibodies, possible to the cell wall M protein, that crossreact with structures in the nervous system. The exact neuronal antigens involved are not known; however, pathological findings and neuroimaging studies suggest that the basal ganglia may be sites rich in target antigens. This would correlate well with symptoms and would account for fluctuations in disease activity related to recurrent streptococcal infections that would presumably trigger a renewed antibody response. The B-cell alloantigen D8/17 is a genetic marker for increased susceptibility to rheumatic fever, Syndenham's chorea, and PANDAS and may confer a propensity to form a crossreactive antibody.

Therapy

Drug therapy with neuroleptics, psychotropics, and/or muscle relaxants may reduce symptom severity but is not curative for either Sydenham's chorea or PANDAS. The autoantibody hypothesis suggested that immunomodulatory therapies, particularly those that might lower antibody levels, would help. Both TPE and IVIG have been shown to reduce symptoms in Sydenham's chorea. In a randomized, double blind trial in PANDAS, both TPE and IVIG brought improvement not seen in the placebo group. Symptomatic improvement with TPE occurred sooner and was felt to be quantitatively more impressive. Improvement also occurred in control patients who later received open treatment with TPE or IVIG. Improvement persisted at one year of follow-up in many treated patients. Both Syndenham's chorea and PANDAS are category II indications for TPE. In contrast, TPE is of no value in children with OCD who do not have evidence of a prior streptococcal infection.

RASMUSSEN'S ENCEPHALITIS

Clinical Features

Rasmussen's encephalitis, first described by Theodore Rasmussen in 1958, is a rare acquired neurological disorder that begins in childhood. Although a seizure is frequently the presenting symptom, the disease differs from typical epilepsy in that seizures become difficult to control and are accompanied by progressive neurological dysfunction in the affected cerebral hemisphere that may eventually result in hemiparesis and/or dementia.

Pathogenesis

Pathological studies show inflammation and atrophy expanding from a central focus. Recent studies have shown that patients with Rasmussen's encephalitis have autoantibodies to the GluR3 receptor for the CNS neurotransmitter glutamate that could account for a humoral attack on neurons bearing this receptor at a site where the blood-brain barrier has broken down. GluR3 antibodies may arise as a crossreactive response to a microbial antigen. Antibody to glutamic acid decarboxylase 65 (anti GAD65) has also reported in association with epilepsia partialis continua.

Therapy

Antiseizure medications do not halt the progression of Rasmussen's encephalitis. Excision of the inflammatory focus has been tried, as well as immunosuppressive drugs. No controlled trials have been reported, but both TPE and IVIG have been reported to reverse inexorable progression, at least temporarily, in patients with GluR3 and GAD65 antibodies. Rasmussen's encephalitis is a category III indication for TPE (Andrews et al. 1996).

THROMBOTIC THROMBOCYTOPENIC PURPURA

Clinical Features

Thrombotic thrombocytopenic purpura (TTP) is a disorder characterized by consumptive thrombocytopenia and microangiopathic hemolytic anemia. Patients with advanced disease may also have fever, renal abnormalities, and/or CNS dysfunction. Some of these features overlap with the hemolytic uremic syndrome (HUS), in which renal failure is more prominent while platelet counts tend to be higher. There is a rare congenital form of TTP that becomes evident in early childhood and typically exhibits a relapsing-remitting course. Acquired TTP is usually seen in adults but a recent review (Brunner et al. 1999) identified 35 cases in children, roughly half of which occurred in the setting of systemic lupus erythematosus (SLE).

Pathogenesis

Recent publications have provided a coherent account of the etiology of TTP. The von Willebrand factor (vWF) multimers secreted by endothelial cells are considerably larger than those found in normal plasma and have greater prothrombotic activity. These ultralarge multimers (UL-vWF) are normally cleaved into an assortment of smaller multimers by a plasma metalloprotease. Children with congenital TTP have a defective gene for the vWF-cleaving enzyme, while adults with acquired TTP have made an autoantibody that inhibits it. In either instance, UL-vWF persist in plasma and cause inappropriate adherence of platelets to microvascular endothelium, perhaps in areas of high shear stress. This consumes platelets and creates myriad microvascular stenoses that either damage red cells mechanically as they pass by, causing hemolytic anemia, or prevent their passage entirely, causing ischemia that results in renal and CNS dysfunction.

Therapy

The vWF-cleaving protease has a rather long half-life, and normal plasma levels apparently far exceed those required to prevent accumulation of UL-vWF. Hence children with congenital disease can usually be treated satisfactorily with monthly infusions of normal plasma. In acquired disease, daily TPE with plasma replacement gives better results than daily plasma infusion, presumably because it removes antibody as well as supplying the missing enzyme.

No reports of acquired TTP in the pediatric age group have included measurement of vWF-cleaving enzyme activity. It is therefore not clear whether the above mentioned association of pediatric TTP with SLE is due to formation of autoantibody to the enzyme in the context of SLE or to production of a similar symptom complex in some other way; for example, severe vasculitis. The observation that many SLE-related cases required six to eight weeks of treatment with daily TPE and other measures to obtain a response would suggest the latter possibility to the present author, since most adults with anti-vWF-cleaving enzyme antibody respond to daily TPE with a significant increase in platelet count within a week in his experience.

Daily TPE is the primary therapy for acquired TTP due to anti-vWF-cleaving enzyme antibody, which is a category I indication for TPE (Robson 2002; Grima 2000). Corticosteroids, other immunosuppressive drugs, and splenectomy have all been advocated as adjunctive therapies; all are supported by the anti-vWF-cleaving enzyme autoantibody proposal.

HEMOLYTIC UREMIC SYNDROME

Clinical Features

HUS is characterized by microangiopathic hemolytic anemia, renal insufficiency, and some degree of thrombocytopenia. As mentioned before, distinction from TTP on clinical grounds alone is occasionally difficult; however, this syndrome can arise in children in several ways that are distinct from TTP and from each other.

The most common form of HUS in pediatric patients follows a significant (typically bloody) diarrheal illness caused by *Shigella* species or, more often in developed countries, by the O157:H7 strain of *Escherchia coli*, which also elaborates a shiga-like exotoxin. This type of HUS may be mild, but even severe cases that require temporary hemodialysis and/or red cell transfusions usually resolve with only supportive therapy, albeit with residual renal damage in some instances.

A second category of childhood HUS is the rare congenital form. It may be evident from infancy and is not associated with diarrhea or with any other infectious illness.

A third subgroup of children and the majority of adults with HUS have an acquired illness that is not preceded by dysentery. This form of HUS has been called "atypical" or "idiopathic" and may have a relapsing-remitting course.

Pathogenesis

HUS associated with dysentery is most likely due to direct endothelial damage caused by shiga toxins that are secreted by the infecting organism and find their way into the bloodstream. Thromboses that develop on damaged renal epithelium consume platelets and produce microvascular obstructions that result in mechanical hemolysis and ischemia (Baker 2000).

Congenital HUS in some families has been shown to be due to an inherited deficiency of factor H, an inhibitor of activation of the alternative pathway of complement. Presumably inappropriate complement activation leads to endothelial damage in these patients.

In the past it has been widely assumed, based on clinical and pathological similarities, that atypical HUS in children and adults was a part of the spectrum of TTP. However, measurements of vWF-cleaving protease in adults with an HUS symptom complex have repeatedly revealed normal levels, and hence no evidence for an autoantibody inhibitor. Thus, most cases of atypical HUS are not related to TTP and their etiology remains unknown.

Therapy

Treatment of diarrhea-associated HUS is usually supportive. Although TPE has sometimes been used in the most severe cases, no compelling clinical evidence supports efficacy (Robson 1991). Furthermore, there is no knowledge concerning the distribution or production kinetics of shiga toxins or any other pathogenic agent in this illness that could be marshaled to suggest a rationale for TPE.

Replacement therapy with plasma is easily justified in congenital HUS due to factor H deficiency; however, it is not very effective in preventing symptoms. Possibly the half-life of infused factor H and/or the levels needed to prevent endothelial damage are not as conducive to replacement therapy with plasma as they are for the vWF-cleaving protease that is deficient in congenital TTP.

TPE has been advocated and used in adults and children with atypical HUS (Gianviti 1993). Most evidence for efficacy comes from older series that included patients with both TTP and HUS symptom complexes based on the assumption that both had the same pathogenesis. There are no reports concerning the efficacy of TPE in adults or children that specifically address patients labeled clinically as HUS who are not deficient in the vWF-cleaving enzyme. Absent compelling clinical data of this sort, one can only say that a firm rationale for TPE is lacking in such cases. HUS has been rated a category III indication for TPE.

FAMILIAL HYPERCHOLESTEROLEMIA

Clinical Features

As the name implies, familial hypercholesterolemia (FH) is an inherited disorder. Patients who are homozygous for the defective gene develop signs of illness during childhood. Plasma low density lipoprotein (LDL) levels are very high and total plasma cholesterol may approach 1000 mg/dL. Accumulation of cholesterol in cutaneous or tendinous xanthomas may be evident before age five years. Atherosclerotic coronary vascular disease also develops prematurely, and untreated patients often die of myocardial infarction before age 20 years.

Pathogenesis

FH is caused by a mutation in the gene encoding LDL receptors on cell surfaces. Cells lacking these receptors cannot internalize the cholesterol carried by LDL, and it accumulates in plasma and in extracellular sites, causing the clinical manifestations described previously.

Therapy

Dietary and drug therapies for hypercholesterolemia are usually ineffective in lowering LDL in homozygous FH patients. More drastic measures, such as portacaval shunt, ileal bypass, and biliary diversion are also of limited benefit in the pediatric population. Liver transplantation permanently corrects the metabolic deficit but carries considerable morbidity and mortality.

LDL are large molecules that are predominantly confined to the intravascular space and have a relatively long half-life. They are therefore amenable to meaningful depletion by TPE. TPE carried out every one to two weeks will significantly lower time-averaged LDL cholesterol levels and lead to resolution of cutaneous xanthomas and coronary atherosclerosis.

Because TPE also depletes high density lipoproteins (HDL), which have a cardioprotective effect, there has been interest in developing extracorporeal systems to specifically deplete LDL. Four approaches have been tried: (1) immunoadsoprtion with anti-LDL antibodies, (2) precipitation of LDL with heparin and removal by filtration, (3) cascade filtration systems that deplete the largest plasma macromolecules, including LDL, and (4) adsorption of LDL by dextran sulfate columns (Demetriou 2001; Al-Shaikh 2002). All these systems require an LDL depletion module in addition to the apheresis instrument that separates the plasma to be LDL-depleted. They therefore entail a greater extracorporeal volume than TPE alone. Nonetheless, with appropriate adjustments in technique as described in Chapter 29 the latter two have been used in pediatric patients and have been effective in maintaining lower time-averaged LDL levels and preventing cardiovascular complications. The dextran sulfate adsorption system has the best selectivity for LDL versus HDL. FH is a category I indication for selective LDL depletion therapy (Zwiene 1995).

REFSUM'S DISEASE

Clinical Features

Refsum's disease, also called heredopathia atactic polyneuritiformis, is an inherited disorder that can result in multiple neurological abnormalities including retinitis pigmentosa, peripheral neuropathy, cerebellar ataxia, anosmia, and sensorineural deafness. Other findings may include ichthyosis of the skin, skeletal abnormalities, and cardiac arrhythmias. All of these clinical manifestations may develop during childhood.

Pathogenesis

Patients with Refsum's disease are genetically deficient in the enzyme needed for α-oxidation of phytanic acid, a branched-chain fatty acid derived entirely from dietary sources, mostly ruminant fats. Lack of this enzyme blocks the normal pathway for metabolism of phytanic acid, which then accumulates in lipoproteins in the blood to levels that are neurotoxic. Acute exacerbations may occur in association with concurrent illnesses or other causes of weight loss, presumably because phytanic acid stored in adipose tissue is released into the bloodstream as fat is catabolized.

Therapy

A diet low in phytanic acid is the mainstay of treatment. A slow ancillary oxidation pathway exists that will eventually eliminate accumulated phytanic acid once ingestion has ceased. Good overall nutrition should be maintained to prevent exacerbations associated with weight loss. TPE has been used to alleviate severe disease in patients with very high blood phytanic acid levels, either at diagnosis or during periods of exacerbation. Maintenance TPE may also hasten depletion of body stores. Because it is bound to plasma lipoproteins, phytanic acid can also be removed by the selective LDL depletion devices described in the section on FH. These have been used to good effect in a few patients. Peripheral neuropathy, ataxia, ichthyosis, and cardiac problems usually improve with TPE and/or dietary therapy; cranial nerve defects usually do not. Refsum's disease is a category I indication for TPE.

FOCAL SEGMENTAL GLOMERULOSCLEROSIS

Clinical Features

Focal segmental glomerulosclerosis (FSGS) is an important cause of renal failure in children and adults.

Most cases are primary, but a few appear to be secondary to other conditions such as human immunodeficiency syndrome (HIV) and chronic ureteropelvic reflux. Dialysis-dependent children with FSGS are candidates for renal transplantation. An unusual feature of FSGS is a propensity to recur in a transplanted kidney. Recurrence is seen in about 30% of transplanted patients, sometimes so rapidly that patients have nephrotic range proteinuria within 24 hours after transplantation.

Pathogenesis

As the name implies, FSGS is a pathologically bland process. Fusion of glomerular epithelial foot processes is an early change evident on electron microscopy, but scarring visible on light microscopy eventually ensues without much evidence of inflammation. Scarring is initially focal and segmental but eventually becomes diffuse. The propensity to recur in transplanted kidneys, sometimes quite rapidly, suggested the possibility of a circulating pathogenic factor. A 50 kD substance has in fact been isolated from the plasma of FSGS patients that will increase the permeability of isolated glomeruli to albumin in a tedious bioassay. The molecular weight of the permeability-enhancing factor is considerably lower than that of IgG; however, like IgG it binds to staphylococcal protein A.

Therapy

Immunosuppressive therapy with corticosteroids, azathioprine, cyclosporin, and/or cyclophosphamide has been tried in FSGS, but no drug regimen has been found to reliably halt progression in either primary FSGS in native kidneys or recurrent FSGS in a renal transplant. TPE has been tried because of the evidence for a circulating pathogenic factor. There is evidence that factor levels are lower at the end of a TPE treatment than at the beginning; but specific metabolic data that would bear on the feasibility of TPE therapy, such as half-life and extent of extravascular distribution, have not been published. There are a number of reports of children whose primary FSGS failed to respond to various drug therapies but improved when TPE was started. Other reports describe a lower incidence of recurrent FSGS in living related transplants where children receive "prophylactic" TPE before transplantation. Still others describe higher rates of remission in posttransplant FSGS when TPE is included in the treatment regimen (Zimmering 2000). Protein A immunoadsorption has also been reported (Belson 2001). None of these conclusions is supported by any controlled data in children

or adults, so their status remains anecdotal. Recurrent FSGS in a renal transplant is a category III indication for TPE.

SOLID ORGAN TRANSPLANTATION

Transplantation of solid organs is an immunological tour de force that can have a tremendous positive impact on longevity and quality of life for recipients. The bulk of transplantation immunology concerns cellular immunity. There are, however, certain circumstances in which circulating antibodies pose, or may be thought to pose, an obstacle to receipt or retention of a transplanted organ. TPE has been used extensively in such circumstances in both adults and children (Shishido 2001; Grauhan 2001).

Humoral Rejection

Transplantation of a kidney or heart into a recipient with pre-existing antibody to donor antigens is doomed to failure. Hyperacute rejection becomes evident within hours (sometimes within minutes) after recipient blood begins to flow through the transplanted organ. An intense inflammatory attack is directed mainly against the donor blood vessel cells that first encounter antidonor antibody; thus pathological findings are primarily vascular. TPE has been tried in hyperacute rejection but, like other therapies, it is virtually always ineffective. Fortunately, hyperacute rejection is rare with modern antibody detection and crossmatching methods.

There is a long-held belief that later rejections in which vascular pathology is prominent are also likely to be mediated, at least in part, by antidonor antibodies. Although pathology texts now emphasize that this is an oversimplification, the belief led to extensive use of TPE for rejection of renal transplants in the years following widespread availability of apheresis instruments. Eventually, however, controlled trials showed that TPE conferred no advantage, even in patients with vascular pathology (Winters 2000). Thus, renal transplant rejection is a category IV indication for TPE. There are anecdotal reports of efficacy in heart transplantation but most of these lack any documentation of circulating antidonor antibodies. Heart transplant rejection is a category III indication for TPE. TPE may still be considered as an antirejection therapy for pediatric organ transplant recipients who have not responded promptly to standard measures. It seems especially plausible for those who have circulating donor specific antibodies documented by retrospective crossmatching. It should be borne in mind, however, that controlled trials proving efficacy in patients with donor specific antibodies have not been conducted in children or adults.

Presensitization

Patients who have preexisting antibody to HLA antigens are at high risk for hyperacute rejection of an organ transplant bearing the antigen. Modern crossmatching techniques can ensure that such transplants are not done. A corollary, however, is that it is difficult to find a compatible organ for patients with multiple HLA antibodies, and such patients incur excess morbidity and mortality associated with long waiting times for transplantation.

Pretransplant immunosuppression combined with TPE to lower antibody levels can sometimes produce negative crossmatches with available organ donors and thereby facilitate transplantation for highly alloimmunized patients. Subsequent humoral rejection is rare with modern posttransplant immunosuppressive regimens, and children who qualify for kidney transplants in this way do almost as well as those with no pretransplant antibodies. Controlled trials are lacking, however, and presensitization is rated a category III indication for TPE.

Antibodies in the ABO system are also considered a barrier to transplantation. Pretransplant immunosuppressive regimens including TPE to lower isoagglutinin titers have been reported to allow successful transplantation of ABO incompatible livers and kidneys to pediatric patients. In particular, a transplant can sometimes be obtained from a willing living donor, with all the advantages that implies, for a child who could not otherwise be transplanted. Survival of these organs is reported to be excellent in some series.

Liver Failure

TPE with plasma replacement has been advocated as a treatment for severe liver failure that might at least prolong life until an organ for transplantation becomes available (Kawagishi 2001). Apparent success has been reported anecdotally in both children and adults (Singer 2001). There is no question that this approach can temporarily improve coagulation factor levels and ameliorate bleeding without producing volume overload. On the other hand, rapid infusion of the large quantities of citrate this entails can pose new metabolic challenges for the failing liver. There is little evidence that TPE lowers intracranial pressure or improves neurological status in encephalopathic patients, and in small controlled series it has not seemed to improve outcomes.

Thus liver failure is at best a speculative indication for TPE. It is rated a category III indication.

OTHER DISEASES

Use of TPE in pediatric patients has been reported in a number of other illness in addition to those discussed previously (Ohta 2001; Olson 2002; Pahl 2000). Several of these are discussed briefly in this final section.

Myasthenia Gravis

Myasthenia gravis (MG) is a disease of weakness and fatigability. There is a congenital form that presents in infancy for which TPE is not indicated. Another form, called juvenile MG, is similar to MG in adults, being caused by an autoantibody to the acetylcholine receptor on muscle cell motor endplates that interferes with neuromuscular transmission. MG is a category I indication for TPE, and TPE should be used in recalcitrant juvenile cases (Andrews 1998).

Goodpasture's Syndrome

Goodpasture's syndrome (GS) classically presents with renal failure and pulmonary hemorrhage. It is caused by an autoantibody to a noncollagenous domain in the α-chain of type IV collagen. Glomerular and pulmonary basement membranes are unusually rich in type IV collagen. GS is a category I indication for TPE, and TPE should be done on the rare occasions in which GS occurs in pediatric patients.

Systemic Lupus Erythematosus

SLE is characterized by a variety of autoantibodies including antibodies to double-stranded DNA. This provides a strong rationale for TPE, and it has been reported to be beneficial in pediatric SLE patients. Controlled trials in adults, however, have been disappointing. SLE nephritis is now a category IV indication for TPE based on a large controlled trial showing no benefit. Another large controlled trial that enrolled patients with a variety of severe manifestations of SLE also failed to show any added benefit for patients receiving TPE. Other manifestations of SLE are still rated as category III indications for TPE, however.

Autoimmune Hemolytic Anemia and Immune Thrombocytopenic Purpura

Autoimmune hemolytic anemia (AIHA) and immune (or idiopathic) thrombocytopenic purpura

(ITP) are caused by autoantibodies to red cells and platelets, respectively. They may occur in children, and TPE might seem a reasonable therapeutic option. TPE has been reported in persistent cases that do not respond promptly to immunosuppressive drug therapy. Limited experience with TPE in adults with AIHA or ITP has been controversial at best, however, and AIHA is currently considered a category III indication for TPE, mostly on the basis of old anecdotal reports. ITP has been treated with protein A immunoadsorption columns that deplete IgG and some immune complexes. This has been reported effective in open trials in ITP associated with HIV. The protein A column is approved by the Food and Drug Administration (FDA) for this indication, and ITP is rated a category II indication for protein A immunoadsorption.

SUMMARY

In summary, TPE has been tried in a number of childhood illnesses and seems to have a beneficial effect in many of them. The most compelling evidence for clinical benefit is found in diseases that also provide a strong rationale, being caused, or thought to be caused, by circulating antibodies of other toxic plasma macromolecules that have a long half-life. These diseases include GBS, CIDP, MG, GS, FH, and Refsum's disease. TPE can be used in children with these illnesses with reasonable assurance of a favorable risk:benefit ratio. The rationale for, and value of, TPE is less clear in other disorders such as SLE, organ transplant rejection, and atypical HUS. Readers are urged to consult the latest clinical data before using TPE in these latter disorders.

References

Al-Shaikh AM. 2002. Impact of the characteristics of patients and their clinical management on outcomes in children with homozygous familial hypercholesteremia. *Cardiology in the Young* 12:105–112.

Andrews PI et al. 1996. Plasmapheresis in Rasmussen's encephalitis. *Neurology* 46:242–246.

Andrews PI. 1998. A treatment algorithm for autoimmune myasthenia gravis in childhood. *Annals New York Academy of Sciences* 841:789–802.

Baker KR and Moake JL. 2000. Thrombotic thrombocytopenic purpura and the hemolytic-uremic syndrome. *Curr Opin Pediatr* 12:23–28.

Belson A et al. 2001. Long-term plasmapheresis and protein A column treatment of recurrent FSGS. *Pediatric Nephrology* 16:985–989.

Brunner HI et al. 1999. Close relationship between systemic lupus erythematosus and thrombotic thrombocytopenic purpura in childhood. *Arthritis and Rheumatism* 42:2346–2355.

Demetriou K et al. 2001. Familial homozygous hypercholesterolemia: effective long-term treatment with cascade double filtration plasmapheresis. *Blood Purification* 19:308–313.

Eder AF and Kim JC. 2002. Pediatric therapeutic apheresis. In *Pediatric transfusion therapy*. JH Herman, CS Manno, eds. Bethesda, MD: AABB Press.

Gianviti A et al. 1993. Plasma exchange in children with hemolytic-uremic syndrome at risk of poor outcome. *American Journal of Kidney Diseases* 22:264–266.

Grauhan O et al. 2001. Plasmapheresis and cyclophosphamide in the treatment of humoral rejection after heart transplantation. *J Heart and Lung Transplantation* 20:316–321.

Grima KM. 2000. Therapeutic apheresis in hematological and oncological diseases. *J Clinical Apheresis* 15:28–52.

Jones HR, Jr. 1995. Guillain-Barré syndrome in children. *Current Opinion in Pediatrics* 7:663–668.

Kawagishi et al. 2001. Experience with artificial liver support in 16 living related liver transplant recipients. *Therapeutic Apheresis* 5:7–11.

Kim HC. 2000. Therapeutic pediatric apheresis. *J Clin Apheresis* 15:129–157.

McLeod B. 2003. Therapeutic plasma exchange in neurological disorders. In *Apheresis: principles and practice*, 2nd ed. BC McLeod, TH Price, R Weinstein, eds. Bethesda, MD: AABB Press.

McLeod BC. 2002. An approach to evidence-based therapeutic apheresis. *J Clin Apheresis* 17:124–132.

Nevo Y and Topaloğlu H. 2002. 88th ENMC International Workshop: childhood chronic inflammatory demyelinating polyneuropathy (including revised diagnostic criteria), Naarden, The Netherlands, December 8–10, 2000. *Neuromuscular Disorders* 12:195–200.

Ohta T et al. 2001. Effect of pre- and postoperative plasmapheresis on posttransplant recurrence of focal segmental glomerulosclerosis in children. *Transplantation* 71:628–633.

Olson JA et al. 2002. Type 1 diabetes mellitus and epilepsia partialis continua in a 6-year-old boy with elevated anti-GAD65 antibodies. *Pediatrics* 109:E50.

Pahl E et al. 2000. Reversal of severe late left ventricular failure after pediatric heart transplantation and possible role of plasmapheresis. *Am J Cardiol* 85:735–739.

Renard TH and Andrews WS. 1992. An approach to ABO-incompatible liver transplantation in children. *Transplantation* 53:116–121.

Robson WL and Tsai HM. 2002. Thrombotic thrombocytopenic purpura attributable to von Willebrand factor-cleaving protease inhibitor in an 8-year-old boy. *Pediatrics* 109:322–325.

Robson WLM and Leung AKC. 1991. The successful treatment of atypical hemolytic uremic syndrome with plasmapheresis. *Clin Nephrology* 33:119–122.

Rogers R and Cooling LLW. 2003. Pediatric apheresis. In *Apheresis: principles and practice*, 2nd ed. BC McLeod, TH Price, R Weinstein, eds. Bethesda, MD: AABB Press.

Shishido S et al. 2001. ABO-incompatible living-donor kidney transplantation in children. *Transplantation* 72:1037–1042.

Singer AL et al. 2001. Role of plasmapheresis in the management of acute hepatic failure in children. *Annals of Surgery* 234:418–424.

Swedo SE. 2002. Pediatric autoimmune neuropsychiatric disorders associated with streptococcal infections (PANDAS). *Molecular Psychiatry* 2:S24–S25.

Weinstein R. 2000. Therapeutic apheresis in neurological disorders. *J Clin Apheresis* 15:74–128.

Winters JL et al. 2000. Therapeutic apheresis in renal and metabolic diseases. *J Clin Apheresis* 15:53–73.

Zimmering FD et al. 2000. Treatment of FSGS with plasma exchange and immunoadsorption. *Pediatric Nephrology* 14:965–969.

Zwiene RJ et al. 1995. Low-density lipoprotein apheresis as long-term treatment for children with homozygous familial hypercholesterolemia. *The Journal of Pediatrics* 126:728–735.

A P P E N D I X

Transfusion Formulas

CASSANDRA D. JOSEPHSON, MD

APPROXIMATE BLOOD VOLUMES (BV)

1. Preterm neonate = 100 ml/kg
2. Term neonate = 85 ml/kg
3. >1 month = 75 ml/kg
   (Pisciotto, 2002)

PLASMA VOLUME (PV)

$$PV = BV \times (1 - \text{Hematocrit [Hct]})$$

(Pisciotto, 2002)

RED BLOOD CELL (RBC) TRANSFUSION

1. Grams hemoglobin (Hgb) in the intravascular space = total blood volume (TBV) × Hgb concentration
2. Volume of RBCs

$$= \frac{TBV \times \left(\begin{array}{l}\text{posttransfusion Hgb} \\ - \text{pretransfusion Hgb}\end{array}\right)}{\text{Donor blood Hgb concentration}}$$

$$= \frac{TBV \times (\text{Hgb desired} - \text{Hct observed})}{\text{Hgb donor unit}}$$

3. Using Hct instead of Hgb yields:

$$= \frac{TBV \times (\text{Hct desired} - \text{Hgb observed})}{\text{Hct donor unit}}$$

If packed red blood cells (pRBC) are used,
Hgb donor unit = 22 to 24 g/dL
Hct donor unit = 70% (65–80%)
With additive solutions, Hct = 55 to 65%.
If whole blood is used,

Hgb donor unit = 11–12 g/dL
Hct donor unit = 35%
4. 3 ml/kg pRBC will raise Hgb by 1 g/dL
5. 10 ml/kg pRBC will raise Hct by 10
6. In an individual with a blood volume of 70–75 ml/kg

Volume pRBC = weight (kg) × change in Hct desired

Volume of whole blood = weight (kg) × change in Hct desired × 2

(Kasprisin, and Luban, 1987)

ESTIMATION OF BLOOD VOLUME EXCHANGED

1. Percent (%) $R = [(V_T - V_E) \div V_T]^n$,

   R = percent of blood remaining in patient
   V_T = total blood volume
   V_E = aliquot size for infusion/removal
   n = number of cycles of infusion/removal

Use of this equation helps determine that a single volume exchange removes approximately 65% of the patient's blood volume, double volume exchange removes 88%, triple-volume exchange removes 95%, and a quadruple volume exchange removes 98%.

Double volume performed to remove plasma bound substances: preterm = 2 × 100 ml/kg
full term = 2 × 85 ml/kg

Single volume for correction of coagulopathies or anemia: preterm = 100 ml/kg
full term = 85 ml/kg

2. Red cell mass = $\dfrac{(\text{blood volume} \times \text{Hct})}{100}$

(Kasprisin, and Luban, 1987)

Handbook of Pediatric Transfusion Medicine

PARTIAL EXCHANGE TRANSFUSION FOR POLYCYTHEMIA

Volume replaced (ml)* =

$$\frac{BV \times (\text{Hct observed} - \text{Hct desired})}{\text{Hct observed}}$$

(performed to change Hgb level)

*replacement fluid will depend on clinical situation (Pisciotto, 2002)

EXCHANGE TRANSFUSION FOR ANEMIA

Partial volume to change hemoglobin level in anemia

Volume RBC exchanged (ml) =

$$\frac{BV \times (\text{Hgb desired} - \text{Hgb initial})}{\text{Hgb RBC (approximately 22 g/dL)} - \text{Hgb initial}}$$

(Pisciotto, 2002)

PREPARATION OF BLOOD PRODUCT WITH SPECIFIC HCT USING FFP

1. $W_{FFP} = [(\text{Hct}_{prbc}/\text{Hct}_{RB}) - 1] \times W_{prbc}$
   W = weight in grams of product
   Hct_{RB} = hematocrit of red cell product
   Hct_{prbc} = hematocrit of pRBCs
2. If one has a known exchange volume needed and wants to use pRBCs reconstituted to a lower hematocrit by using FFP or albumin, the following equations apply:

Volume of pRBC =

$$\frac{\text{Exchange volume} \times \text{reconstituted Hct}}{\text{Hct}_{prbc}}$$

Volume of FFP = exchange volume
   − volume of pRBC

(Kasprisin and Luban, 1987)

PLATELET TRANSFUSION

Corrected count increment (CCI)

CCI =

$$\frac{\text{Posttransfusion count} - \text{Pretransfusion count}}{\text{Number of platelets per unit} \times N} \times BSA$$

Subsequent to platelet transfusion, 50% to 70% of platelets are recovered in the peripheral circulation. The remainder of the platelets are sequestered in the spleen. The recovered proportions of peripheral platelets are decreased further if splenomegaly is present.

PLATELET SOURCE	PLATELET COUNTS	VOLUMES
Whole blood derived	$\geq 5.5 \times 10^{10}$ platelets*	50 ml plasma
Apheresis	$\geq 3.0 \times 10^{10}$ platelets‡	250–300 ml plasma

*There should be at least $> 5.5 \times 10^{10}$ in at least 90% of the units tested

†There should be at least $\geq 3.0 \times 10^{10}$ platelets in at least 90% of the units tested.

‡Dose equivalent to 6–10 units of whole blood derived platelets.

(Brecher M, Combs, MR, et al., 2002; Kasprisin and Luban, 1987)

COAGULATION

1. Fresh frozen plasma contains all clotting factors in a concentration of 1 unit/ml. The dose is 10–15 ml/kg and may be used to attain a minimal hemostatic level of 20% to 30% for many coagulation factors.
2. Cryoprecipitate contains a higher concentration of Factor VIII (≥ 80 units) and a much higher concentration of fibrinogen (≥ 150 mg). The dose is 1 to 2 units/10 kg and volume will vary, but each unit can contain a maximum volume of 15 ml. The expected increment should be 60 to 100 mg/dL rise in fibrinogen with a proper dose for body weight.

   Number of units cryoprecipitate needed for fibrinogen replacement:

$$\frac{\left(\begin{array}{l}\text{desired fibrinogen level} \\ -\text{initial fibrinogen level}\end{array}\right) \times PV \div 100\,\text{ml/dL}}{250\,\text{mg fibrinogen/unit (content of 75\% of units)}}$$

3. Factor VIII Dosing

   Factor VIII (units) = (desired Factor VIII
      − initial Factor VIII) × PV

   EXAMPLE: initial Factor VIII level = 0, desired Factor VIII level is 50%, weight = 40 kg, Hct 30

   Factor VIII (units) = (0.5 units/ml − 0) × 75 ml/kg
      × 40 kg × (1−0.3)
      = 1050 units

(Kasprisin and Luban, 1987)

MASSIVE TRANSFUSIONS

Once the blood volume of a patient has been completely replaced, cross-matching RBCs is not usually performed because the cross-match would reflect the transfused blood and not the patient's blood type. However, a new cross-match is required if the blood type were to be changed. The following chart provides approximate cutoffs for determining when to discontinue cross-matching.

AGE (years)	WEIGHT (kg)	TOTAL BOOD VOLUME	UNITS
2–3	10	700	2–3
4–5	20	1400	4–5
6–7	30	2100	6–7
8–9	40	2800	8–9
10–12	50	3500	10–11
Adolescent	60	4200	12–13

Massive bleeding transfusion guidelines:

1. Transfuse 1 unit of FFP and 1 platelet apheresis for each 5 units of pRBCs administered.
2. Transfuse type-specific (ABO and Rh type) products as long as possible.
3. ABO/Rh type-specific blood or O negative packed cells must be used when transfusing neonates for hyperbilirubinemia.

(Kasprisin and Luban, 1987)

COMPOSITION OF ANTICOAGULANTS

	CPD, CPDA-1	AS-1, AS-3, AS-5
Hematocrit	70%–75%	55%–65%
Expected Hgb rise post transfusion of 10 to 15 ml/kg	3 g/dL	2 g/dL
Glucose	+	+++
Mannitol	0	AS-1, AS-5
Sodium	Minimal	++
Adenine	CPDA only	Yes
Shelf-life	21–35 days	42 days

(Pisciotto, 2002)

References

Brecher M, Combs MR, Drew MJ, et al., eds. 2002. AABB Technical manual, 14th ed. Bethesda, MD: AABB Press.

Kasprisin DO, Luban NL. 1987. Pediatric transfusion medicine. Boca Raton, FL: CRC Press, Inc.

Pisciotto P. 2002. Pediatric hemotherapy data card. Bethesda, MD: AABB Press.

Index